Advances in Science, Technology & Innovation

IEREK Interdisciplinary Series for Sustainable Development

Advances in Science, Technology & Innovation (ASTI) is a series of peer-reviewed books based on important emerging research that redefines the current disciplinary boundaries in science, technology and innovation (STI) in order to develop integrated concepts for sustainable development. It not only discusses the progress made towards securing more resources, allocating smarter solutions, and rebalancing the relationship between nature and people, but also provides in-depth insights from comprehensive research that addresses the **17 sustainable development goals (SDGs)** as set out by the UN for 2030.

The series draws on the best research papers from various IEREK and other international conferences to promote the creation and development of viable solutions for a **sustainable future and a positive societal** transformation with the help of integrated and innovative science-based approaches. Including interdisciplinary contributions, it presents innovative approaches and highlights how they can best support both economic and sustainable development, through better use of data, more effective institutions, and global, local and individual action, for the welfare of all societies.

The series particularly features conceptual and empirical contributions from various interrelated fields of science, technology and innovation, with an emphasis on digital transformation, that focus on providing practical solutions to **ensure food, water and energy security to achieve the SDGs.** It also presents new case studies offering concrete examples of how to resolve sustainable urbanization and environmental issues in different regions of the world.

The series is intended for professionals in research and teaching, consultancies and industry, and government and international organizations. Published in collaboration with IEREK, the Springer ASTI series will acquaint readers with essential new studies in STI for sustainable development.

ASTI series has now been accepted for Scopus (September 2020). All content published in this series will start appearing on the Scopus site in early 2021.

More information about this series at http://www.springer.com/series/15883

Sara Eloy · David Leite Viana ·
Franklim Morais · Jorge Vieira Vaz
Editors

Formal Methods in Architecture

Proceedings of the 5th International Symposium on Formal Methods in Architecture (5FMA), Lisbon 2020

Springer

Editors
Sara Eloy
Instituto Universitário de Lisboa
(ISCTE-IUL), ISTAR
Lisbon, Portugal

David Leite Viana
Instituto Universitário de Lisboa
(ISCTE-IUL), ISTAR
Lisbon, Portugal

Franklim Morais
Laboratório de Investigação em Arquitetura e
Design (LIAD)
Escola Superior Artística do Porto (ESAP)
Porto, Portugal

Jorge Vieira Vaz
Laboratório de Investigação em Arquitetura e
Design (LIAD)
Escola Superior Artística do Porto (ESAP)
Porto, Portugal

ISSN 2522-8714 ISSN 2522-8722 (electronic)
Advances in Science, Technology & Innovation
IEREK Interdisciplinary Series for Sustainable Development
ISBN 978-3-030-57508-3 ISBN 978-3-030-57509-0 (eBook)
https://doi.org/10.1007/978-3-030-57509-0

This Springer imprint is published by the registered company Springer Nature Switzerland AG
The registered company address is: Gewerbestrasse 11, 6330 Cham, Switzerland

Preface

The 5th International Symposium Formal Methods in Architecture (5FMA) follows a successful series of symposiums dedicated to the debate in the fields of architecture and urbanism on the application of new formal methods to emerging societal and technical problems. Previous editions have been in 2011, 2013, 2015 and 2018 in Escola Superior Artística do Porto, Portugal. For the edition of 2020, Iscte-Instituto Universitário de Lisboa (Iscte) hosted the symposium for the first time outside Porto.

Formal Methods in Architecture's (FMA) symposiums are focused on methodological advances based on new developments coming from collaborative work with Mathematics and Computer Sciences which enables several different grades of abstraction and formalization. From the millennial geometry to current shape grammars, several formal approaches to architecture and urbanism take part of the debate during these events. The aim of these symposiums is to discuss, disseminate and promote the use of formal methods in the creation of new explicit languages for problem-solving in architecture and urbanism. In them, we discuss current problems in the field and the potentials and drawbacks of the use of formal methods to address them. These problems range from production, sustainability, representation, communication, heritage among others, never ceasing to see architecture and urbanism as technological activities as well as artistic ones. The FMA symposiums focus on scientific fields whose areas of application use methodologies that stem from the mathematical and computer sciences, especially those that have witnessed recent developments. Areas that have been represented in these symposium are the ones related to: collection of information; semantic organization of information; syntactically and semantically formal languages; representation, visualization and interaction; architectural design automation; building performance analysis; SCAVA-Space Configuration, Accessibility and Visibility Analysis; automated manufacturing and construction; active management of the built environment.

Due to the unfortunate times of the global outbreak of COVID-19, the 5FMA initially scheduled to May 2020 was postponed to 13–16 October 2020 and some changes occurred regarding the initial plan. Even though we face troubled times, this volume presents a very interesting set of papers that focus on the main topics in stake in the global agenda and to which formal methods in architecture have a relevant role by responding with innovative, creative and efficient approaches.

Chapters in this volume present the use of formal methods in architecture to respond to social-ecological problems that are currently discussed in the field. Approaches to problems as cities renewal, biodesign, climate change, participatory design, ethics and architecture education are part of this volume. The chapters in this volume are organized in four parts.

Firstly, an Introduction to the problematic involved in 5FMA which includes statements by keynotes and the chairs of the previous FMA Symposiums.

José Pinto Duarte refers to the world context marked by asymmetrical development and impactful events as the COVID-19. Duarte questions the societal value of using formal methods in architecture and urban design and discusses the ethics involved in the work of architects.

Maria Lopez from MVRDV explores on how the office thinks about how spaces and cities can be adapted to future users and how formal methods take part on their design methodology. Lopez highlights the fact that architects design according to the parameters defined by social patterns and human desires and the combination of the socially driven wills with the architecture knowledge is fundamental and enhanced using formal methods.

The beauty of space is a topic not frequently addressed by formal methods. Tasos Varoudis (UCL) brings a vision on this topic that combines machine intuition, applied to design and learning, and neural networks. Varoudis explores how a neural model could be trained to learn inherent morphological characteristics of the urban street network and extract knowledge completely unsupervised.

David Viana, Franklim Morais and Jorge Vaz, chairs of the first four editions of FMA symposiums, bring us their view on the global and disciplinary frameworks for formal methods in architecture.

The papers in this volume are organized in four parts. The first part includes seven chapters dedicated to the topic of design, construction and management of the built environment. This part starts with a chapter by Inês Caetano and António Leitão that explores the topic of algorithmic design within the design practice and proposes a framework for the design of buildings' facades. The chapter presents a mathematics-based framework to support architects with the algorithmic development of designs by following a continuous workflow embracing the main design stages. The second chapter, by Ekin Unlu and Sema Alaçam, presents an experimental curriculum in the field of digital modeling and discuss the students' learning and making processes. Shota Tsikoliya et al. uses algorithms simulating differential growth for additive fabrication of molding of architectural structures. The effort of the chapter to demonstrate the structural, tectonic, spatial and architectural advantages of such formal production machines is welcome.

Ítalo Guedes et al., develops an analysis method of improving airport design using definitions of regulations in a BIM checker. Authors shows the importance of creating a standardization process in the modeling of information during the development of airport designs in software authored by BIM, considering the export to an integrated IFC model. Moving toward the next chapter of this part of the book, Robert Neumayr, uses agent-based simulations to analyze offices occupation. The chapter mentions a methodology based on agent-based simulation, focusing on agent behavior rather than on space morphology to assess the social performance of spaces. Alexandros Peteinarelis and Nikolaos Patsavos explore a semi-formal design methodology, in which user and machine (parametric design machine controlled by the designer at every step) interact permanently. The presented digital tool can calculate directly output data for every design choice and therefore help to inform the designer in the decision-making process.

The last chapter of this part is by Aniruddha Mukherjee who presents a culturalist assumption of production methods and affirms its autonomy. The chapter is linked to the problems addressed by the traditional studies within the disciplines of History, Theory and Critique of Architecture and brings other languages (philosophical, ideological, epistemological, aesthetic, sociological, etc.) to architecture.

The second part includes six contributions on the topic of social and ecological concerns in the built environment. Canan Colaço and Zeynep Mennan present models of users' participation first by non-digital methods and later using parametric generative grammars of simple access to non-expert users. The authors state that computational design proves to be beneficial for design participation since it allows users to co-design in respect with formalized design rules. Sara Eloy et al. contribution is more operational. The authors present generative grammar tools to enable users to change their houses' design. It proposes human–machine interaction that allows an active participation of non-expert users through visualizations in augmented and virtual realities.

Maria João Oliveira and Filipa Osório report a workshop dedicated to teaching formal methods of producing architectural forms. The aim was to design a self-supporting structure based on the terrestrial plants' biological properties and on origami geometric principles.

The methods taught are suitable for a general project production methodology with some automation. The study of space geometry related to thermal behavior of buildings is common in thermal analysis. Leen Hiasat and Pedro Januário make an unusual use of space syntax techniques, not especially concerned with space configuration but rather with space accessibilities of circulation spaces.

Luísa Almeida et al. proposes an approach based on artificial intelligence so that buildings respond to our moods, although still only in a methodological framework, providing the various components that are part of the project. In the same line of research, Beatriz Couto and Sara Eloy present concrete case studies of active systems—walls—which respond interactively to the presence and activity of humans, who also have the possibility to guide these reactions.

The third part is dedicated to the topic of adaptability and resilience of the built environment and includes eight chapters. In the contribution by Carolina Coelho, the traditional methods of space syntax and agent-base analysis are semantically developed for the analysis of adaptability of a School. Camilla Pezzica et al. bring us methodological ways of dealing with problems of urban management related to disaster contexts. The most relevant is a comparative study between two known representations of space syntax graphs—primal and dual, also generally associated with metric and non-metric spaces, and even the introduction of others when needed. Chiara Chioni et al. focus is on the same situation of catastrophe but deals with different problems. These authors study the production of measures to obliviate the immediate consequences of catastrophes, e.g., by the design of new provisional settlements. Space syntax and public life studies are used to analyze diverse situations.

Ricardo Correia et al. proposed a sound methodology to solve the problem of measuring theoretical quantities in a space other than the usual Euclidean space. The authors also bring the functional application of that methodology for the study of aspects of the city renewal, clarifying the potential uses and advantages of new and previous methodologies, while providing useful advances in the management of the city. Silvia Spolaor and Vítor Oliveira speak of the need to find solutions for cases where the organic and informal settlement and growth of the city is the norm. In these places, the participation of citizens in the production of construction is very intense and has economic, social, political and cultural problems that are not found in communities living in older urban settlements.

David Pereira-Martínez based solely on the design of the facades, uses shape grammars to create close-to-reality models of existing buildings without the need for intrusive interior visits. Shape grammars are created and tuned in a process like supervised learning, in this case only with heuristically constructed cases and not automatically. Olha Thikonova uses space syntax and space pattérns analysis in heritage buildings, inferring new historic information, not known or discovered by other methods, about the spaces and their temporal uses. Finally, Momen FoadMarashi and Francisco Serdoura analyze a historical Bazaar, also with space syntaxes, providing useful information about its temporal transformations and desirable future urban transformations to meet the intents of the city management for those public spaces.

The fourth part includes seven chapters dedicated to the topic of design tools and technics introducing the 'hard' technological innovations in formal methods in architecture, as languages or digital tools. These are generic methodologies, whose main concern is not the presentation of use cases, but the technology or tool itself.

In the field of analysis, Tasos Varoudis and Alan Penn presented innovations for space syntax tools, deepening the analysis of graphs to more powerful techniques and crossing it with unsupervised AI learning based on the data obtained by space syntax, which allows a deeper penetration into the 'essence' of the reality of the built environment. In the same field—SCAVA (space configuration, accessibility, and visibility analysis) the DepthSpace3D (a digital tool for 3D space syntax) production team, Ruivo et al. present a new digital tool that integrates several types of space syntax analysis. The goal is to provide architects and city planners with a digital tool closer to their normal language.

In the field of automated architecture design Pedro Veloso and Ramesh Krishnamurti present an interesting proposal, still in the development phase, of producing the design project through a production environment with multiple agents (with competing, collaborative or other rules of behavior aimed at spatial synthesis). This method is extremely generic, since the semantics imbued with the agents can be semantically established in a very generic way, suitable for any type of architectural space, reflecting a more or less complex theory of space.

Wassim Jabi and Aikaterini Chatzivasileiadi present a systematization of language that intends to integrate simultaneously geometry, topology and semantics—therefore good for conceptual pre-BIM design. This linguistic syntax, in the field of declarative languages, has advantages over traditional BIM. Andrew Malcom et al. participate in the international effort to democratize BIM through non-proprietary dissemination tools and via the WEB—for the available information bases, interoperability and the possibility of viewing all available information, through a web digital tool.

Finally, two chapters on tools for capturing semantic meaning from binary large objects. June-Hao Hou and Chi-Li Cheng propose AI learning tools in photogrammetry, on photos of drones (but applicable to other cases), reconstructing 3D models, more effectively, in terms of speed, effectiveness, meaningfulness and usability in other domains than the most used in this specific field. Egemen Kizilcan also uses learning and associates it to a deeper semantics based in visual intelligence.

Lisbon, Portugal Sara Eloy
sara.eloy@iscte-iul.pt
Lisbon, Portugal David Leite Viana
david.viana@iscte-iul.pt
Porto, Portugal Franklim Morais
franklim-esap@cyberoikos.com
Porto, Portugal Jorge Vieira Vaz
iorge@hotmail.com

List of Reviewers

Andrew I-kang Li (Kyoto Institute of Technology)
Anetta Kepczynska-Walczak (Lodz University of Technology)
Anette Kreutzberg (The Royal Danish Academy of Fine Arts, School of Architecture)
Antonio Fioravanti (Sapienza University of Rome)
António Menezes Leitão (Instituto Superior Técnico, Universidade de Lisboa)
Armando Trento (BIMTEGRA SRL)
Bob Martens (TU Wien)
Canan Colaço (Instituto Universitário de Lisboa (ISCTE-IUL), ISTAR)
Catarina Ruivo (CIAUD, Lisbon School of Architecture)
David Leite Viana (Instituto Universitário de Lisboa (ISCTE-IUL), ISTAR)
Eva Castro (Singapore University of Technology and Design)
Gonçalo Castro Henriques (LAMO, Rio de Janeiro Federal University)
Gulen Cagdas (Istambul Technical University)
Henri Achten (Czech Technical University in Prague)
Ioanna Symeonidou (Department of Architecture, University of Thessaly)
Isabel Carvalho (CIAC-UAb Arts and Communication Research Centre)
Joachim Kieferle (Hochschule RheinMain)
João Rocha (Architecture Department, University of Évora)
Joaquim Flores (Arts School of Superior Education, Porto-ESAP)
Jorge Gil (Chalmers University of Technology, Department of Architecture)
Jorge Vieira Vaz (Arts School of Superior Education, Porto-ESAP)
José P. Duarte (The Pennsylvania State University)
José Pedro Sousa (Architecture School of the University of Porto)
José Beirão (Architecture School of the University of Lisbon)
Leandro Madrazo (School of Architecture La Salle, Ramon Llull University)
Marcel Tramontano (Institute of Architecture and Urbanism, University of São Paulo)
Maria Ângela Dias (PROARQ, Rio de Janeiro Federal University)
Maria João Oliveira (Instituto Universitário de Lisboa (ISCTE-IUL), ISTAR)
Maurício Breternitz (Instituto Universitário de Lisboa (ISCTE-IUL), ISTAR)
Michelle Cannatà (Arts School of Superior Education, Porto-ESAP)
Nancy Diniz (Central Saint Martins, University of the Arts London)
Nuno Montenegro (CIAUD, Architecture School of the University of Lisbon)
Pablo C. Herrera (Universidad Peruana de Ciencias Aplicadas)
Paolo Marcolin (Arts School of Superior Education, Porto-ESAP)
Pieter Pauwels (Eindhoven University of Technology)
Ricardo Pontes Resende (Instituto Universitário de Lisboa (ISCTE-IUL), ISTAR)
Rosalia Guerreiro (Instituto Universitário de Lisboa (ISCTE-IUL), CRIA)
Sara Eloy (Instituto Universitário de Lisboa (ISCTE-IUL), ISTAR)

Sema Alaçan (Istambul Technical University)
Sérgio Mendes (Arts School of Superior Education, Porto-ESAP)
Vasco Rato (Instituto Universitário de Lisboa (ISCTE-IUL), ISTAR)
Vitor Oliveira (Universidade do Porto)

Introduction

The Ethics of Our Work

In a World context marked by asymmetrical economic development, strong populations growth, fast urbanization, climate change, and impactful events, such as the migrant crisis and the COVID-19 pandemic, one may question the societal value of using formal methods in architecture and urban design. These methods, I would argue, can empower designers and provide them with the means to develop innovative solutions for some of the big challenges affecting humanity today. However, they require the traditional disciplines that revolve around the construction sector to embrace non-traditional methods and keep pace with the rest of the World.

In 2015, the United Nations adopted the 2030 Agenda for Sustainable Development, which includes 17 Sustainable Development Goals (SDGs). Based on the principle of 'leaving no one behind,' this agenda emphasizes a holistic approach to guaranteeing sustainable development for all.[1] These goals are interrelated and mutually dependent, but for the sake of this discussion, let us focus our attention on Goal 11, Sustainable Cities and Communities. The World is experiencing significant growth in population and fast urbanization at such a scale that we will need to build over the next twenty years as many houses as we have built in the past two thousand years. Overcoming this challenge requires new approaches. We will need not just to build houses in great quantity in a short time but do it also in a way to take into account the unique features of each site and household. This way we can guarantee that we consume the least material and energy resources too. This approach corresponds to solving the problem of mass customization.

Mass customization implies an apparent paradox, the production of unique objects on a large scale, where uniqueness stems from adaptation to context. It can therefore be described as context-sensitive design, where the configuration of designed object results from an improved response to various contextual features, including physical, social and individual. This paradox can be resolved by recurring to computational design and production systems—formal methods. The design system couples parametric or rule-based generative systems with performance simulation and optimization, and it inputs contextual site and user data to output a 3D model of a matching design solution. The production system may recur to different fabrication technologies to materialize the design into a physical object. However, additive manufacturing, also called 3D printing, is a particularly effective technology because it simplifies construction, decreasing time and cost, while increasing safety and performance. For instance, it may avoid the use of formwork, which can account for up to 30% of the construction costs. It may also lead to significant energy savings and a decrease in CO_2 emissions in the construction and operation of buildings. For instance, 3D printing permits the extrusion of concrete where traditional aggregates are partially replaced by light aggregates in varying degrees, leading to heterogeneous materials with different properties. By manipulating

[1] https://www.un.org/development/desa/disabilities/envision2030.html.

extrusion, we can make building parts with optimized performance from a variety of viewpoints, including environmental and structural, leading also to optimized building performance. We would call these buildings smart and the use of similar methods at the larger, urban scale, will lead to smart cities.

The lecture will briefly describe this theory on mass customization and then illustrate it with several examples of design and production systems for housing developed over the years. It will also show how innovative design and construction technologies developed to overcome the situation on Earth were used to design a habitat to support the human exploration of Mars. This effort took place within the context of the NASA 3D-Printed Mars Habitat Challenge, and it represented an extreme case of the application of the proposed mass customization theory, where even basic features of the context, such as gravity, atmospheric pressure and temperature changed. The lecture will also show how the lessons learnt from this Martian effort may impact the way we design and build cities on Earth, making them more resilient, including to pandemic threats like COVID-19.

The use of formal methods in architecture and urban design will progressively lead to changes in the construction sector, eventually changing how we design and build. Instead of design a building, architects will design a design system, which can be used to design buildings. Construction workers will be required to have a different set of skills as they will need to learn how to operate sophisticated machines. At the end of the day, we will have to weigh the advantages against the disadvantages to society, and if the former exceeds the latter, then the paradigm shift is ethically justified.

Ethics, I believe, is of paramount importance for society, in general, and for the design professions, in particular. A society where people comply with the basic principles of ethics is a more just society; one in which people will perceive the value of their work and will, therefore, feel compelled to participate. Ethics builds trust, and trust means less conflicts and increased economic development, which will lead to a better quality of life for all. This ethics-based virtuous cycle is key to solve the big challenges affecting humankind today. It is also necessary for the design professions to maximize their contributions, particularly through the use of formal methods. In fact, such powerful tools may aggravate the situation if their use is not bound by strong ethical principles.

José Pinto Duarte
The Penn State Stuckeman School
The Pennsylvania State University
University Park, PA, USA

From Datascape to Formal Methods

As a practice, MVRDV is always eager to explore and expand on existing typologies. We like to think about how spaces and cities can be adapted to future users; to think of what society demands and what the main challenges are in creating a design that is ambitious, rethinking the typology. How do we achieve that?

Since MVRDV was founded, the office has been fascinated by data-driven parametric design, a concept that results in something we called a Datascape. Central to MVRDV's method is that the design is literally defined by the given parameters so that rather than trying to reach a formal ideal or goal, architects are able to create unexpected forms that go beyond artistic intuition or known geometry. It is architects' assignment to design according to given parameters that are defined by social patterns and human desires. Our work draws from the cutting-edge of architectural knowledge, incorporating research and thinking in urbanism, sustainability, sociology, materials development and more.

For us, design is therefore determined to a large degree by quantifiable parameters that influence or even define the work of architects and urbanists. In this endeavor, the use of formal methods, of purposeful techniques, is key to our design methodology.

This analytic approach from MVRDV was deepened by the development of several design tools and interactive software packages—for example the Functionmixer, the Regionmaker, the Climatizer, the Villagemaker and The (Green) City Calculator—created by both MVRDV and The Why Factory, the think tank led by Winy Maas at TU Delft. These self-built computer programs are tools for development and visualisation, evaluation and comparison, testing and optimization. All aim to avoid static analytical models and enable designers to interact with the constraints and the opportunities exposed by the design parameters. They enable discussion and are constructed on the basis of objective argumentation, bridging the gap with other disciplines and the wider public.

The development of these computational tools has strongly influenced MVRDV's design process, just as MVRDV's design process informed the creation of these tools; technological evolution allows for this development, while the need for insight creates the impulse to develop new tools.

Parametric models, driven by BIM software, are the next technological step in this evolution of the process of spatial design and planning for MVRDV, and BIM has become a true design tool within the office. MVRDV have made a significant commitment to BIM over the past eight years. BIM, as a process and as a tool, essentially helps designers to harness data in order to enrich human experience.

BIM technology has the capacity to connect many disciplines. We like to collaborate, and with BIM, we create a virtual building model where all parties—from designers to contractors to managers—can participate in a model environment, with comments, amendments and other features.

The design team is able to achieve a unified flow of information and thus effectively address the many aspects of design in a cohesive way. This information flow is dynamic, as all building components within the model are intelligently connected to one another. This powerful and flexible dynamic model gives us more space to design, process, test, evaluate and optimize the conditions and demands generated by our data analysis.

The friendly visualization of complex data from BIM also introduces inhabitants to urban participatory processes and mediates the need for collectiveness. Different parties confront their desires in a process of negotiation through real-time collaboration, in a faster and more accurate way, simulating the design at an earlier stage—a design that consists of data.

Through BIM, we design in a more intelligent manner by the use of parametric change engine that allows us to manipulate data, sculpt and explore design options, for problem-solving in architecture and urbanism.

Maria Lopez
MVRDV
Rotterdam, The Netherlands

Variational Beauty of Space: Machine Intuition and Urban Networks

There is a long tradition for humankind debating beauty, design, machine thinking and creativity. While I do not pretend to have a complete answer, one thing that we can see is the relation between a Platonic 'generalized description' of objects, or the 'essence' of them and the way that modern 'thinking machines' work [1–3].

My motivation for following this new path of research was my own fascination for how our brain, after years of urban analytics and design, subliminally generates a number of urban networks, some of them 'optimized' for space syntax performance, every time we look at a road central line map. I wanted to break down this intuitive process and with the help of modern artificial intelligence try to reconstruct it from simplified parts.

Working with spatial networks you intuitively associate yourself with Bill Hillier's idea of internal spatial order or beauty [4] of urban environments, something that has inspired many in the field of spatial computation and analytics to employ a number of graph-based methodologies in order to understand or extract intrinsic attributes of the urban fabric around us. My view and vision though is to take a completely different approach by looking at urban structure through the use of deep convolutional variational autoencoders.

Autoencoders are an unsupervised learning technique in which we employ neural networks for the task of representation learning. Specifically, an autoencoder network architecture imposes a bottleneck in the network which forces a compressed and generalization of the knowledge representation of the structure of space. This nonlinear compression and subsequent reconstruction creates a unique set of features that encapsulate inherent properties of urban space. Using networks that are multiple layers deep, we will be able to encode basic spatial complexity through convolutional network architectures which are inspired by biological processes similar to the connectivity pattern between neurons that resembles the organization of the animal visual cortex. Artificial neurons respond to real urban networks of London in a restricted region of the visual field, which partially overlaps such that they cover the entire convolutional visual field. Convolutional layers apply a convolution operation to the input, passing the result to the next layer. Each convolutional neuron processes data only for its receptive field but the cascading nature of our network build a knowledge of more complex spatial relations as data move deeper. While convolutional neural networks are extensively used in supervised image classification, the focus of my work is completely unsupervised.

My main vision for this stream of work is to demonstrate how a deep convolutional variational neural model could be trained to learn inherent morphological characteristics of the urban street network. Starting with images of London's street network broken up in small neighborhoods, I have already trained a model that was able to 'see' by imitating neural processes similar to our visual system and extract generalized representations, or knowledge, completely unsupervised. The first steps of our vision detailed in Varoudis and Penn [5] focus on the potentials of spatial neural-based clustering in latent space and the experimentations with generative machine intuition. They both give us insights to two unexplored areas in the field of spatial morphology; first is in experimenting with machine intuition in urban morphology and design, and second the research of how performance-based space syntax analytics can be hybridised with the help of neural networks. These focused explorations will help to overcome and understand better the future of spatial computing with neural networks.

Tasos Varoudis
Bartlett School of Architecture
University College London (UCL)
London, UK
e-mail: t.varoudis@ucl.ac.uk

References

1. Sparkes, B. (1996). *The red and the black: Studies in Greek pottery*. Routledge
2. Tandy, D. W. (1997). *Works and days: A translation and commentary for the social sciences*. University of California Press.
3. LeCun, Y., Bottou, L., Bengio, Y. & Haffner, P. (1998). Gradient based learning applied to document recognition. Proceedings of the IEEE.

4. Hillier, B. (2007). *Space is the machine: A configurational theory of architecture*. Space Syntax.
5. Varoudis T., Penn A. (2019). *Variational beauty of space: The probability model perspective of urban morphology*. Beijing, China: Space Syntax SSS12.

Global and Disciplinary Frameworks for Formal Methods in Architecture

The Global Human Path Towards Formalization

Formalization as Mathematical Paradigm

If it were not for the '*Formal Methods in Architecture*' symposia series, since 2011, googling 'formal methods architecture' would result in millions of references in the field of computer sciences. The idea of calling formal methods to a whole new generation of architectural methodologies started right there—in that analogy with the evolution of computer sciences. The attempt to mimic this formalization trends also had to take into account that these were processes in very different phases. Computing was already mathematical before the current stage of formal methods and is already at a later stage than that in which architecture is now attempting to formalize. Computer science itself also sought inspiration for this formalization in the mathematical formalization process itself, whose main steps had taken place much earlier.

The ultimate reference to mathematics, but through the path of computer science, also reflects the current situation in which most of the concrete tools that use formal methods in architecture, are tools of information and communication technologies (ICT), in such a way that many times the formal is wrongly confused with the digital. Nor should the formal be confused with the quantitative. Many formal methods in architecture deal with non-quantifiable concepts and semantics, as math does itself.

When we first claimed this designation of formal methods for the field of architecture, we wanted to emphasize the import of methodologies from the mathematical sciences, without much demand for a formal definition. Mathematics itself does not have a definition accepted by all. It is even discussed whether there is a mathematics or if there several different mathematics exist. And even a formal mathematics suffer from framing by another, even more formal meta-mathematics. So, formalization is not just an import of know-how coming from elsewhere. It is a continuous and endless process, involving the discipline and its methods.

The Path of Formalization

If the biological strategy of *homo sapiens* is to transform the world, this possibility is conferred on him by his ability to produce an explicit mental picture of that world and its transformation. In other words, its behavior (human-matter interaction) is not carried out exclusively by a simple mechanism of stimulus-action. Human actions are not simple physiological reactions to material stimuli. Many other biological species also have more evolved and effective mechanisms (this effective term is perhaps unfortunate, when you see the resilience of entities as simple as viruses). Human beings also have these basic, unconscious physiological mechanisms, reflexes, instincts, habits and others, which in fact take over most of their behavior. But *homo sapiens* is referred to as the rational animal, precisely because of this additional capacity—this explicit mental representation of the world in the form of languages.

Languages Languages are mechanisms, specific to the homo sapiens species, that intermediates action from circumstance in the process of human behavior in a material (natural and social) environment. Languages are expressions of the mental representations of the material world accomplished by the human mind. Unlike other species, humans are called 'rational animals' because they have a very complex mediation process 'between the organism and the external environment… [In humans,] another system of signalization is added: it can be assumed that this system relates to the frontal lobes, which in animals are much less developed than in man. It represents a signalization of the first signaling system by means of speech and of its basis or basal component—kinesthetic stimulations of the speech organs. In this way, a new principle of nervous activity arises—abstraction and at the same time generalization of the countless signals of the first signaling system which is again accompanied by analysis and synthesis of the new generalized signals—a principle which ensures unrestricted orientation in relation to the surrounding world and ensures the highest degree of adaptation, namely science…This second system of signalization and its organ, representing the latest acquisition in the process of evolution…' [1]. Nevertheless, first human languages were not speech like. Imitation of real behaviors in abstract performances (such as hunting training practices or mating rites) or objects (paleolithic paintings and sculptures) constituted an astonishing new way to nature domination through representational (and not immediately real) means, such as in a comparable civilization achievement—the use and production of tools. Those languages, after losing (toward speech) their early importance, evolved to and influenced rites, games, performative and visual arts and also physiognomic actions accompanying speech acts and ideographic written languages.

Those early languages already revealed some of the characteristics that later semi-formal, symbolic, discursive, natural languages would accomplish:

The first is that their function is solving problems of the primordial relation between man and environment, through two main features: They organize knowledge, and they allow communication between humans.

Second is displacement. The speech refers to situations with which the thinker has no immediate contact. The problems are solved firstly in idea, through the language. And only later, the 'solution' is transferred to reality. With that mechanism, humans have not to deal directly and at every moment with the material world. Instead, they define and give solution to the problems, acting first in the world of ideas.

Third is semantic universality and abstraction. Human language acquired its specificity over animal languages when it acquired the capacity to organize information on aspects, domains, properties, places and events from the past, the present or the future, real or possible, true or false, near or far.

Fourth is productivity. The creation of new phrases in the language is accomplished with a limited set of elements. Those new phrases have an informational content that cannot be deduced from the ancient phrases. Other animals can do this in a very limited way.

Fifth is arbitrariness. The signs used in the language are not programmed in the genes of the species. The decoding codes of other animals are genetically programmed. All elements of the species participate in the same language. Nothing in the genes determines the 'English.' It does not appear in the majority of the humans, near populations talk differently, every guy can speak English. And also, there is no physical relation between sign and meaning [2].

Formal Languages Mathematics, if it was not born formal, was christened formal as early as in classical Greek culture. Perhaps we can set the date of its birth registration (which had occurred earlier) in Euclid's 'Elements', c. 300 BC. With the logical versions of Aristotle or (apparently) geometric of Euclid, the logical-deductive sciences have been presented for centuries as synonyms of formal methods.

Euclid's 'Elements' condensed in a text the mathematical knowledge of the epoch, but also the introduction of deduction, a powerful meta-mathematical method of producing new knowledge. With those production algebras, men had no longer to confront permanently with reality. Understood the problem, solutions were provided mentally, and only then have men to act materially to fulfill the desired goals.

Formalization attained formal languages themselves. More recently, at the end of the nineteenth century, Cantor, Piano, Frege, Hilbert and other mathematicians carried out the task that the development of mathematics required—the further formalization of their own formal methods, through foundation, axiomatization or in a generic way through meta-theories of mathematical theories themselves. Interestingly, these studies remain within the mathematical sciences and are still in full development.

Can we not then define mathematics for this millennial process of formalization?

Since then languages have acquired very precise structural definitions: lexicons, syntaxes, grammars, (static, denotation, connotational, operational) semantics, pragmatics and production algebras. They also acquired qualitative (and quantitative) parameters (such as expressivity, completeness, simplicity, complexity, determinism, decidability, soundness, finiteness, recursion) that permit to assure meta-theoretically the desirability of their use in the set of problems we have to deal with. In the same way, formal rules of acceptability of sentences in the chosen language do an active depuration over acceptable knowledge. Maintenance of formal coherence became a powerful tool of knowledge.

Theories Theoretic knowledge, much more than others, is not a passive container, but mainly an active process of creation of more knowledge, primarily by its structured integration in later conceptual concrete knowledge and also by means of its own organization. Total explicitness of knowledge needs the inclusion of the interpretative code to be used by other languages, making clear assumptions. And theories also include production rules of new knowledge.

The historical importance for human kind of the production of theories cannot be denied, constituting one of its most important achievements to date.

Greek formal geometry, beyond being a language (the 'mother' of formal languages), is also a theory—the earlier theory of space. Twenty centuries later, Newton's *Principia Mathematica* not only established the second theory—of mechanics, this time. It also established the possibility to expand the use of formal methods, the Theory as civilization achievement, something to be attempted by every new science.

Theory is one of the forms of knowledge—a mental representation of the material reality, explicit and with self-conscience. Although, as a special form, theory has some unique characteristics, not only as a product, but also in its production processes. As product, theories are formalized linguistic structures, originated in logical-deductive sciences, populated by connotation semantics that represent the specific domain to which they refer/denotate. These structures allow the generalization of the understanding of empirical facts in a very condensed form that reflects material reality and its reciprocal action and universal connection.

The concepts of those structures (such as symmetry, synchrony, intelligibility or synergy in space syntax theories, for example) are becoming more and more diverse from those immediately tied to empirical perceptions of the objects and try to search for essences of the reality, far away from our senses.

Research Methodologies R&D is the production process of theories and is becoming a very well-settled methodology, with increased formality. It has a great commitment with a set of empirically observable facts and phenomena.

The definition of the domain is one of the kernel problems of the establishment of the theory. The simple collection of samples and facts must obey to a previous hypothesis, which is followed by a very interactive and mobile definition. Even the acceptance of empirical facts must overcome a process of formal validation. We are far away from the empirical mechanical transition from facts to theories.

To arrive to the formalized linguistic structures, a concatenation of data is not enough. Those concepts and that structure are not at all immediately given by the phenomenon of the domain. Abstraction and generalization are necessary to create the above-cited structure of concepts. Those Cartesian methods have the very bad reported consequence of narrowing the meaning of the knowledge. In the normal agenda, to generalize to further classes of phenomena, we have to abstract, i.e., to let fall more and more properties of those base entities. And then, we must opt from either a very vague knowledge about a large domain, or a great set of diversified rules to attain to multiple particular cases. But the real 'magic' of theory is that, through the establishment of a convenient structure of relations, abstraction can lead simultaneously to a very deep penetration in the essence of the material domain without any loss of generalization. In a single theory, we can explain deeply a large slice of reality.

It must be said that for that purpose, modern formal methodologies of R&D can help to reach that goal. But really, that is never enough, and all the skill and genius of the researchers are fully needed, in a rather informal way. And that is particularly true in the major scientific advances.

Scientific praxis does not culminate in the production of the theoretical abstract. They are indeed fundamental elements in the overall evolution of the behavior of the primordial relationship between man and reality, which is not a mental and abstract relation, but an active, material and concrete one. There is no such thing as an abstract action. So, theories must go back to reality.

What proves the value of the theory is not the formal coherence or the scientific esprit, but the tight adjustment to reality and the capacity of solving problems. Many theories did not survive the rigors of the confrontation to reality or the lack of operational capacity.

Before being 'thrown' to the real world, theories have also to pass a severe set of tests in controlled environment: scientific experimentation. They must have the capability of 'previewing the future,' previewing a set of results for a set of tests. All these operations are controlled via social mechanisms, including advertising of results, the explanation of how to reproduce results and discussion inter pares.

Finally, this idea of permanent opening to confrontation and refutation by reality and by debate is an integrant part of the scientific process. There are no immutabilities based on faith, dogmas and '*magister dixit*'. And, also, that mistakes and errors happen and are not dramatic, because they can be corrected.

This is the human path that architecture is currently traveling.

Main Areas of Dissemination of Formal Methods in Architecture

Which domains of architecture do formal and digital methods have most insistently penetrated?

The Digitalization of the Construction Sector and its Influence on Architecture

Before analyzing this impact on the specific disciplinary work of architecture, let us allow ourselves to analyze what happens in the broadest area of human activities linked to the constructive transformation of the natural environment.

As is known, *homo sapiens* has a biological strategy that is not that of its own adaptation to the environment, as other species do. Instead, he proceeds to massive modification of the environment, adapting it to human characteristics and needs, in the global anthropological process of humanization of the world. In this process, the construction component is the one that brings the greatest consequences in changes in mass and volume, and thus becomes one of the most visible transformations in the natural environment by human beings. In this global process, we highlight the following aspects in which formal methodologies and information and communication technologies (ICT) are marking a great influence on the work of architects.

ICT revolution Although smart city and smart building are still in an incipient stage, compared to other sectors of human activities, the truth is that the global trend of the ICT revolution is bound to affect the built environment. Think of the radical and inevitable transformation that autonomous vehicles will introduce in the configuration of cities. Or the consequences of the widespread use of telework in buildings and also in cities.

Active and intelligent built environment ICT is beginning to profoundly change basic characteristics of the built environment. What is still essentially a set of blockish, static and passive elements and devices is giving way to the appearance of active and changeable systems, responding quickly to changing situations, adjusting in real time to the needs of the customer. Dynamic spatial configurations, mobility management, active environmental responses by constructive elements are beginning to invade buildings and cities. We still have to consider that active systems are machines that can and are already being managed by artificial intelligent systems.

Interface between humans and the built environment The interaction between the built environment and its users is being transformed, still in an incipient way for now, but where radical changes are expected, as indeed observed in other human activities (the main one being the automation of productive activities). Humans then act differently—they provide real-time motivations to the built environment, delegate (albeit in an assisted manner) the execution and management of the built environment on intelligent machines, maintain an increased exchange of information with those machines, aiming to know the reasons for automatically assumed actions. With new forms of interaction, new forms of mental perception and the attribution of meanings to constructive entities by humans are necessarily created.

Automation in construction processes The automated production processes of the building materials themselves, and even of entire buildings are changing the productivity of the construction systems. What is economically more sustainable is no longer the same as before. In addition, there is the technological possibility of creating new forms, which were previously very difficult to implement.

Participatory design processes New possibilities for the participation of the final users of the built environment are also created, at the stage of its design and planning. These interactions are enhanced by several digital instruments such as those underlying participatory urbanism and new methods of representation, visualization and interaction, for example through mixed, virtual and augmented realities.

Interaction between stakeholders In addition to social users, and even more intensely, digital technologies have been changing the interaction between stakeholders in the building production process. Let us put it simplistically: The first generation of CADs was aimed at the interaction between architect and drawer. The D in CAD was more for drawing than design. The interaction between producers was then established (through blueprints—the project) between skilled professionals trained in the use of the codes for reading those blueprints—nothing new about it. The second generation of CAD and associated technologies, generally under the name of BIM, is dedicated precisely to the communication between all stakeholders: other designers (such as engineers or landscape architects), contractors, builders, manufacturers of building materials and even building managers during the entire life cycle of buildings and constructions. To these human actors, it is necessary to add a completely different class of players—the digital machines that already perform autonomously part of the tasks and speak another type of languages that should be included in the general process. These and other changes in the architectural work environment have profound consequences for the architect's professional practice, in addition to the specific digital transformations within his professional domain, things that we will analyze below.

A Framework of Formal and Digital Trends in the Disciplinary Work of Architecture

After mentioning some domains outside architecture in which digital transformations have important consequences in the specific domain of architecture, what we intend to do now is to situate the main collaborations of formal methods in architectural work itself.

An alert should be made What follows may seem naive to some and definitive to others. But simultaneously, it is and is not naive, and it is certainly not definitive. The editors are perfectly aware of the debates (involving even the definition of architecture itself) that take place in that area of architectural studies that is called 'history, theory and critique' and that involve autonomous scientific and ideological areas, such as philosophy, social sciences, psychology and even politics, which dialogue with architecture. And it will become clear that starting points and non-consensual positions (a consensus that, by the way, does not exist) are assumed and will not deserve everyone's support. The positions are therefore presented in a positive way, without discussion, but taking into account these debates.

Many of the architects, researchers and practitioners, dedicated to the formal methods, carry out their work with some detachment from these issues, and there are also those who are even in favor of these methods as opposed to the more traditional methods and also to the discussions of 'history, theory and critique,' that they believe useless. For this reason, what we are going to put in place rides this 'spontaneous philosophy of scientists' (paraphrasing Althusser). For this reason, this alert becomes important, to make it clear to the least attentive that there is a meta-debate that is here suppressed but not forgotten.

The work of architecture Architecture is essentially the accomplishment of the mental processes that lead to the production of plans for the transformation of material reality through the construction of the built environment. In a developed economy, with a great specialization of work, the architect's job goes mainly in the opposite direction, being responsible for the integration of all the expertise involved. In addition to coordination, the architect is also charged with carrying out these mental processes in some specialized areas. For example, is normally associated with the architect the spatial organization of buildings, taking into account their adequacy to the needs of human activities and lives. Another aspect that architects also take as their own is that of making buildings to be vehicles of communication between human beings of a wide range of cultural meanings. In this sense, we will try to formalize the architect's own work, finding a set of different mental processes assembled in a formal structure. And then we will see where the formal and digital methods in use fit into that structure.

Structured model of the architect's work In the model, we have adopted, stimuli are a certain part of the global state of the world, limited but complex, and actions are constructive activities. The mental processes that mediate between each other are grouped into four zones: world learning, society learning, action plan and decision processes.

The **world learning process** reaches a body of socially consolidated knowledge, which includes three main pathways: inductive, through the treatment of empirical data that are acquired through the senses; the deductive, which transforms established knowledge into new knowledge; and acquired, which stores the exchanges of knowledge between the various actors. The results are the referred bodies of knowledge expressed in more or less formal languages. This knowledge can have different characteristics, according to its duration of veracity (persistent or volatile), abstraction (abstract or concrete), actuality (current or potential), declarative or operational, etc., generating typologies such as events, states, theoretical methods, policies or laws. Generically, knowledge will be placed between the extremes of the momentary state of the world and the stable theoretical model of the world.

The **learning process of society** reaches a set of motivations. These motivations are usually expressed in the 'customer' program, or 'the brief.' But architecture does not just accept this program. First, because the 'client's brief' is generally very incomplete, it can be expressed in multiple ways (for example and in order of complexity: emotions, objectives, utilities, policies). Then, it is necessary to frame individual motivations in social motivations, which can be defined practically in regulations, but which have many other ways of expressing themselves, for example ideologies. If we add the fact that societies are not consensuous, but conflicting, in which there are different motivations in dispute, it is concluded that architecture must define many of the undeclared motivations and take sides between them. Just as an example of what we say, the choice between an architecture for the 'society of waste' or a sustainable architecture is on the table of contemporary ideological discussions.

The **action plan** is embodied in what is usually called the project, which is intended to be a plan of actions to be carried out by its executors, expressed in sufficiently consolidated languages and dominated by the various stakeholders.

Between the motivations and the state of the world on the one hand and the action plan on the other, there is no direct line, and it is precisely here that the most complex process appears—that of **decision**. In a rational decision process, it is intended that the material action to be developed (based on the project) changes the external material environment to one that satisfies human motivations. Rationality is thus here defined as a methodology that effectively controls the production of futures.

Framework of formal methods in architectural work The various formal methods in use in architecture fit into several points of this model:

Two of them have already been mentioned above, because they are on the frontier of architecture, at the **interface between all stakeholders** and the mental design processes: CAD, BIM (especially for the other technicians in the design and construction process); its extensions of virtual presentation in near-reality renderings and mixed realities as communication tools with the users of the built environment; CAM is also used to produce reduced models as a communication language between the various actors. All of these methods are widely used as means of communication between the various digital processes and the architect himself.

In the area of **learning**, we have for example:

Sensory acquisition and first semantic treatment of empirical data: tracking, mapping, computer video, photogrammetry, several sensor technologies and devices—photo, video, IoT, GPS, WiFi, social networks, beacons, biometric, climatic.

Semantic treatment of empirical big data to more or less deep theoretical levels of knowledge: AI—machine learning techniques.

More formal languages for storing information about the state of the world: BIM, GIS, various mapping technologies, ontologies. These languages can be semantically flatter (with a grammar that allows simple semantics) or can extend to languages based on logic and allowing the organization of semantically advanced and complex theories (for example, ontologies).

Languages suitable for certain theories—these languages simultaneously combine knowledge of the world (theories) and adapted decision machines: declarative semantics + production machine: SCAVA, space syntaxes, agent-based, BIM checker, structure, climatic, fluid mechanics.

In **action** process, we have CAM that can transform a plan into concrete material construction actions, from building materials to large robotic building construction systems.

The **decision-making process** is certainly the most important and complex and deserves a little more analysis effort.

In the decision-making process, we essentially have two different cycles—the synthesis and the validation cycles.

In the synthesis cycle, there are machines that have production grammars for the project, such as: processing, generative programming, parametricism, shape grammars, artificial intelligence machines. However, they are very different in terms of semantic capacity.

At the lower limit is, for example, the technique behind parametricism—the ability to automatically generate architectural forms. These automated shape production processes were a fashion within the architectural poetics of a few years ago and still retain their cultivators (e.g., blobitecture). Although they produce architectural forms, their decisions are not semantically oriented. The generated forms do not tend to respond to human motivations or states of the world and are not theoretically founded. They are practically only syntactic languages, without semantic settlement.

Shape grammars also produce architectonic forms. They are also a set of rules for producing projects. However, these rules are already semantically founded. They try to mimic in some explicit way what a good professional would do. The production rules are heuristically defined, or use the best practices of the profession. In terms of language theory, it has broad operational semantics and small denotational semantics. In operational semantics, knowledge of the world and motivations are concatenated without explaining what is what. This means that normally the expressiveness of shape grammars is small; that is, it is necessary to make a shape grammar for each class of problems. When shape grammars are very expressive, in other words, very generic, they suffer from the same problem as parametricism, their semantics are very weak and the projects produced are not the answer to material, human and social problems.

There are currently other attempts to implement synthetic methodologies, combining an explicit declarative semantics and a production machine that are independent of each other. For example, through the declaration of theories in ontologies with first-order logics, which allows complex theories. And then a concrete problem is introduced in the production machine, which is an AI logical inference machine, which queries the ontology and reaches the solution (or solutions). These models are still under development.

Whatever the model in the synthesis process, the tendency is to produce a wide range of possible solutions. This is where the analysis cycle comes in. These solutions can be analyzed in terms of their quality, based on theories that are much better founded than those of the synthesis processes, and thus limit the number of projects to the best ones. These analysis methodologies are also suitable for validating the solutions proposed by the traditional manual work of architects.

As mentioned above (in the learning process), there are already well-established theories of construction. For example, those that are generally the responsibility of engineering—structures, climatic-environmental, networks, etc. But, also, some specifically architectural ones, like SCAVA, space syntaxes, agent-based.

The big problem is that theories only say how the world behaves, demonstrating fixed relations between diverse scientific quantities representing the world

Without great difficulty, we can even put theories in the form:

Future state of the world = T (current state of the world, action)

But our big problem is deciding what our action will be and has the following form:

Our Action = M (current state of the world, future desired state of the world)

T and M are transformations or functions on the considered domains. T is our theory. M is some kind of inverse transformation of T, which we call methodology.

Our big problem is that, starting with T, we do not easily get M. Usually, it is a very complicated problem. There is no easy way to obtain a formal methodology from a formal theory, and without a lot of effort, the best theory explains everything but solves nothing. This is well known in the mathematical theory of functions. For example, analytical solutions for integral or differential functions are very difficult to find (if possible) and this is what made it impossible to apply elasticity theory immediately after its invention. Another example is that of first order logic (FOL), a language with widespread use in ontologies. Although the solution to a query (an inverse function of a declaration) in the Horn clauses (a subset of FOL) was easy to create, it was difficult to find a complete solution for any query in this logic.

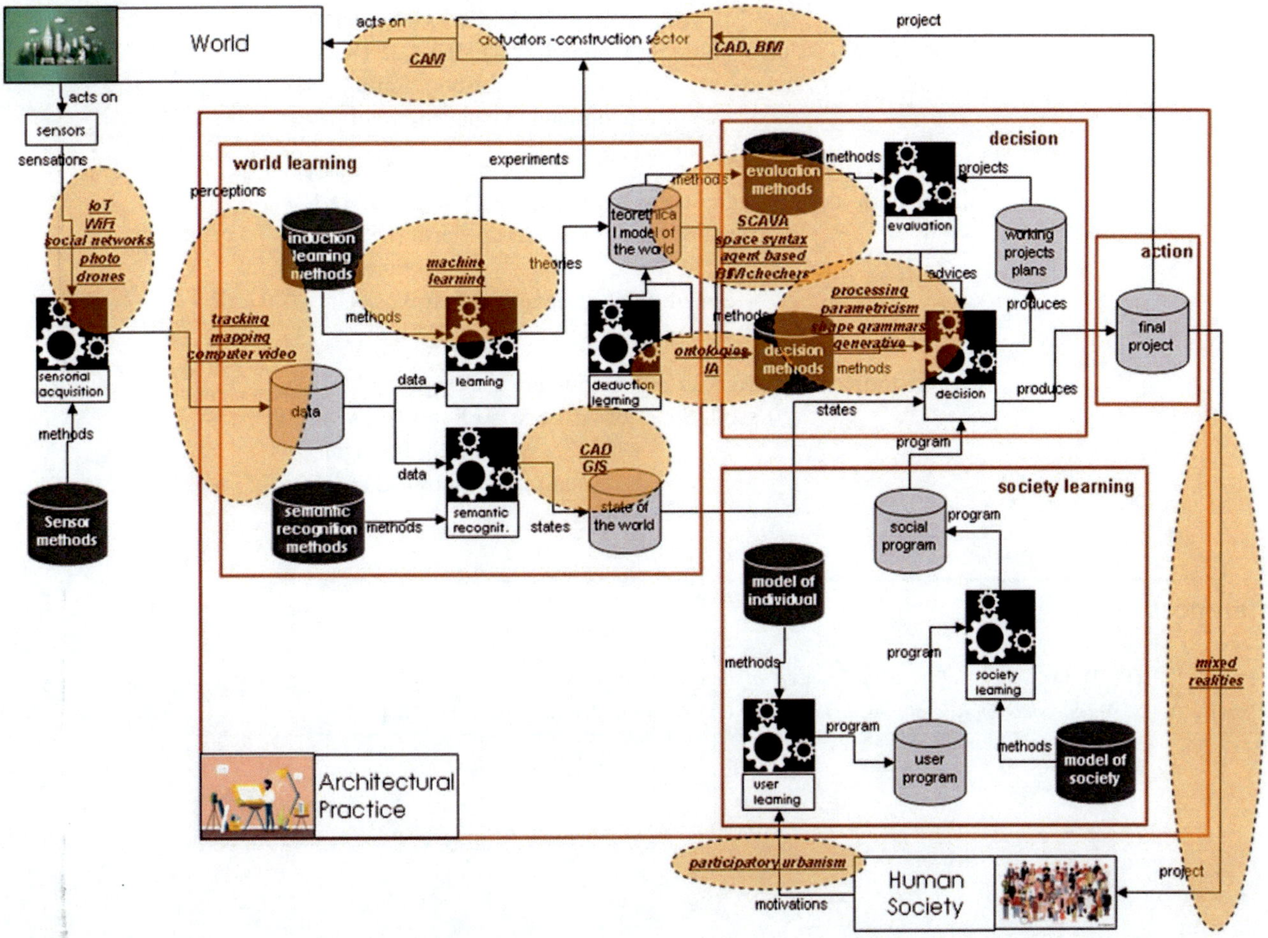

Fig. 1 Framework of formal methods in architectural practice

So, these theories cannot be applied in the synthesis process, but they have plenty of success in the analysis process. This process has two distinct components: the explicit theory, and the analysis machine, for example, in the agent-based analysis method or the theory of structures. Theoretical semantics is the behavior of agents in various generic situations and their consequences, or else the theory of elasticity, in the case of structures. To this theoretical semantics is associated a calculation machine that analyzes the project applying the rules of behavior of the agents to the project in question and arriving at the concrete consequences of that behavior, or else the theory of elasticity to a pre-dimensioned structural geometry. The results obtained are analyzed and can follow as recommendations for the author of the project (architect or automatic process). Some of the projects analyzed are not viable (the structure collapses or traffic is subject to traffic jams or bottlenecks), others are not recommended (e.g., due to costs), and there are some that have the best optimization indicators.

A preliminary scheme We tried to represent more visually what we said earlier. The four different processes of architectural practice are represented, with their mental procedures (the gear wheels), their acquired architectural methodologies—the know-how (the black cylinders), their knowledge about the material world implied in the process—natural, individual or social (the grey cylinders). The various formal methods in use today in architecture fit into several points of this scheme and are represented by cream dashed ellipses. Of course, this is a preliminary, over-simplified model, but it is intended to clarify ideas. May this scheme be useful, is our best wish (Fig. 1).

David Leite Viana
Instituto Universitário de Lisboa (ISCTE-IUL), ISTAR
Lisbon, Portugal
e-mail: david.viana@iscte-iul.pt

Franklim Morais
Laboratório de Investigação em Arquitetura e Design (LIAD)
Escola Superior Artística do Porto (ESAP)
Porto, Portugal
e-mail: franklim-esap@cyberoikos.com

Jorge Vieira Vaz
Laboratório de Investigação em Arquitetura e Design (LIAD)
Escola Superior Artística do Porto (ESAP)
Porto, Portugal
e-mail: iorge@hotmail.com

References

1. Pavlov, I. P. (1932). *Essay on the physiological concept of the symptomatology of hysteria*. Myr, Moscow.
2. Harris, M. (2004). *Introducción a la antropologia general* (7th edition). Madrid: Alianza Editorial.

Contents

Editors and Contributors

About the Editors

Sara Eloy is an architect, Ph.D. in Architecture. Eloy main areas of research include digital technologies applied to architecture, shape grammars, virtual and augmented reality, analysis of the building space namely considering space perception and housing rehabilitation. She is the Director of the Information Sciences Technologies and Architecture Research Center (ISTAR), Iscte, where she is a fellow researcher. She has curated several exhibitions being the latest 'Artificial Realities: Virtual as an Aesthetic Medium for Architectural Ideation,' in the scope of Lisbon Architecture Triennale 2019. She has participated in national and international research projects and published her work in several journals.

David Leite Viana holds a post-doc. in Urban Morphology/Civil Engineering (FEUP), a Ph.D. in Urban and Spatial Planning (IUU-UVa), a DEA in City Planning (ETSA-UVa) and Dipl. Arch. (ESAP). He develops his professional activity in the Municipal Planning Department at Porto City Council and at the Digital Living Spaces Research Group at ISTAR, Iscte. He is visiting fellow in the Theories of Architecture Fellowship Program at TU Delft, Ph.D. supervisor in the doctorate program on Architecture for the Contemporary Metropolitan Territories (Iscte) and external reviewer at Chalmers University of Technology (Architecture and Urban Design Program). He is professor in the specialization program on Collaborative Territories (IPPS, Iscte) and in the Master Program on Geographic Information Systems Applied to Spatial Planning, Urbanism and Landscape at UPV.

Franklim Morais is a Portuguese civil and electronic engineer, Ph.D. professor and researcher. He was the leader of development teams of private commercial software for scientific applications: pioneering, in the early 1980s, the use of finite element methods for structural engineering on micro-computing; and, since the early 1990s, developing COTS private hardware and software systems for smart buildings, urban spaces and cities. He is an invited professor in ESAP's Architecture Program since 2001. He participates in LIAD, an academic research team, in a group dedicated to Formal Methods in Architecture. A major achievement is the production of DepthSpace3D, a digital tool for space syntax analysis. Since 1991, he participated in the organization (scientific and operational committees) of 11 conferences on his scientific fields of interest, including the 5 symposia of the series Formal Methods in Architecture (since 2011).

Jorge Vieira Vaz is a registered architect at Portuguese Architectural Board since 1991, creative designer and technologist. He has a Diploma in Architectural Design from the University of Porto and an Advanced Studies Diploma in Architecture from University of Valladolid. He is professor in Construction and Technologies at the Integrated Master Program in Architecture, ESAP, since 1993; Building Thermal Behavior qualified expert of the Portuguese Energy Agency; Virtual Design to Construction Project Manager (AAVV) and BIM (Building Information Modelling) methodologist. Develops research in taxonomies, ontologies and classification systems, in the scientific field of architecture and construction technology, at LIAD, ESAP.

Contributors

Sema Alaçam Istanbul Technical University, Taskısla, İstanbul, Turkey

Canan Albayrak de Brito Colaço Instituto Universitário de Lisboa (ISCTE-IUL), ISTAR, Lisbon, Portugal

Luísa Almeida Instituto Universitário de Lisboa (ISCTE-IUL), ISTAR, Lisbon, Portugal

Max Andrade Universidade Federal de Pernambuco, Recife, Brazil

Clarice Bleil de Souza Welsh School of Architecture, Cardiff University, Cardiff, UK

Filipe Brandão Instituto Universitário de Lisboa (ISCTE-IUL), ISTAR, Lisboa, Portugal

Inês Caetano INESC-ID/Instituto Superior Técnico, University of Lisbon, Lisbon, Portugal

Adriana Carvalho Faculdade de Engenharia da Universidade Do Porto, Porto, Portugal

Isabel Carvalho Centro de Investigação em Artes e Comunicação (CIAC), Universidade Aberta, Lisbon, Portugal

Aikaterini Chatzivasileiadi Cardiff University, Cardiff, UK

Chi-Li Cheng Graduate Institute of Architecture, National Chiao Tung University, Hsinchu City, Taiwan

Chiara Chioni D.E.S.Te.C, University of Pisa, Pisa, Italy

Carolina Coelho Centre of 20th Century Interdisciplinary Studies, FCTUC Department of Architecture, University of Coimbra, Coimbra, Portugal

Ricardo Mendes Correia Instituto Universitário de Lisboa (ISCTE-IUL), CRIA, Lisboa, Portugal

Fábio Costa Instituto Universitário de Lisboa (ISCTE-IUL), ISTAR, Lisbon, Portugal

Beatriz Couto CGI, Lisbon, Portugal

Filipa Crespo Osório Instituto Universitário de Lisboa (ISCTE-IUL), ISTAR, Lisboa, Portugal

Valerio Cutini Department of Energy, Systems Territory and Construction Engineering, University of Pisa, Pisa, Italy;
D.E.S.Te.C, University of Pisa, Pisa, Italy

Ana de Almeida Instituto Universitário de Lisboa (ISCTE-IUL), ISTAR, Lisbon, Portugal; Centre for Informatics and Systems, University of Coimbra, Coimbra, Portugal

Maria João de Oliveira Instituto Universitário de Lisboa (ISCTE-IUL), ISTAR, Lisboa, Portugal

Sara Eloy Instituto Universitário de Lisboa (ISCTE-IUL), ISTAR, Lisbon, Portugal

Momen FoadMarashi Lisbon School of Architecture, CIAUD, Universidade de Lisboa, Lisbon, Portugal

Ítalo Guedes Universidade Federal de Pernambuco, Recife, Brazil

Rosália Guerreiro Instituto Universitário de Lisboa (ISCTE-IUL), CRIA, Lisboa, Portugal

Leen Hiasat Lisbon School of Architecture, CIAUD Research Center, Universidade de Lisboa, Lisbon, Portugal

June-Hao Hou National Chiao Tung University, Hsinchu City, Taiwan

Wassim Jabi Cardiff University, Cardiff, UK

Pedro Januário Lisbon School of Architecture, CIAUD Research Center, Universidade de Lisboa, Lisbon, Portugal

Egemen Berker Kızılcan Department of Architecture, Middle East Technical University, Ankara, Turkey

Ramesh Krishnamurti Carnegie Mellon University, Pittsburgh, PA, USA

Gisela Lameira Center for Studies in Architecture and Urbanism (CEAU), University of Porto (UPorto), Porto, Portugal

Plácido Lizancos Group of Territorial Studies (GET), University of A Coruña (UDC), A Coruña, Spain

António Leitão INESC-ID/Instituto Superior Técnico, University of Lisbon, Lisbon, Portugal

Andrew Malcolm Ghent University, Ghent, Belgium

Zeynep Mennan Department of Architecture, Middle East Technical University, Ankara, Turkey

Franklim Morais Laboratório de Investigação em Arquitetura e Design (LIAD), Escola Superior Artística do Porto, Porto, Portugal

Aniruddha Mukherjee Universität Innsbruck, Institute of Urban Design (IOUD), Innsbruck, Austria

Robert R. Neumayr University of Applied Arts Vienna, Vienna, Austria

Vítor Oliveira Centro de Investigação do Território, Transportes e Ambiente, Faculdade de Engenharia, Universidade do Porto, Porto, Portugal

Nikolaos Patsavos University of Ioannina, Athens, Greece

Pieter Pauwels Eindhoven University of Technology, Eindhoven, Netherlands

Alan Penn Bartlett School of Architecture, University College London (UCL), London, UK

Virgílio Borges Pereira Center for Studies in Architecture and Urbanism (CEAU), University of Porto (UPorto), Porto, Portugal

David Pereira-Martínez Center for Studies in Architecture and Urbanism (CEAU), University of Porto (UPorto), Porto, Portugal;
Group of Territorial Studies (GET), University of A Coruña (UDC), A Coruña, Spain

Alexandros Peteinarelis University of Ioannina, Ioannina, Greece

Camilla Pezzica Department of Energy, Systems Territory and Construction Engineering, University of Pisa, Pisa, Italy;
Welsh School of Architecture, Cardiff University, Cardiff, UK

Isabel Raposo Research Centre for Architecture, Urbanism and Design (CIAUD), University of Lisbon (ULisboa), Lisbon, Portugal

Micaela Raposo Instituto Universitário de Lisboa (ISCTE-IUL), ISTAR, Lisbon, Portugal

Catarina Ruivo Centro de Investigação em Arquitetura, Urbanismo e Design (CIAUD), Faculdade de Arquitetura da Universidade de Lisboa, Pólo Universitário, Lisbon, Portugal

Simone Rusci D.E.S.Te.C, University of Pisa, Pisa, Italy

Alessandro Santucci Department of Energy, Systems Territory and Construction Engineering, University of Pisa, Pisa, Italy

Francisco Serdoura Lisbon School of Architecture, CIAUD, Universidade de Lisboa, Lisbon, Portugal

Petra Sochůrková Academy of Arts Architecture and Design in Prague, Prague, Czechia

Silvia Spolaor Centro de Investigação do Território, Transportes e Ambiente, Faculdade de Engenharia, Universidade do Porto, Porto, Portugal

Daniel Sviták Academy of Arts Architecture and Design in Prague, Prague, Czechia

Olha Tikhonova Faculty of Architecture, University of Lisbon, Lisbon, Portugal

Shota Tsikoliya Academy of Arts Architecture and Design in Prague, Prague, Czechia

Ekin Ünlü Istanbul Technical University, Taskısla, İstanbul, Turkey

Tasos Varoudis Bartlett School of Architecture, University College London (UCL), London, UK

Jorge Vaz Laboratório de Investigação em Arquitetura e Design (LIAD), Escola Superior Artística do Porto, Porto, Portugal

Imro Vaško Academy of Arts Architecture and Design in Prague, Prague, Czechia

Pedro Veloso Carnegie Mellon University, Pittsburgh, PA, USA

David Leite Viana Instituto Universitário de Lisboa (ISCTE-IUL), ISTAR, Lisbon, Portugal

Pieter E. Vermaas Instituto Universitário de Lisboa (ISCTE-IUL), ISTAR, Lisbon, Portugal;
Philosophy Department, TU Delft, Delft, The Netherlands

Jeroen Werbrouck Ghent University, Ghent, Belgium

Design, Construction and Management
of the Built Environment

Mathematically Developing Building Facades: An Algorithmic Framework

Inês Caetano and António Leitão

Abstract

The importance of Algorithmic Design (AD) is growing due to its advantages for the design practice: It empowers the creative process, facilitating design changes and the exploration of larger design spaces in viable time, and supports the search for better-performing solutions that satisfy environmental demands. Still, AD is a complex approach and requires specialized knowledge. To promote its use in architecture, we present a mathematics-based framework to support architects with the algorithmic development of designs by following a continuous workflow embracing the three main design stages: exploration, evaluation, and manufacturing. The proposed framework targets the design of buildings' facades due to their aesthetical and environmental relevance. In this paper, we explain the framework's structure and its mathematical implementation, and we describe the predefined algorithms, as well as their combination strategies. We focus on the framework's algorithms that generate different geometric patterns, exploring their potentialities to create and modify different facade designs. In the end, we evaluate the flexibility of the framework for generating, modifying, and optimizing different geometrical patterns in an architectural design context.

Keywords

Algorithmic Design • Mathematical framework • Higher-order functions • Facade design

1 Introduction

Algorithmic Design (AD) is a design approach based on algorithms [1]. Compared to manual approaches, AD provides greater flexibility, supports more complex geometries, and handles larger amounts of information. AD also facilitates design changes and automates repetitive tasks, therefore enabling design optimization by automating the search for better-performing design solutions [2]. Still, AD requires specialized knowledge that most architects do not have, namely programming experience. To make AD more attractive to the architectural community, we need to make it more accessible by providing ready-to-use algorithms that can be combined in arbitrary ways. We focus on building facades, and we present a mathematics-based framework for the design of facades.

Building facades are the outer layer of buildings, separating the indoor spaces from the outside ones and, therefore, having many associated functions, namely environmental performance [3, 4], structural behavior [5], cultural identity [6, 7], and urban communication [8, 9]. Lately, facade design has become an increasing complex task due to the current design trend to create intricate geometries/patterns and the growing design constraints, namely environmental requirements, economic limitations, and tight deadlines. AD helps dealing with these constraints, reducing the time and effort needed to explore different design solutions. To facilitate the adoption of AD in architecture, we propose a flexible mathematics-based framework that tames the complexity of AD techniques in the design exploration, analysis, and optimization of facade solutions. The framework is structured in a fivefold classification inspired by previous

I. Caetano (✉) · A. Leitão
INESC-ID/Instituto Superior Técnico, University of Lisbon, Lisbon, Portugal
e-mail: ines.caetano@tecnico.ulisboa.pt

A. Leitão
e-mail: antonio.menezes.leitao@tecnico.ulisboa.pt

S. Eloy et al. (eds.), *Formal Methods in Architecture*, Advances in Science, Technology & Innovation,
https://doi.org/10.1007/978-3-030-57509-0_1

research [10–13], containing several algorithms addressing the facade design process.

2 Methodology

The idea of an algorithmic framework had as inspiration previous works [14–16] proving that sets of algorithms can be generalized and reused in the exploration of new designs. These strategies, known as *modular programming* and *design patterns* [14, 15], promise to mitigate the limitations architects still face when using AD, especially to reduce the time and effort spent with the algorithmic task: They avoid writing algorithms from scratch for each new design, while preventing extensive and potentially error-prone programming efforts. However, when the predefined algorithms are not well-structured, it becomes difficult to combine them. To solve this problem, we propose a mathematics-based algorithmic framework for facade design that provides, for each scenario, a set of algorithms and combination strategies suiting the different design stages. Despite not considering all possible scenarios, which would be an unviable task, the framework addresses the most common ones and can be adapted to more specific ones.

The framework development was twofold: (1) defining a mathematical theory for facade design and (2) implementing the theory in a framework containing predefined algorithms targeting building facades that can be combined in arbitrary ways. The resulting framework promises to (1) promote the use of AD in facade design, (2) solve interoperability issues between design and analysis tools, and (3) guide the selection of the algorithms/strategies that best suit a design scenario.

3 Mathematics-Based Framework

The architectural practice depends on several external factors like the design brief's specificities and both environmental and economical requirements. This means different projects require different approaches either using an AD or a non-AD approach. As we address the AD one, we must handle design in a way that a computer understands. Given that computational tools are based on instructions transmitted through programming languages (PLs) and that most PLs are inspired by the universal language of mathematics, we consider the latter's formalism to:

1. structure the AD theory;
2. implement the different algorithms;
3. organize both (1) and (2) in an algorithmic framework specialized in facade design.

The mathematics-based framework must be able to handle the design (1) *variability*, adapting to the ever-changing design requirements, (2) *diversity*, embracing and evaluating diverse design problems and scenarios, and (3) *coherency*, correctly integrating the design information in a single workflow. Regarding its structure, the framework organizes the predefined algorithms in a fivefold classification, which we describe in the next sections.

4 Framework Implementation

The framework organizes the algorithms based on their type and role in the facade design process in the following categories: *geometry*, *pattern*, *distribution*, *optimization*, and *rationalization*. For each one, it provides $\mathbb{R} \rightarrow \mathbb{R}$, $\mathbb{R} \rightarrow \mathbb{R}^2$, $\mathbb{R}^2 \rightarrow \mathbb{R}^2$, and $\mathbb{R}^2 \rightarrow \mathbb{R}^3$ algorithms that can be then combined with one another through function composition.

In general, the framework handles all surface-related functions $S(u, v)$ within the domain $0 \leq u \leq 1, 0 \leq v \leq 1$ and provides operators that can be arbitrarily combined, namely the one-dimensional linear variation function $linear(a, b) = \lambda(t).a + (b - a)t$ and the (paradoxical) constant "variation" function $constant(c) = \lambda(t\ldots).c$. Here, we employ the λ-calculus notation for an anonymous function with parameter t [17]: The result of the function *constant* is therefore an anonymous function that can be combined with functions of any number of arguments, adapting its number of arguments according to those of the combined function.

The next sections explain the implementation of the framework's categories.

4.1 Geometry

This category contains algorithms to define the facade geometry. As the latter's domain is often two dimensional, we need to extend the one-dimensional variations' domain into $\mathbb{R}^2$. We use *higher-order functions* (HOFs) [17] to define two functions, $dim_u(f) = \lambda(u, v).f(u)$ and $dim_v(f) = \lambda(u, v).f(v)$ that make a function f vary only in one dimension, i.e., u or v accordingly. To generalize function composition operations, we provide the operator

$$\circ(f, g_1, \ldots, g_n) = \lambda(x_1, \ldots, x_m).f(g_1(x_1, \ldots, x_m), \ldots, g_n(x_1, \ldots, x_m)) \tag{1}$$

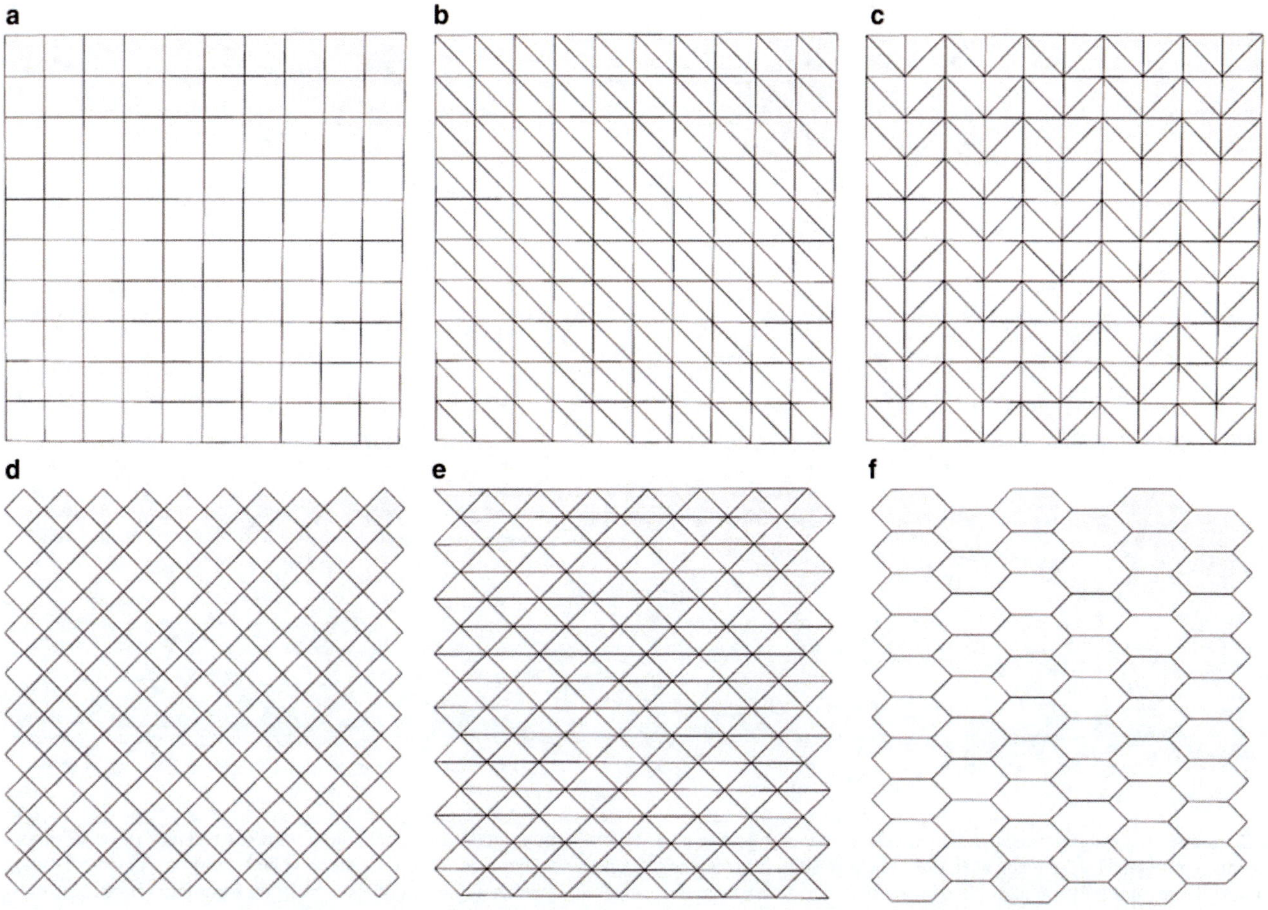

Fig. 1 Some *distribution* algorithms: **a** $\text{grid}_{\text{square}}$; **b** $\text{grid}_{\text{triangles}}$; **c** $\text{grid}_{\text{triangles2}}$; **d** $\text{grid}_{\text{rhombus}}$; **e** $\text{grid}_{\text{triangles3}}$; **f** $\text{grid}_{\text{hexagons}}$

and, to simplify the notation used, we define $u \otimes lim = dim_u(linear(0, lim))$ and $v \otimes lim = dim_v(linear(0, lim))$, wherein lim is the domain's upper limit. We also treat all numbers in a function context as $constant(n)$ and we represent any first-order function that receives functional arguments $g_1, \ldots, g_n$ as $\circ(f, g_1, \ldots, g_n)$, thus being $f \times g$ the same as $\circ(\times, f, g)$.

With HOFs, we can move from the numeric space into the functional space and combine functions using functional operators rather than simply combining numbers using numeric operators. In a functional space, the algorithm $straight(w, h) = \lambda(u, v).XYZ(u \times w, 0, v \times h)$, which represents a $w \times h$ planar parametric surface, has the equivalent representation $straight(w, h) = XYZ(u \otimes w, 0, v \otimes h)$. The other predefined surface geometries follow the same logic, which is further detailed in [13].

Having the surface's algorithmic description, we can now explore its geometric pattern by using the algorithms of both *pattern* and *distribution* categories, which we explain in the next sections: The former create the shape(s) composing the pattern and the latter distribute those shapes on the surface domain.

4.2 Distribution

This category contains algorithms to distribute elements on a surface. These algorithms receive a matrix of the surface points on which the distribution will be made, as provided by the *geometry* algorithms, returning another matrix with the same points rearranged in different distribution configurations. As an example, the algorithm $grid_{squares}$ rearranges the points in sets of four points describing a squared area on the surface. Regarding the algorithm $grid_{hexagons}$, it reorganizes the points in sets of six points representing hexagonal areas. Figure 1 illustrates some of the predefined distributions.

4.3 Pattern

This category contains algorithms to create different geometric patterns. Generally, a geometric pattern results from repeating an element either along one or both dimensions of a two-dimensional domain. In the first case, the result is a pattern continuous in one of the domain's directions but

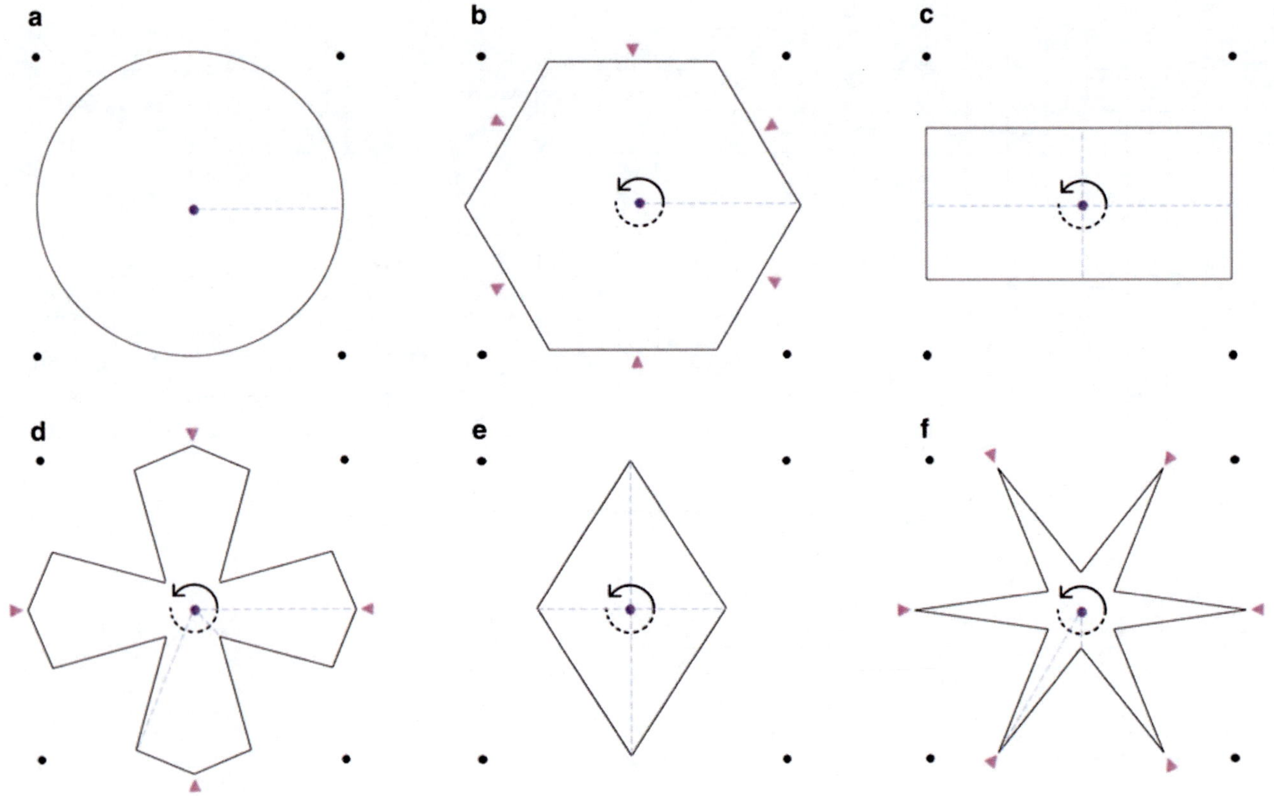

Fig. 2 Some *shape* algorithms and their inputs: In addition to the set of points represented by the black dots, the circle algorithm (**a**) receives a radius; the regular polygon algorithm (**b**) receives a radius, an angle, and a number of sides; the rectangle algorithm (**c**) receives a length, a width, and an angle; the rosette algorithm (**d**) receives three radii, an angle, and a number of vertices; the rhombus algorithm (**e**) receives two diagonals and an angle; and the star polygon algorithm (**f**) receives two radii, an angle, and a number of vertices

discrete in the other; whereas in the second, the pattern is discrete in both directions. We name the former as *continuous* and the latter as *discrete* pattern. In both cases, the repeated element can be kept unchanged along the facade domain or can suffer some transformations regarding its shape, size, etc. Therefore, this category provides algorithms to generate different geometric shapes, as well as to apply different geometric transformations to them, which are organized in two groups: *shape* and *transformation*. Each group contains algorithms handling both *discrete* and *continuous* patterns but, for the scope of this paper, we focus on the former ones. The latter ones are further detailed in [18].

Shape. This group contains algorithms to create different 2D and 3D geometric shapes, including polygonal, ellipsoidal, and spherical ones, among others. The available algorithms all receive the set of points where to center the geometric shape; an information provided by the *distribution* algorithms. The remaining parameters, in turn, depend on the characteristics of each geometric shape. Figure 2 illustrates some *shape* algorithms and their corresponding parameters.

As an example, consider the function that creates star polygons: besides the set of points (*pts*), it receives the number of vertices ($n_{vertices}$), the inner and outer radii (r_{inner} and r_{outer}), and an angle (α):

$$shapeStar(pts, n_{vertices}, r_{inner}, r_{outer}, \alpha) \qquad (2)$$

To create a geometric pattern based on star polygons, we need to define (1) the surface on which to apply the pattern and (2) the type of elements' distribution. We choose a straight facade and a rhombus distribution, which requires combining three algorithms:

1. one shaping the straight surface where to create the pattern—*straight*;
2. another creating a rhombus grid of points—$grid_{rhombus}$;
3. a last one producing the star polygons—*shapeStar*.

Figure 3a–c illustrates the result of this composition using different inputs.

We can simplify the composition of the first two algorithms by writing $grid_{rhombus}(straight(w, h))$ and, to facilitate the mathematical representation of algorithms dealing with matrices, we can take advantage of *broadcasting*, i.e., the

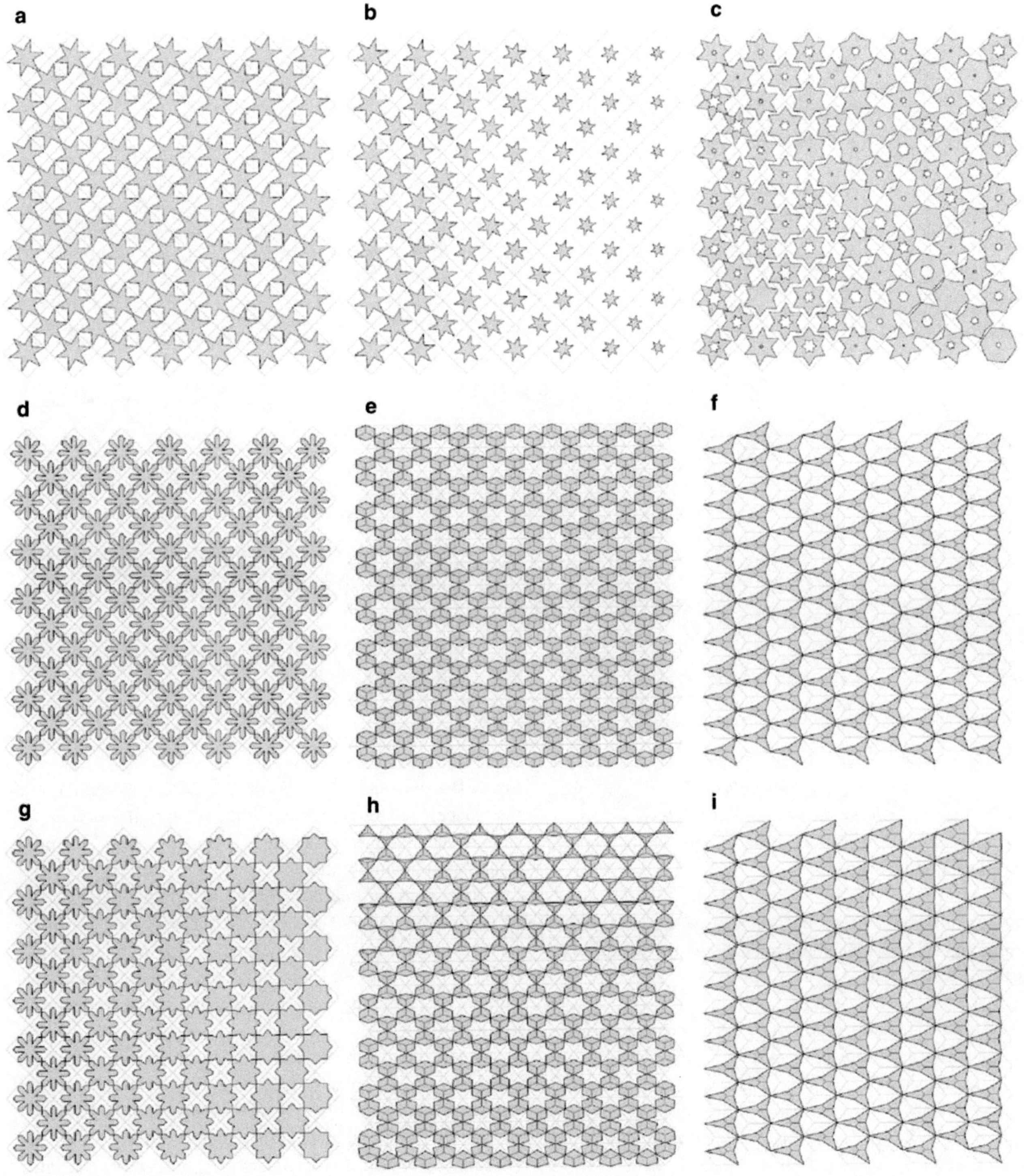

Fig. 3 Examples of *shape* algorithms combined with different *geometry* and *distribution* algorithms

application of a function to an array of elements, even if the latter has a different number of dimensions from the other received arguments. *Broadcasting* is represented by the *dot syntax* $f.(args...)$ and it can be applied in single or nested calls $f.(g.(args...))$. This means we can simplify

$$shapeStar\begin{pmatrix} \circ(grid_{rhombus}, straight)(w, h), \\ constant(n_{vertices}), \\ constant(r_{inner}), \\ constant(r_{outter}), \\ constant(\alpha) \end{pmatrix} \quad (3)$$

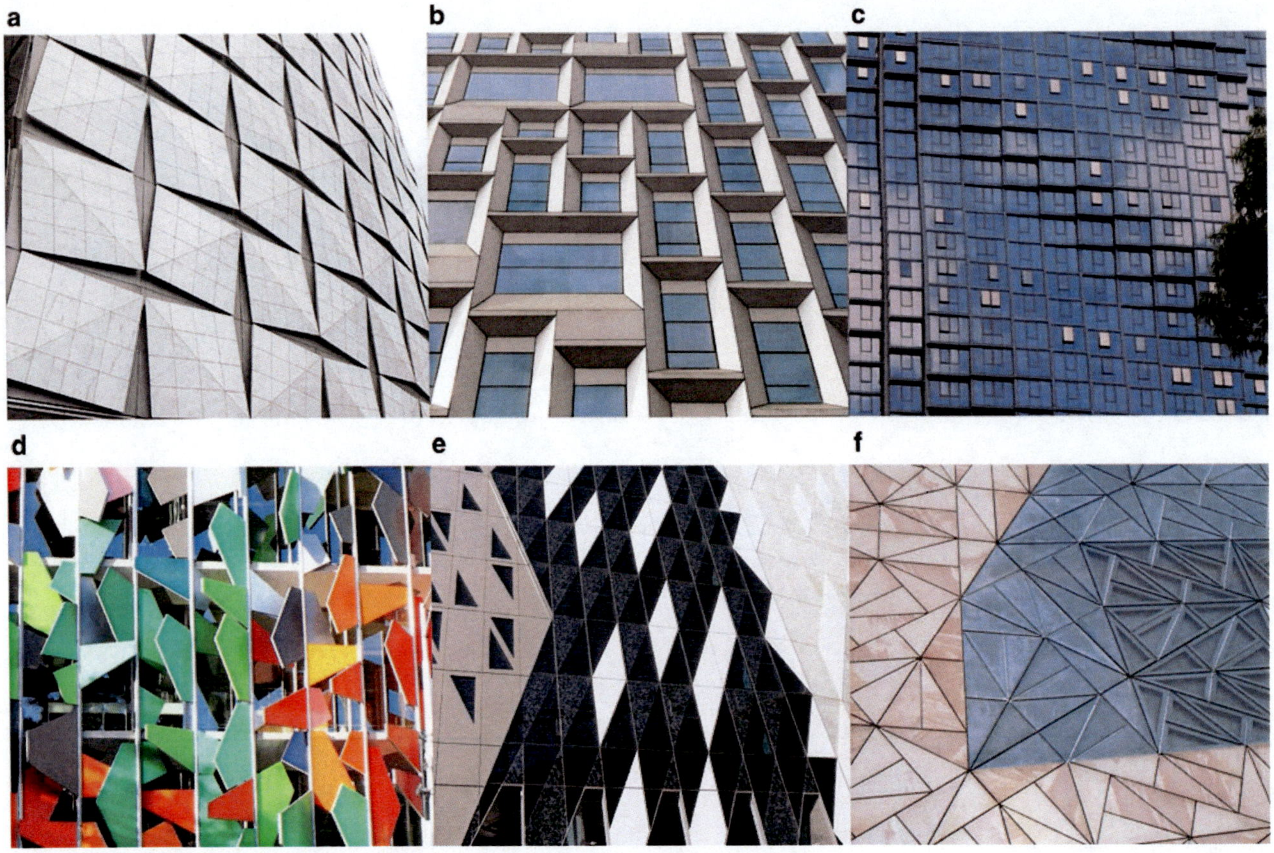

Fig. 4 Examples of building facade patterns (Photos ©The authors)

into $shapeStar.(grid_{rhombus}(straight(w,h)), n_{vertices}, r_{inner}, r_{outer}, \alpha)$ $regularPolygon.(grid_{square}(straight(w,h)), n_{sides}, r_{size})$. This is valid to all *shape* algorithms, being some of them illustrated in Fig. 3.

Transformation. This group provides geometric transformations algorithms, which are organized into *affine* and *rule-based* transformations.

The former includes *scaling* (T_{scale}), to change distances between points according to a constant factor k, compressing or stretching shapes if $k < 1$ or $k > 1$, correspondingly; *reflection* (T_{mirror}), to invert shapes with respect to an axis or point; *rotation* (T_{rotate}), to rotate shapes around an axis; *shearing* (T_{shear}), to distort shapes parallel to an axis or plane; and *translation* ($T_{translate}$), to move shapes according to a displacement vector.

The latter contains transformations resulting from the application of rules inspired by real examples (Fig. 4), including *shape substitution* (T_{shape}), to replace shapes with new ones; *color application* (T_{color}), to apply different colors to the shapes; *protrusion creation* ($T_{protrude}$), to move the shapes perpendicularly to the surface; *recursive subdivision* ($T_{subdivide}$), to subdivide shapes by recursively applying a geometric rule; and *edge deformation* (T_{edge}), to bend or fold the shapes' edges. Combining these algorithms with the *shape* ones allows generating a wider variety of more dynamic geometric patterns.

Affine Transformations. These algorithms receive a surface ($ptss$), a specific position on the surface (pt), and a factor controlling the transformation effect intensity (k), being 0 and 1 the null and maximum effects, correspondingly. The remaining arguments depend on the transformation to apply.

As an example, the algorithm T_{scale} allows scaling a shape according to different criteria, including (1) its position, (2) its distance to one or more *attractor* points/curves, (3) being or not contained in certain surface areas, and (4) random rule(s). For each case, T_{scale} receives the information needed to perform the transformation, e.g., the direction of the effect to produce (option 1); a set of *attractor* points or curves (option 2); a set of surface areas (option 3); and random values (option 4). We describe this algorithm as $T_{scale}(ptss, pt, k, args\ldots)$, being $args\ldots$ the supported optional arguments. This logic applies to all *affine* transformations.

To illustrate the practical application of these algorithms, we start with a pattern resulting from horizontally and

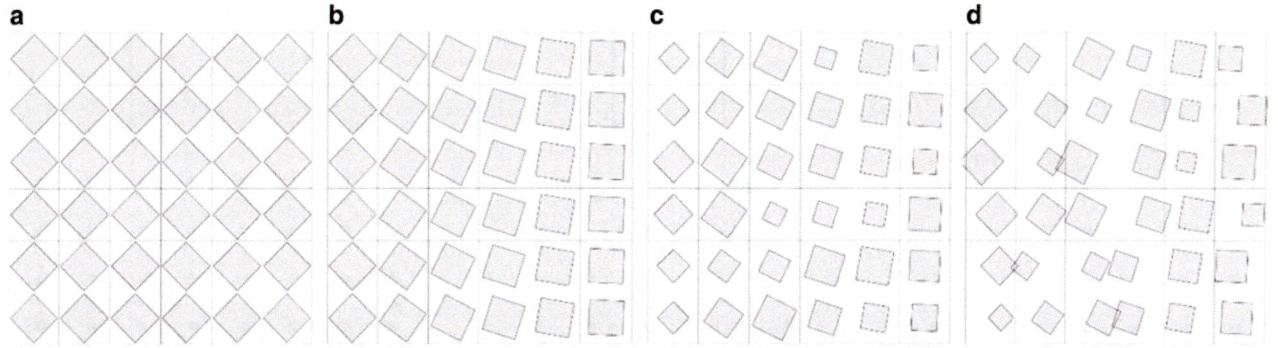

Fig. 5 Pattern of squares distributed in a squared grid (**a**), whose angle varies horizontally (**b**) and whose radius (**c**) and center position (**d**) change randomly

vertically aligned squares. We select the algorithms *shapePolygon*, to create the squares, and $grid_{squares}$, to distribute them in a squared grid (Fig. 5a). To increase the squares' rotation in the u direction, we combine both algorithms with the algorithm T_{rotate}, which, in this case, returns a factor that increases with the surface length. The latter, in turn, affects the *shapePolygon*'s parameter angle (α), producing squares whose angle increases with the u dimension (Fig. 5b):

$$shapePolygon.\left(grid_{squares}(ptss), n_{sides}, r_{outer}, \alpha \times T_{rotate}\left(ptss, pt, k, \underset{v_u}{\rightarrow}\right)\right) \tag{4}$$

wherein $\underset{v_u}{\rightarrow}$ is the transformation effect direction.

In this composition, $grid_{squares}$ and T_{rotate} inform the two parameters *pts* and α of *shapePolygon*, correspondingly: The former provides the position of each polygon and the latter changes its angle according to a factor. In turn, both algorithms are informed by the algorithm *straight* but with a small difference: while $grid_{squares}$ takes all surface points at once, T_{rotate} receives a surface point at a time, corresponding to the position of the element to rotate. The latter case can therefore benefit from *broadcasting*:

$$T_{rotate}.\left(straight(w,h), grid_{squares}(straight(w,h)), k, \underset{v_u}{\rightarrow}\right) \tag{5}$$

The resulting composition therefore applies *broadcasting* techniques in nested function calls:

$$shapePolygon.\left(\begin{array}{c} grid_{squares}(straight(w,h)), \\ n_{sides}, \\ r_{outer}, \\ \alpha \times T_{rotate}.\left(straight(w,h), grid_{squares}(straight(w,h)), k, \underset{v_u}{\rightarrow}\right) \end{array}\right) \tag{6}$$

Following this logic, we can apply multiple transformations to this same pattern by combining more algorithms in sequential function compositions. For instance, to randomly change both the squares' radius (Fig. 5c) and center position (Fig. 5d), we select the algorithms T_{scale}, to control the parameter radius (r_{outer}), and $T_{translate}$, to change the *pts* position.

Figure 3 illustrates more examples combining T_{scale} with different *shape* algorithms: In pattern B, T_{scale} makes both *shapeStar*'s parameters r_{inner} and r_{outer} uniformly decrease along the u dimension; in pattern C, T_{scale} controls only the *shapeStar*'s parameter r_{inner} to randomly increase along the u dimension; and, in pattern G, T_{scale} controls the *shapeRosette*'s parameter r_{inner} to uniformly increase along the u dimension.

Rule-based Transformations. Rule-based transformations are more complex than *affine* ones. Therefore, their mode of combination differs from what we have seen so far. Before explaining their application, we introduce the concept of *matrix of functions* (MF), i.e., a matrix containing different functions. We use MFs to group the transformation algorithm(s) to apply, while defining their pattern of application. We represent a MF as

$$MF = \begin{bmatrix} f1 & f2 \\ g1 & g2 \end{bmatrix}, \quad \text{where} \left| \begin{array}{l} f_1 \in \mathcal{L}(U,U), f_2 \in \mathcal{L}(U,V), \\ g_1 \in \mathcal{L}(V,U), g_2 \in \mathcal{L}(V,V). \end{array} \right. \tag{7}$$

MFs are useful to iterate along two-dimensional domains, which is the case of our surface-related algorithms: *Geometry* and *distribution* algorithms produce two-dimensional *matrices of points* (MP). To deal with possible size differences between matrices, *rule-based* algorithms map the smallest size matrices along the largest size ones, iteratively applying the former to a submatrix of the latter of the same size. For instance, consider Fig. 6: When mapped along the MP below, the 2 × 2 matrix A affects 2 × 2 submatrices of

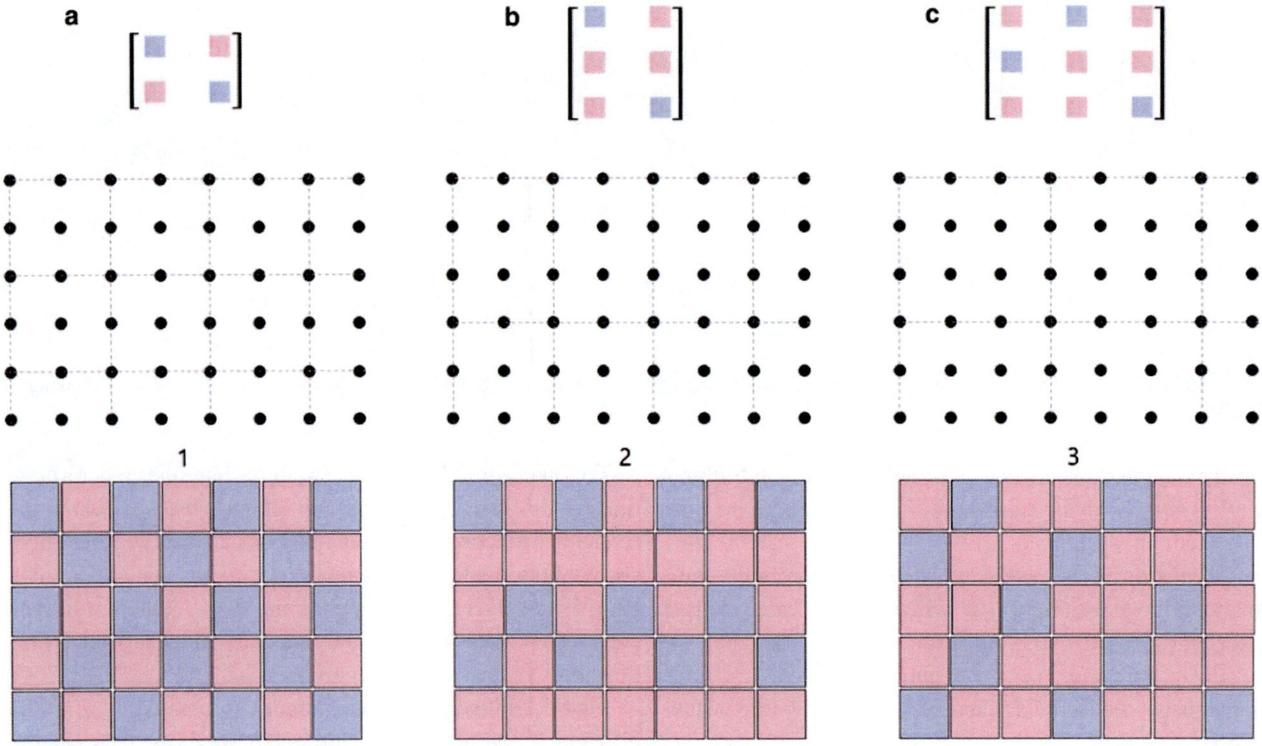

Fig. 6 Examples of *matrix of functions* of different sizes applied to the same *matrix of points*

the MP at a time, producing pattern 1; the same happens with the examples B and C. Note that, the framework supports MFs of any size, including non-squared matrices, row matrices, columns matrices, unit matrices, or even matrices covering the entire surface at once. Having this knowledge, we can now focus on these algorithms' application.

Rule-based algorithms receive tree matrices: one with the surface points (*ptss*), another one with one or more algorithms to apply (M_{rules}), and a last one describing their order of application ($M_{pattern}$):

$$T_{ruleBased}\left(ptss, M_{rules}, M_{pattern}\right) \tag{8}$$

Mathematically, these matrices are different, being *ptss* a MP; M_{rules} a MF; and $M_{pattern}$ an integer matrix. The result is a new matrix merging both M_{rules} and $M_{pattern}$ information, which organizes the algorithms of M_{rules} according to the sequence set by $M_{pattern}$: The latter's integers identify which algorithm of M_{rules} to apply at each position. For instance, consider both matrices $M_{rules} = [f \quad g]$ and $M_{pattern} = \begin{bmatrix} 1 & 2 \\ 2 & 1 \end{bmatrix}$. As the latter's integers correspond to the former's algorithms indexes, integer 1 matches the same index algorithm, i.e., f, and integer 2 the algorithm of index 2, i.e., g, being the result $\begin{bmatrix} f & g \\ g & f \end{bmatrix}$. As this new matrix size often differs from that of *ptss*, it is then resized to be a matrix of the same size as *ptss*, containing the algorithms of M_{rules} arranged according to the pattern set by $M_{pattern}$.

We illustrate their application with some examples combining *rule-based* algorithms with those previously presented. For example, to replace the squares in Fig. 5d with circles at every four elements (Fig. 7a), we select the *shape substitution* algorithm T_{shape} and we provide it with three matrices:

1. one containing the surface points: *ptss*;
2. one identifying the shapes composing the pattern: M_{shapes};
3. a last one describing their mode of application: $M_{pattern}$.

In this case, *ptss* are the same as in Fig. 5d, M_{shapes} is a matrix containing both algorithms *shapePolygon* and *shapeCircle*, and $M_{pattern}$ a matrix describing their order of application:

$$T_{shape}(grid_{squares}(straight(w,h)), [shapePolygon \quad shapeCircle], [1 \quad 1 \quad 1 \quad 2]) \tag{9}$$

As visible in Fig. 7a, we can apply different transformations to each *shape* algorithm in M_{shapes}: while the squares are affected by both algorithms T_{scale} and $T_{translate}$, the circles are not. The example in Fig. 7b, in turn, already applies both transformations to both *shape* algorithms.

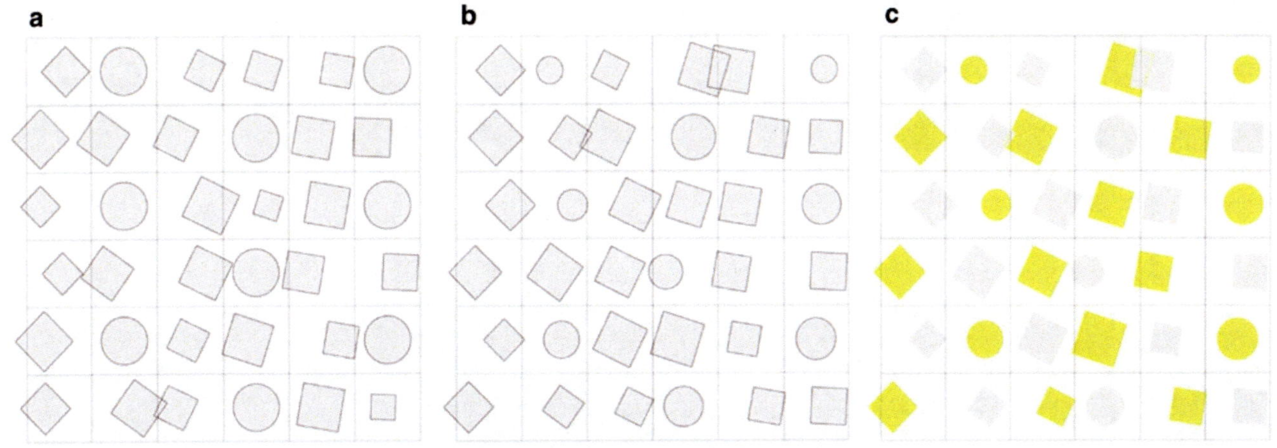

Fig. 7 Pattern in Fig. 5d combined with *rule-based* algorithms to create a circle for every four squares with **a** a constant size and position or **b** a size and position that varies in the same way as the squares, and to **c** color both squares and circles in a yellow and white chess pattern

We follow the same logic in the remaining *rule-based* algorithms. As an example, to apply different colors to the pattern in Fig. 7b, we select the *color application* algorithm T_{color}. We provide it with the same *ptss* plus two new matrices, one with the colors to apply and another with their pattern of application (Fig. 7c):

$$T_{color}\left(ptss, [\blacksquare\ \square], \begin{bmatrix}2 & 1\\ 1 & 2\end{bmatrix}\right) \tag{10}$$

In this composition, T_{shape} and T_{color} originate a new matrix of the same size as *ptss*, merging the information of the following matrices $[shapePolygon\ \ shapeCircle]$, $[1\ 1\ 1\ 2]$, $[\blacksquare\ \square]$, and $\begin{bmatrix}2 & 1\\ 1 & 2\end{bmatrix}$: as the first two produce $[shapePolygon\ shapePolygon\ shapePolygon\ shapeCircle]$ and the last ones return $\begin{bmatrix}\square & \blacksquare\\ \blacksquare & \square\end{bmatrix}$, the resulting matrix therefore is

$$\begin{bmatrix} shapePolygon & shapePolygon & shapePolygon & shapeCircle \\ shapePolygon & shapePolygon & shapePolygon & shapeCircle \end{bmatrix} \tag{11}$$

To move the shapes perpendicularly to the surface and create different three-dimensional effects like the one in Fig. 4c, we use the algorithm $T_{protrude}$. Besides the surface points *ptss*, it receives the protrusion movements to apply, which we represent with vector functions, and their order of application:

$$T_{protrude}\left(ptss, M_{vectors}, M_{pattern}\right) \tag{12}$$

Like the previous *rule-based* algorithms, $T_{protrude}$ maps the protrusion movements along the surface's shapes as set in $M_{pattern}$.

As another example, to recursively subdivide a triangular tiling [19] using the geometric rule in Fig. 8a, we select the algorithm $T_{subdivide}$ and provide it with the subdivision rule to apply ($M_{subdivide}$). As subdivision rules are recursively applied to shapes, we have to prevent them from being endlessly executed by controlling the number of iterations. Therefore, $T_{subdivide}$ receives an additional input, the subdivision rules' level of recursion ($l_{recursion}$):

$$T_{subdivide}\left(ptss, M_{subdivide}, M_{pattern}, l_{recursion}\right) \tag{13}$$

As an example, in Fig. 8b, rule A is applied twice, corresponding to $l_{recursion} = 2$; whereas, in Fig. 8c and d, it is applied three and four times, respectively. Figure 8e illustrates rule A applied to a triangular tiling with different levels of recursion. Figure 8f and g result from the same rule and $l_{recursion} = 3$ on different tilings.

We can also add some randomness to the subdivision rules, as illustrated in Fig. 9(I.C and II.C). This is triggered by the additional input *random*, which receives a *Boolean* value: when true, it adds some randomness to the subdivision rule's level of recursion.

Finally, to deform the edges of a square tiling (Fig. 10a), we select the algorithm T_{edge}. Mathematically, we represent the edge deformation movement with a single vector function v, being its amplitude the vector's length and its direction the vector's sign (Fig. 10(1)). We can apply multiple deformation movements to the same edge and obtain deformations with different amplitude and directions: Fig. 10(2) illustrates the result of two deformation movements of opposite directions. To deal with multiple deformations, T_{edge} needs to be informed about the edge's subdomain to which each deformation movement will be applied, which we represent with different t factors (Fig. 10(2)). We can also control the maximum curvature

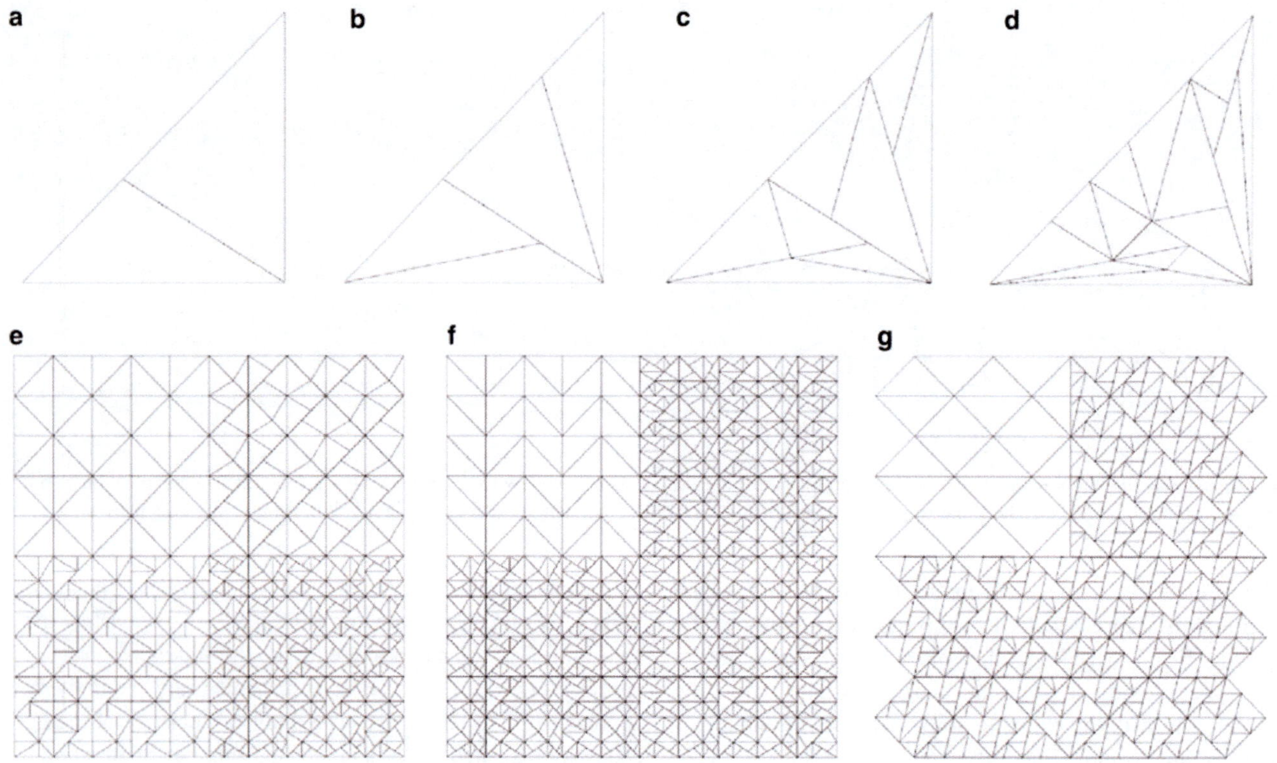

Fig. 8 **a** subdivision rule; **b–d** the rule applied twice, thrice, and four times; **e–g** rule **a** applied in three triangular tilings (the latter are illustrated in the top-left of each example)

position of each deformation movement with different k factors: In Fig. 10(3), the left deformation has $k = 0.5$ (the middle position); whereas, the right one has $k = 0.3$.

Considering this, T_{edge} receives two additional inputs: a matrix containing the t factors and another one with the k factors. In sum, T_{edge} receives (1) the surface points (*ptss*), (2) the vector functions organized in sets $\{v_1, ..., v_n\}$ ($M_{vectors}$), (3) the latter's order of application ($M_{pattern}$), (4) the t factors $\{t_1, ..., t_n\}$ ($M_{subdomain}$), being $t = 0$ the edge's starting point and $t = 1$ its ending point, and (5) the k factors $\{k_1, ..., k_n\}$ ($M_{curvature}$):

$$T_{edge}(ptss, M_{vectors}, M_{pattern}, M_{subdomain}, M_{curvature}) \quad (14)$$

In practice, $M_{pattern}$ dictates where each set of vector functions is used over the *ptss* and, in each position, the applied vector functions receive the respective factors in both matrices $M_{subdomain}$ and $M_{curvature}$.

Figure 10 illustrates some edge deformations examples, being B, for instance, the result of applying T_{edge} to A with the following inputs:

- $M_{vectors} = [\{v_1, v_1, v_2, v_2\}]$, wherein v_1 and v_2 have the same amplitude but opposite directions;
- $M_{pattern} = \begin{bmatrix} 1 & 1 \\ 1 & 1 \end{bmatrix}$, meaning the set $\{v_1, v_1, v_2, v_2\}$ is applied to all squares;
- $M_{subdomain} = [\{0, 1\}]$, which indicates the deformation occurs in the entire edge;
- $M_{curvature} = [\{0.5\}]$, informing the maximum curvature matches the edge's middle.

4.4 Optimization and Rationalization

These categories contain algorithms to analyze, optimize, and rationalize designs, by either driving the changes made to the geometric pattern or controlling the number of different shapes composing the pattern.

For the former case, we can select algorithms like $opt_{daylight}$, to optimize natural daylight illumination, or $opt_{ventilate}$, to improve the natural ventilation, among others. Generally, *optimization* algorithms receive (1) the surface domain *ptss* and (2) the pattern to optimize *shapes*. The other arguments, in turn, depend on the type of optimization to perform. As an example, $opt_{daylight}$ receives information

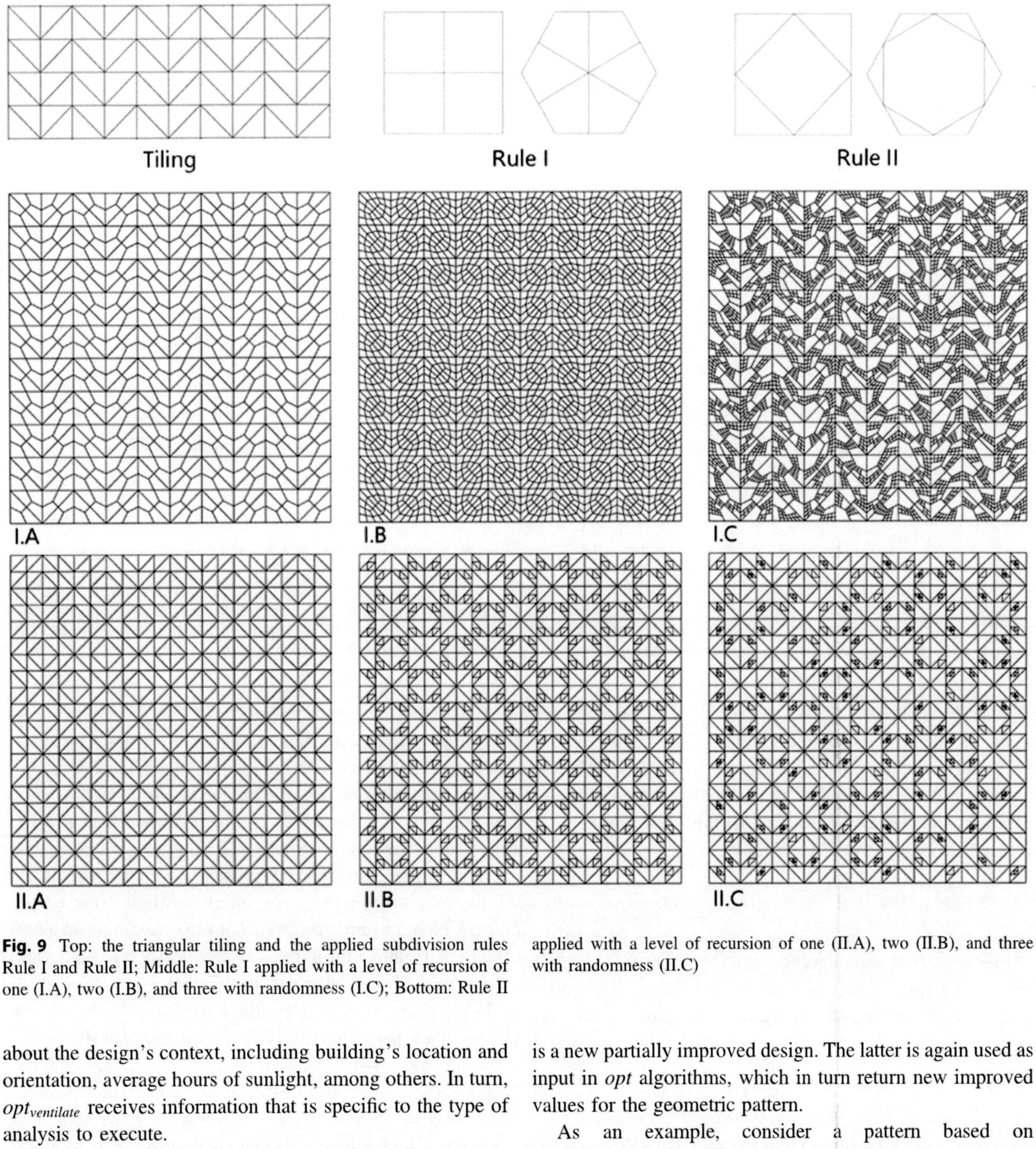

Fig. 9 Top: the triangular tiling and the applied subdivision rules Rule I and Rule II; Middle: Rule I applied with a level of recursion of one (I.A), two (I.B), and three with randomness (I.C); Bottom: Rule II applied with a level of recursion of one (II.A), two (II.B), and three with randomness (II.C)

about the design's context, including building's location and orientation, average hours of sunlight, among others. In turn, $opt_{ventilate}$ receives information that is specific to the type of analysis to execute.

Based on the received information, *opt* algorithms analyze the design and change it according to the results obtained. Then, they rerun the analysis, repeating the cycle until reaching a solution with the desired performance requirements. In practice, *opt* algorithms return one or more values to be then used as input in the final pattern, which may result in shapes of different sizes (if affecting size-related parameters), positions (if altering translation-related values), deformations (if changing deformation-related inputs), or even geometries (if modifying shape-related functions). The result is a new partially improved design. The latter is again used as input in *opt* algorithms, which in turn return new improved values for the geometric pattern.

As an example, consider a pattern based on rhombus-shaped openings whose size needs to be optimized to match certain performance requirements: The more intense the red tone is, the smaller the aperture size should be (Fig. 11). We select the rhombus-shaped algorithm

$$shapeRhombus\left(pts, u_{diagonal}, v_{diagonal}, \alpha\right) \quad (15)$$

wherein $u_{diagonal}$ and $v_{diagonal}$ are its horizontal and vertical diagonals and α its rotation. To optimize their size, we select an *opt* algorithm, which we generically represent as

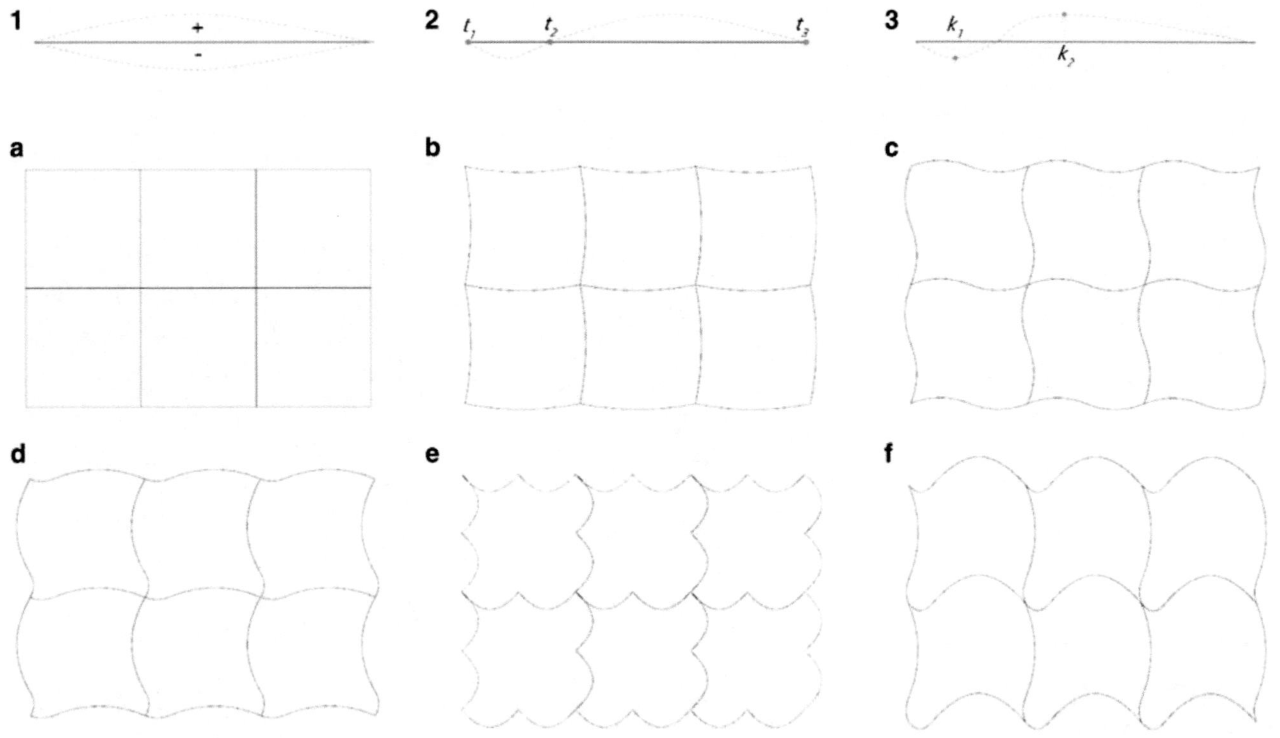

Fig. 10 Top: (1) edge deformation movement sign and amplitude; (2) edge subdomain factors; (3) deformation maximum curvature factors; bottom: **a** a square tilling; **b–f** the same square tilling after applying the algorithm T_{edge} with different input values

$opt(ptss, shapes, args\ldots)$, being $args\ldots$ the information related to the performed optimization. We combine both algorithms so that the latter controls the size-related parameters of the former, which, in this case, are $u_{diagonal}$ and $v_{diagonal}$. In case the *opt* algorithm optimizes the $u_{diagonal}$, the pattern converges toward example A:

$$shapeRhombus(pts, opt(ptss, shapes, args\ldots) \times u_{diagonal}, v_{diagonal}, \alpha) \tag{16}$$

Otherwise, if the *opt* algorithm improves the $v_{diagonal}$, the pattern converges toward example B. Finally, if both parameters are optimized, the pattern becomes similar to pattern C:

$$shapeRhombus\begin{pmatrix} pts, \\ opt(ptss, shapes, args\ldots) \times u_{diagonal}, \\ opt(ptss, shapes, args\ldots) \times v_{diagonal}, \\ \alpha \end{pmatrix} \tag{17}$$

To control the number of different shapes composing the pattern, we select the algorithm *discretize*, which is important to minimize the design manufacturing costs, while increasing its construction viability. When applied, *discretize* iteratively reduces the pattern's diversity of shapes, while making the balance between its design intent and performance. Finally, to identify and count the different existing shapes, we use the algorithm *tallying*, which returns each shape typology locations and quantities; an information that is critical to proceed to the ensuing manufacturing stage.

Mathematically, *discretize* follows a discretization process that aims at reducing the number of values accepted by a continuous variable, converting a continuous set of values into a finite range grouped into n intervals. This allows us to set the maximum number of intervals of the discretized range, which, in turn, corresponds to the maximum number of different elements in the design ($n_{typologies}$). Since the existence of different elements in a pattern is directly related to the number of *shape* algorithms used and to the values given to their parameters, it is then important that *discretize* constrains these two ranges. In practice, *discretize* collects the values that differ from element to element, producing a new range of values varying between the maximum and minimum values found and containing the same number of values as $n_{typologies}$. Then, it replaces the parameters' original values with their closest match in the new interval, guaranteeing the final design contains $n_{typologies}$ different shapes.

Regarding the algorithm *tallying*, it extracts/stores the design information that is important to manage its construction process on site, including the existing element typologies' quantities and positions. Mathematically, *tallying* receives a pattern, returning a list with its shapes organized by type, with their quantities, geometric

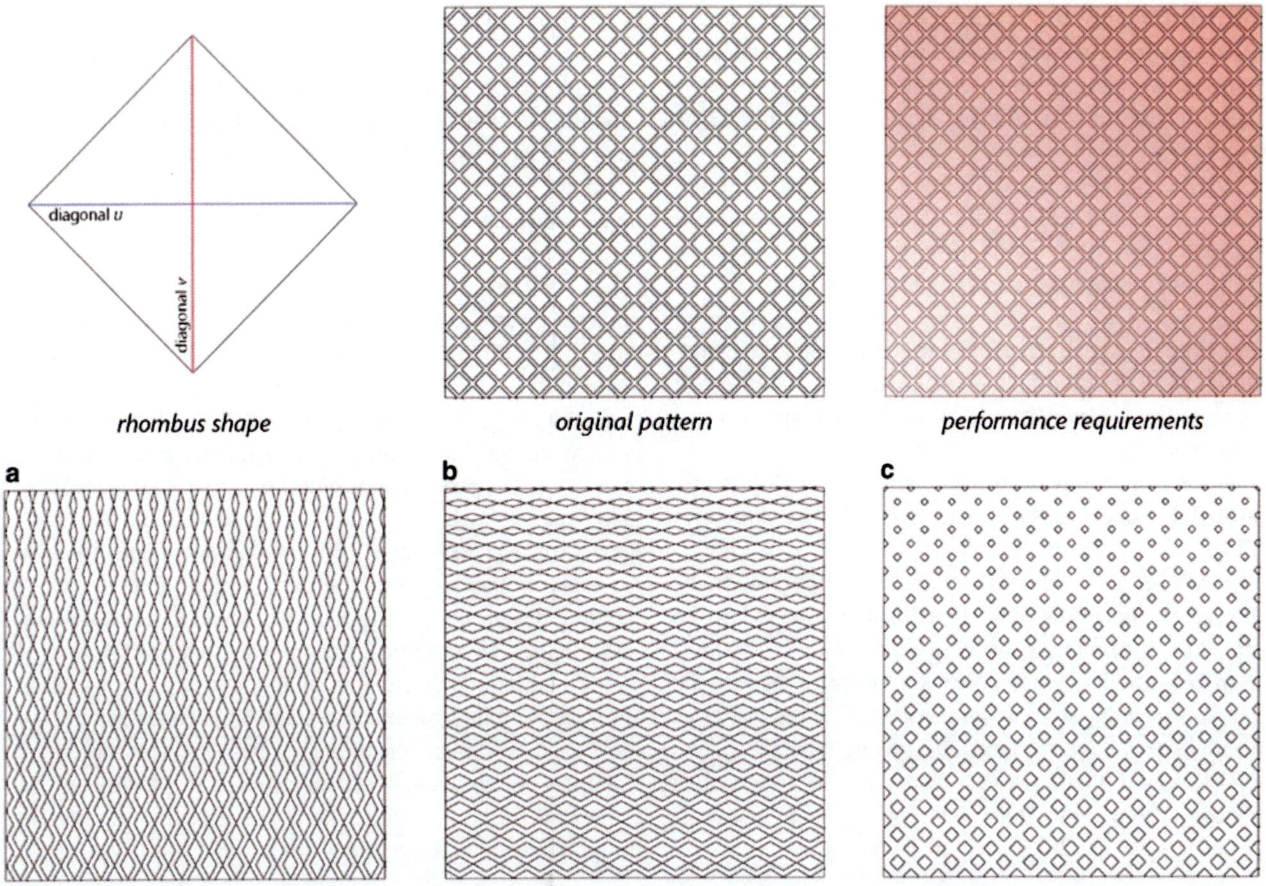

Fig. 11 Top: the rhombus shape's diagonal parameters (left); the rhombus-based pattern to optimize (middle); a conceptual representation of the optimization requirements (right); Bottom: the original pattern with their rhombuses' size optimized regarding their **a** diagonal *u*, **b** diagonal *v*, and **c** both diagonals

information, and spatial locations. It also enables the visualization of this information through the 3D model in the design tool, applying different colors to each shape typology. Further details on this algorithmic category can be found in [20].

5 Evaluation

To evaluate the framework's application in an AD context, we selected the AD tool *Khepri* [21], a textual programming tool tailored for architectural design, based on the *Julia* PL [22]. Although we evaluate the framework on a particular tool using a specific PL, it can be applied in any other AD tool using its own PL because the proposed theory is described in mathematical terms.

As the framework follows an AD approach, it therefore benefits from the latter's advantages. First, it allows the transition from a manual design process to a fully automated one. Moreover, the resulting process is both flexible and parametric, facilitating the integration of design changes, supporting the generation of more complex designs, enabling the exploration of wider design spaces, and allowing for the automation of the *generation-analysis-regeneration* cycle typical of optimization processes. Contrastingly, manually using design tools requires considerable time and effort to change the designs, which not only limits the complexity of the obtained solutions, but also narrows the design space explored. Finally, the proposed approach is also highly configurable due to allowing the available algorithms to receive functions as input.

Moreover, the framework embraces all designs stages in a single and continuous workflow: The resulting function compositions integrate algorithms addressing different stages, meaning that modifications made to one algorithm automatically propagate to the other algorithms. Also, its implementation in the AD tool *Khepri* allowed the framework to benefit from some of its capabilities, namely the algorithms portability among different CAD, BIM, and analysis tools. In practice, the same algorithmic description generates identical models in the different supported tools by adapting the embedded information according to the tool:

While in CAD tools, it produces a simple geometric model, in BIM tools it enriches the model with building semantics, and, in analysis tools, it generates a simplified version of the model with only the information needed for each specific analysis.

There are already some tools that provide functionalities similar to those available in the framework. These include *ParaCloud Gem* [23], a 3D pattern modeler, and some plug-ins for *Grasshopper* and *Dynamo*, including:

1. *PanelingTools* that contains surface paneling functionalities supporting grid manipulation and the morphing of patterns, and rationalization techniques for analysis and fabrication [24].
2. *LunchBox* that provides functionalities to explore mathematical shapes, surface paneling, and wire structures [25].
3. *Weaverbird* that includes mesh subdivision and transformation operators and functionalities to help the preparation of meshes for fabrication [26].
4. *Parakeet* that has functionalities to generate algorithmic patterns resulting from tilings, geometric shapes and grids subdivisions, edge deformations, among others [27].
5. *SkinDesigner* that provides functionalities to generate facade geometries from building massing surfaces repeating panels [28].

Despite addressing the same problems, these tools are limited by the available predefined operators [29], not allowing the latter's configuration so as to respond to more specific design contexts or intents. Also, they require the user to directly interact with the design tool, hindering the automation of the design process. This results in iterative design exploration processes that are tiresome and error prone. Finally, most tools resort to visual PLs, e.g., *Grasshopper* and *Dynamo*, that are hard to apply to complex problems [30, 31]. Our framework, in turn, overcomes these shortcomings by (1) structuring and providing algorithmic strategies that adapt to diverse design scenarios and requirements and to different design tools and workflows; (2) promoting an entirely algorithmic use of the functionalities available, despite also allowing user interaction with the design tool, e.g., to take advantage of visual inputs [21]; and (3) being designed for textual PLs implementations and, thus, benefiting from their expressive power [30, 32, 33]. Nevertheless, in its current state, the framework requires more programming experience than the previous tools and is less visually attractive and intuitive. Future work will focus on improving these shortcomings.

6 Conclusions

Algorithmic Design (AD) is a powerful design approach that supports the exploration of more complex designs and wider design spaces, as well as the search for better-performing solutions. Unfortunately, architects still face several limitations when adopting AD techniques: (1) the transition from a purely visual to a more complex and abstract design process and (2) the need for specialized knowledge, including programming experience. The challenge, then, is to make AD more accessible and attractive to the architectural community.

Based on previous studies proving that the provision of predefined algorithms and strategies facilitates the development of algorithmic solutions, we proposed a mathematics-based framework designed to tame the complexity of AD techniques. As domain of application, we focused on building facades due to their major role in building design, presenting a mathematical theory for them. In the paper, we described the proposed mathematics-based theory and its implementation in an algorithmic framework. We explained its structure and available algorithms, and we illustrated its application through a sequence of conceptual examples. In the end, we demonstrated the ability of the framework to be generalized and to be applied in different design contexts and workflows.

Acknowledgements This work was supported by national funds through *Fundação para a Ciência e a Tecnologia* (FCT) with references UIDB/50021/2020 and PTDC/ART-DAQ/31061/2017, and by the Ph.D. grant under contract of FCT with reference SFRH/BD/128628/2017.

References

1. Caetano, I., Santos, L., & Leitão, A. (2020). Computational design in architecture: Defining parametric, generative, and algorithmic design. *Frontiers of Architectural Research,* 287–300.
2. Aguiar, R., Cardoso, C., & Leitão, A. (2017). Algorithmic design and analysis fusing disciplines. In *Disciplines and disruption—Proceedings of the 37th annual conference of the association for computer aided design in architecture*, pp. 28–37. Cambridge, Massachusetts, USA.
3. Trubiano, F. (2013). Performance based envelopes: A theory of spatialized skins and the emergence of the integrated design professional. *Buildings, 3,* 689–712.
4. El Sheikh, M. M. (2011). *Intelligent building skins: Parametric-based algorithm for kinetic facades design and daylighting performance integration*. PhD thesis. University of Southern California, Los Angeles, USA.
5. Al-Kodmany, K., & Ali, M. M. (2016). An overview of structural and aesthetic developments in tall buildings using exterior bracing

and diagrid systems. *International Journal of High-Rise Buildings, 5*(4), 271–291.
6. Schulz, C. N. (1971). *Existence, space & architecture*. New York: Praeger. Stamps.
7. Schittich, C. (2006). *Building skins*. Birkhäuser.
8. Stojšić, M. (2017). (New) media facades: Architecture and/as a medium in urban context. *AM Journal, 12,* 135–148.
9. Venturi, R., Brown, D. S., & Izenour, S. (1972). *Learning from Las Vegas*. MIT Press.
10. Moussavi, F., & Kubo, M. (2006). *The function of ornament*. Actar.
11. Otani, M., & Kishimoto, T. (2008). Fluctuating patterns of architecture façade and their automatic creation. In: *CAADRIA 2008—Proceedings of the 13th international conference on computer aided architectural design research in Asia*, pp. 375–382. Chiang mai, Thailand.
12. Pell, B. (2010). *The articulate surface: Ornament and technology in contemporary architecture*. Germany: Birkhäuser GmbH.
13. Caetano, I., Santos, L., & Leitão, A. (2015). From idea to shape, from algorithm to design: A framework for the generation of contemporary facades. In G. Celani, D. Sperling, & J. Franco (Eds.), *Computer-aided architectural design: The Next City—New technologies and the future of the built environment 16th international conference, CAAD futures 2015, selected papers* (pp. 527–546). Berlin: Springer.
14. Woodbury, R., Aish, R., & Kilian, A. (2007). Some patterns for parametric modeling. In *Expanding bodies: Art • cities• environment—Proceedings of the 27th annual conference of the association for computer aided design in architecture*, pp. 222–229. Halifax, Nova Scotia.
15. Qian, Z. C. (2009). *Design patterns: Augmenting design practice in parametric CAD systems*. PhD thesis. School of Interactive Arts and Technology: Simon Fraser University. Burnaby, Canada.
16. Chien, S., Su, H., & Huang, Y. (2015). PARADE: A pattern-based knowledge repository for parametric designs. In *Emerging experience in past, present and future of digital architecture—Proceedings of the 20th international conference of the association for computer-aided architectural design research in Asia (CAADRIA)*, pp. 375–384. Daegu, Korea.
17. Leitão, A. (2014). Improving generative design by combining abstract geometry and higher-order programming. In *Rethinking comprehensive design: Speculative counterculture—Proceedings of the 19th international conference on computer-aided architectural design research in Asia (CAADRIA)*, pp. 575–584. Kyoto, Japan.
18. Caetano, I., & Leitão, A. (2019). Weaving architectural façades: Exploring algorithmic stripe-based design patterns. In *Hello, culture—Proceeding of the 18th international conference on computer aided architectural design futures*, pp. 1023–1043. Daejeon, South Korea.
19. Grünbaum, G., & Shephard, G. C. (1987). *Tilings and patterns*. New York, USA: W.H. Freeman.
20. Caetano, I., & Leitão, A. (2018). Algorithmic patterns for facade design: merging design exploration, optimization and rationalization. In *Facade tectonics 2018 world congress conference proceedings* (Vol. 1, pp. 413–422). Los Angeles, USA.
21. Sammer, M., Leitão, A., & Caetano, I. (2019). From visual input to visual output in textual programming. In *Intelligent & informed—Proceedings of the 24th international conference of the association for computer-aided architectural design research in Asia* (Vol. 1, pp. 645–654). Wellington, New Zeland.
22. Bezanson, J., Edelman, A., Karpinski, S., & Shah, V. B. (2012). *The Julia language*. https://julialang.org/, last accessed Mach 28, 2020.
23. ParaCloud GEM. (2011). https://paracloud-gem.software.informer.com/, last accessed March 29, 2020.
24. Issa, R. (2013). *PanelingTools for rhino and grasshopper*. https://www.food4rhino.com/app/panelingtools-rhino-and-grasshopper, last accessed March 27, 2020.
25. Miller, N. (2011). LUNCHBOX. https://www.food4rhino.com/app/lunchbox, last accessed March 21, 2020.
26. Weaverbird—Topological Mesh Editor. (2009). http://www.giuliopiacentino.com/weaverbird/, last accessed March 19, 2020.
27. Esmaeil. (2019). *PARAKEET*. https://www.food4rhino.com/app/parakeet, last accessed March 23, 2020.
28. sgaray. (2017). *SKINDESIGNER*. https://www.food4rhino.com/app/skindesigner, last accessed March 21, 2020.
29. Zboinska, M. A. (2015). Hybrid CAD/E platform supporting exploratory architectural design. *CAD Computer Aided Design, 59,* 64–84.
30. Leitão, A., Santos, L., & Lopes, J. (2012). Programming languages for generative design: A comparative study. *International Journal of Architectural Computing, 10*(1), 139–162.
31. Janssen, P. (2014). Visual dataflow modelling: Some thoughts on complexity. In *Fusion—Proceedings of the 32nd eCAADe conference* (Vol. 2, pp. 305–314). Department of Architecture and Built Environment, Faculty of Engineering and Environment, Newcastle upon Tyne.
32. Wortmann, T., & Tunçer, B. (2017). Differentiating parametric design: Digital workflows in contemporary architecture and construction. *Design Studies, 53,* 173–197.
33. Celani, G., & Vaz, C. (2012). CAD scripting and visual programming languages for implementing computational design concepts: A comparison from a pedagogical point of view. *International Journal of Architectural Computing, 10*(1), 121–138.

Inês Caetano Caetano has a M.Sc. in Architecture from Instituto Superior Técnico (IST) of the University of Lisbon. She has also studied at Politecnico de Milano during her masters. She is currently a junior researcher at INESC-ID and a Ph.D. student at IST, exploring the integration of algorithmic design methods in the design, analysis, and optimization of facade designs.

António Leitão Leitão has a B.Sc. in Mechanical Engineering, a M.Sc. in Electronics Engineering, and a Ph.D. in Computer Science and Engineering, all from Instituto Superior Técnico of the University of Lisbon. Currently, he is Assistant Professor at the same university, Senior Researcher at INESC-ID, within the Software Engineering Group, and Coordinator of ADA, teaching, lecturing, and researching on bringing together the fields of Computer Science and Architecture.

Design to Experiment—Experiment to Design: Tool- (User, Breaker, Designer)

Ekin Ünlü and Sema Alaçam

Abstract

This study aims to present a curriculum with threefold experimental modules in the field of digital modeling and discuss the reflection of the module on students' learning and making processes. The paper proposes an experimental approach to address introduction of CAD/CAM tools in design education, enhanced by constructivist theories of teaching. The submodules we call tool- (user, breaker, designer) allows students to make digital modeling by experiencing a variety of tools. We introduce the outcomes of the course and interrelations between submodules and students' engagement in students' learning process. Initial findings suggest that the students' personalization of the design tools in the tool-breaker phase and associating them with their own experiences contribute positively to the development of a unique perspective in the design process.

Keywords

Design process • Design pedagogy • Learning by doing • Experimental methods • Constructivism

1 Introduction

The creativity and experimentation are often considered as main elements of the design process, while analyzing the nature of creativity remains as a debatable topic. There are remarkable number of studies in the field of design studies which attempt to decode the design process through various concepts such as diverging and converging processes [1], see-move-see schema [2] elaboration and reduction loop [3], stages in the development of expertise [4], a product development knowledge model [5], design process as iteration of divergence, and convergence steps or prototype cycles [6]. The reciprocal relation between design and experiment provides an unfinished ground for exploring new possibilities in the edge of each field. This study investigates the pedagogical methods and approaches of the design process regarding the interrelationship between design and experiment. The constructivist learning theory and its related concepts are a starting point for the theoretical considerations of this study.

Constructivist theories consider learning as an information building process. In this construction process, every individual builds a different meaning or concept based on her/his own perceptions and understandings [7, 8]. When it comes to teaching methods and approaches of CAD/CAM in design education, teaching what the digital tools provide and associating the gained skill with the design problems lead to a twofold axis. In most of the cases, learning the tool and its functions requires more time and dedication in comparison to learning how to apply these CAD/CAM techniques in a design context. Learning how to use a tool does not always mean one can implement the tool in the design process in a creative way. Therefore, it becomes critical to seek a way to connect the learning process and designing process, while providing students to generate personalized dialogues with tools, approaches, and design. In other words, this study aims to examine pedagogical methods to support students' creativity in the process of integrating CAD/CAM in the design process and developing their own approaches rather than using already given computational ones. In this context, it is considered that constructivist theories of learning have potential to provide insights for design pedagogy. John Dewey's concepts of 'arts as experience' [9],'learning by doing' [10], and Piaget's 'learning by discovery' [11] are considered in the proposed curriculum development.

E. Ünlü (✉) · S. Alaçam
Istanbul Technical University, Taskısla, 34347 İstanbul, Turkey
e-mail: unlue19@itu.edu.tr

S. Alaçam
e-mail: semosphere@gmail.com

S. Eloy et al. (eds.), *Formal Methods in Architecture*, Advances in Science, Technology & Innovation,
https://doi.org/10.1007/978-3-030-57509-0_2

This paper is framed by two approaches outlined above, and it explores the impact of experiments in terms of construction, observation, analysis cycles on students' designing with CAD/CAM tools. The experiments are expected to include a change of an initial form and transformation of it through physical or chemical processes. The aim is to reveal potential connections between observed physical experiments and the design process through informing the design process with one's own observation. In that sense, the role of digital modeling tools is examined in relation with the experiment and design processes. This study is based on a case of design practice and teaching, from physical experiments to the development of the use of digital modeling tools for a specific design project and tasks. An experiment was designed under three main submodules namely 'tool-user,' 'tool-breaker,' and 'tool-designer.' The research is carried out with video, model, and outputs of eight groups created with 24 graduate students. Initial results suggest that the analysis diagrams created during the physical experiments supported the establishment and development of students' own algorithms with a bottom-up approach during the design process.

2 Design to Experiment: Designing an Experimental Digital Modeling Course

The constructivist learning theory is an educational strategy that is open to change. The constructivist theory, which forms the basis of the experimental course, provides a theoretical basis for rethinking and redesigning learning strategies. Constructivism, which is the result of cognitive science, is a process where learning knowledge is structured by the individual [8, 12]. Piaget uses the concepts of assimilation, adaptation, and balance from biology to explain the development of these cognitive structures [13]. It emphasizes that experience is assimilated within the framework of existing structures and that mental balance can be maintained in this way [13, 14]. According to Piaget during 'learning by discovering' in an environment that supports open-ended learning, the existing mental schemes are reworked [11]. As well as, according to John Dewey, discovery is never made, it always continues [15, 16]. The 'learning by doing' theory developed by Dewey in the twentieth century is part of modern education. According to John Dewey, learning, it is possible by doing and living. In the constructivist approach, the student actively constructs and reproduces the information [13, 17]. 'Learning by doing,' which supports teamwork, has been developed with the reinforcement of practical experiences. Questioning and experimentation bring critical thinking habits [10]. Afterward, a learning model in which the nature of the application will be explored was examined by Schön [18] as 'reflection in action.' Design course takes its teaching power from experiential learning or learning by doing method [19]. Schön described it as a different and unique way of knowing; 'learning by doing' settings are suitable settings for a process of creating customizable information [20]. The course process, which is an experimental study, is designed with inputs that trigger the thinking mode that comes out when designing. The education process of design, which is a discovery process, is also designed with the reference of constructivism theory. In view of all that has been mentioned so far, the constructivist approaches, learning by doing and learning by discovering come from the learning experience.

Hofstadter introduces an abstract and narrative diagram to illustrate conceptual parts of an articulating structure [21]. Köknar and Erdem [22] propose using Hofstadter's 'chunking' idea in the context of design [21] and they argue that the chunking analogy might be helpful to gain a better understanding about the act of design. Derived from Hofstadter's [21] 'chunking' idea, Köknar and Erdem's [22] discussion but different from them, a diagrammatic illustration of chunking is developed and presented in Fig. 1 in terms of adapting the chunking concept to the design domain. Keeping these conceptual foundations in mind, weekly tasks are defined in consideration of the 'chunk'-based modular structure in the proposed digital modeling course. The tasks are expected to involve links with other tasks. Apart from the emphasis on gaining new skills in the field of CAD/CAM, the course also involves readings and discussion sessions in terms of gaining a critical understanding about the available technologies. The physical/chemical experiments firstly provide an experiential ground to support students' constructing an internalized dialogue between the given task and their own observations, with Hofstadter's terminology [21] we call it as 'internal relations' of a chunk. Secondly, supported with the weekly instructions and tasks, physical/chemical experiments provide a trackable and a common denominator for different tasks. In that sense, experiments contribute to creating 'external relations' among chunks. The learners are encouraged to work in groups to construct common meanings in constant interaction also to identify ways of thinking and learning alternative measurement tools and evaluation methods. In terms of learning from experience, the focus of the proposed pedagogy can be considered as learning by interacting with materials, machines, and tools during the construction phase [16, 23].

Generally speaking, the students who take the digital design and modeling courses at the master level mostly start with different knowledge levels. Even if the entire course is operated in the same way, the learning outcomes of the course may differ due to the differences in the individual learning styles of the students [24]. During the semester, the previous experiences (background) of the students and actual

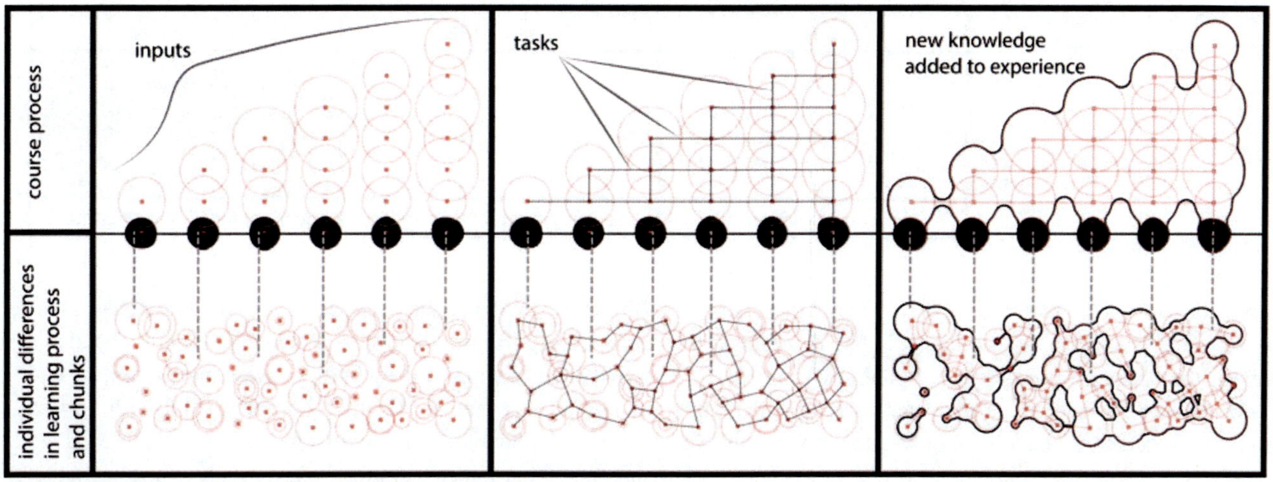

Fig. 1 Threefold modulation of the course regarding pedagogical objectives

experiences (learning by doing) gained in the process are bound together continuously. While some of the students tend to learn via visuals, others might prefer verbal content; some students might prefer more open-ended design formulations or vice versa. How to respond to all those variants becomes a critical issue. The expectation of the experiment is that the tasks given to the students trigger gaining knowledge in chunks to be reflected in new contexts. As a result, experience and knowledge are structured by the learner and are constantly changing.

3 What Is Tool- (User–Breaker–Designer)?

Dewey [9] states that 'there is no way to discover what is "more truly educational" except by the continuation of the educational act itself. The discovery is never made; it is always making'. Today, Dewey's emphasis on the action of making rather than the noun version of 'made' is still a valid debate in the context of tools and tools' learners. Aranda and Lash invite us to use the term 'tooling,' as an attempt to externalize a process with a finite number of steps, namely algorithms [25]. On the other hand, it is not easy and possible to externalize every aspect of the design process, particularly tacit and embodied knowledge that designers gain during the process. Considering the inseparable relationship between design and construction, the designing subject and the designed, the maker and the object, El-Zanfaly [16] mentions 'craft knowledge' in reference to Dewey [9] and Eisner [26]. El-Zanfaly suggests intertwining the actions of 'Imitation–Iteration–Improvisation,' namely I3, through a modulated making process [16]. In today's architectural design practice, the tool-user, tool-designer, and tool-implementer are becoming the same person. Thinking about not only the development of design ideas but also the processes of developing design ideas is an integral part of the design process. Now architects, makers of notations, can be used immediately and directly for physical fabrication [27]. With the expansion of the design repertoire of designers, the traditionally used tool is now designable. To get more freedom in design, keeping a critical distance to the predefined or previously known usage strategy of existing tools is recommended [28, 29].

In this study, the tool is described as a physical or mental mechanism that enables the design process to take place. In the context of this mechanism, the relationship between the tool and the subject differs. Füssler argues that the tool can become a dominant factor in determining the formal features of the product by considering the tool as a part of the design process, the tool can add content to the design, the tool itself can be the producer of an idea, and the tool can become creative [30]. However, he also argues that the tool may not benefit from the design idea [30]. The tool is in a complementary and variable relationship with its user. Three different experimental sets have been set up to examine the relationship of the subject with the tool when designing: tool-user, tool-breaker, and tool-designer [Fig. 2].

Tool-user: It is a skill-oriented learning process for the first three weeks of the module. In this phase, students are encouraged to use a variety of digital modeling and fabrication tools such as 3d printer, CNC, laser cutter, grasshopper, and agisoft. The weekly given tasks involve constraints regarding the specific tool to be used.

Tool-breaker: Students are encouraged to develop their own experiment setups consisting of observation of a physical or chemical process. The physical/chemical processes are expected to cover some of the following key concepts: differentiation, variation, change from one state to another, ability to change in the volume, porosity, repeatability with changing one parameter at each time. Students are

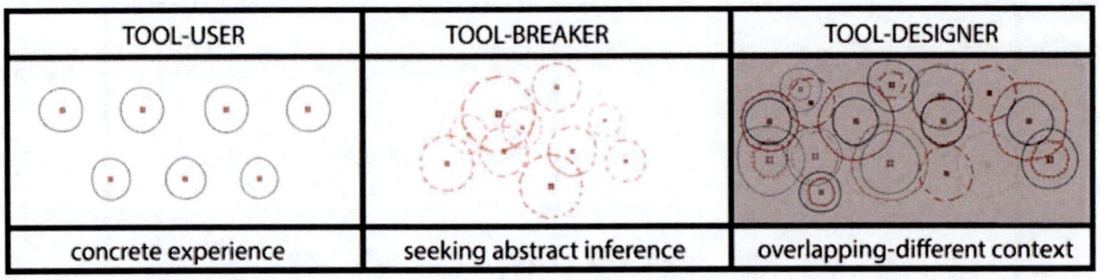

Fig. 2 Tool- (user–breaker–designer) and tasks relationships

asked to document the observed process, make diagrams to analyze, and develop a bottom-up rule-based strategy to represent the observed process in a digital environment.

Tool-designer: Students are expected to adapt their previously generated from making strategies to a given context.

4 Experiment to Design: Process of an Experimental Digital Modeling Course

In the scope of this study, the digital modeling and fabrication tools are introduced to architecture students through form and material experiments in a seven-week master-level course module. To schematize the teaching process, Buxtons' [3] elaboration–reduction concept in the design process is borrowed (Fig. 3). Buxton introduces the 'Overlapping Funnels diagram' in the context of design creativity and its process [3]. In Buxton Overlapping Funnels diagram, a design process is described between elaboration–reduction [3]. Elaboration in Buxton's diagram refers to opportunity-seeking and reduction refers to decision making [3]. But the narrative shown in Fig. 3 focuses on the design process. The course process diagram that we created shows that each task creates a new focal point (Fig. 3). The inputs of the design to be produced as a result of the lesson are not given to the students at once. Different tasks going on the same plane contain hidden learning objectives, which are introduced task by task.

During the first three weeks, model making exercises, mold models, and material experiments were performed. In the first week, students are asked to model the lost–missing–broken parts of an object on their campuses with a digital tool and produce them with a 3D printer. In the third week, students are asked to rescale their previous mode 2.5 times and reproduce it via using various digital manufacturing tools such as rapid prototypes, laser cutters, and CNCs. During the following two weeks, exploratory material testing tasks based on 'porosity' research were given. In the experiment set called tool-designer, students' tasks include digitalization, digital model, optimization, and algorithmic definition.

Students were asked to document their observations made in physical and chemical experiments. During the 6th and 7th week, students were expected to adapt their 'growth-porosity algorithm' to a spatial context. Besides, students were asked to support their findings with spatial context and a narrative scenario. Abstract concepts, ability to grow, degree of porosity were to inspire scenarios designed by students. This paper examines the data up to the students' midterms, a period of seven weeks. After the 7th week, the course was studied on an urban scale. Students continued to work in groups focusing on a slum analysis of a given city. And the students were asked to develop the growth-porosity algorithm of each group sent as the midterm exam in their 15th week final assignments. The tasks given during the course process consist of hidden learning objectives that follow and complement each other. All the assignments in the experiment sets were recorded by each group in the form of videos, sketches, diagrams. Successive and triggering exercises and tasks are successive sets of exercises and constitute our experiment sets for research.

5 Tasks and Results

The course process consists of six different tasks. Tasks are grouped in three different modules. The first module is tool-user, weeks 1 and 2, the second module is tool-designer 3–4 and 5 weeks, and the third module tool-designer is 6 and 7 weeks (Fig. 4). Task-1 and task-2 (missing part and mold design shown in Fig. 3) were asked to do exactly the same with the measurements of a lost–missing–broken part of an object on their campus. In task-3, students are expected to reflect to optimize the process through a series of performance goals.

- How can the mold model be made cheaper and lighter?
- If the mold making method is changed, can the amount of material used change as well?
- What other materials can we use besides the selected materials?
- How can the topography in the mold model be abstracted?
- How can the resulting model be lighter?

In the tool-breaker section, in task-4, students were asked to investigate the concept of 'porosity' and conduct a

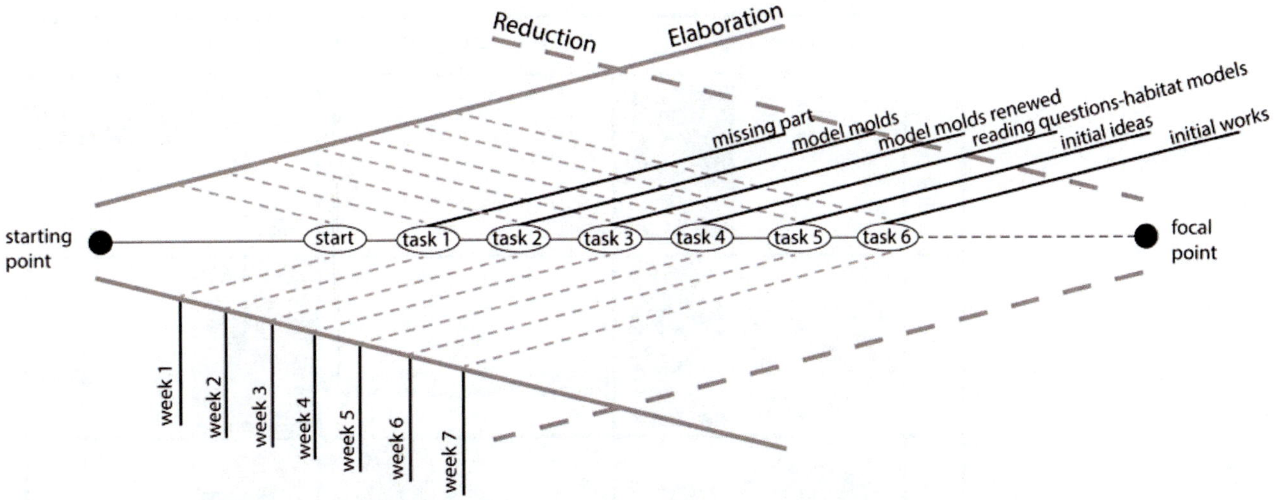

Fig. 3 Tool- (user–breaker–designer) and tasks relationships. The illustration utilizes and is inspired by Buxton's [3] elaboration–reduction separation as a conceptual base

material experiment. Afterward, they were expected to relate to the concepts of 'growth-porosity' and 'habitat' in task-5. In the tool-designer part, namely task-6, students' habitat models were translated into an algorithm (Fig. 4).

Group-1: The first group replaces the mold production method in task-3 with two different stages and presented in Fig. 4. The topographic surface of the model (missing-part model) is revised by triangulation. In order to convert the surface into mold shape, a base is created with a waffle structure in grasshopper. It was decided to use epoxy resin and foam like materials and the model to be formed was added by adding a sponge. In task-4, form tests and observations are made with an ice–sugar experiment. As a result of the observations made in task-5, they have created a rule with the variables 'boundary-gravity-viscosity-gravity-freezing point' (Fig. 4). In task-6, a simulation was done with grasshopper–kangaroo using these rules. Grasshopper–anemone plug-in was used in the habitat model. The effect of the research on the tool-breaker experiment set has shown its effect on the habitat design. The effects of the tool-breaker test set are observed in the tool-designer section. In the last study, algorithm rules were created by using the tools in the tool-user.

Group-2: The second group creates surfaces with triangulate and fold techniques in revised models in task-3. Mold designs consist of two parts as base and surface. They carried out their experiments separately from three different materials. In lightness and porosity studies, foam, plaster + styrofoam, plaster + balloon were used, respectively (Fig. 4). The experiment was continued with sourdough in task-4 and task-5. The group's experiments consist of three steps: theory of sourdough formation, shape and growth analysis, and behavior analysis and rules. In task-6, manipulate volumetric growth and porosity ratio of sourdough and behavior modeling were done.

Group-3: In the third group, while producing molds with surfaces in task-3, triangulation method was used. Mold is produced with a similar logic as other groups. The material poured into the mold is plaster. To lighten the general material, materials such as styrofoam and garbage were added to the plaster. In the experiments with sugar–ice in task-4, different forms and behaviors were examined. In task-5, model trials were made with Voronoi (Fig. 4). The experiment was simulated in task-6. The observations obtained in the tool-breaker experiment set to form the basis of the simulation made in the tool-designer section. It is observed that the sugar form obtained in the tool-breaker section comes from the porosity experiences obtained in the tool-user section.

Group-4: In the fourth group, task-3, two-piece mold model consisting of a mesh model and the base model was produced. Stone powder, textile, soft soap, and tape materials were used (Fig. 4). In task-4, a chemical reaction was observed by burning sugar and soda. A new model has not been created by removing rules from this reaction. However, a model similar to the test result was made in the form. The model is not produced from rule sets created through observations, but by productive systems in the literature. Reflection points in the tool-designer section are more than the tool-breaker experiences in the tool-user section.

Group-5: The fifth group has simplified their task by revising their models in the triangular method in task-3. The material they use to fill the mold is plaster. To lighten the material used, he has experimented with ice + plaster, dry ice + plaster, water-filled bag + plaster, and balloon + plaster (Fig. 4). Plaster–dry ice experiment is continued in task-4. In this task, the reflection of the learned

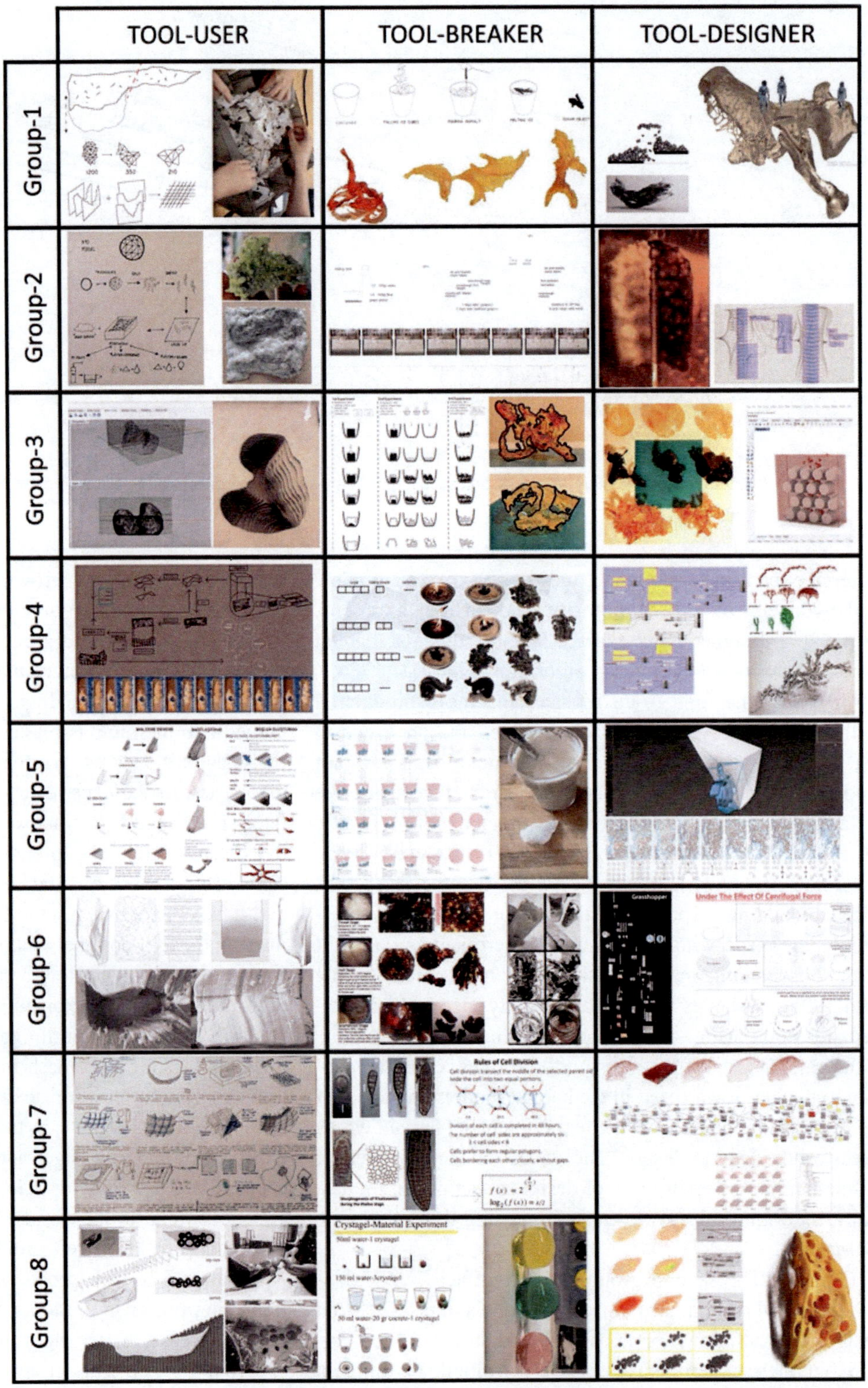

Fig. 4 Outcomes of the given tasks

information in the tool-user set is reflected in the tool-breaker set. In task-5, the experiment is continued with dry ice and liquid glue. In task-6, a simulation model is made with the rules obtained in the experiment. The rules in the tool-breaker section have created data for the tool-designer, but the habitat model has not progressed further. The habitat model was only a simulation of the material behavior in the experiment.

Group-6: The sixth group reused their molds. But they changed the material. They filled the mold using silicone and

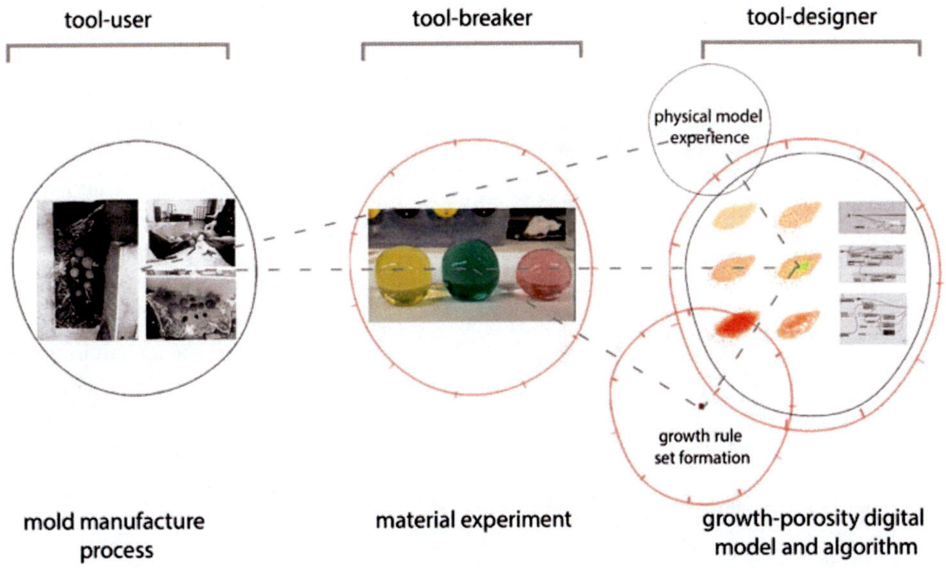

Fig. 5 Results of Group 8's tool- (user, breaker, designer) tasks and details of the relationships between them

then covered the mold with plaster and plaster. Then, with the removal of the silicone, a model was created from hollow surfaces. In task-4 and task-5, sugar was caramelized, and a mixture of small foam was prepared together (Fig. 4). In task-6, this mixture is modeled by analogy with the bacterial habitat. The information learned in the tool-breaker section remained only the concept in the tool-designer section. However, the traces of the tools used in the tool-user section are also seen at the last stage.

Group-7: The seventh group renewed their mold designs with a waffle structure system. The foam was used as the material. Pyropia (red algae) was examined in task-4–5. The group has not conducted a new material experiment (Fig. 4). However, they continued their research by making use of the research in the literature. In task-6, the growth behavior of Pyropia is modeled by analogy by creating a cells population. The information in the tool-breaker section does not provide enough data to create a new situation for the final stage. However, the effect of the tools learned in the tool-user section is seen more in the tool-designer section.

Group-8: The eighth group used mold designs in the same way and did not revise them. But they made a change of material. They carried out material experiments such as plaster and concrete + balloon. An experiment to examine the growth of crystagel material was carried out in task-4. Growth rules of crystagel have been established in task-5. Moreover, the experiment was repeated in different contexts. Mold making in the tool-user experiment set was repeated on a small scale. Crystagel has also been observed to grow in concrete. The effect of the material types in the tool-user experiment set is a reflection point affecting the toll breaker (Fig. 5). The growth design in task-6 is modeled with grasshopper. Behavior models learned in the tool-breaker experiment set were used when creating the habitat model.

6 Conclusions

Computer-aided design (CAD)/computer-aided manufacturing (CAM) is becoming a part of the design process rather than being merely tools. As Oxman indicates digital techniques in design not only change our design representations but also create new bases for design thinking [31]. The design theory that emerges as it moves away from the traditional transforms the concept of form into the concept of formation [31]. The tool- (user–breaker–designer) course setup created in this context supports the digital modeling process with experiments. It has the potential to encounter changing and transforming material behaviors with experiments to discover physical events using digital modeling and analog computation methods. The effect of learners' experiences in design processes was achieved by comparing the results of each task and presented in Fig. 6.

- After the chemical–physical experiments, tectonic analysis, and abstractions, the qualifications related to the model are reconstructed by the students. In the tool-user experiment set, the tools, construction techniques and materials preferred by the students include similarities. In mold model revisions, 75% of the groups used the triangulation method while 25% (group 6–8) reused the mold but changed the spilt material. 37.5% of the experiment groups (group 4–5–8) continue their materials used in the tool-user section to create spaces in the tool-breaker section in the experiment sets (Fig. 6). In the mold designing process, students have to develop their personal methods and approaches according to their own experience and interests. In this context, the software used in the representation and digitization processes is not the

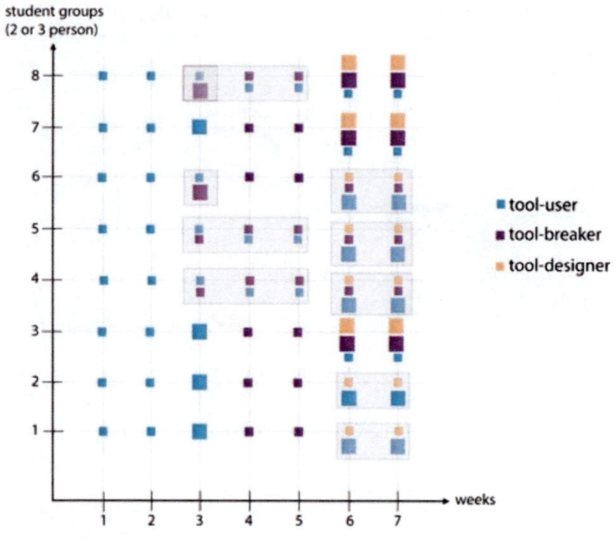

Fig. 6 Relationship of tool- (user–breaker–designer) experiences obtained by comparing task results

focus, but similar constructions can be repeated in different software and platforms.

- In all experimental groups, the effect of grasshopper-script experiences in the tool-user section is observed in the tool-designer section. It is seen that students who are in search of abstract inference are looking for more original material in the tool-breaker section. The discovery, research, and learning process of each student is different as it covers the re-representation and conversion processes of the rules obtained in chemical experiments. The resulting student's modeling approach is supported by creative processes and unpredictable results are encountered.
- Our relationship with digital modeling tools changes in the design process. Although chemical experiments are expected to be stimulating in this experimental proccss, the effects of digital modeling thinking can differ in future studies. It is observed that one experience is reflected in another experience/context while using CAD/CAM tools while designing. In the tool-designer section, 62.5% of the students progressed by extracting the rule set of material experiments and making them an analogy to the experiment–result form. Experiment parameters were transferred to the digital model by analyzing, abstracting, or by developing an algorithm. However, 37.5% (group-4–5–6) compared the existing productive system or grasshopper plug-ins to their experiments. In these groups, the experiment parameters were transferred to the digital model by using the existing algorithms and adapting them with interpretation (Fig. 6).
- The first two stages, which overlap in a different context with the habitat concept, were turned into scenarios by 75% of the students and evaluated as space. However, 25% (group 1–2) only simulated the results of the experiment (Fig. 6). In these two groups, it is observed that CAD/CAM tools are used only for representation purposes in design processes.

In the process from experiment to design, from design to experiment, every student is expected to develop their methods according to their own experience and interests. Similarities may be reduced in studies conducted by course mentors considering previous digital modeling experience processes. In an environment where tools, interfaces, and representation are constantly diversifying, pedagogical methods and approaches that will gain the digital modeling skills to design students by exploring the design process with experiments is an issue that needs to be researched and discussed.

Acknowledgements We would like to thank research assistant Burak Delikanlı, and students: Nisa Çelebi, Betül Ünlü, Gözde Küçükoğlu, Aysun Demirel, Zeynep Candar, Sevgi Altun, Pelin Arman, Betül Uçkan, Gülbin Lekesiz, Nurdan Akman, Nisanur Özdemir, Gizem Özpolat, Can Ozmen, Deniz Tutucu, İrem Konur, Tahsin Oğuz Koç, Ayşenur Baltacı, Ekin Kılıç, Erhan Sevinç, Atra Haddur, Dilara Gökçe, Necmettin Sancak, Bilal Togrul, Serhat Gülmüş.

References

1. Brown, T., & Katz, B. (2009). *Change by design: How design thinking transforms organizations and inspires innovation*. New York: HarperCollins.
2. Schön, D. A. (1992). Designing as reflective conversation with the materials of a design situation. *Knowledge-Based Systems, 5*(1), 3–14.
3. Buxton, B. (2007). *Sketching user experiences: getting the design right and the right design*. San Francisco: Morgan kaufmann.
4. Cross, N. (2011). *Design thinking: Understanding how designers think and work*. New York: Berg Publishers.
5. Eris, O., & Leifer, L. (2003). Faciliating product development knowledge acquisition: Interaction between the expert and the team. *International Journal of Engineering Education, 19*(1), 142–152.
6. Leifer, L. J., & Steinert, M. (2011). Dancing with ambiguity: Causality behavior, design thinking, and triple-loop-learning. *Information Knowledge Systems Management, 10*(1–4), 151–173.
7. von Glasersfeld, E. (1995). A constructivist approach to teaching. In L. P. Steffe & J. Gale (Eds.), *Constructivism in education* (pp. 3–15). Hillsdale: Erlbaum.
8. Chen, C. A. (2003). Constructivist approach to teaching: Implications in teaching computer networking implications in teaching computer networking. *Information Technology, Learning, and Performance Journal, 21*(2), 17.
9. Dewey, J. (1934). *Art as experience*. New York: Berkley Pub. Group.
10. Dewey, J. (1933). *How we think. A restatement of the relation of reflective thinking to the educative process*. Boston: DC Heath and Company.
11. Piaget, J. (1973). *To understand is to invent: The future of education*. New York: Grossman Publishers.
12. Fosnot, C. T., & Perry, R. S. (1996). Constructivism: A psychological theory of learning. Constructivism: Theory,

perspectives, and practice. New York and London: Teachers College Press, Columbia University.
13. Aydınlı, S. (2015). Tasarım Eğitiminde Yapılandırıcı Paradigma: 'Öğrenmeyi Öğrenme'. *Tasarım + Kuram, 11*(20), 1–18.
14. Philips, D. C., & Soltis, J. (2005). *Öğrenme Perspektifler*. Ankara: Nobel Yayın Dağıtım.
15. Dewey, J. (1929). *The sources of a science of education*. New York: Horace Liveright.
16. El-Zanfaly, D. (2015). [I3] imitation, iteration and improvisation: Embodied interaction in making and learning. *Design Studies, 41*, 79–109.
17. Milne, C., & Taylor, P. C. (1995). Metaphors as global markers for teachers' beliefs about the nature of science. *Research in Science Education, 25*(1), 39–49.
18. Schön, D. A. (1984). The architectural studio as an exemplar of education for reflection-in-action. *Journal of Architectural Education, 38*(1), 2–9.
19. Schön, D. A. (1987). *Educating the reflective practitioner: Toward a new design for teaching and learning in the professions*. New Jersey: Wiley.
20. Schön, D. A. (1985). *The design studio: An exploration of its traditions and potentials*. International Specialized Book Service Incorporated.
21. Hofstadter, D. R. (1979). *Gödel, Escher, Bach: An eternal golden braid*. New York: Basic Books.
22. Köknar, S. A., & Erdem, A. (2011). Tasarlama eyleminin tasarım araçları modeli üzerinden bir okuması. *İTÜDERGİSİ/a, 9*(2), 51–62.
23. Dewey, J. (1986). Experience and education. In *The educational forum* (Vol. 50, No. 3, pp. 241–252). Taylor & Francis Group.
24. Cross, N., & Nathenson, M. (1981). *Design methods and learning methods. Design: science: method*, pp. 281–296.
25. Aranda, B., & Lasch, C. (2006). *Pamphlet architecture 27: Tooling* (No. 27) (2nd ed.). New York: Princeton Architectural Press.
26. Eisner, E. W. (2002) The arts and the creation of mind. Yale University Press.
27. Youtube, Mario Carpo. (2008). "The Second Digital Turn"|Talks at Google https://www.youtube.com/watch?v=UVerq5DSdKU&list=LLuGEMrl4tBY_7UOqPwaR-iQ&index=6&t=0s (Access time: March 03, 2020).
28. Fischer, T. (2008). Designing (tools (for designing (tools for …))). Doctoral dissertation, School of Architecture and Design, Faculty of the Constructed Environment, RMIT University.
29. Loh, P., & Leggett, D. (2017). *Tools as agents in design and making processes*.
30. Fussler, K. (2008). Design by tool design. In *Proceedings of Advances in Architectural Geometry*.
31. Oxman, R. (2006). Theory and design in the first digital age. *Design Studies, 27*(3), 229–265.

Ekin Ünlü Ünlü received her Bachelor of Architecture degree from Eskişehir Osmangazi University. After graduating, she worked with the PET-MAT team at Czech Technical University for a few months during 2018–2019. She worked as a designer and researcher involved in all aspects of the design process with 3D printing and recycled material. She is currently completing her Masters in Architectural Design at Istanbul Technical University.

Sema Alaçam Alaçam is Associate Professor at Istanbul Technical University (ITU), Faculty of Architecture, where she also is Coordinator of the Architectural Design Computing Graduate Program. She received her B. Arch. degree from ITU, M.Sc. degree from Architectural Design Computing Graduate Program of ITU and Ph.D. degree from Architectural Design Computing Graduate Program of ITU. She has research experience from TU Delft, Hyperbody Research Group, ETH Zurich, Chair of Structural Design, University of Oregon, Department of Philosophy and Université de Versailles Saint-Quentin-en-Yvelines.

Tectonics of Differential Growth. Folds in Additive Fabrication and Moulding for Architectural Design

Shota Tsikoliya, Imro Vaško, Petra Sochůrková, and Daniel Sviták

Abstract

Differential growth is a process of volumetric expansion with spatially uneven intensity, which results in appearance of folds and is an infinite source of patterns in nature. Examples of differential growth are particularly prevalent in epithelial tissue. The paper proposes to use algorithms simulating differential growth for additive fabrication of moulding of architectural structures. The project starts with a research of folding at the level of epithelia cells and proposes a geometrical model simulating the process. Chosen geometrical approach has the advantage of continuity and closeness of any horizontal cross section, which allows for rapid additive fabrication as well as for use as a mould. Since one of the main issues of additive fabrication at construction scale is the problem of keeping production time down while keeping the detail scale required for the finish, the authors propose 3D printing moulds with extremely thin walls. These moulds are later filled with liquid material after hardening of which the structural qualities are reached. Small thickness of the walls is essential for the production speed, and its vulnerability to deformation is solved by simultaneous filling of the mould from both sides, with materials of similar density. The outer material is later extracted and recycled. Project results in production of several prototypes, which demonstrate the structural, spatial and aesthetical qualities of the approach. Project unravels and explores the potential of the surface maximization in relation to environmental impact of the architecture, in particular heat dissipation.

Keywords

Differential growth • Folds • Fabrication • Moulding

S. Tsikoliya (✉) · I. Vaško · P. Sochůrková · D. Sviták
Academy of Arts Architecture and Design in Prague, Prague, Czechia
e-mail: stsiko@vsup.cz

1 Introduction

Differential growth is a process of volumetric expansion with spatially uneven speed and intensity, which often results in appearance of folds and wrinkles and is an infinite source of shapes and patterns in nature. Recent research proves geometrical exploration as essential for understanding of growth and folding in nature. The paper proposes to use algorithms simulating differential growth for additive fabrication or moulding of architectural structures. The algorithms used in the project are based on random insertion of rapidly growing elements within the structure, which reflects the natural process of apico-basal cell intercalations, common in curved epithelia. Resulting geometries are used for FDM fabrication of thin-walled moulds and subsequent concrete prototypes. Within those prototypes, the authors explore range of architectural typologies from furniture to a structural element such as column to a semi-enclosed space. Project further unravels the potential of the surface maximization in relation to environmental impact of the architecture, in particular energy management and heat dissipation.

2 Context

Examples of differential growth could be found in various areas, but are particularly prevalent and visible in epithelial tissue. Tissue bending contributes to shaping the organs into complex three-dimensional structures. There the control of differential growth resides in the level lower, than that in which it is expressed, therefore resulting in non-determined character of the form [2]. Patterns of differential growth

S. Eloy et al. (eds.), *Formal Methods in Architecture*, Advances in Science, Technology & Innovation,
https://doi.org/10.1007/978-3-030-57509-0_3

present a result of complex interactions of growing cells. Within the process, various elements cooperate to achieve qualitative difference and play different roles, ranging from energy minimization to creation of tubular or folded structures, presenting an emergent phenomenon [3].

The existing application of various differential growth algorithms in the field of design or architecture among others has been applied to achieve biologically inspired aesthetics [6], to differentiate mechanical properties of the structures [7] or to be able to design across multiple resolutions [1]. While taking the previous applications in consideration, project focuses on applying differential growth algorithms to a narrow field of additive fabrication of moulding for architectural design using conventional FDM printing technology. In addition, this project tries to implement recent foundings of the geometrical consequences of differential growth on cellular level [4].

Research of the cellular organization within epithelial tissue has shown that the apical surface of cells is assumed to be a faithful proxy for their three-dimensional (3D) shape. The arrangement of cells in the apical surface of the epithelium has been extensively analysed from biophysical, mechanical, and topological viewpoints [5]. Topologically, the apical surface of epithelial sheets is arranged similarly to Voronoi diagrams. However, cells in bent or folded epithelia can undergo intercalations along the apico-basal axis. This phenomenon forces cells to have different neighbours in their basal and apical surfaces. As a consequence, epithelial cells adopt a novel shape, recently termed "scutoid". Scutoids, in contrast to prisms, are not necessarily convex and present non-planar lateral faces. This is due to the fact that the boundary between cells follows geodesic trajectories (the shortest path connecting two points). This ensures that scutoids display concave surfaces that allow the 3D packing of cells in a curved tissue [4].

There have been several approaches of computational generation of scutoid-based epithelia representations. Among others lofting of the several layers of Delaunay diagrams, described as Delaunay lofts, proved to be a simple and intuitive approach for designing space filling scutoid-shaped tiles in 3D space. [9] The approach suggests performing Voironoi subdivisions of apical and basal surfaces of the structure as well as of multiple virtually interpolated surfaces within it. Lofting process of the Voronoi cells results in the scutoid shape.

3 Methods

The project starts with a research of folding at the level of epithelia cells and proposes a geometrical model simulating the process. The model uses the geometry of scutoids, which, according to the previous research, seem to be essential for three-dimensional packing of folded epithelia [4]. These geometries make possible the minimization of the tissue energy and stabilize three-dimensional packing—qualities, not only essential in biological systems such as epithelial organs, but also having potential in the field of architectural design.

In this part of the research, the authors expand on the examples found in nature and create series of possible geometrical solutions to the problem of uneven growth and folding. The solutions are generated using spring-based engine. Among the tools, used by the authors, is physics-simulation engine Kangaroo as well as custom-made digital tools within the environment of Grasshopper and Rhinoceros software. The catalogue of the proposed geometrical solutions is approximated into a common digital model. Authors apply this differential growth model on closed two- and three-dimensional geometrical shapes, different stages of transformation of which are later interpolated between, and used as an input for lofting operation, which defines final structure.

4 Design Process

Within the first approximation, geometrical model presents two identical offset surfaces (representing basal and apical surface of the epithelial tissue). The surfaces are evenly and identically populated with points and subsequently subdivided using Voronoi diagrams. Cells are modelled by the way of lofting Voronoi diagrams, and since those are identical, cells are represented by prisms. Additional set of points is inserted at either of the surfaces to cause folding process, the inserted points being drivers of the growth. This insertion transforms Voronoi diagrams and gives cells, neighbouring the insertions the shape of scutoids. The resulting model is being iteratively transformed using spring engine Kangaroo, with Delaunay edges between the points defining attracting forces and maximization of the volume of the cells defining repulsive forces. The iterative transformation results in the creation of folds in places of inserted points (see Fig. 1).

While the geometrical model is aimed on representing the growth and creation of folds within epithelial tissue, it presents several challenges, when applied to creation of folds in additive fabrication and moulding for architectural design. One of those challenges is overall computational redundancy of the model, which slows down the process and limits the possibilities of its design applications. Another challenge is based on the fact, that while, in nature, growth of the shape and its materialization are inseparable, in case of architectural application, the fabrication process follows different rules than the process of generating the geometry.

The second approximation of the model was based on the fact that the authors planned to use a particular fabrication

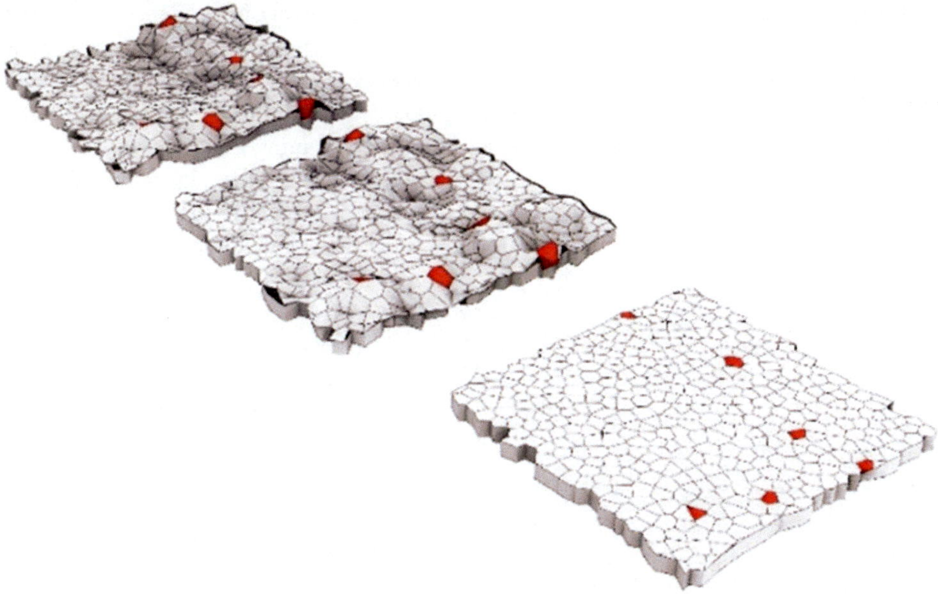

Fig. 1 Iterative transformation of epithelial tissue model—first approximation

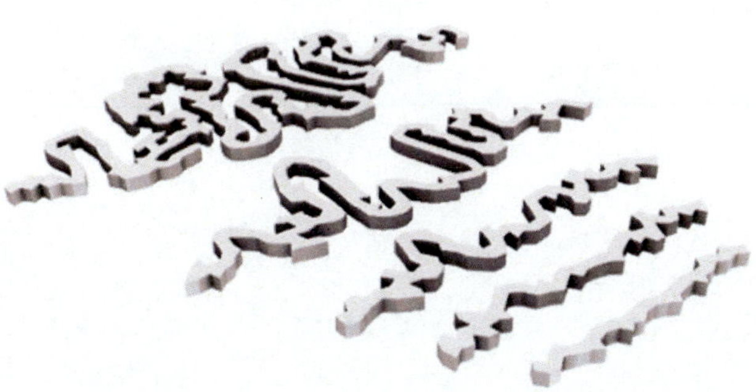

Fig. 2 Iterative transformation of epithelial tissue model—second approximation

process, i.e. FDM 3D printing of the mould, which is characterized on materialization of the shape layer by layer. Therefore, the geometrical model reflects a section of virtual epithelial tissue. In this approximation, it presents two identical offset curves, evenly and identically populated with points. The lofted segments of the curves represent single cells. Additional set of points is inserted at either of the curves to cause folding process, the inserted points being drivers of the growth. The model is being iteratively transformed using custom-made spring engine, with curve segments defining attracting forces and maximization of the area of the cells defining repulsive forces. The iterative

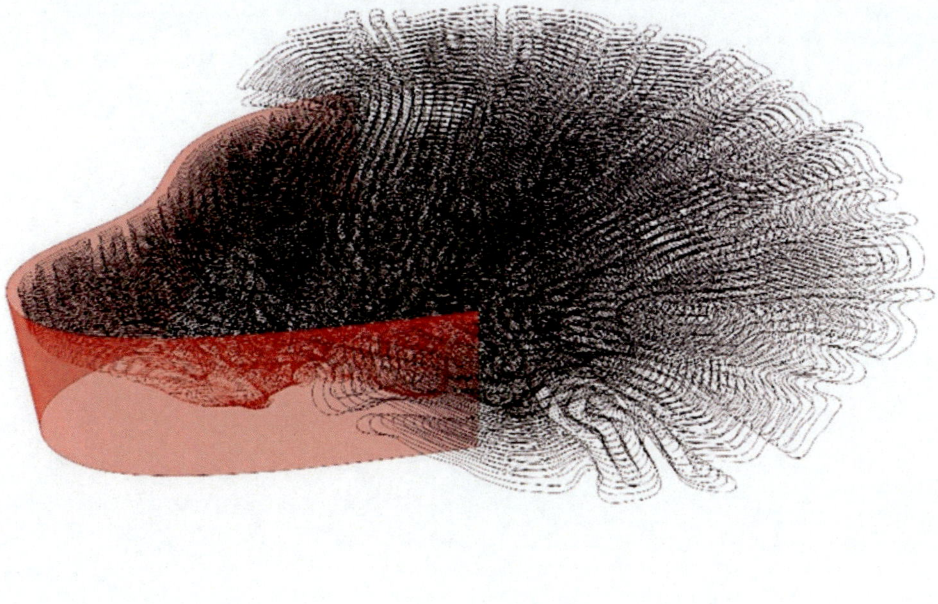

Fig. 3 Design through constraints

transformation results in the creation of folds and branches, in places of inserted points (see Fig. 2).

This approach allows to precisely define the thickness of the section, an important parameter of the fabrication process. To achieve three-dimensional structure, the authors loft different stages of section transformation positioned above each other at regular intervals. The above-described method leaves control of differential growth in the level lower than that in which it is expressed, therefore resulting in non-determined and emergent character of the form. In order to provide top-down defined limits to this bottom-up approach, the predesigned original structure (later "footprint") and the use of constraints have been implemented into the algorithm. The process of growth starts at the footprint and stops or resumes where and when it fits the particular aim of the design, and therefore, the footprint and

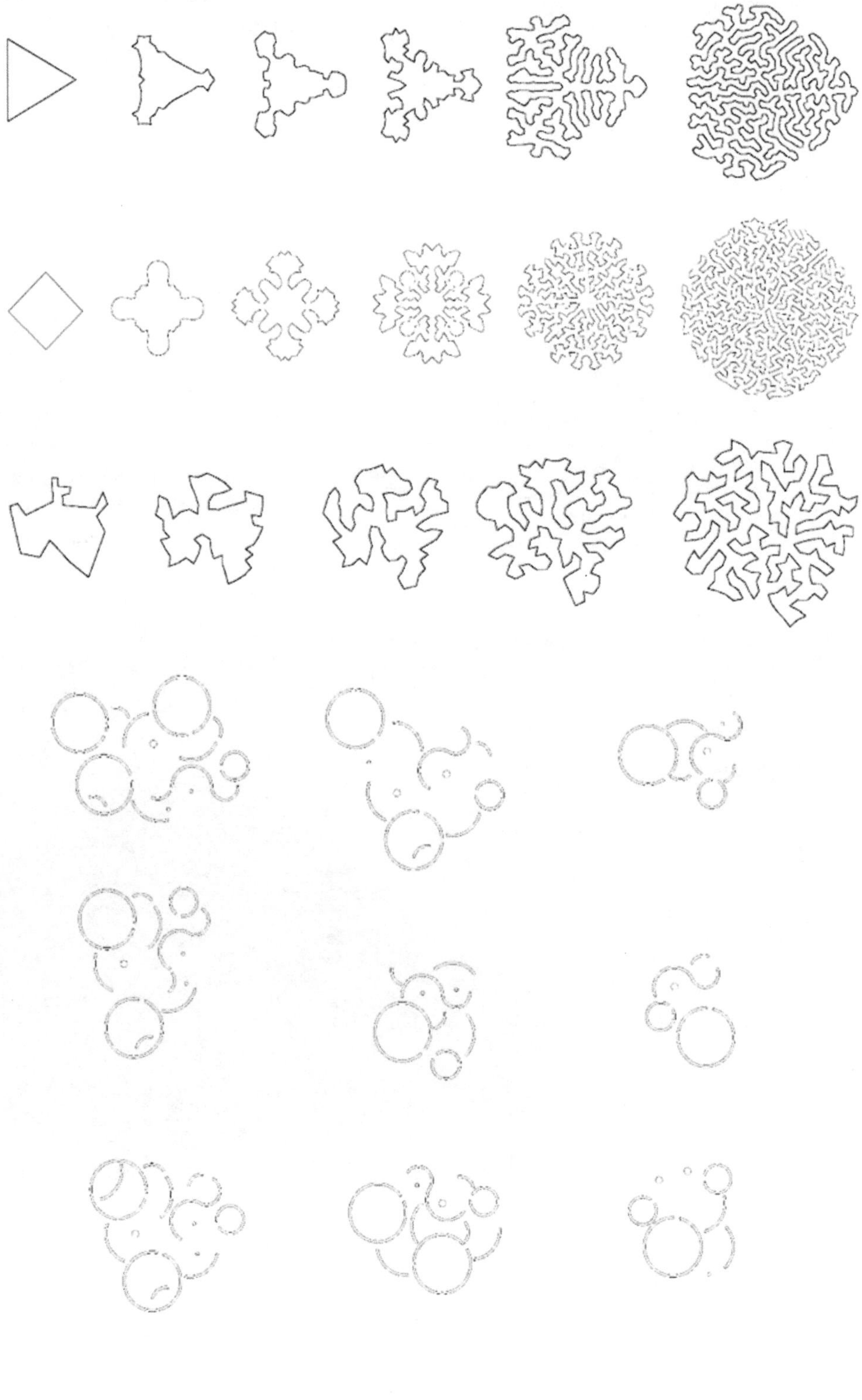

Fig. 4 Design through footprint

the spatial and temporal constraints become the main tools of the designer (see Figs. 3 and 4).

The spatial constraints are defined through modelling of geometry, in which the structure is not allowed to reach or intersect. The temporal constraints define the speed of growth by defining the number of inserted cells as well as varying the ratio of attracting and repelling forces between the cells. The spatial constrains allow to model predesigned shapes while the control of the speed of the growth defines parameters like density of the structure or the angle of cantilevered elements.

The process has been applied to the creation of various architectural typologies from furniture to a structural element such as column to a semi-enclosed space. Basic typology was a mushroom-shaped element, which is naturally generated by vertically arranging various stages of growth, starting at the bottom and ending at the top (see Fig. 5). This element was later modified or transformed using constraints. Of particular, interest was parallel growth of several mushroom-shaped elements, which, eventually meeting, allowed to achieve more complex typologies, such as vaults.

5 Fabrication

The approach has the advantage of continuity and closeness of any horizontal cross section of the structure, which allows for rapid additive fabrication as well as for use as a mould. Since one of the main issues of fabrication at construction scale is the problem of keeping production time down while keeping the detail scale required for the finish [8], the authors propose 3D printing moulds with extremely thin walls. These moulds are later filled with liquid material (concrete, epoxide, or metal) after hardening of which the structural qualities are reached. Small thickness of the walls is essential for the production speed, and its vulnerability to deformation is solved by simultaneous filling of the mould from both sides, with materials of similar density. The outer material is later extracted and recycled. Another advantage of the approach is controllable character and size of created cantilevers, an issue particularly important in most of the widespread techniques of 3D printing. The authors use variable growth rates for different parts of the input curves to achieve variable density and structural properties. Minimal and maximal sizes of the folds are explored to achieve successful moulding from materials of various viscosities, i.e. concrete, epoxide, or metal.

For the fabrication of prototypes, the authors created moulds using FDM 3D printers with the size limitation of 300 mm × 305 mm × 457 mm (see Figs. 6 and 7). The produced moulds can be stacked together in horizontal layers to produce a bigger prototype. The authors used expanding fibre-reinforced concrete as a main construction material, its expansion being advantageous in filling the small deformations of the mould (see Fig. 8). The outside of the mould was simultaneously filled with sand, which resulted in easy and energetically non-demanding recycling process.

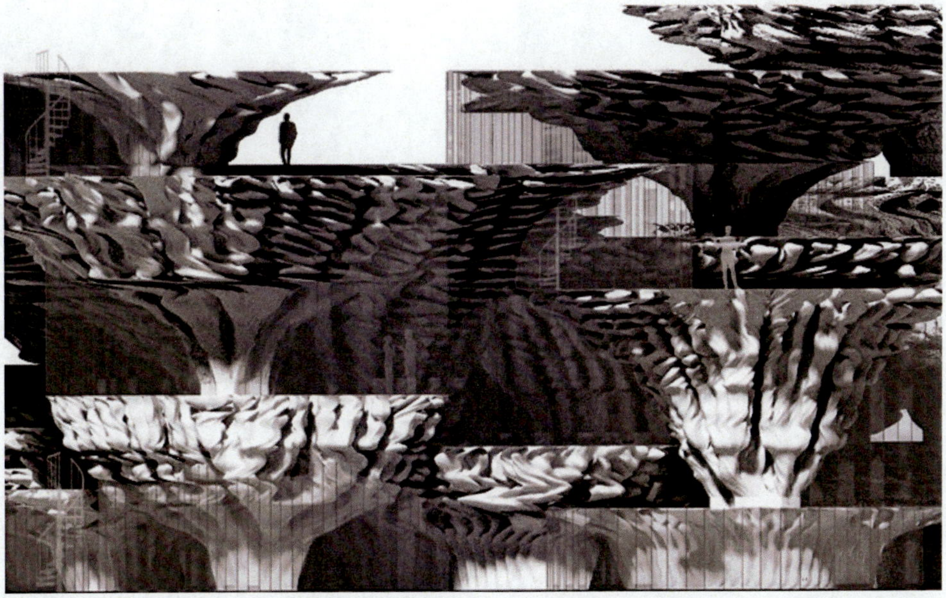

Fig. 5 Basic typology—a mushroom-shaped element

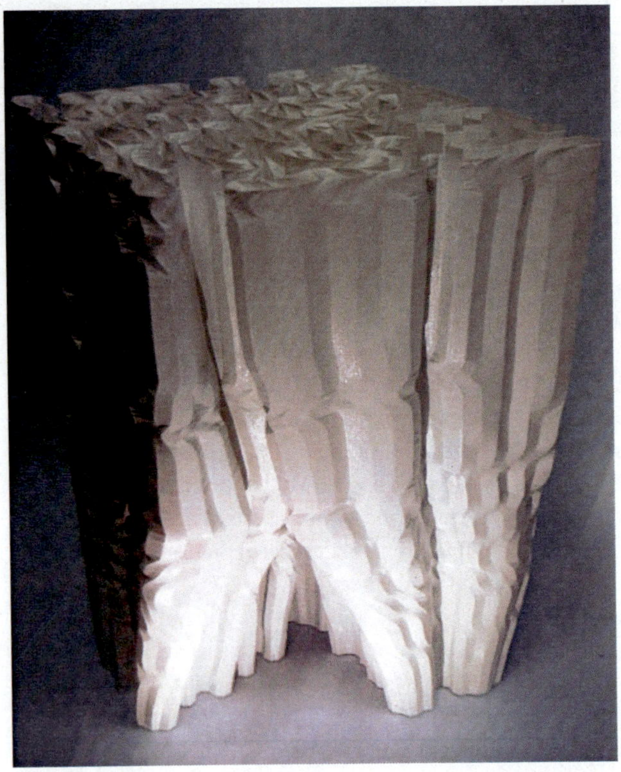

Fig. 6 3D printed mould

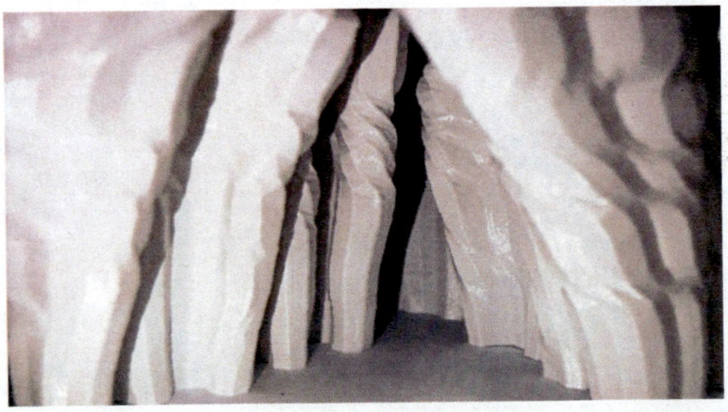

Fig. 7 3D printed mould—detail

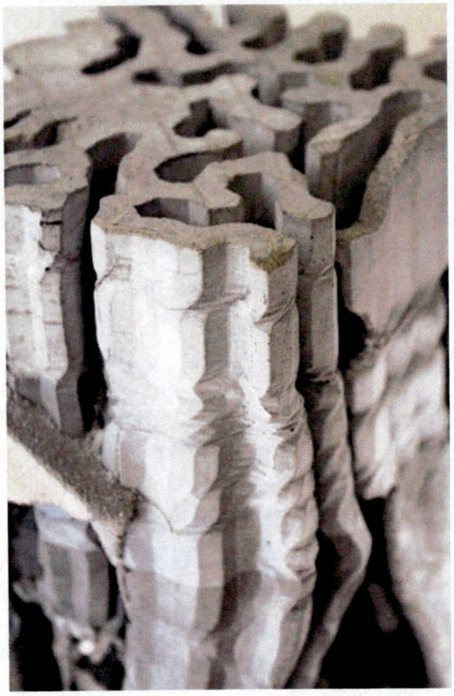

Fig. 8 Prototype detail

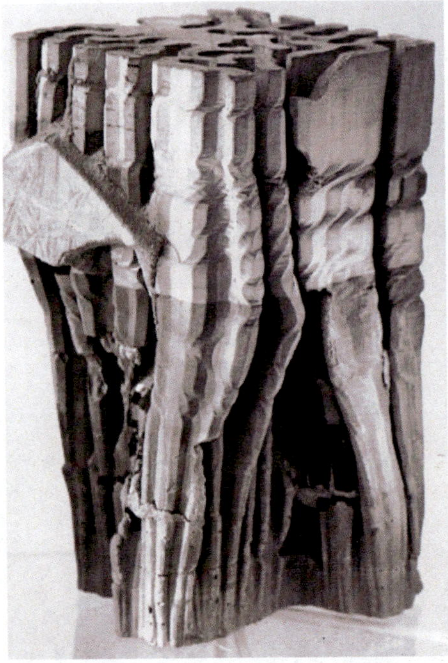

Fig. 9 Prototype

6 Outlook

The project resulted in production of several prototypes of various scales, which demonstrate the structural, tectonic, spatial, and architectural qualities of the approach (see Figs. 8, 9 and 10).

The project questions the potential of the surface maximization in relation to environmental impact of the architecture, in particular energy management and heat dissipation. In the tested case of the single living unit with the built-up area of 54 m^2 and the concrete shell thickness of 100 mm, the approach allows to maximize surface as much as by 1400%. The predominantly vertical orientation of the folds is convenient for dissipating the heat through natural air ventilation caused by chimney effect. However, further research is necessary to explore practical consequences of

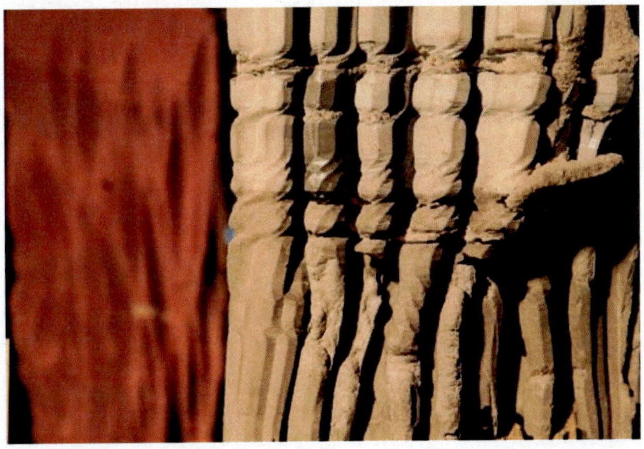

Fig. 10 Prototype detail

implementation of differential growth generated structures to heat dissipation in architectural scale.

References

1. Andrasek, A. (2016). Xenocells: In the mood for the unseen. *Architectural Design, 86*(6), 90–95.
2. Barlow, P. (1989). Differential growth in plants—A phenomenon that occurs at all levels of organization. *Environmental and Experimental Botany, 29*(1), 1–5.
3. Corning, P. (2002). The re-emergence of emergence: A venerable concept in search of a theory. *Complexity, 7*(6), 18–30.
4. Gómez-Gálvez, P., Vicente-Munuera, P., Tagua, A., Forja, C., Castro, A., Letrán, M., et al. (2018). Scutoids are a geometrical solution to three-dimensional packing of epithelia. *Nature Communications, 9*(1).
5. Lecuit, T., & Lenne, P. (2007). Cell surface mechanics and the control of cell shape, tissue patterns and morphogenesis. *Nature Reviews Molecular Cell Biology, 8*(8), 633–644.
6. Louis-Rosenberg, J., & Rosenkrantz, J. (2020). Anti-entrepreneurs using computation to unscale production. *Architectural Design, 90* (2), 112–119.
7. Oxman, N. (2015). Templating design for biology and biology for design. *Architectural Design, 85*(5), 100–107.
8. Soar, R., & Andreen, D. (2012). The role of additive manufacturing and physiomimetic computational design for digital construction. *Architectural Design, 82*(2), 126–135.
9. Subramanian, S., Eng, M., Krishnamurthy, V., & Akleman, E. (2019). Delaunay lofts: A biologically inspired approach for modeling space filling modular structures. *Computers & Graphics, 82,* 73–83.

Shota Tsikoliya M.Sc., Ph.D., holds a doctoral degree in Architecture from the Academy of Arts Architecture and Design in Prague, with focus of his doctoral research in Computational Design and Emergent Architecture. He graduated with Master of Science degree from University of Stuttgart and a Master's degree in architecture from the Academy of Arts Architecture and Design in Prague. Currently, he holds a position of Assistant Professor at the Academy of Arts Architecture and Design in Prague and is co-founder and partner at architecture studio COLLARCH.

BIM-Based Airport Design Project Standardization (IFC) for Use of Code Checking

Ítalo Guedes, Max Andrade, and Adriana Carvalho

Abstract

Building information modeling (BIM) in the AECO industry is a fact. Users have overcome the initial learning curve and perceive a new potential for using BIM, such as rule checking and automatic design validation, which it is possible through dissemination and use of Industry Foundation Classes (IFC). This paper assesses the main properties that must be inserted in BIM models of airport design projects so that they can be evaluated using code checking. In this sense, the properties of the information standards will be analyzed and which IFC entities should be linked to those standards. The method is based on design science research. This consists of two macro-stages: construction and evaluation of the artifact. The stage of construction, discussed in this paper, consists of the theoretical foundation and construction of the artifact, which is a method to evaluate airport design using code checking. For evaluation, it is necessary not only to create consistent rules that can be implemented in BIM evaluation software but also to create BIM models that are consistent during the assessment. This research shows the importance of creating a standardization process in the modeling of information during the development of airport designs in software authored by BIM, considering the export to an integrated IFC model. In view of the inexistence of manuals dealing with this subject in Brazil, it is possible to observe the potential that this approach has and may, shortly, benefit all teams of airport designs and analysts.

Keywords

Building information modeling • Airport • Code checking • IFC

1 Introduction

Airport design projects, by nature, are complex because of the series of design requirements to be met in the design's documentation phase. In Brazil, the evaluation and approval of airport design project are carried out by INFRAERO (Brazilian Federal Public Company responsible for the administration of the country's airports). Currently, design projects are evaluated predominantly using CAD software. In this scenario, the designers develop the design projects, print the documents and submit these printed papers for a manual evaluation and validation process carried out by INFRAERO analysts. Such processes tend to have a prolonged evaluation. In some cases, it may last months or even years, considering the entire process. Besides, the evaluation processes allow margins for ambiguity in the evaluation, due to the different perspectives and specific knowledge of each analyst, with a lack of standardization in the design evaluations.

As an answer to optimize the process, minimize errors in the evaluation of airport design projects and reduce the evaluation time, we consider in this paper the use of Building Information Modeling (BIM). The use of BIM linked to this type of activity is called code checking or checking of rules. According to Estman et al. (2009), code checking is defined as a process of evaluating relationships or design attributes based on the configurations of its objects. Benning et al. [1] classify automatic detection based on interference rules in two categories: interference based on geometry (clash detection) and interference based on code checking. The checking of rules in BIM models allows a new scenario for analysts to automate the evaluation process,

Í. Guedes (✉) · M. Andrade
Universidade Federal de Pernambuco, 1235 Cidade Universitária, Recife, Brazil
e-mail: italoguedesarq@gmail.com

A. Carvalho
Faculdade de Engenharia da Universidade Do Porto, 4200-465 Porto, Portugal

S. Eloy et al. (eds.), *Formal Methods in Architecture*, Advances in Science, Technology & Innovation,
https://doi.org/10.1007/978-3-030-57509-0_4

reducing the analysis time and increasing efficiency concerning the validation process.

The key points discussed in this paper are not only to identify the rules and the properties that must be associated with the BIM models but also how these properties should be transcribed to IFC and exported. Thus, regardless of the BIM authoring software used for the design project or of the evaluation software to be used by INFRAERO, the BIM model must be consistent enough and exported in the right way, so that it can be evaluated correctly and precisely based on predefined rules.

An example of a rule tested and demonstrated by Guedes and Andrade [2] is the rule that establishes the width of the service circulation between the check-in queues and check-in counters. For this case, it was not only necessary to create a rule in a BIM evaluation software (SOLIBRI Model Checker), but also, it was necessary that the model that carries this information had been created in a BIM authored software and had been exported bringing the correct information, such as Environment Name (IfcSpace) and Bouding Box Width (IfcSpace) referring to the circulation between check-in queues and check-in counters so that the analysis software could validate the rule.

Finally, this paper proposal is to discuss and evaluate what are the minimum necessary properties to be hung in a BIM design project, as well as, which IFC entities that must be linked to the building architectural design of airports (passenger terminal, runway, taxiway, apron, etc.) so that they are evaluated consistently by INFRAERO.

2 Method

The methodological process of the present work is based on design science research. This, according to March and Smith [3], consists of two macro-stages: **(1) Construction**, a stage referring to the development of artifacts, which in this research consists of the construction of a specialized method for code checking and identifying properties inserted in BIM models of airport design projects, to demonstrate the feasibility of acting as a solution to the proposed problem, stage in development; and **(2) evaluation of the artifact**, a step to measure the efficiency of the artifact, based on the checking of how it behaves in its environment or context, a stage yet to be developed.

The stage of construction of the artifact, discussed in this paper, consists of two parts: foundation and construction of the artifact:

Part (1)—**theoretical foundation**—consists of two phases. The first phase consisted of the theoretical foundation on the theme. It started with a systematic literature review (RSL). International and national publications on the subject have been identified in the past ten years. The second phase consisted of carrying out case studies in processes of analysis and validation of airport designs in Brazil, with INFRAERO.

Part (2)—**construction of the artifact**—still under development. It consists of creating rules for checking the architectural design based on Brazilian legislation in the area to identify the minimum properties of the IFC model, and how it should be exported, to be used in the rule checking process. This step aimed to translate the main Brazilian Regulations for airports design into criteria that can be inserted in automatic assessment software based on BIM with the function of code checking and validation [4]. From some software tests and literature experiences, which points to more systematic use of Solibri Model Checker (SMC) in places like Singapore (Corenet), Norway (Stasbygg), Australia (CRC), USA (ICC and GSA) [5], this research chose to carry out the experiments using this software. Once the software has been defined, the SMC checking rules are being created and tested in a Brazilian architectural design project for an airport design project modeled on a software authored by BIM.

The types of rules that can be implemented are analyzed as well as how to implement them in BIM assessment software. Also, the analysis includes how the model should be built by the designers so that it may be accurately evaluated. In this part, the characteristics that need to be incorporated in the BIM models (modeling characteristics) are also being studied so that they can be evaluated in the SMC and which are the terminologies for the types of rules and their classification using a publicly available open standard (in this work, it is being used based on the IFC format). Industry Foundation Classes (IFC) have been accepted as the only open and relatively mature standard currently supported by major BIM applications. Finally, the last part of the first stage consists of the automatic generation of the analysis report based on the rules template created in the SMC, verifying whether the properties of the IFC model comply with the pre-established design criteria.

Figure 1 presents the synthesis of this ongoing research, graphically. Since the design of an airport is conceived in a BIM logic, with the creation of the construction information model in BIM authorship software and subsequent evaluation in BIM analysis software, this implies a change in the evaluation of the design process that passes to be automated, based on guidelines established in the design. Within this design logic, the design revision time is reduced, with automation in the evaluation phase. This paper investigates more specifically the relevance of the guidelines manual (based on regulations for airports design) in the interrelationship between the airport designer team and the airport design analysis team, suggesting the transposition of data from standards, manuals, ordinances, among other documents for a standard document that allows the creation of the

BIM model containing all the pertinent and necessary information for the verification of automatic rules (code checking) in the evaluation design phase. Thus, it is expected, greater efficiency and reliability in the design process, with reduced time for the approval of an airport design project.

3 Airport Design

A few years ago, the design of an airport was conceived according to the specific demand of the aircraft that would operate on it. Currently, as a result of several factors, such as proximity to demand-generating centers, multifunctionality of services at a passenger terminal, among other points, the concept of airport design has transformed into a complex endowed with its own life that has correlations with its immediate surroundings, as well as the adaptation to parameters, such as service levels, security levels, adaptation to the advancement of new technologies, making the program of need for a passenger terminal somewhat complex. Something more than guidelines defined in the functional and operational limits.

Here, it is important to make a distinction between airport and aerodrome. According to Art.27 of the Brazilian Aeronautical Code (1986), "aerodrome is an area destined for landing, takeoff, and movement of aircraft" and can be classified as civil and military. The difference between aerodromes and airports is that airports have specific systems and facilities for basic infrastructure for aircraft operations, loading and unloading of people and cargo; in short, the difference between aerodrome and airport is the existence or not of the basic systems of infrastructure and facilities. The object of study of this work is the airport; therefore, the evaluation approach faddressed is for airport and does not apply to aerodrome (Table 1).

About the development of an airport design project in Brazil, there are a series of normative prerequisites that guide the designer when developing an airport design. Among the documents, we can highlight: (a) RBAC (Brazilian Civil Aviation Regulations), (b) Ordinance 1141/GM5, c) Annex 14, (d) Airport Planning Criteria and

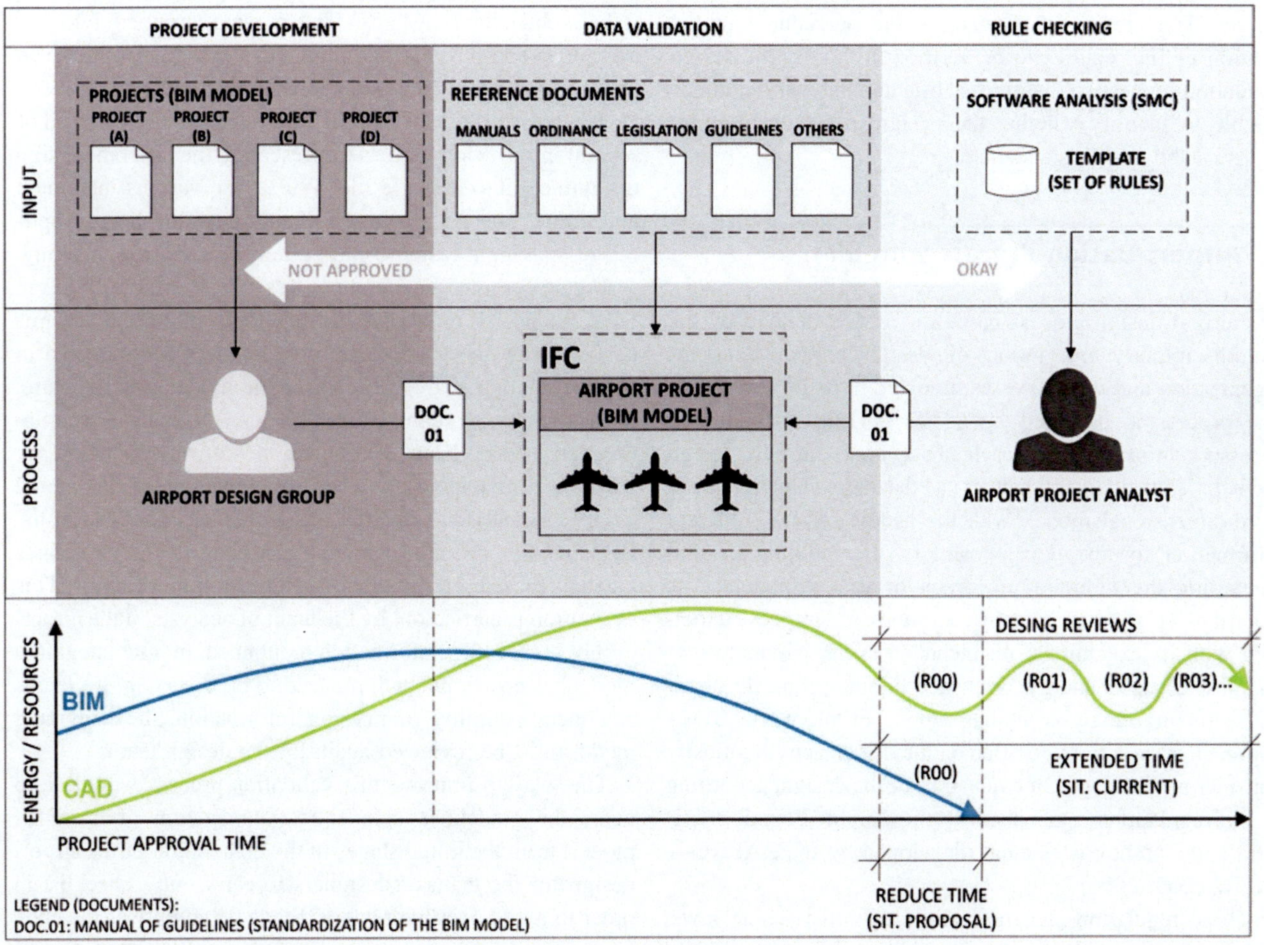

Fig. 1 Synthesis of the research proposal

Table 1 The difference between aerodrome and airport

Description	Airstrip and Takeoff	Taxiway	Apron	Facilities for loading, unloading people and cargo
Aerodrome	(Y)	(Y)	(Y)	(N)
Airport	(Y)	(Y)	(Y)	(Y)

Legend (Y) yes / (N) no.

Conditions Manual (INFRAERO), among others of documents of references. The team of designers must have the expertise to design a project in compliance with all current regulations.

With the incorporation of code checking in the airport design process, the team of design analysts in Brazil can take advantage of an automated code checking, through software that evaluates a set of rules on the information model of the airport design. However, to do so, it is necessary that this model contains the necessary and sufficient information as well as uses an information exchange protocol capable of bringing the right information and in the right place, using a scheme internationally accepted. This work will base the studies on the use of the IFC scheme as an information exchange protocol between the synthesis stage (development of the design solution) and evaluation of the architectural artifact. The aim is to standardize the procedures in the creation of IFC entities to be verified in the airport design evaluation process so that the use of automatic verification is reliable to identify whether the design meets the requirements established in standards.

4 Airport Design Process with BIM

The airport design process based on BIM is defined by the multidisciplinary integration of design project data, its organization and the processes involved in its development. The process is designed by adding all the relevant and necessary information in each phase of the airport design project. The final product is a database hanging in a three-dimensional model, with the necessary and sufficient information for the development of the building design, containing the defining properties of its components, its materials, specific characteristics, codes of services associated with its executions, maintenance cycle, parameters of quantities, costs, energy, acoustic, luminotechnical, financial, structural analysis, among others. In this work, BIM's approach is essentially focused on the design activity, mostly considering the stages of evaluation of the design, according to the regulations established by the Airport Planning Criteria and Conditions Manual (developed by INFRAERO—INFRAERO, 2006).

These regulations are normative texts that define laws, codes, specifications, standards and so on. These generally provide restrictions for solutions or technical activities that occur under defined conditions. The use of automated or semi-automated verification may contribute to the checking of regulations that have not been verified due to a lack of knowledge of the code specifications or time as well as competence limitations. The main objective of a method such as the one proposed in this work is to offer computable models with clear syntax and semantics that can be used in the creation of requirements for the evaluation of airport design projects with regard to meeting the existing regulations.

Guedes and Andrade [2], from the analysis of the Criteria and Conditioning Manual for Airport Planning, managed to identify a series of categories of possible components and their respective parameters that can be translated into rules checking of regulatory code for the passenger terminals (PT) architectural designs. Those were grouped into six categories and have 119 components (only for PT) that can be implemented in an automatic BIM rule checking process.

The basic process flow in BIM assumes the integration of several information models (representing the different design disciplines) into a single file. With these various integrated and unified design projects, the coordination effort is optimized, making it easier to carry out automated rule checking.

Figure 2 shows a general process map of airport design in BIM. Stage 01 refers to the development of the designs conceived by the team of airport designers and modeled in software authored by BIM. Once the models (architecture, infrastructure, structures, installations, mechanical ventilation, etc.) have been developed in software authored by BIM, they are exported via IFC to an integrated BIM model (in IFC format). In stage 02, the properties contained in the BIM model (.ifc) are validated according to the guidelines manual (based on regulations for airports design). This evaluation is carried out by the team of analysts of the airport design. Once the information contained in the integrated model has been validated, proceed to the next step. If there is any incompatibility of missing information, the integrated model must be reviewed again by the design team.

One way to improve this validation process would be to make the guidelines manual (standardization of the BIM model) from the initial stages in the elaboration of the airport design for the team of designers to carry out a checklist in order to avoid possible rework. Stage 03 comprises the code checking step in the analysis software SOLIBRI Model

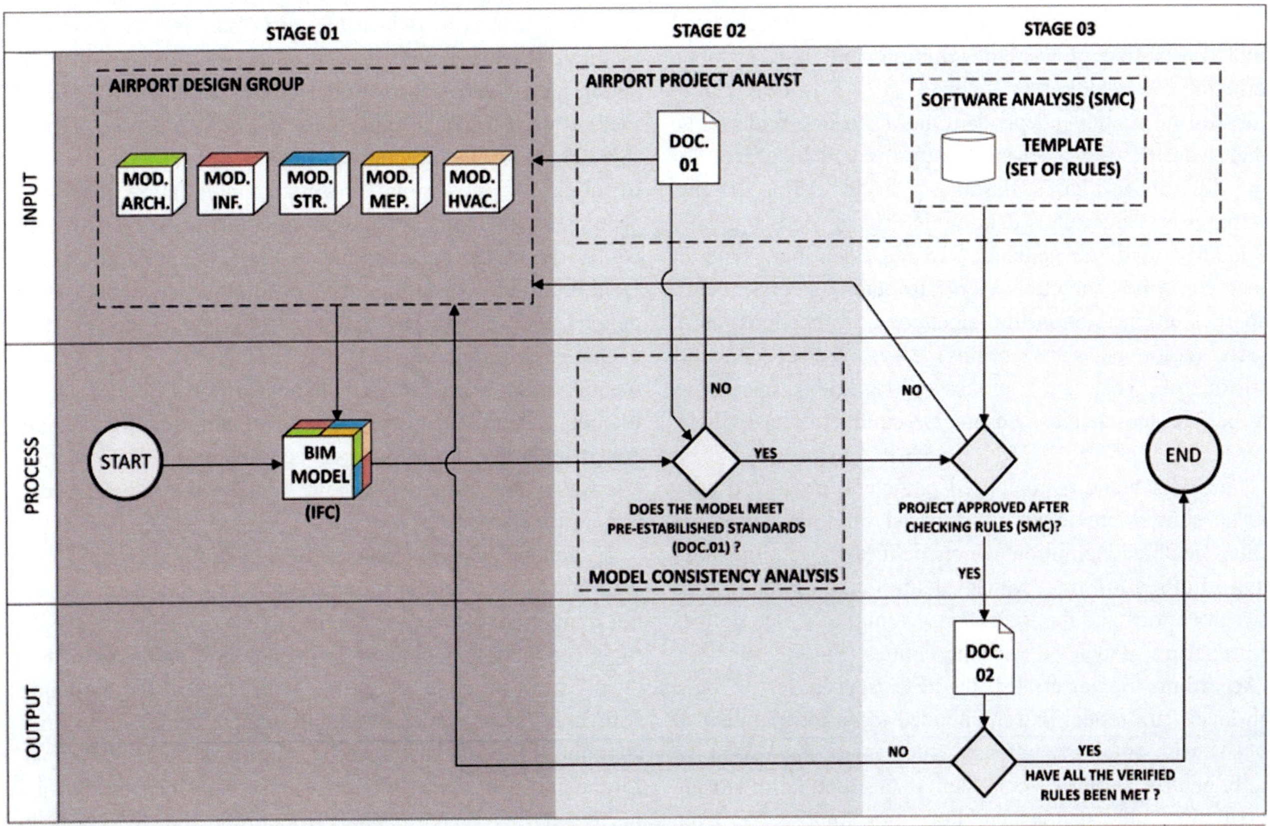

LEGEND (DOCUMENTS):
DOC.01: MANUAL OF GUIDELINES (STANDARDIZATION OF THE BIM MODEL)
DOC. 02: AUTOMATIC VERIFICATION REPORT (RESULTS)

SMC.: SOLIBRI MODEL CHECKER

LEGEND (BIM MODELS):
BIM MODEL (.ifc): PROJECT FILE - AIRPORT PASSENGER TERMINAL - EXPORTED (.ifc)
MOD. ARCH.: ARCHITECTURE BIM MODEL / MOD. INF.: INFRAESTRUCTURE BIM MODEL
MOD. STR: STUCTURE BIM MODEL / MOD. MEP.: MECHANICAL, ELECTRICAL AND PLUMBING BIM MODEL
MOD. HVAC.: HEATING, VENTILATING AND CONDITIONG BIM MODEL

Fig. 2 BIM model validation process map

Checker (SMC). In this software, the automatic checking is performed based on a rules template (created by the team of airport project analysts), automatically generating a report with the results obtained. If all the verified rules are met, the process is completed; otherwise, the report is sent to the team of airport designers for a review of the detected problems or a request for information to be redefined in the model. In this way, it can be accurately evaluated. This report (document 01) should be presented as a written report (in PDF) or digitally communicated to the design team using the BIM Collaboration Format (BCF). Both stage 02 and stage 03 will be coordinated by the team of airport design analysts.

This paper brings some initial investigations into how to document 01 should be created and organized. This proposes guidelines for the standardization of properties and data to be hung in the BIM model of airport designs, allowing teams of designers to design their respective designs projects using BIM authored software and that when these project information models are sent for evaluation, they can bring the necessary and sufficient information for an accurate assessment following the criteria established by the Airport Planning Criteria and Conditions Manual, i.e, to enable documentation 01 to be well validated in stage 02, thus enabling the checking of rules efficiently in stage 03.

5 IFC

To make it possible to exchange information between different BIM models, according to Sacks et al. [6], there are three types of formats for data exchanges between BIM applications: (1) direct link; (2) proprietary exchange format; (3) public product data model exchange formats. The third type comprises an open standard of information exchange models. It has properties of objects, measures, parameters, materials in addition to geometric properties. They are environments suitable for use in construction analysis and management applications, for example, IFC. This paper investigates the applicability and standardization of the third type of data exchange through the IFC scheme in airport designs.

According to buildingSMART, the IFC is a standardized digital description of the built environment. It is an open international standard (ISO 16739–1: 2018 [7]), designed to be neutral or vendor-independent and used in a wide range of hardware devices, software platforms and interfaces for many different use cases. Eastman et al. [8] define the IFC scheme as a standardized data model that encodes logically the identity and the semantics (name, identifier, type of object, etc.); the characteristics or attributes (material, color, properties, etc.); relationships (locations, connections, etc.); objects (columns, slabs, beams, etc.); abstract concepts (performance, cost, etc.); processes (facilities); operations and people (companies, designers, contractors, suppliers, etc.).

The IFC scheme is composed of several organized entities, as shown previously, and based on three structural points, IfcObjectDefinition which defines the entities (objects), IfcRelationship which defines the relationships between entities and the IfcPropertyDefinition which defines the properties associated with the entities (Fig. 3).

According to Nawari [3], the IFC is proposed to be a high-level data model that is intended to be independent of any software implementations, which is to stay, must be strictly neutral. It offers a consistent data structure for storing construction information, but does not impose, or even allow, any particular method of implementing it in a software application. Currently, the Industry Foundation Classes (IFC) scheme is used to exchange information from one party to another for a specific business transaction.

Due to the wealth of information that can be encapsulated in attributes, properties and geometries throughout the life cycle of a building and its environment, the amount of IFC data is very large. To avoid the difficulty of transferring information in an IFC model, it is important to define uniform means and content specification standards to be used in stages of the building's life cycle. Thus, it is important to establish what information should be delivered, by whom, when and to which recipient [9].

To solve this, buildingSMART developed the structures of the Information Delivery Manual/Model View Definition [10]. The idea is to reduce the space for interpretation and facilitate the implementation of use cases in specific application areas. The Information Delivery Manual (IDM) is a method used to consistently describe BIM processes (IDM, ISO 29481). The predefined IDM structure and uniform methods for presenting process models allow users to accurately develop, agree and document their BIM processes. From the data defined in the IDM, a Model View Definition (MVD) defines the specific sub-elements of the general IFC data model that can support the specific data exchange requirements [11]. That is, an MVD is a recommendation for what data and elements the IFC model should include, depending on the purpose of the model exchange. As an example of MVD, we can mention: Coordination View 2.0, FM Handover View, Structural Analysis View, among others.

In general, data exchanges used in rule verification systems must meet certain requirements. For this work, essential requirements are considered: (1) use of appropriate BIM tools for modeling objects. For example, the creation of a "wall" must be done using the "wall" tool of the modeling software; (2) guarantee that all nomenclatures are corresponding to those established in the manuals. For example, for the names of the environments, use the names established in the INFRAERO manuals (boarding lounge, check-in queue, etc.); and (3) ensure that the IFC mapping and its native BIM creation tools are correctly configured to export the IFC model that complies with the MVD defined by the rules verification systems to be used for airport projects. In this case, Coordination View 2.0 will be used.

6 Results

The initial studies of this research, presented in this paper, point to the importance of standardizing the BIM information models of airport designs. In order to arrive at these initial results, simulations were carried out on a passenger terminal design modeled on BIM authoring software. These simulations show, even in a preliminary way, the feasibility

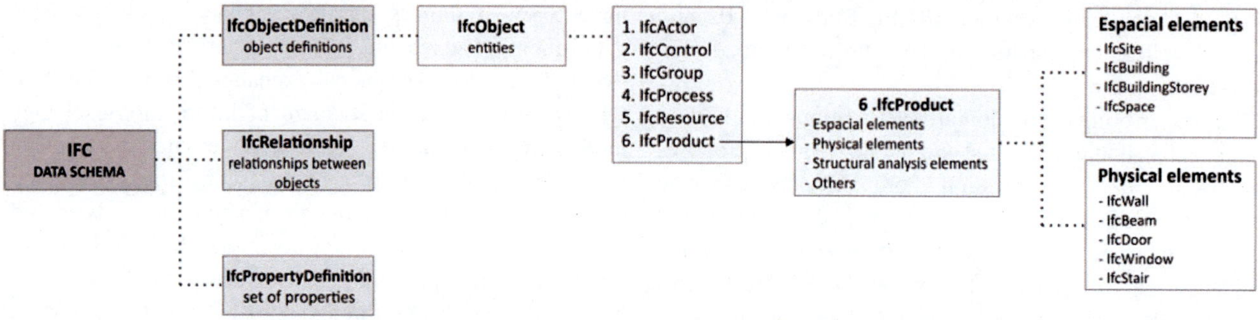

Fig. 3 Simplified example part of the IFC hierarchy classes

of standardizing certain airport design parameters to check compliance with requirements (Stage 03—Fig. 2). These studies also indicate the need for without creating protocols for modeling so that the information is hung following a standard coding in line with the rules created for evaluation (Stage 02—Fig. 2).

The starting point of the research was to analyze the evaluation process developed by teams of airport design analysts in Brazil. The results indicate that it was possible to identify some important points regarding the project evaluation process currently in force (analog): (1) the quality of the analysis is directly associated with the analyst's maturity, (2) ambiguity in the process of analyzing airport designs, having in view of the lack of standardization of analysis, with each analyst responsible for a specific criterion, (3) extended analysis time considering that the majority of Brazilian airport designs are still submitted for analysis in two-dimensional computer-aided drawings (CAD).

On the other hand, the first studies have shown that through the use of checking through regulatory codes and the standardization of the information model in BIM software (using the IFC format), it is possible to identify the possibility of improving the current process of airport designs. But, for this, there is a need to develop the manual for standardizing guidelines for modeling designs in BIM software. Through the application of the rules established in the manuals, which may be incorporated in the templates of the designs teams, the building information models produced by the designers may be able to check the automatic rules carried out by the analysts of the competent bodies responsible for such activities (Stage 03—Fig. 2).

The research current stage consists of checking automatic evaluation rules and identifying the minimum IFC properties to be hung in the BIM model. Two evaluation situations were simulated in existing components in the Airport Planning Criteria and Conditions Manual (INFRAERO, 2006). Each simulated rule was classified according to the general classes of rules recommended by Solihin and Eastman [12], being: class 01—rules that require a single or a small number of explicit data; class 02—rules that require simple derived attribute values; class 03—rules that require extended data structure; and class 04—rules that require a "proof of solution." The following will show examples, already implemented (see Figs. 4 and 5), which demonstrate the description of the rule checking, the classification of the rule and the minimum properties required in the design's BIM IFC model.

Table 2 describes the simulation rule 01 presented in this work. The condition of acceptance of this rule is a minimum distance of $\geq$ 10.00 m between the limits of the aircraft envelopes in the apron. If the IFC model is compliant with, the analysis software (SMC) will show the graphical result in green (compliant with); otherwise, the graphical result in the verified component is not compliant with, SMC will show the graphical result in red (not compliant with), for rule 01, the condition was compliant, so the SMC identified compliance with the established rule, without the need to change the design.

Table 3 describes the simulation of rule 02 presented in this work. The condition of acceptance of this rule is: the width of the checking service desk must be $\geq$ 3.00 m. If the IFC model is compliant, the analysis software (SMC) will show the graphical result in green (compliant); otherwise, the graphical result in the verified component is not compliant with, SMC will show the graphical result in red (not compliant), for rule 02, the condition was not fit, so the SMC identified the non-compliance with the established rule, with the need to resend the airport design model to the team of designers responsible for the design review.

What can be seen with these two examples is that just as it is important to establish rules for the evaluation of information models of buildings, the creation of rules alone is insufficient for the evaluation process to be efficient. For the rules to be effectively read and validated or not (true or false), the information models of the designs must be organized in such a way that the information present in the IFC properties is standardized in the creation of the models according to the parameters established in the rules checking the model. Also, it must be exported in IFC using the protocols set out in the manuals (for example, using the appropriate MVD). In the rule presented in Fig. 5, for example, entities such as IfcSpace, IfcQuantities, and Boundign Box Widht were considered. For the information model to be used to evaluate the rule transcribed in Table 3, the information model would need to contain these entities. IfcSpace, for example, would need to be defined as "service desk." Any other word defining this space would result in the impossibility of evaluating this rule.

These two examples described in this paper demonstrate, even preliminarily, the importance of standardizing IFC entities within the BIM information models during the design process. That is, in order for the evaluation team to be able to successfully carry out the automated rule checking process using BIM analysis software, the BIM model must contain the standardized IFC entities, incorporated into the BIM model. These entities must be correctly incorporated in the design process, according to the parameters created in the rules template implemented in the analysis software. For this, it is necessary to know the organization structure of the information present in the analysis software, so that it is possible to overlap the IFC data with the set of rules created. It is also necessary that the BIM information models, developed by the designers, have incorporated the IFC entities that are following the rules established in the modeling software, as well as their terminologies (for example, "service desk").

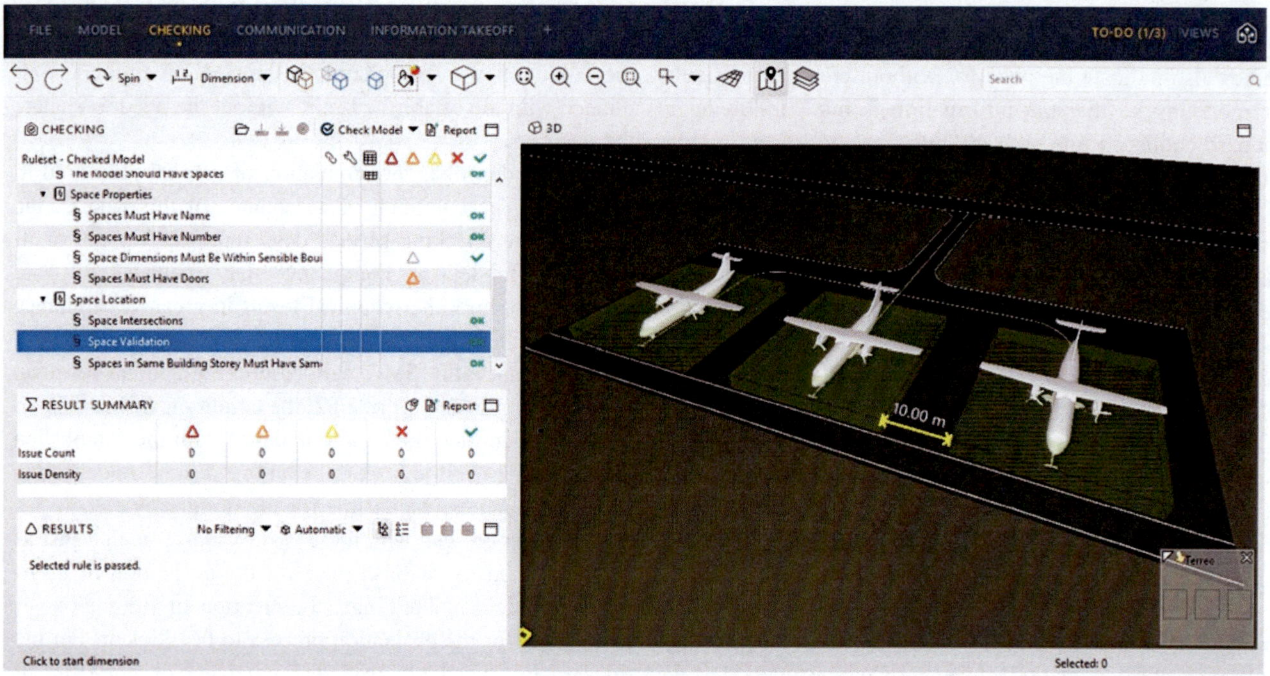

Fig. 4 A view representing the compliance with rule 01, automatically evaluated in SMC

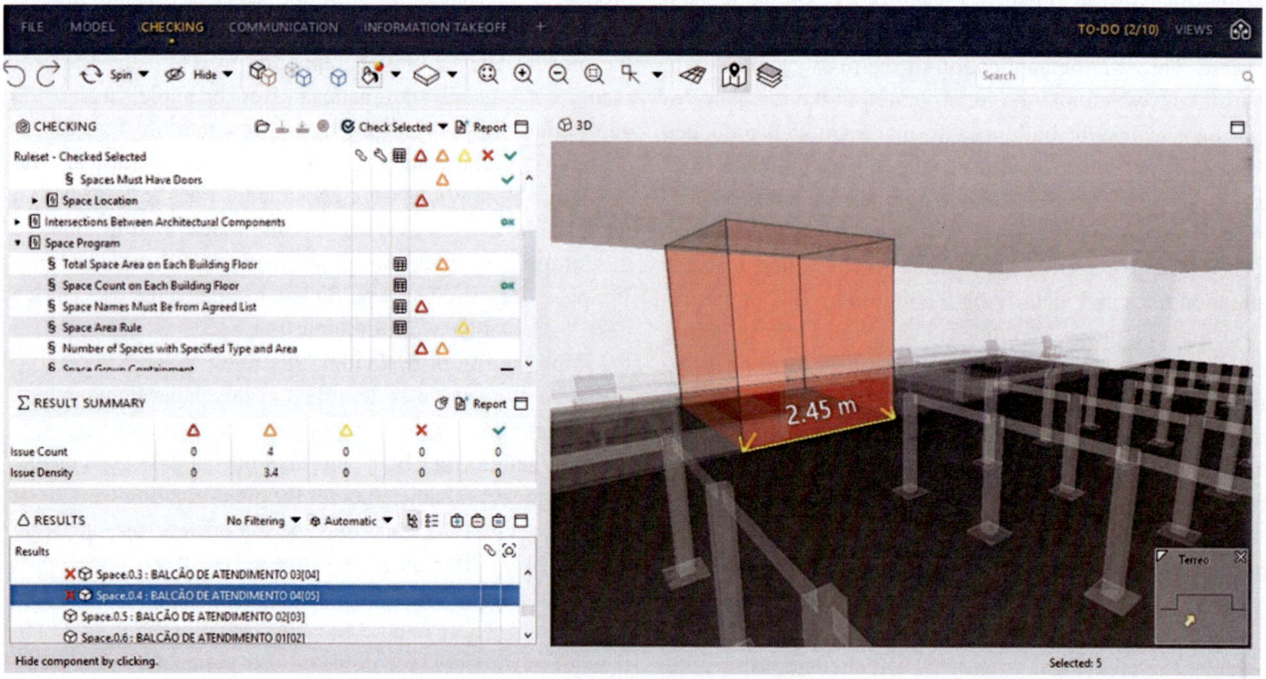

Fig. 5 View representing the non-compliance with rule 02, automatically evaluated in SMC

It is important to note that the examples of the rules presented in this research refer to class 02 rules [12], that is, they are rules that require attribute values derived from simple; this does not mean that this method cannot benefit analysts in the validation of rules classes 03 and 04. A classic example of the use of these last rules is the so-called design assessment tool (DAT). This tool, developed by the Georgia Institute of Technology, used the circulation and safety rules found in the US Courts Design Guide [13] encoding them in 302 computable parametric rules [5]. Another example of a class 04 rule, for the case of PT, would be for the analyst to assess how an airport worker

Table 2 Rule description 01 verified on BIM model

Reference document: manual of criteria and conditioning of airport planning—INFRAERO (2006)
Rule description: minimum distance between an aircraft envelope area in the apron
Condition: the distance between wingtips ≥ 10,00 m
Classification of the rule—according to Solihin and Eastman [12]: Class 02
Verified IFC properties | Verified SMC (ruleset)
IFC entity: IfcSpace/IfcProperties: Name = ENVELOPE|SMC Parameter: Tolerance = 10,00 m

Table 3 Rule description 02 verified on BIM model

Reference document: manual of criteria and conditioning of airport planning—INFRAERO (2006)
Rule description: minimum width of the check-in desk
Condition: minimum width ≥ 3.00 m
Classification of the rule—according to SOLIHIN, EASTMAN [12]: Class 02
Verified IFC properties | Verified SMC (ruleset)
IFC Entity: IfcQuantities/Boundign box width/constain: ≥ 3.00 m

would make an access path through the service area to a certain system within the PT with his equipment to perform maintenance or repair in a specific system. In this case, it would be necessary for the analyst to have knowledge regarding the building maintenance processes, as well as, to have knowledge of lessons learned from other projects; thus, this type of analysis could be semi-automated.

From the moment it is possible to automate the analysis process in a set of rules, be they simple or more complex, consequently this automation will allow more time for analysts to deal with more pertinent issues regarding efficiency and the design process, such as form analysis, functionalities related to operational activities, building maintenance, and operation processes, among other relevant points. In the current evaluation process (manual), analysts spend a lot of time in the evaluation phase, checking a series of components that have the potential to be implemented in a code checking process. An example of possible components to be implemented in an automatic verification process are the items of architectural design for an Airport Passenger Terminal [2]. In this case, mentioned above, the authors identified 119 components grouped into 06 categories [14].

Figure 6 demonstrates the time and energy ratio that the analyst currently has to verify compliance with the rules. In

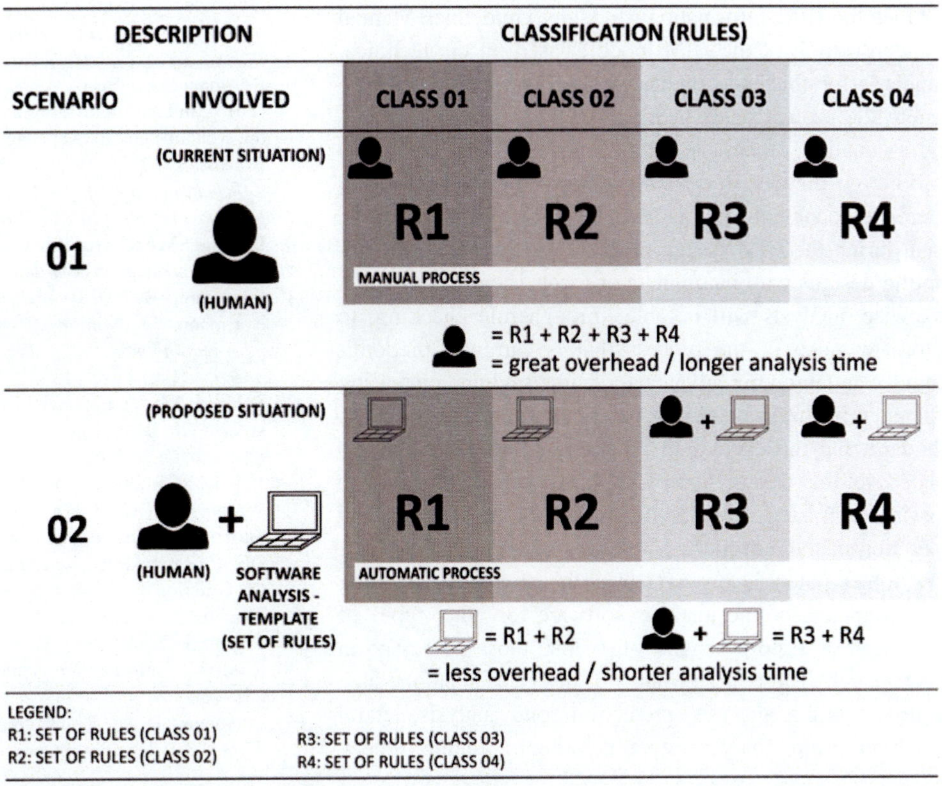

Fig. 6 Relationship between project analyst activities

scenario 01, the analyst performs a manual analysis process, devoting more time to check the low complexity rules, leaving more time to analyze the more complex rules. On the other hand, in scenario 02, with the aid of a computational system, the analyst optimizes his time in the analysis of the simplest rules (and that takes less time for implementation) and starts to have more time to analyze more complex rules in a process manual and automated hybrid. This does not mean that complex rules cannot be implemented, but that the implementation of simple rules (classes 1 and 2), at an early stage of transforming design evaluation practices, is easier to implement and at the same time can bring great productivity gain to the project evaluation activity.

7 Conclusion

This research, even though it is still in the development phase, demonstrates the importance of creating a standardization process in the modeling of information during the development of airport designs in software authored by BIM, considering the export to an integrated IFC model. In view of the inexistence of manuals dealing with this subject in Brazil, it is possible to observe the potential that this approach has and may, shortly, benefit all teams of airport designs and analysts.

Once the BIM Information Modeling Guidelines Manual (standardization of the BIM model) based on the technical standards for this type of design has been created, it will benefit the professionals involved in the designs' synthesis and evaluation process at various points. Thus, the teams working on the design development (synthesis) will have a reference document for standardizing IFC information to be used during the BIM design process, which can avoid submitting designs with incomplete or inconsistent information. Likewise, analysts will be able, through rule checking, to automate much of the process that is currently predominantly manual, taking advantage of time savings, increasing efficiency in the evaluation process, minimizing ambiguity, standardizing the evaluation process, among other benefits related to the design approval phase. In addition to the possibility of later use of the IFC model for maintenance and operation of the airport.

Another potential observed was the possibility of using the parameters of the analysis software for other types of evaluation of airport designs, other than those addressed in this research. There are elements of design evaluation, such as noise zoning analysis, protection zone analysis, lateral transition ramp analysis, approach ramp, among others, which can also be customized in rules, or new ones can be created rules for analyzing these parameters that in many cases go beyond the airport site boundary.

References

1. Benning, P., Dumolin, C., Dehlin, S., Tulke, J. Âberg, P., & Ryd, N. (2010). Collaboration Processes: A State of the Art.
2. Guedes, Í., & Andrade, M. (2019). Automatic rule-based checking for the approval of building architectural designs of airport passenger terminals based on BIM. In *Proceedings of the 37th eCAADe and 23rd SIGraDi Conference* (Vol. 2, pp. 333–338). Porto, Portugal: University of Porto (11–13).
3. Nawari, O. N. (2018). *Building information modeling: Automated code checking and compliance processes*, 1st edn. Boca Raton; CRC Press.
4. BIM ThinkSpace. (2015). Episode 24: Understanding Model Uses, v 0.73.
5. Eastman, C., Lee, J., Jeong, Y., & Lee, J. (2009). Automatic rule-based checking of building designs. *Automation in Construction, 18*(8), 1011–1033.
6. Sacks, R., Eastman, C., Lee, G., & Teicholz, P. (2018). *BIM handbook: A guide to building information modeling for owners, managers, designers, engineers, and contractors* (3rd ed.). New Jersey: Wiley.
7. ISO—International organization for standardization. ISO 16739-1: 2018—Industry Foundation Classes (IFC) for data sharing in the construction and facility management industries.
8. Eastman, C., Telcholz, P., Sacks, R., & Liston, K. (2008). *BIM Handbook—A guide to building information modeling for owners, managers, designers, engineers, and contractors*. New Jersey: Wiley.
9. Borrmann, A., Beetz, J., Koch, C., Liebich, T., & Munic, S. (2018). Industry foundation classes: A standardized data model for the vendor-neutral exchange of digital building models. In A. Borrmann, M. Köning, C. Koch, & J. Beetz (Eds.), *Building information modeling: Technology foundations and industry practice* (1st ed., pp. 81–126). Cham: Springer.
10. BuildingSMART. (2113). Construction operations building information exchange, MVD definition for IFC4.
11. Steinmann, R. (2018). IFC certification of BIM software. In A. Borrmann, M. Köning, C. Koch, & J. Beetz (Eds.), *Building information modelling: Technology foundations and industry practice* (1st ed., pp. 139–154). Cham: Springer.
12. Solihim, W., & Eastman, C. (2015). Classification of rules for automated BIM rule checking development. *Automation in Construction, 53*, 69–82.
13. Judicial Conference of the US, US Courts Design Guide, Washington DC (2007).
14. Superintendência de Empreendimentos de Engenharia (2006). DEEP Gerência de Planejamento de Engenharia—EPPL. Manual de Critérios e Condicionantes de Planejamento Aeroportuário. Infraero.
15. March, S. T., & Smith, G. F. (1995). Design and natural science research in information technology. *Decision Suport Systems, 15*, 251–266.
16. International Air Transportation Association (IATA). (2019). *Airport development reference manual*, 11th edn.

17. Garber, R. (2009). *Closing the gap information models in contemporary design practice* (Architectural Design March April 2009 Vol. 79 No. 2). Wiley.
18. Eastman, C., Telcholz, P., Sacks, R., & Liston, K. (2014). *Manual de BIM—Um Guia de Modelagem da Informação da Construção para Arquitetos, Engenheiro, Gerentes*. Porto Alegre: Construtores e Incorporadores. edn. Bookman Editora Ltda.

Ítalo Guedes is an architect and urban planner, specialist in BIM Platform, Master in Architecture and Urbanism at the Federal University of Pernambuco. He has professional experience in the development of BIM projects in infrastructure and architecture of airports in Brazil, currently developing a master's research project on Verificação automática de regras em projetos de aeroportos no Brasil com base em BIM, professionally linked to the Works and Projects Directorate—Secretariat of Infrastructure, City Hall of Ipojuca/PE, researcher and teacher graduate course.

Agent-Based Semiology: Optimizing Office Occupation Patterns with Agent-Based Simulations

Robert R. Neumayr

Abstract

Measuring the performativity of office space has been a long-standing topic of research. With the emergence of knowledge economy, the nature of work has changes considerably, foregrounding personal interaction and information interchange. Consequently, traditional tools of space evaluation, such as space syntax, have become increasingly difficult to apply. This research, therefore, uses a methodology based on agent-based simulation, focusing on agent behavior rather than on space morphology to assess the social performance of spaces. The research process is conducted in two phases: In a first research phase, simple social models are developed for the agent population in order to set up simulations that show differentiated agent behavior toward other agents and architectural frame dependency. A series of simplified yet plausible life-like office event scenarios with strategic changes to the furniture layout is used to evaluate and calibrate the simulation's social performance. Based on these simulations and the social algorithms derived from them, in a second research phase an experimental setup that follows the generative logics of evolutionary design solving is devised to identify office layouts with the highest social performativity. Methodically searching the design space for performance peaks, algorithmic design is used to generate, simulate, and test an initially large number of random scenarios against a set of predefined success criteria in order to obtain a subset of the most successful configurations.

R. R. Neumayr (✉)
University of Applied Arts Vienna, Vienna, Austria
e-mail: robert.r.neumayr@gmail.com

Keywords

Agent-based semiology • Human space design • Work and office environments • Social performance simulation • Algorithmic design

1 Introduction

Measuring and improving the performativity of office space and the workflow within it have been a long-standing topic of research. However, since office space environments first emerged as a distinct typology in the second half of the nineteenth century, their layouts have changed and evolved constantly. Initially widely used Taylorist office space layouts, whose design was based on a strictly linear organization of work, at some point proved to no longer capable of proficiently emulating the increasingly intricate systems of co-operation that had developed over time.

Consequently, innovative designs, that were rejecting the then prevailing rigid structures of large bureaucratic organizations, were developed, most notably the German Quickborner team's Bürolandschaft concept. Its typically open-plan office configurations were based on a two-dimensional spatial distribution logic, which in turn was based on the different modes of interaction, which could be detected between the various groups and teams of workers within the heterogenous types of office organization in the 1960s.

While Bürolandschaft concepts at that time reflected the relational qualities within a workplace, their conception remained largely on a purely diagrammatic level. The first concept bringing the esthetics of design and the need for a more dynamic office environment into the modern workplace was Robert Propst's and George Nelson's Action Office. However, while initially designed to provide the flexible and autonomous environment much needed in a Bürolandschaft, economics eventually turned the Action

S. Eloy et al. (eds.), *Formal Methods in Architecture*, Advances in Science, Technology & Innovation,
https://doi.org/10.1007/978-3-030-57509-0_5

Office into the monotonous cubicle farms that came to largely dominate the 1990s office landscapes, compressing a lot of employees into small spaces (see e.g., [1: 205–220]).

Within the last years, knowledge economy has taken over as the most important factor in almost every developed country's service sector. In knowledge economy, processes are no longer organized along the transfer of materials or goods but focus on the constant exchange of information and knowledge. Contemporary office space layouts, accordingly, have become more informal and multi-functional, as their performance depends on their capability to facilitate the interactions between different actors in various formal and informal configurations.

2 Office Space Analysis and Simulation

2.1 Scientific Theory

All innovations in office space design have been, therefore, informed by two key factors: an increase in complexity of the patterns of workflow and a profound shift toward immaterial work. Changes in the workflow's underlying organizational principles have resulted in innovative spatial configurations and consequently necessitated more advanced tools and methods to evaluate the performance of different office environments.

For a long time, traditional work environments were designed according to the theory of scientific management, a concept developed by Frederick W. Taylor in the 1880s to improve labor productivity in workspaces that mostly rely on a linear transfer of work and goods. The performance of any such space could be easily assessed, for example, by counting the units assembled or tracking the amount of paperwork done.

2.2 Bürolandschaft

While the Bürolandschaft concept still followed a linear understanding of workers' interaction, albeit arriving at spatially more complex office solution, it presented two major innovations for future office space design strategies: First of all, it used the result of an analysis of the flow of communication as a generative tool. At the same time, it overcame restrictive hierarchical systems, thus fostering informal conversations, which were considered essential to their cybernetic organizational model [2]. The performance of such multi-directional open office spaces that were designed to accommodate more complex patterns of social interaction, however, have proven much more difficult to evaluate.

2.3 Space Syntax as an Analytical Tool

The most successful tool today to understand the relationship between the configuration of an office space and the spatial occupation patterns of its users is space syntax. It was initially developed by Bill Hillier and Julienne Hanson, who suggested to look at the built environment "*[...] at the level of [a] system of spatial relations*" [3: 3] in order to understand its societal implications. Quickly space syntax developed into a tool to analyze various indoor social spaces, most notably offices environments. In space syntax research, the relation between a space's morphological characteristics and the user interactions therein is established by means of applying space syntax' analytical concepts to assess a space's geometrical attributes and subsequently correlating these findings with information about the social interaction within it, which is normally collected from questionnaires taken by the employees who work in that space (see e.g., [4].

The premise of this entire analytical procedure then is "*[...] that the sociologically relevant aspects of configured space can be captured at the level of topological description.*" [5: 19]. However, recent studies suggest that in confined spatial systems, such as typically sized office spaces, social factors, such as employees with their respective position within an organization's hierarchy, expertise, or personal relationships outweigh spatial specifications (see e.g., [6]. In these cases, space syntax analysis generates inconclusive results. The quantitative description of the spatial layout in question can no longer be correlated to the interaction patterns observed in the space, as space syntax cannot metricize the environment's social qualities (see e.g., [7].

2.4 Agent-Based Modeling

To understand and simulate spatial occupation patterns in contemporary office environments, one therefore needs to focus the research on the social interactions between the actors themselves, rather than looking at office geometries. Contrary to earlier office layouts, which were based on fixed furniture layouts and static behavior of their users, contemporary patterns of user interaction can no longer be considered permanent, but temporal and emergent in nature and are the result of a series of simple user interactions, which over time amount to a sequence of complex ephemeral configurations. They can, therefore, be simulated using agent-based modeling (ABM), which in its most concise definition is "*a computational method that enables a researcher [to] experiment with models composed of agents that interact within an environment.*" [8: 2].

Agent-based modeling was first introduced into computer science in 1987, when Craig Reynolds for the first time successfully digitally reproduced the flocking behavior of birds in his computer simulation "*Boids*" [9]. Since then, the field of application for agent-based modeling has broadened considerably and now encompasses various fields of research from physics to social science nowadays.

2.5 Tools for Agent-Based Simulations

Over the last years, as agent-based modeling gains popularity across different disciplines, its scientific community has developed a wide range of agent-based modeling program packages that can be used to develop agent-based simulations with different scientific focus and at various levels of complexity (see e.g., [10]).

For the research of social phenomena, NetLogo is the most popular software toolkit. It is particularly widespread within the scientific community, and at this point, "*[...] there are a large number of agent-based models written in NetLogo in a wide variety of domains*" [11: xiv]. Therefore, its users can rely on a large and active user community and a wide range of online tutorials and code libraries. Created for agent-based simulations only, NetLogo already comes with built-in processes designed to solve typical agent-based simulation scripting problems. Due to its rather simple code structure, it is easily accessible, fast, scalable, and its list-based processes makes data extraction easy.

While these software properties allow for the quick development of a first simulation setup and subsequently for the evaluation of the resulting data and therefore suggest the use of NetLogo for this strand of research, its representational capacities are basic, and visual output is limited to simple 2.5 D graphics (Fig. 1).

At the same time, with the recent advances in software technology, readily accessible crowd simulation tools have become part of almost every commercial animation program, such as Autodesk's Maya, Alias, or 3ds Max, or game engine software package, such as Unreal or Unity. These tools initially created for the animation or gaming industry, focus on the visualization of digital crowds and can therefore be considered a possible alternative when looking for simulation software. However, the development of social agent models proves to be difficult, because more complex behavioral patterns cannot be scripted and developed properly, but need to be assembled as a sequence of ready-made basic behaviors (such as attraction, repulsion, or collision avoidance), which themselves are hard to customize or change. In some instances, data handling and collection is rather opaque, the number of free variables is limited, and data export is not possible.

After testing four different software environments (NetLogo, Maya Miarmy, Unreal Engine, and Unity) at the very beginning of the research by setting up and simulating a basic office scenario, and evaluating the coding process, simulation results and data output, it was decided to use NetLogo for the research in question.

3 Constructing Social Models for Agent-Based Simulations

3.1 Basic Considerations

Architecture has only recently started using agent-based simulations to investigate the intrinsic logic of architectural crowds. Based on the assumption that human crowds—quite similar to a flock of birds or a school of fish—show emergent behavior caused by a multitude of recurring interactions that add up to the complex overall state of a system, engineers began to develop simulations for circulation, congestion, and building evacuation. Like flocks in nature, these kind of models only operationalize the spatial relations between agents, as they computationally negotiate agent properties such as cohesion (stay close to the other agents), distance (keep a minimum distance to the neighboring agents), and alignment (adjust your direction toward the overall direction of the crowd).

It is safe to assume, however, that the occupational patterns in contemporary office spaces heavily depend on a number of different non-spatial factors, such as fellow agents, objects of interaction and architectural features (quantitative factors) or light, temperature, cultural context, and work atmosphere (qualitative factors). Considering that, this research aims to develop a more sophisticated behavioral model in which an agent's actions and responses are no longer triggered by a space's morphology but by the social interaction with other agents and specific spatial or environmental features. In order to achieve this, any agent-based simulation needs to include a series of novel features that go beyond the processing of space [12: 51], of which the two most important ones are agent differentiation and architectural frame dependency.

Agents need to be differentiated by their position, status, group membership, or importance within the social network, resulting in behavioral differences as they interact with each other. But agents also need to show architectural frame dependency, allowing for varying behavioral patterns depending on their location within a space and the objects of interest within it.

The aim is to develop a population of agents with individual behavioral rules that enable agent-based simulations to "*[...] move from the current evacuation- and traffic-engineering crowds to architectural and semiological crowds as the basis for generalized life process simulation*" [13: 112].

In general, all agent-based simulations share the same set of characteristic features: Ontological correspondence, a

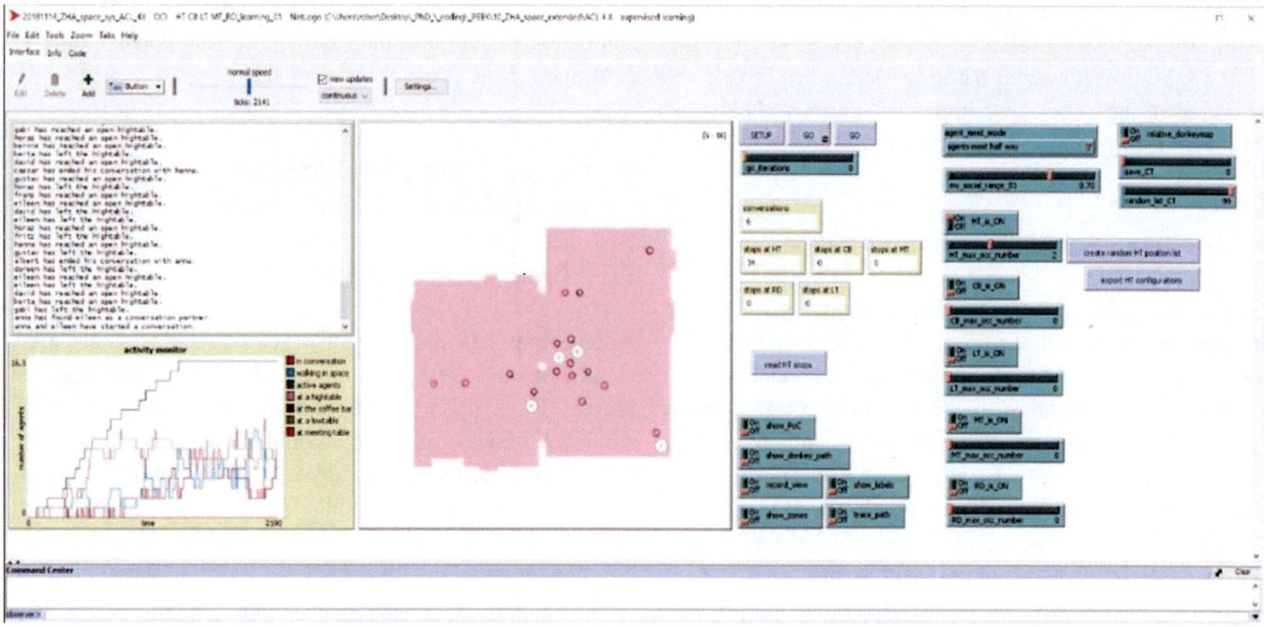

Fig. 1 The NetLogo interface for the simulation. Columns from left to right: data readout windows, graphic representation of the simulation space, control panel. Image license: CC BY-SA 4.0

representation of the environment in question, a set of heterogeneous agents, and agent interactions based on bounded rationality [14].

In this simulation, the morphology of an existing office environment in London is used as an experimental setup, focusing on its most informal area, the breakout space, where it is fair to assume that face-to-face interaction will most likely occur. The space is furnished with several typical office furniture elements, such as high tables, low tables, reception desks, coffee bars, and meeting tables, that allow for spontaneous encounters as well as for scheduled meetings in different constellations. The simulation space will be populated by sixteen agents, entering, navigating, and leaving the space, while interacting with other agents and semiological objects (points of interest that drive specific types of behavior) in that space.

3.2 Simplification and Bounded Rationality

Social science research is conducted by devising simplified representations of social phenomena. In agent-based simulations, computational agents are digital entities, which are set up to model human capabilities in a much simplified way and are, therefore, always limited in their cognitive abilities. They are modeled to act with bounded rationality.

The concept of bounded rationality was first introduced by Simon [15] who suggested that rather than assuming that an individual's choices are perfectly rational, one should understand the rationality within any decision-making process to be limited for various reasons.

In general, agent-based models are developed gradually, starting from basic rules of interaction. The challenge is therefore "*[…] not to limit the rationality of agents, but to extend their intelligence to the point where they could make decisions of the same sophistication as is commonplace among people*" [8: 16].

3.3 Simplified Social Models

Knowledge economy, which in Western European countries at this point represents around one third of all economic activities [16: 115], has led to the emergence of various new kinds of knowledge work and patterns of interactive workflow, focusing on the exchange of knowledge in different types of conversations and interactions [17].

The simplified social model developed for this research, therefore, is based on operationalizing the basic parameters of agent-to-agent interaction. It is the starting hypothesis of this research that the dynamics of emerging conversation patterns in social spaces can be described by focusing on two main questions of social interaction: "*Who interacts with whom?*" and "*How long does this interaction last?*".

It is safe to assume that the social parameters that add up to the selection of a conversation partner and determine the duration of the ensuing conversation, are manifold, including an agent's social status, position in the company, hierarchy, affiliation, or useful knowledge and information, but also properties such as popularity or physical attractiveness. These attributes will be subsumed and quantified in a variable called "*social attractiveness*".

We can also assume that, in general, any actor in a space will try to engage in a conversation with the actor with the highest social attractiveness available. However, we can observe restrictions to this behavioral rule in real-life scenarios, where spatial distance or big differences in status or hierarchy usually preclude social interaction. Actors therefore always operate within a certain physical and social range, which describes the *relation* between communication patterns and the spatial and social environment they are embedded in. Following the observation that communication patterns will be considerably different in culture groups with divergent concepts of hierarchy, agents in the simulation will be introduced to a physical and a social range that they need to operate in. In this simulation, agents will therefore always look for and try to interact with an available agent with the highest social attractiveness within its social and physical range.

The duration of any social interaction is determined by the differences in value of a variable called "*social alignment*," which represents an agent's personal characteristics, such as personality, profession, expertise, fields of interests, or level of acquaintance with other agents. The more the values of two interacting agents align (i.e., the more they have in common), the longer the interaction will last (Fig. 2).

It can be observed, that in contemporary office spaces not only fellow agents, but also objects of interest in the space will influence the spatial occupation pattern of its users. Steen and Markhede, for example, also register this impact in their space syntax analysis of spatial and social configurations in offices [6: 123]. Consequently, the simplified social model developed for agents is equally applied to the objects in an office space, as similar logics will influence the attractiveness of a specific furniture object in space, such as the type of object (high table, coffee bar, etc.) or its location in the space (e.g., next to the entrance vs. in a remote corner).

Systematic variations of the physical and social ranges of agents and objects will create distinctively different patterns of spatial occupation and interaction (Fig. 3).

At this point of the research, all values of these variables that determine the agents' behavioral properties are assigned randomly as placeholders that can later—once the algorithms of interaction have been refined—be substituted by more viable social data that can, for example, be extracted from social network analysis.

Note: The modes of operations of a simplified social model have been documented in greater depth in a recently published paper by the author [18].

3.4 Assembling Comprehensive Agent Models

First, the simple social model described above is implemented on top of a generic agent model, which already contains essential procedures for spatial navigation, field of vision, collision avoidance, and object and agent recognition. Then, the agents' capabilities are systematically extended step by step to develop and test patterns of interaction with a number of common office furniture elements taken from the office layout developed earlier on, such as high tables, low tables, a meeting table, a reception desk, and a coffee bar. In this process, the agents' capabilities are systematically extended until a level of sophistication is reached that allows for the simulation of a simplified, yet credible, life-like office scenario, including agent-to-agent interaction as well as interaction with a number of common furniture elements, namely a reception desk, a coffee bar, high tables, low tables, and a meeting table.

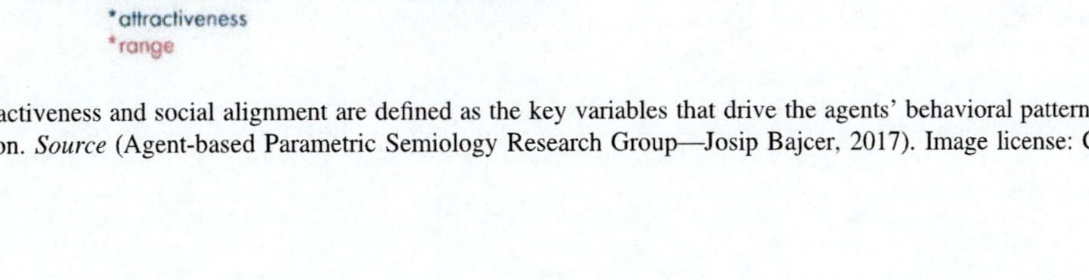

Fig. 2 Social attractiveness and social alignment are defined as the key variables that drive the agents' behavioral patterns in a simplified social model of interaction. *Source* (Agent-based Parametric Semiology Research Group—Josip Bajcer, 2017). Image license: CC BY-SA 4.0

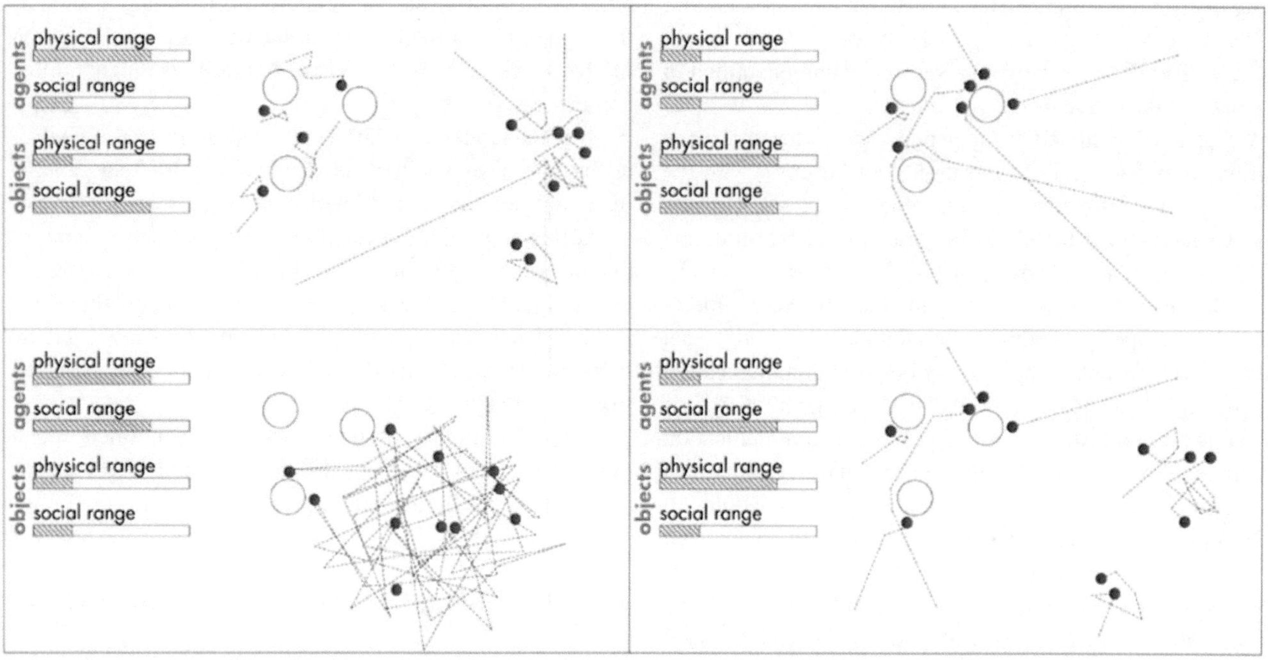

Fig. 3 Different values for social and physical ranges of agents and objects will generate a wide variety of different spatial occupation patterns. Source: (Agent-based Parametric Semiology Research Group – Josip Bajcer, 2017). Image license: CC BY-SA 4.0

4 Assessing the Simulation of "Life-like" Office Scenarios

4.1 Setting Up a Research Matrix

The simulation setup is tested for general plausibility and frame dependency through a matrix of nine similar yet slightly different spatial configurations. While the space morphology and the number and type of furniture elements stay the same, their locations are altered systematically, generating nine discrete plausible scenarios. Furniture elements are meaningfully distributed according to functional and well-established office layout patterns, with proper attention to location, spacing, and distances. Each simulation is run a hundred times to check for consistence, minimum and maximum values, average, standard deviation, and outliers.

Apart from calculating and comparing statistical numbers, the data collected is also used to generate a number of different visual quantifiers, most notably heat maps, which show the agents' occupation patterns over time, and location maps, tracking the positions of each individual agent in regular intervals.

4.2 Assessing the Simulation's Accuracy and Realism

As this research aims to establish a relation between a space's semiological features (agents and objects of interest) and its users in a life-like scenario rather than examining its spatial morphology, all simulation results can be assessed for basic plausibility by looking at their visual output first and comparing it to what we perceive as resulting patterns in similar real-life scenarios.

Apparently, agent show differentiated behavior toward each other as well as to semiological objects in the space depending on their respective social properties. It can be deducted from the agents' apparent behavior that procedures for interaction follow intricate patterns rather than being based on simple parameters, such as distance or random selection. The resulting heatmaps (which depict the agent positions over time superimposed in one single image each) generated by the different simulation setups indicate that occupation patterns shift and adapt according to various office furniture layouts, thus demonstrating architectural frame dependency.

Next, to check for the simulation's consistency, the analysis of the numerical data recorded throughout the nine simulation scenarios serves as an important indicator. Investigating the series of nine simulations in an identical concise simulation space over a fixed length of time with a constant number of agents and objects in varying positions in the space, one would reasonably expect to discover a similar but not identical number of overall interactions, but at the same time different values for particular interactions between agents and the objects in the scene. The numbers extracted from the nine simulations confirm this expectation: While the total number of interactions does not change significantly, values for agent-to-agent conversation and agent-to-object interaction differ considerably depending on the simulation scenario (Fig. 4).

At this point, the overall realism of these simulations is hard to assess. Additional simulations and the subsequent calibration of their results with sensor data collected in the actual office space will clearly be necessary. However, in order to compare and identify the most performative design proposal from within a set of different options no accurate performance simulation is necessary. The empirical finding, that a specific layout's relative advantages in performance can be sufficiently specified, even if absolute performance values might be inexact, nevertheless seems a legitimate first step that calls for further investigation into agent-based semiology.

5 Optimizing Spatial Layouts with Evolutionary Design Strategies

5.1 General Considerations

So far, we have compared the interaction values of a set of different, plausible office scenarios to understand their respective performance properties. Thus, the procedure allows us to identify those configurations out of the nine contenders that hold a relative performance advantage over others in respect to specific performance criteria, for example, furniture layouts that foster free-floating agent-to-agent communication or others that increase the number of encounters at meeting tables. At this point, the simulation is therefore used as an analytical tool that enables us to compare and evaluate existing design solutions.

Based on these findings, we are interested in the question, whether the algorithm can suggest a configuration, different from the ones already tested, with an even higher performativity in respect to the same set of performance criteria. For this, we need to turn the simulation from an analytical tool into a generative tool.

Traditionally design decisions are taken deterministically, mostly relying on intuition and previous experience. Research indicates that successful intuitive decisions rely on

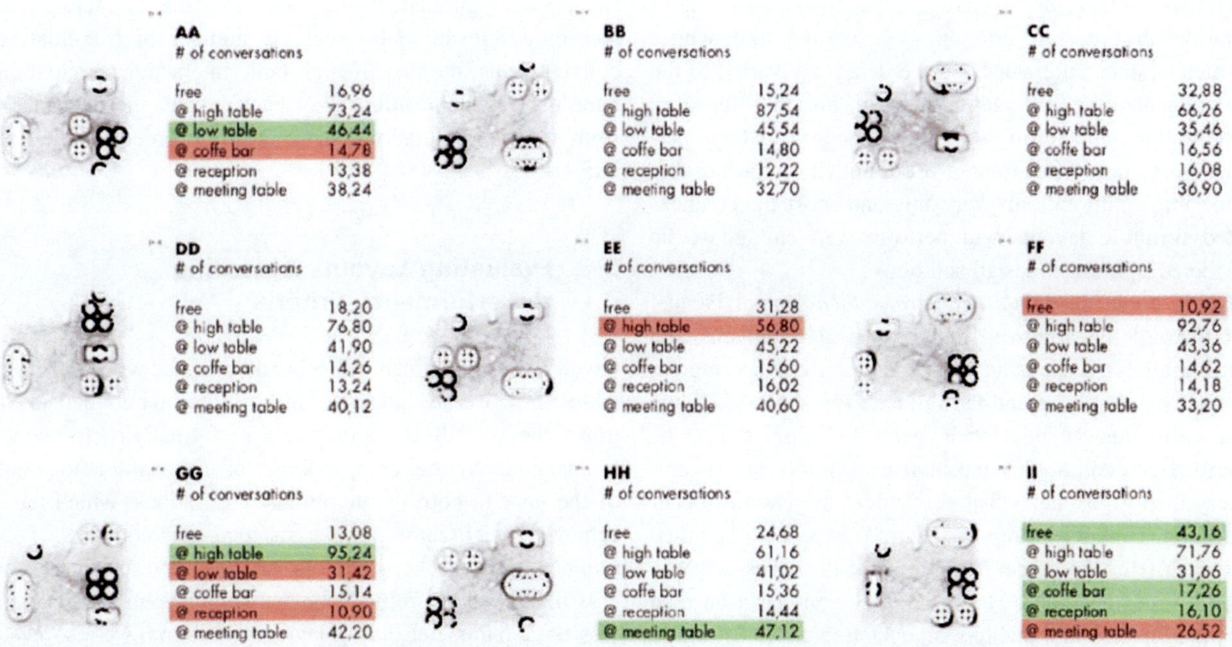

Fig. 4 Table of interaction values for nine different office space layouts (AA to II). Green and red indicate maximum and minimum values for the interaction with specific objects within the entire set of simulations. Images depict the heatmap resulting from each scenario's simulation. Image license: CC BY-SA 4.0

previous experiences which are processed on a subconscious level (see e.g., [19]). In such a process, the designer selects a previously successful solution to a design problem from her repertoire and re-applies it to a similar design problem. While this strategy is useful to develop "satisficing" [20] solutions (in the sense, that a solution is "good enough" for a given problem) to a known design problem quickly, it might not allow the designer to find the optimal answer to the problem in question, as she might be trapped on a local peak in a widely uncharted fitness landscape, unaware of possibly better solutions that lie outside her current design paradigm.

5.2 Evolutionary Design strategies

The use of evolutionary design strategies in the search for optimized architectural solutions allows for the systematic exploration of a much bigger design space, which is no longer limited by the designer's imagination or experience, therefore presenting the designer with a fresh set of additional design options to select from. This is widely considered one of the biggest advantages of digital design processes, as one is able, as Manuel de Landa remarks, to "*[...] explore a space in which it is impossible for the designer to consider all potential configurations in advance, and only if what results shocks, or at least surprises.*" [21].

The task is therefore to change the simulation's mode of operation in order to develop a generative tool whose parameters, once automated, can be iteratively varied to the end of generating a large number of possible furniture configurations and testing them all according to their performative values. After testing, of an initially large number of possible solutions only very few and most likely unexpected furniture layouts will perform well enough to be considered optimized design solutions.

In his book, *Darwin's Dangerous Idea* Daniel Dennet puts forward the argument that biological organisms and human artefacts, such as art, architecture, or engineering are essentially part of one and the same design space [22], and as a consequence, both their generative logics can be described as evolutionary mechanisms. Simply put, evolutionary theory postulates that all changes in genomes occur randomly, without purpose or intent. This process is called genetic drift in biological evolution and describes a lateral motion in design space. The force of natural selection then identifies successful mutations, an operation called lifting. In architectural design, evolutionary design strategies are therefore applied to use the aleatoric potential of automated digital design processes to generate an initially large set of random configurations within the parametric limits determined by the designer. Lifting, then, occurs by testing the results of this generative process against the success criteria defined by the designer, a selection process leading to a smaller subset of the most successful configurations. This subset then in turn serves as the basis for the generation of a new and refined set of configurations, thus starting an iterative process that is stopped as soon as no further improvement in performance is measured.

5.3 Algorithmic Design Setup

The algorithmic design setup is based upon the previous research. For this set of new simulations, the previously developed social model is used for the agents within the same office breakout space and its entry and exit points. It also features the same number of agents and the same runtime length; however, in this first simulation attempt, the simulation's complexity is reduced by removing some of the available furniture elements from the scene in order to get more conclusive results.

The simulation script will automatically generate one hundred office layout variations with four randomly positioned high tables in each of them. As the table location is entirely random (the only stipulation is that a table's center needs to be inside the space), there is no consideration of practicality, purpose, or design intent, thus avoiding any designer's intuition or bias (Fig. 5). Each of these layouts is subsequently simulated ten times, each time extracting and recording relevant data, such as number of free-floating conversations, number of high table interaction, circulation length, or agent positions. For each layout, the data taken from the ten simulations is averaged in order to rule out outliers.

5.4 Evaluating Layouts According to Performance Criteria

Given a clearly defined office breakout space with four high tables in it, we are interested in furniture layouts that maximize the overall configuration's performativity for communication. We are, then, looking for those simulations out of the genetic pool of one hundred variations, which have achieved the highest number of total conversations. This number therefore becomes our criterion for selecting successful candidates from all the available layouts. As realism was not a parameter during layout generation, we also need to eliminate unrealistic layouts, such as tables overlapping or table-wall-collisions. Comparing the remaining successful layout variations subsequently allows for the deduction of general layout patterns that these configurations have in common.

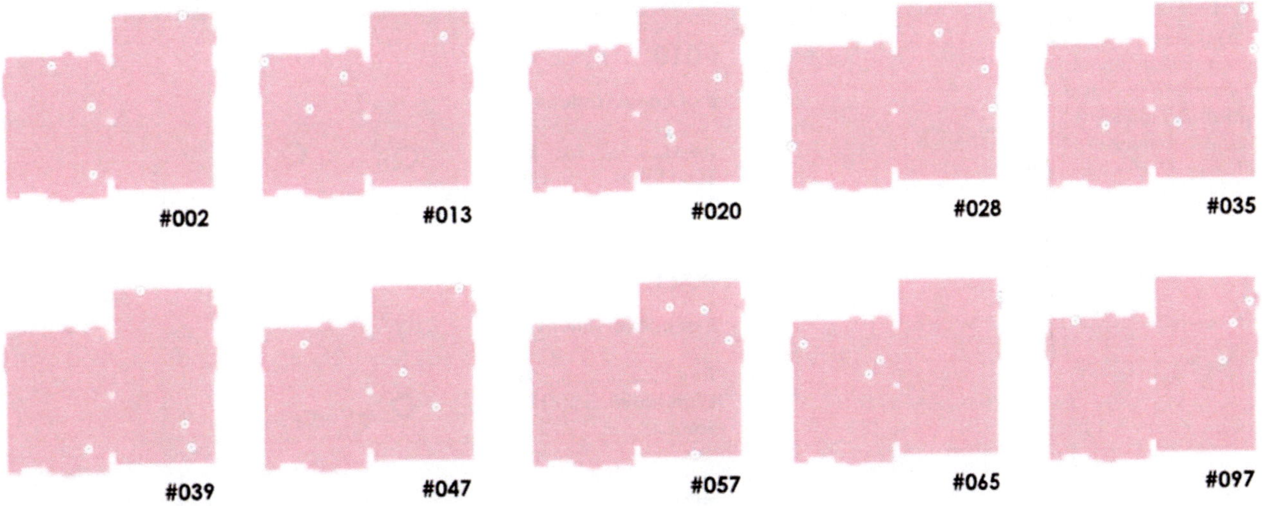

Fig. 5 Exemplary set of 10 office space layouts, randomly selected out of the initial pool of 100 office space layouts randomly generated by NetLogo for algorithmic design purposes. Image license: CC BY-SA 4.0

6 Results and Outlook

6.1 Simulation Results and Findings

Looking at all 100 layouts, the maximum number of total conversations is 216; the minimum number is 202, with an average of 209. The three most successful furniture layouts (#10, #21, and #36) feature 216, 216, and 215 total conversations, respectively. In comparison, the three least successful layouts (#12, #25, and #5) have 202, 203, and 204 conversations in total, establishing a significant difference in the overall number of conversations (Fig. 6).

Comparing the three most successful layouts shows that configurations with high performativity have their four high tables located toward or along the perimeter of the room, sometimes in different corners. This suggests that the lateral positioning of tables allows for enough central room in the center of the space for spontaneous face-to-face interaction between the agents while moving from one table to another. The locations of these conversation recorded in the space confirm this assumption (Fig. 7).

Analyzing the three least successful layouts reveal that in these configurations have the high tables clustered in the middle of the space, close to each other but probably forcing spontaneous conversations to take place in the far corners of the room, rendering them more infrequent. This again is substantiated by looking at the numbers: The three most performative layouts allow for significantly more spontaneous face-to-face interaction (81, 91, and 97) than the three least successful ones (70, 77, and 73).

6.2 Discussion

Building up on an agent-based simulation which is able to generate virtual crowds that exert overall life-like behavior, exhibit differentiated behavior toward other agents and objects in space, and whose cumulative behavior results in different spatial occupation patterns throughout different scenarios, the systematic application of algorithmic design processes allows the researcher to identify layouts with high performativity from a large set of randomly generated configurations and subsequently to determine generalized layout rules that these have in common.

While the result itself might not be that surprising — and in that sense defies de Landa's dictum about "*[...] shock, or at least surprise.*" [21]—the research indicates that out of two different plausible furnishing options that intuitively seem equally reasonable (location of tables centered in the middle of space vs. location of tables at a space's perimeter) those with maximum performance according to the selected performance criteria can be recognized and singled out for further refinement.

Again, at this point it is not possible to assess the simulation's realism in terms of the accuracy of the absolute performance values collected. For this, a comparison of the simulation data with the data collected in the actual space will be necessary in order to further refine and calibrate the algorithmic values.

Successful space layouts will, however, show relative performance advantages over less successful ones. The design rules generalized from analyzing the parameters of those layouts with clear relative advantages in performance

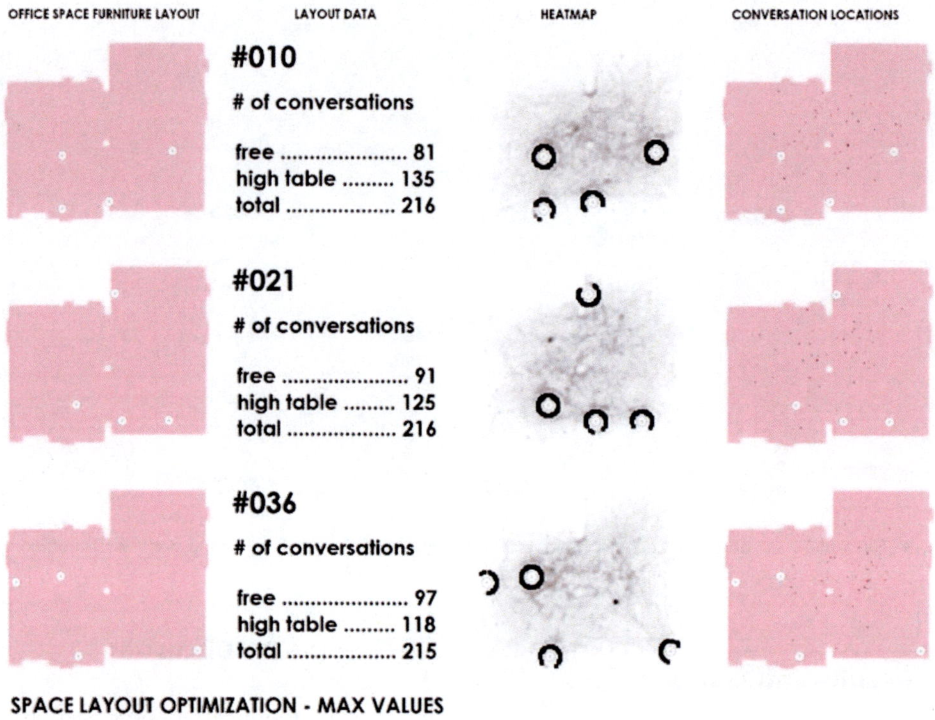

Fig. 6 Diagram showing the three most successful furniture layouts. Columns from left to right: Furniture Layout, Values Recorded in Simulation, Heatmap, Locations of free-floating Conversations in Space. Image license: CC BY-SA 4.0

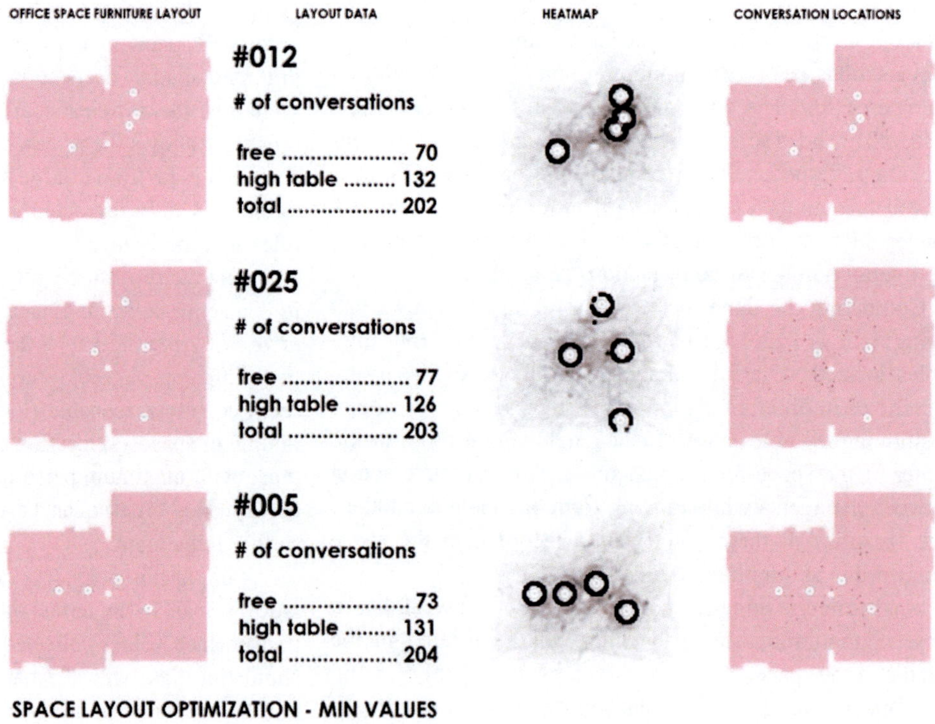

Fig. 7 Diagram showing the three least successful furniture layouts. Columns from left to right: Furniture layout, values recorded in simulation, heatmap, locations of free-floating conversations in space. Image license: CC BY-SA 4.0

can in a next step be used to serve as additional guidelines when approaching similar design challenges.

Acknowledgements This project is part of a broader research program, generously founded by the Austrian Science Fund (FWF) PEEK—project AR 354-G24. My appreciations go to Patrik Schumacher and to my fellow researchers.

References

1. Saval, N. (2014). *Cubed—A secret history of the workplace*. New York: Anchor Books.
2. Kockelkorn, A. (2008). Bürolandschaft—Eine vergessene Reformstrategie der deutschen Nachkriegsmoderne. *ARCH + Zeitschrift für Architektur und Städtebau, 186/187*, 6–7.
3. Hillier, B., & Hanson, J. (2003). *The social logic of space*. Cambridge: Cambridge University Press.
4. Peponis, J., Bafna, S., et al. (2007). Designing space to support knowledge work. *Environment and Behaviour, 39*, 815–840.
5. Bafna, S. (2003). Space syntax a brief introduction to its logic and analytical techniques. *Environment and Behaviour, 35*, 17–29.
6. Steen, J., & Markhede, H. (2010). Spatial and social configurations in offices. *The Journal of Space Syntax, 1*(1), 121–132.
7. Groves, K., & Marlow, O. (2016). Interview with Kerstin Sailer. Spaces for innovation—The design and science of inspiring environments (pp. 61–64). Amsterdam: Frame Publishers.
8. Gilbert, N. (2008). *Agent-based models*. Thousand Oaks: Sage Publications.
9. Reynolds, C. (1987). Flocks, herds, and schools: A distributed behavioral model. In *Proceedings of the 14th Annual Conference on Computer Graphics and Interactive Techniques ACM, 21*(4), 25–34.
10. Abar, S., Theodoropoulos, G., Lemarinier, P., & O'Hare, G. (2017). Agent based modeling and simulation tools: A review of the state-of-art software. *Computer Science Review, 24*, 13–33.
11. Wilensky, U., & Rand, W. (2015). *An introduction to agent-based modeling: Modeling natural, social, and engineered complex systems with NetLogo* (p. xiv). Cambridge, Massachusetts: MIT Press.
12. Schumacher, P. (2019). Digital. *AA Files, 76*, 47–52 (AA Publications, The Architectural Association, London).
13. Schumacher, P. (2016). Advanced social functionality via agent-based parametric semiology. In P. Schumacher (Ed.), *Parametricism 2.0. AD 02/2016* (p. 112). Wiley: London.
14. Fagiolo, G., Windrum, P., & Moneta, A. (2006). *Empirical validation of agent-based models: A critical survey (No. 2006/14)*. Pisa: Sant' Anna School of Advanced Studies, Laboratory of Economics and Management.
15. Simon, H. A. (1957). A behavioral model of rational choice. In H. A. Simon (Ed.), *Models of man*. New York: Wiley.
16. Eurostat. (2013). *Science, technology and innovation in Europe*. Luxembourg: Publication Office of the European Union.
17. Greene, C., & Myerson, J. (2011). Space for thought: Designing for knowledge workers. *Facilities, 29*(1/2), 19–30.
18. Neumayr, R. (2020). *Simulating contemporaray office occupation pattern with simplified social models*. Atlanta: Divergence in Architectural Research, GeorgiaTech University.
19. Gladwell, M. (2005). *Blink*. London: Penguin Books.
20. Simon, H. A. (1956). Rational choice and the structure of the environment. *Psychological Review, 63*(2), 129–138.
21. De Landa, M. (2002). Deleuze and the use of the genetic algorithm in architecture. *Architectural Design, 72*, 9–12.
22. Dennet, C. (1995). *Darwin's dangerous idea*. New York: Simon and Schuster.

Robert R. Neumayr Robert Neumayr is an architect and researcher. He studied architecture in Vienna, Paris, and London where he received his March II with distinction from the AA's Graduate School Design Research Laboratory. He worked with Will Alsop and Zaha Hadid in London and is currently working in Vienna, running his architectural practice hoppe+-partner and his research practice unsquare.at. Robert has been teaching with Patrik Schumacher, Zaha Hadid, and Kazuyo Sejima at the University of Applied Arts in Vienna and was teaching the Post Professional Design Studio at PennDesign. He is currently a Lecturer at the Institute of Design at the University of Innsbruck and a PhD candidate in Vienna, coordinating a research group on Agent-Based Semiology.

Check for updates

Digital Doxiadis: Parametric Thinking for Human Settlements

Alexandros Peteinarelis and Nikolaos Patsavos

Abstract

This paper presents the output of 'Digital Doxiadis: Parametric thinking for Human Settlements', an architectural design workshop that took place in Chania on July 2015 with 25 participants. It was part of a two-year preliminary research project on the future of Doxiadis' unique European industrial settlement and based on the findings of two intensive workshops in Aspra Spitia and Athens on January 2015 and April 2015. One of its main goals was to bring Doxiadis's approach on the study and development of human settlements in the context of contemporary tools and associative parametric modeling. Custom design tools attempt to simulate the function and evolution of Aspra Spitia in Boeotia combined with industrial districts, assuming programmatic benefits through the application of selected cluster theory concepts, expressed as spatial rules. Reflecting on the experience and the workshop results, we suggest that employing procedural parametric master planning offers the possibility of revisiting Doxiadis' pioneering analysis-based and data-driven anthropocentric ekistics, while the advantage of calculating directly output data for every design choice taken augments the design process, shifting beyond the flexible urban configuration parametric model into an informed spatial exploration and quantitative decision-making tool.

Keywords

Parametric modeling • Doxiadis • Ekistics

A. Peteinarelis (✉)
University of Ioannina, Ioannina, 45110, Greece
e-mail: ng5ten@yahoo.com

N. Patsavos
University of Ioannina, Ioannina 45110, Ctrl_Space Lab, Athens, 16345, Greece
e-mail: control.space.gr@gmail.com

1 Historical Identity and Localized Potentiality

1.1 Aspra Spitia

Aspra Spitia, Doxiadis' only European example of complete realization of his ekistic theory, was originally planned and constructed between 1961 and 1964 for French aluminum company Pechiney and its Greek subsidiary Aluminum of Greece (AoG) at Distomitika, Boeotia, Greece [1]. The settlement is a paradigmatic application of Doxiadis' anthropocentric attempt to revive the ancient Greek city in the context of radical critique to modern planning and architecture and the challenges of the future [2]. It was designed for a projected population of 5000 [3, 4], proposing 1100 dwellings for recruited aluminum plant workers, engineers and administrative staff.

After the initial design and construction supervision by Doxiadis Associates (DA), two expansions–modifications followed. Architects Lebessis, Fotiades and Massourides worked on public space plans and dwellings designs. Doxiadis' invisible geometry, expressed with formal and material simplification, aimed to an unplanned organic settlement result and to activate familiarization actions through ephemeral structures on facades[1] [5]. Thus, the dwelling districts would disrupt the sense of continuation of working in a mechanized factory, as typically experienced in uniform industrial settlements of the era. In contrast, the expansions introduced modern high-rise buildings drawing from and cultivating the Athenian suburban lifestyle. At any case, the settlement and the factory it serves should be viewed as an interplaying assemblage, both operating in conjunction, while also preserving their singularity.

[1]The project was initially published by DA in [5], a brochure printed in 1966. This publication was based on the homonymous article [6]. For a critical presentation, see [7, 8].

S. Eloy et al. (eds.), *Formal Methods in Architecture*, Advances in Science, Technology & Innovation,
https://doi.org/10.1007/978-3-030-57509-0_6

The settlement today is not fully occupied, while the adjacent aluminum factory owned by Mytilineos Group (MG) still operates. The concession agreement between the Greek Government and Pechiney specifies that the settlement's land may only be used for the purposes of the factory, excluding tourism and other irrelevant functions. The current condition calls for a redevelopment of the settlement's programmatic and spatial logic to enhance its future social, economic and environmental sustainability.

This topic has been, since October 2014, the object of a preliminary research project directed by Ctrl_Space Lab (Research Director Patsavos, N., Senior Researcher Tsesmetzis, V.) in collaboration with the School of Architecture, Technical University of Crete (Research Director Yiannoudes, S., Senior Researcher Peteinarelis, A.) and AoG. In its first two stages, the project dealt with a fundamental set of historical and theoretical data, detailed analysis of a representative part of the settlement and a pilot strategic investigation on the future of Aspra Spitia, by means of archival research, direct collaboration with MG and two research intensive workshops on January and April 2015. In this paper, we present and discuss the third stage, namely the process and output of 'Digital Doxiadis: Parametric thinking for Human Settlements', an architectural design workshop that took place in Chania on July 13–17, 2015. Doxiadis's approach would be examined this time in the context of contemporary tools and associative parametric modeling.

1.2 Spatial Infrastructure and Social Interaction

Following the first fifty years of their systemic evolution, Aspra Spitia and AoG are now facing a series of challenges:

- the company, having lost its initial connection with Pechiney and due to its relatively small size and impact factor, is facing the risk of isolation;
- technological, environmental and social concerns have been seriously negotiating the settlement's future sustainability;
- deindustrialization in the Western world raises the question of a new industrial paradigm, allowing vertical and horizontal integrated development.

Therefore, Aspra Spitia is at risk of becoming an obsolete and unmanageable monument of twentieth-century industrial heritage. In resolving this, the research team reflected on a strategic recodification of its functional role and organization which would invest in its inherent characteristics and the following assets:

- local knowledge on aluminum;
- emblematic and functional spatial infrastructure;
- unique design with emphasis on public space needed for social interaction;
- extensive and specialized available national human resources.

In short, Aspra Spitia (Fig. 1) possesses the fundamental spatial, social, expertise and productive attributes of what Porter [9] has described as a cluster. A geographic concentration, a network of interconnected businesses, suppliers and associated institutions in a particular field, which can provide with an opportunity to reorganize localities in the context of the global networked society. A concept that aligns with both the Ecumenopolis model, i.e., Doxiadis projection on how in the future all local settlements would end up linked in a global network [10], and Doxiadis' conceptualization of networks, since the 1960s, as the defining parameter for the evolution of human settlements [11]. This was the hypothesis raised at the second relevant workshop and further examined in the third, discussed in this paper.

1.3 The Workshop

Digital Doxiadis workshop was announced as an open call for professionals and students. No prior knowledge of parametric modeling was required. Participants had to submit a cv-portfolio for evaluation of their interests and quality of work. Therefore, the curriculum had to respond to the following challenges:

- fulfill the research agenda objectives with the provided algorithm implementation;
- like Senske [12, 13] suggests, offer computational thinking understanding, thereby participants can grasp the underlying logic of the processes taking place in the provided design tool;
- present the preliminary research findings and previous workshops results as point of departure for the design of new masterplans.

More specifically, participants have been called to design and evaluate Aspra Spitia as the location for a new aluminum R&D cluster, part of a wider European and global network. After three days of six-hour computational thinking and modeling skills sessions, three open to public lectures at the Center for Mediterranean Architecture (CAM) on Doxiadis Ekistics,[2] Aspra Spitia, the preliminary research

[2]Doxiadis had coined Ekistics as a holistic meta-science of human settlements, considering the latter globally and locally via a new synthesis of data provided by the sciences of nature, Anthropos, society, built shells and networks. See [14] for an in-depth set of relative articles.

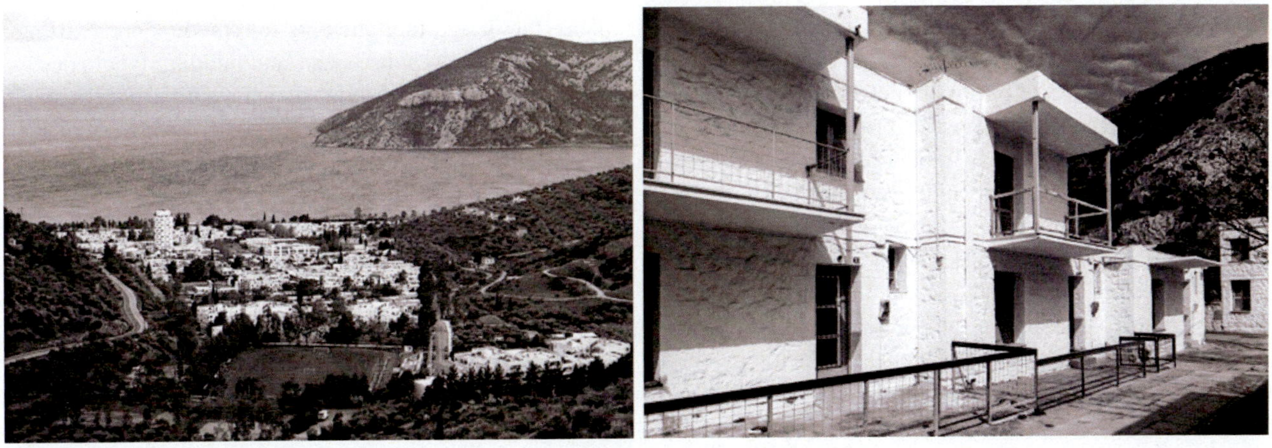

Fig. 1 General view of Aspra Spitia to the south seaside, housing units. Ctrl_Space Laboratory archive, Photographer Drapaniotis Sofianos

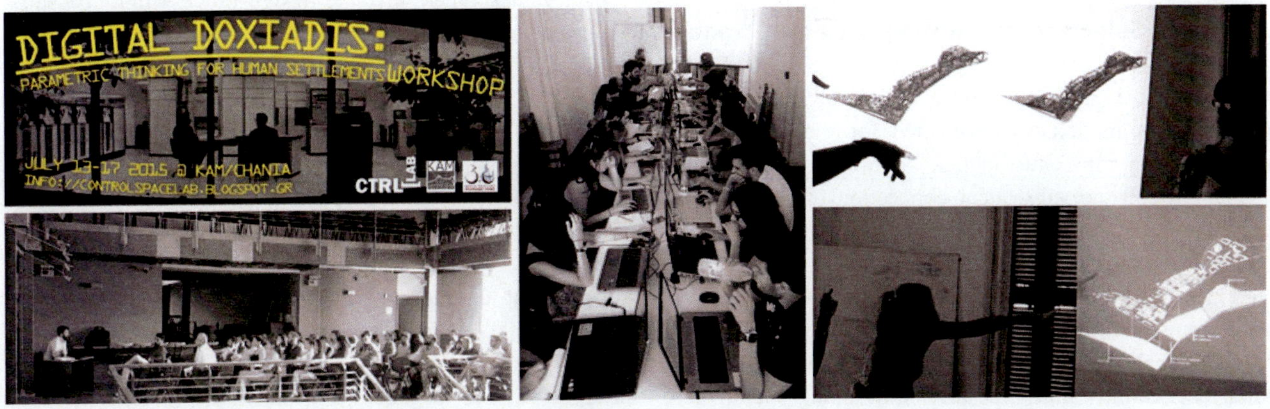

Fig. 2 Workshop call poster, public lecture at CAM, skills sessions, final presentations

findings and a documentary [15] screening, the 25 participants were assigned to four cross-functional teams. Each included members of variable age, interests, skills and design expertise level and in the course of the two remaining days the study and design of the settlement would conclude with a critical review of the design outcomes (Fig. 2).

2 Urban Parametric Design Methods and Systems

Like Beirão et al. [16, 17] argue the design of urban plans considers decisions beyond their formal characteristics. Their research presents an urban design tool with parametric features in interplay with fixed and predefined urban form and use elements. It is confronted as a dynamic design system, steering away from static design of specific layouts, toward a continuous exploration of informed design solutions, derived from parameters adjustments. They opt to conduct the design investigation and decision-making processes simultaneously, to improve the judgment quality of the design outcomes, but also to enhance the designers' understanding of crucial to the design indicators. Visual interfaces, besides geometric transformations, continuously depict updated information on determined components of urban design problems.

Similarly, Pitts and Luther [18] suggest that parametric urban models should be viewed as a support tool that can help inform the designer in the decision-making process, rather than a means of designing on behalf of or removing the architect from the design process. Their proposed method consists of two phases, starting with the diagramming of initial conceptual ideas based on site information analysis, i.e., the collection, compilation and extraction of relevant contextual data and proceeding with custom parametric scripts to translate the diagrams into digital parametric models. Real-time visual analysis, gross floor area values and land use percentages calculations are generated within the same design tool, to communicate the design concept and assess its qualities. The 3D outcomes are used for further computational analytical and transformational processes, specific to the demands of the design intent. Thus, they establish an updating design feedback loop, where ideas are tested and validated up to the point a specific solution is promoted to the next level of a more detailed design phase.

Along this direction, the grasshopper definition implementation for the workshop attempted to parameterize the diagnosed dynamics of Aspra Spitia settlement. The main computational research focus was to examine how its function and evolution, aligned with Doxiadis's initial masterplan design intent and systemic thought, can be embedded into a parametric model, anticipating to go beyond established quasi-representational parametric models into an informed urban design system.

3 Method, Workflow and Custom Design Tool

3.1 Tools Selection and Implementation Goals

For this study, Grasshopper 3D graphical algorithm editor for Rhino 3D modeler was selected for the design tool implementation. In Grasshopper, algorithms are implemented through the subsequent connection of code nodes, the components, forming directed acyclic graphs, the definitions. Three plug-ins were selected: Human UI [19] for the creation of custom user interfaces (UI), Human [20] for on screen numerical and geometry data display and Urban Network Analysis Toolbox (UNA) for Rhino [21, 22] for spatial accessibility and pedestrian flow along spatial networks analysis. The design tool and method considered the following properties and goals:

- the algorithm implementation, being a pedagogical tool too [23], had to accommodate participants of variable expertise, skills and understanding of urban design concepts and programming, if any, skills. Three types of participants were considered, beginner, intermediate and advanced, to be able to navigate through, modify and extend the provided parametric tool, respectively;
- one of the main intents of Aspra Spitia initial planning was to create a sequence of spaces, in which scale, form and character would follow the sequence of importance of events in the life of the town and its inhabitants [5]. Thus, the design tool considers the first original Doxiadis implementation as a hard constraint and attempts to alter the settlement only in the areas of later modifications, through scale and size parameters that create urban gradients of structures and place;
- although purely formal exploration design methods are educational and allow freedom for experimentations [24], steering away from purely stylistic, malleable forms [17] and in line with Doxiadis's framework of the Dynapolis [25], calling for linearly expandable, rectangular dwelling plans, the design tool attempts to reproduce the existing Doxiadis housing units layout and building typologies;
- since clusters are networks of competitive participants, who share resources and have equal access to specialized information, in the end it is human agency that binds clusters together and provides the social glue that facilitates their competitive benefits [9]. The notion of social glue abides to the spatial values of Doxiadis' Gossip Square [26, 27] planning. Therefore, the 'square of implicit knowledge' (SoK) is proposed as the spatial representation of social glue to the master plan, where knowledge and current innovation concepts are exchanged outside the working spaces, and is the key urban attribute for design iterations;
- the design tool aims to accommodate evolving masterplan variations, which can be assessed through fitting target data values. Therefore, the design process must be supported by visual interfaces that offer spatial and numerical updates.

3.2 Workflow—Iterative Bidirectional Analysis-Design-Evaluation

Adopting the definition of design as a negotiation between problem and solution through the processes of analysis, synthesis and evaluation [Lawson cited in 16], where the designer pursues to generate a matching problem–solution pair, through a coevolution of the problem and the solution [28], participants were challenged to apply this iterative bidirectional method (Fig. 3).

Analysis, as the first step, considers the provided preliminary research findings, along with an examination of the current spatial dynamics of Aspra Spitia, to form design questions regarding the future of the settlement. The existing buildings, road network and urban blocks are studied in the given masterplan. Buildings are classified as per type, use and size, while their spatial allocation is examined in correlation with the urban blocks, from district to single block level. The analysis step is the groundwork for the formulation of the working vector master-file, to be edited at any time, matching alternating design intents.

Design begins with strategy diagrams that examine the social and spatial settlement function. Target metrics are specified to resonate with the design intent, and the working master-file parameters are referenced to the design tool. Persistent layers contain non-editable geometry from the original Doxiadis masterplan and dynamic layers are utilized for the design of new urban blocks, buildings and road network. SoK squares are the main urban attribute that transforms the settlement and creates cluster districts. Urban blocks in user given proximity to the square, i.e., the

Fig. 3 Design method—workflow overview

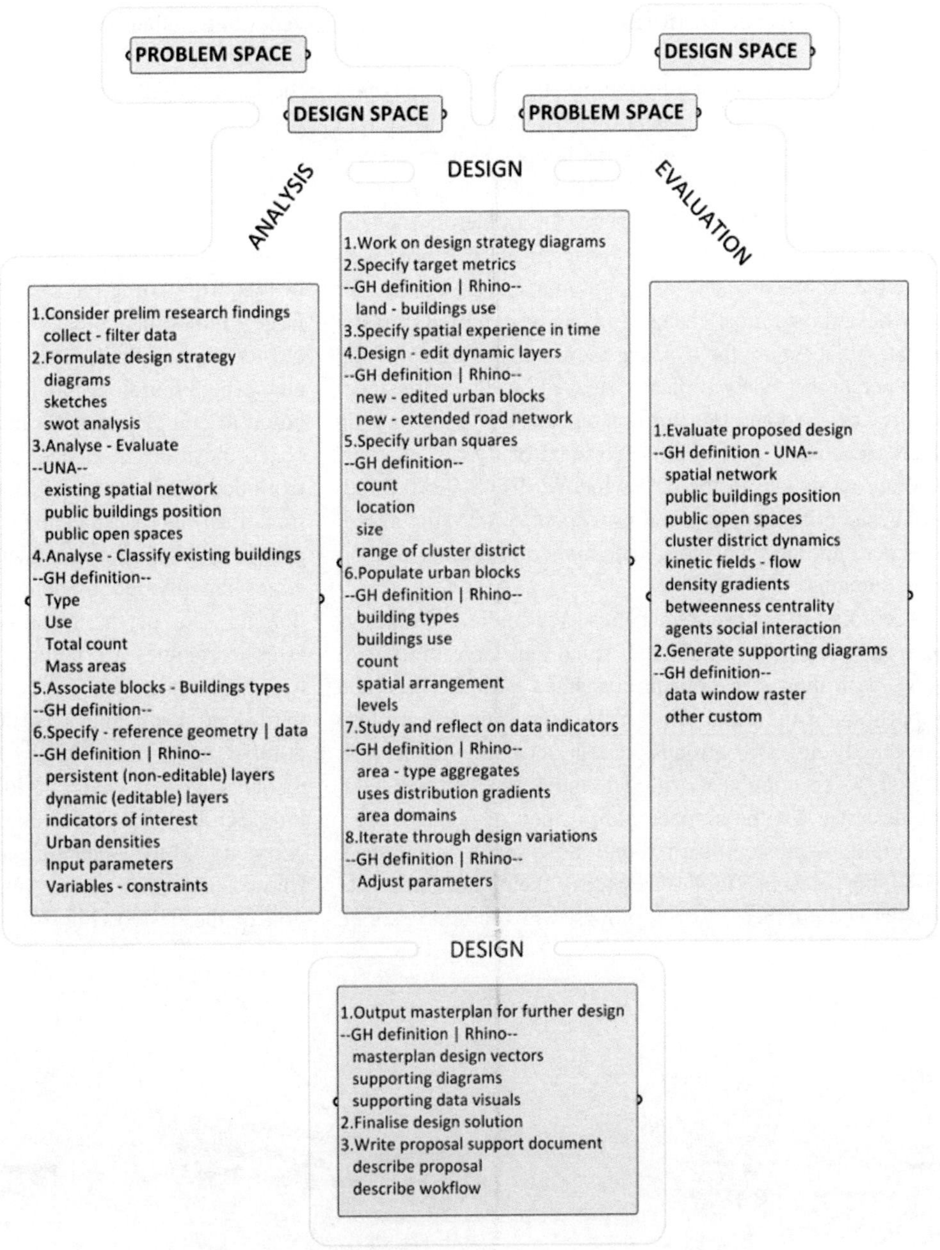

range of the district, are populated with three types of buildings:

1. cluster experts' buildings, where aluminum research and development take place;
2. cluster-type 01 buildings, for pertinent to aluminum research and applied technologies industry experts;
3. cluster-type 02 buildings, short-term dwellings for cluster experts and visitors.

Outside the cluster range are allocated public use buildings and Type 3 long-term dwellings, to complete the settlement masterplan and its private and social function.

The evaluation step examines the proposed solutions assessing the chosen positions of the cluster districts in reference to each other and to urban amenities. The masterplans are evaluated in time, the daily routine of the settlement, during rush hours and leisure time. Pedestrian and vehicle accessibility from/to social–private destinations and working places are quantified using UNA's features, to pinpoint either areas of overcrowding and road network bottlenecks, or secluded and poorly accessible regions. The settlement is further evaluated with the reflection on density indicators. Design aspirations and strategies are validated or reconsidered according to the minimum metrics that every case initially is set as target and priority.

3.3 Grasshopper Definition

The workflow is supported by the interface layout shown in Fig. 4. It has three working Rhino views, one legend view, a custom UI data window and Grasshopper remote control panel (RCP), which hosts slider parameters and Boolean operation buttons. From left to right, the masterplan legend view displays on screen, in fixed positions, non-user-selectable geometry and data, automatically updates for every design parameter change and can be exported in raster format. Adjacent are the working views, an orthographic top and a perspective view vertically stacked, where the designer can preview, add and transform geometry, e.g., move gossip square generating points, edit persistent or dynamic layers, resulting an update to the data window, displaying recalculated residential density, buildings floors count, area aggregates per building type, along with numeric extremes of their areas domains.

According to the preliminary research findings, buildings were clustered into two groups, small and large units (S/L units), with their size domain correlated with the area and proportions of the urban blocks. Therefore, blocks are also dispatched into two groups, small and large area (S/L blocks), to be populated with S/L units, as describer later. The designer sets the number and position of points (GSP), the origin of cluster districts and SoK squares. For each GSP, the design tool calculates their proximity to user-defined number of urban blocks, which are next sorted according to their areas. To create open space for the square, a slider defines the distance that every sorted block, except the largest, will shrink parallel to its edge closest with the GSP.

The cluster district range (CR) is defined by the radius of a circle (Fig. 5). S/L units inside the CR are sorted according to their area and assigned with cluster-type 01 and 02 buildings, with their ratio defined by the user given number of type 01, starting the assignment from the largest unit. The large, unmodified blocks of the SoK squares receive only cluster experts' buildings. Every unit outside CR is considered type 3 building. Finally, the designer chooses the levels count to complete the 3D massing.

To position and size the units, each block curve is exploded to evaluate its segments according to their direction. East and west facing segments are used for the size grouping of blocks, as discussed earlier. North and south edges are divided by length, associated with the local x domain size of the rectangular units. To create variable sizes, rectangles are randomly scaled, in local *y*-axis only, to given scale values domain. Similarly, to create front yards and back alleys, scaled rectangles are moved randomly along their local y-axis. Lastly, the definition resolves cases, where building curves often overlap and intersect with block outlines, by removing the offending geometry. Thus, a clean masterplan result can be baked into Rhino for further edit, evaluation with UNA and finalization of the design (Fig. 6).

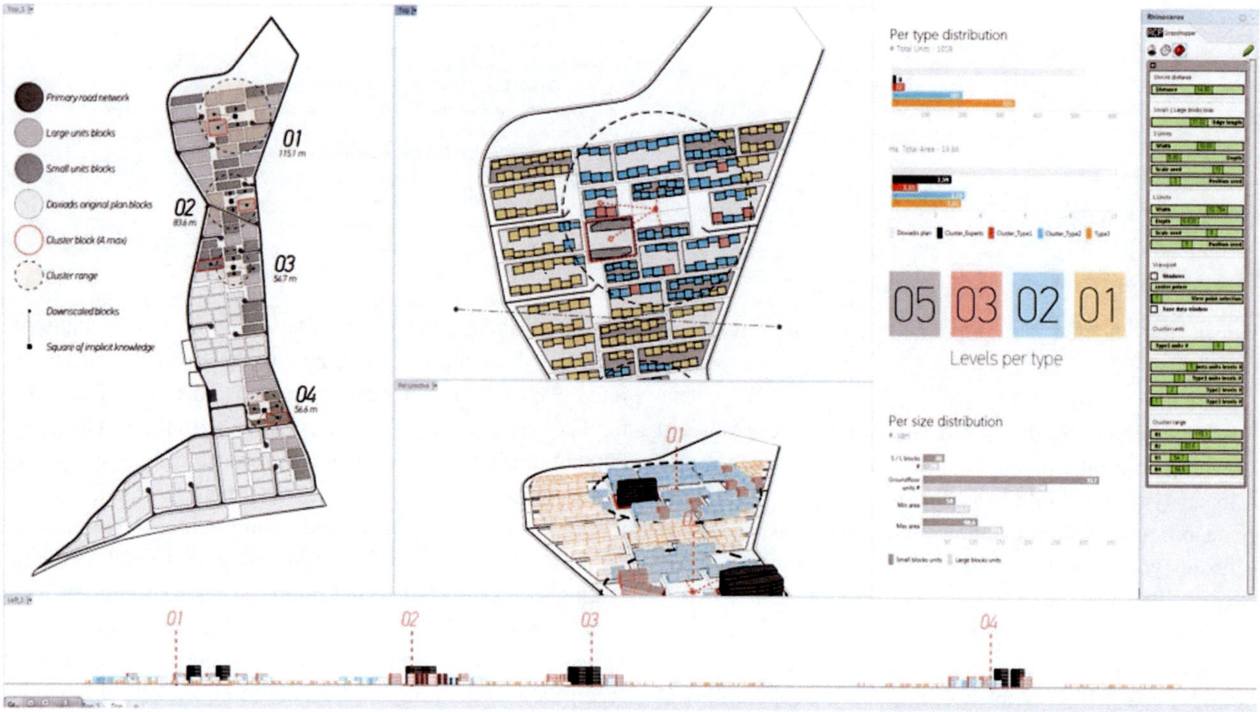

Fig. 4 Interface layout

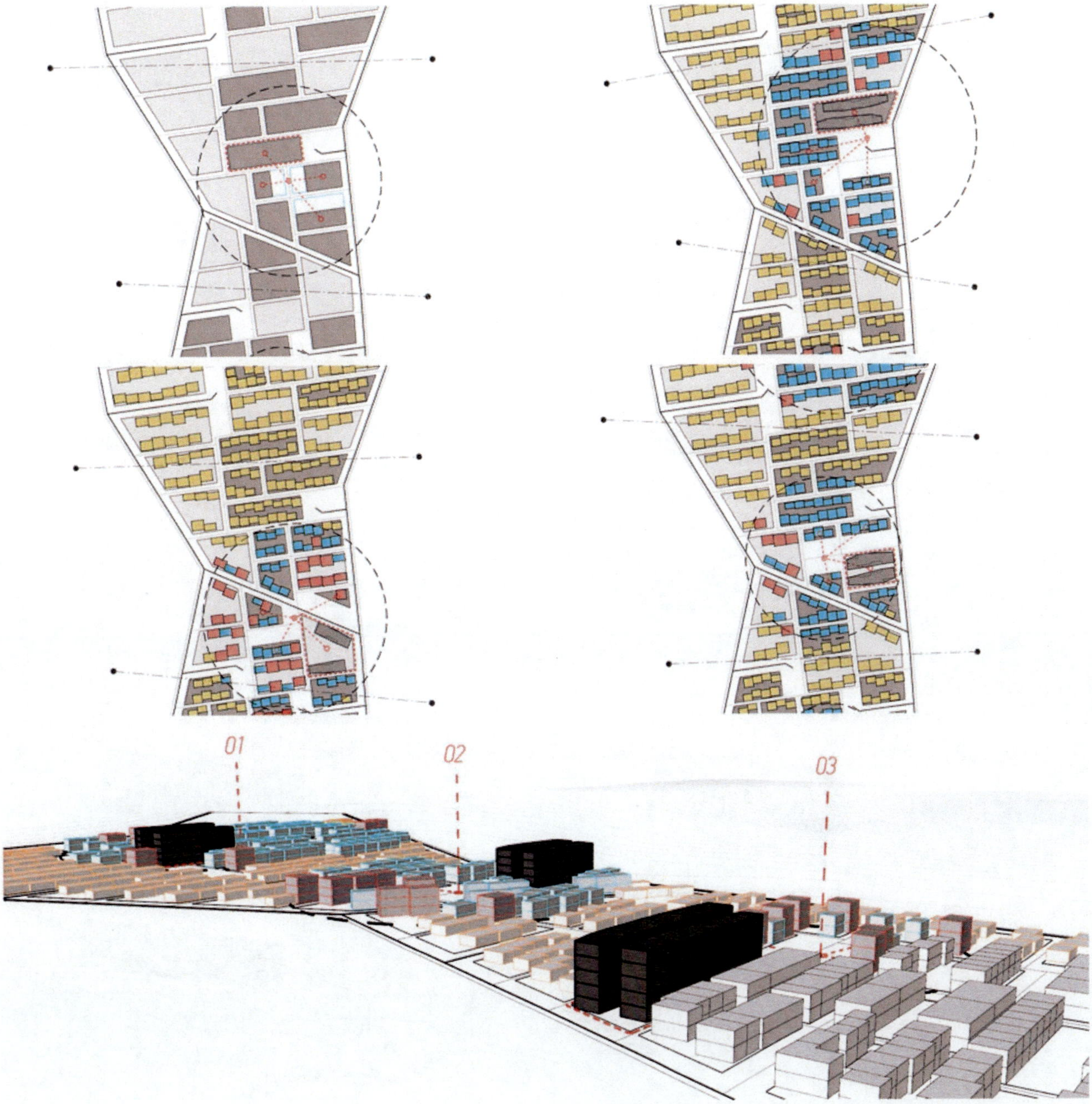

Fig. 5 SoK blocks shrinking, multiple outcomes of variable GSP positions, cluster radii and building types, color-coded 3D massing

4 Workshop Output

The workshop outputs, in terms of the parameters discussed in this paper, are organized in two sets; the first regards the project by Team A (PR01), and the second both Team B (PR02) and Team C (PR03) projects. The fourth team mostly abides to the research patterns better represented by PR01. The main questions driving the final jury were the following:

- How did each team address the provided analytical data?
- How did participants adopt to the design tool and the suggested design method?
- How did they conceptualize into diagrams their design intentions?
- How did these evolve through the feedback of the design tool, the evaluation tools and processes into proposal masterplans?

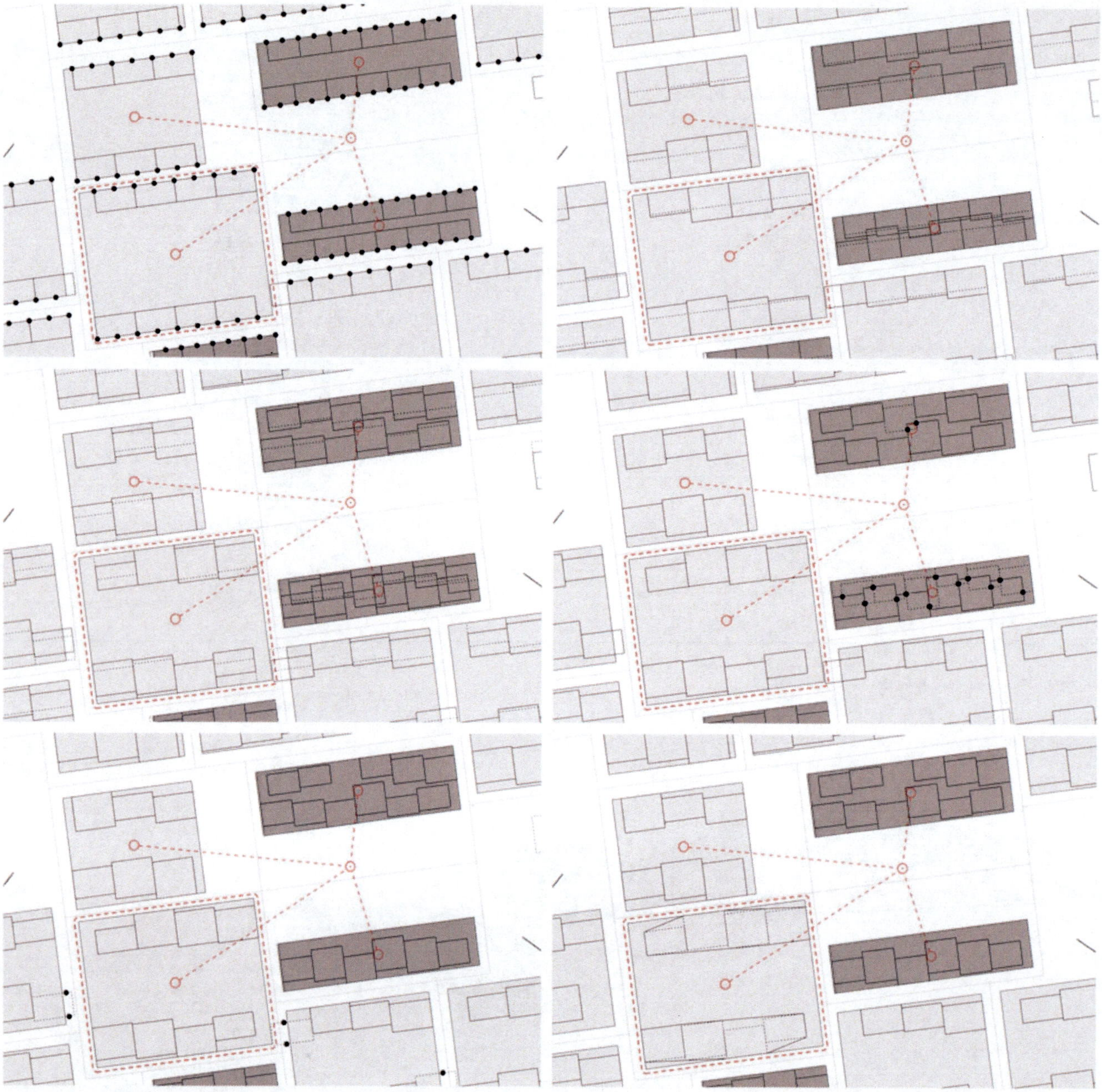

Fig. 6 Position and size, units variability, front yards—back alleys, units self-intersections, units-blocks intersections, cluster experts buildings

4.1 Team A

PR01 (Fig. 7) reframed the project agenda by emphasizing three strategic intentions:

- to optimize the productive and financial performance of the settlement and the factory, by proposing specific models for the relation between the company production and the research conducted in the cluster;
- to enrich the sociocultural mixture of the settlement's visitors, mainly by spatially enhancing the formers' sense of place, and by adding–situating new functions attracting more visitors;
- to promote the new and the historical identity of Aspra Spitia, as an outcome of their comprehensive programmatic and functional reorganization of the settlement.

The team's concept was intellectually based on the premises of different kinds of interconnected networks. Thus, they coined a new sociocultural spatial lace considering different temporalities and a gradual evolution toward the envisioned eventual prevalence of the new knowledge ecosystem.

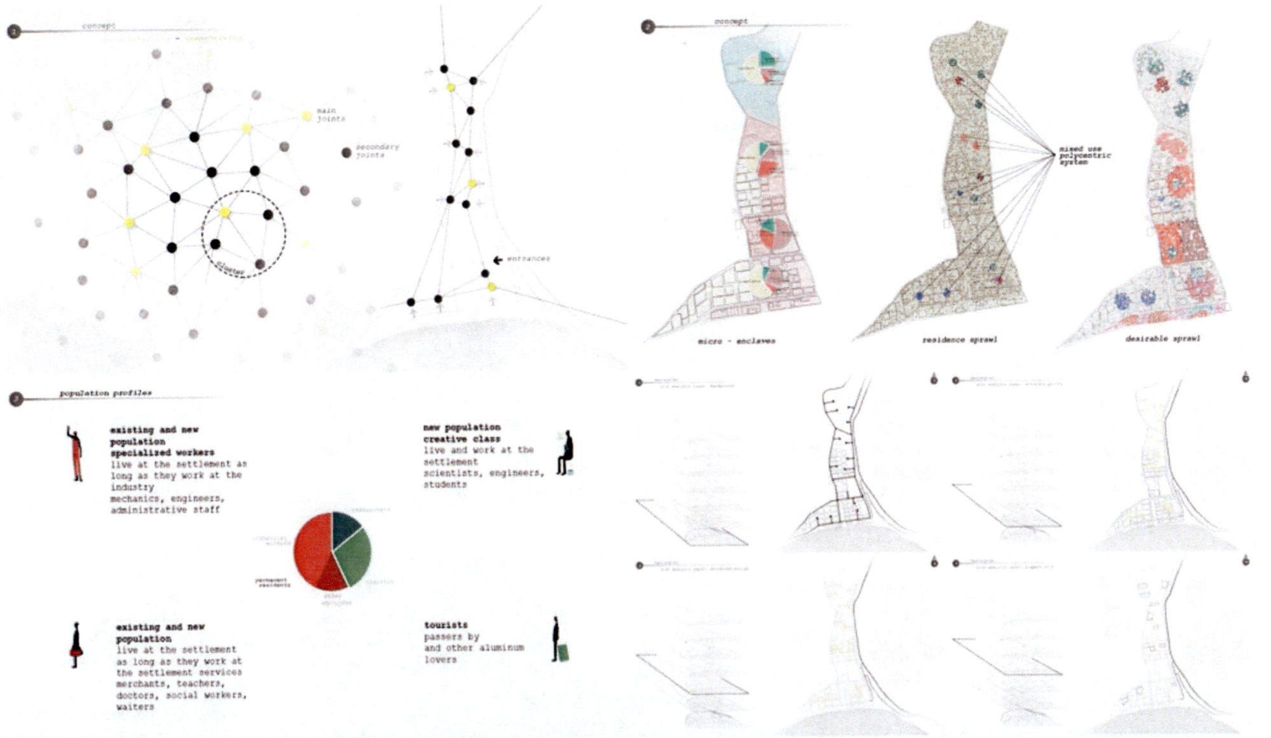

Fig. 7 Concept diagrams, suggested building types, polycentric mixed land use of initial and projected-desirable sprawl, population profiles density, layered site analysis

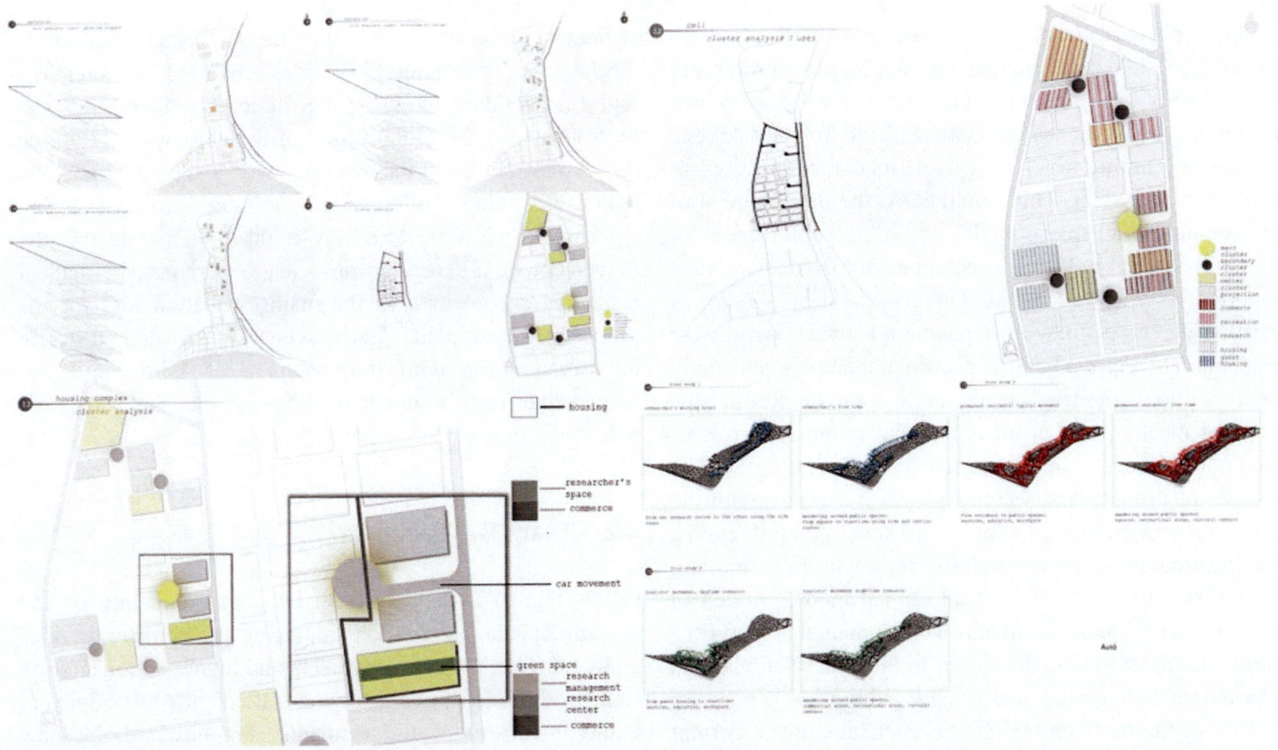

Fig. 8 Cluster districts locations, typical twofold districts, main secondary cluster areas, land use and social interaction, UNA evaluations

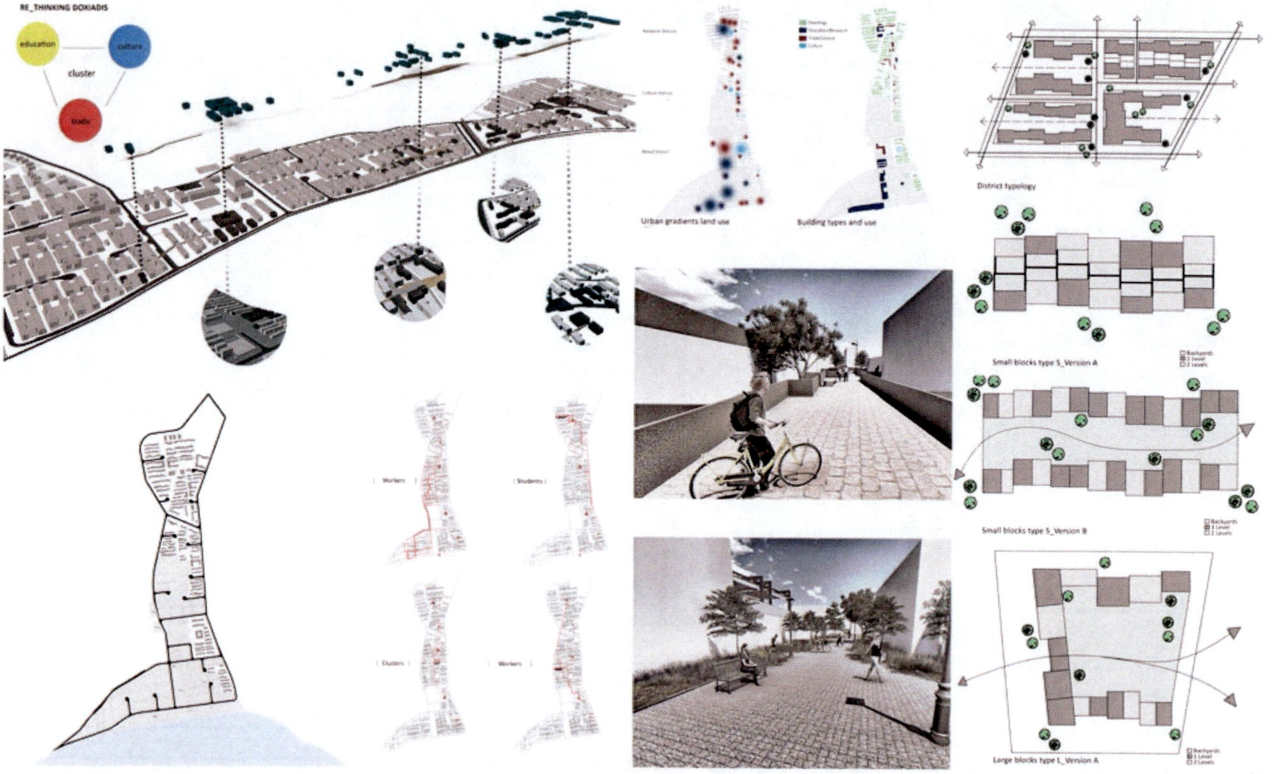

Fig. 9 Concept and land use diagrams, district and blocks typologies, UNA evaluations

On a physical level, they tested possibilities toward re-enhancing Doxiadis' original circulation plan, preserving the residents' and visitors' possibility for unobstructed movement along the whole settlement at different kinetic fields, while further developing its peripheral connections to near towns and areas. This would allow the development of a new multicentral spatial model, spreading public spaces of different scales and character, one of Doxiadis' most important original intentions. On a social level, Team A tested different degrees of separation between permanent residents, short period residing cluster members and daily visitors. PR01 modeled space in a way supporting the informal meeting of the different social groups and researched on the intensity of the possible gatherings (Fig. 8). The new set of primary and secondary nodes, together with the new gates opened to Aspra Spitia, was structured following a multilayered and non-hierarchical logic, in this, moving beyond original Doxiadis' organizational model. Added to that, the team proposed and allocated an annual ALU-FEST event, aiming at linking the cluster to the wider national and international community and general public.

One of the most intriguing characteristics of the current situation is that, since the factory works in three eight-hour shifts, there is always one-third of residents working, one sleeping and one spending free time or taking care of other social and practical necessities. In that sense, the original settlement has been working as a typical modernist-mechanized community. Team A invested on breaking with that tradition by analyzing different temporalities, and by testing how the organization of the cluster would enrich the current settlement's timeline with a more variant and hybrid reasoning. Different temporal scenario, i.e., during worktime, free time and day–night time, simulated the movement of different groups, i.e., researchers, permanent residents and visitors, as the main evaluation tool for the proposed masterplan. Each scenario revealed different intensities of use along different parts of Aspra Spitia and helped the team's understanding of the new system's behavior.

4.2 Team B—Team C

PR02 (Fig. 9) and PR03 (Fig. 10) focused mainly on the existing spatial segregation, reflecting the company's hierarchy between workers, engineers and higher administration staff. Both aimed to enhance the settlement's attractivity to young members of the residents' families and the new researchers. In contrast to the closed society of AoG, being the only factory of its kind in Greece, and since the company needs highly specialized employees and may not easily recruit new staff often, Team B noted that the new

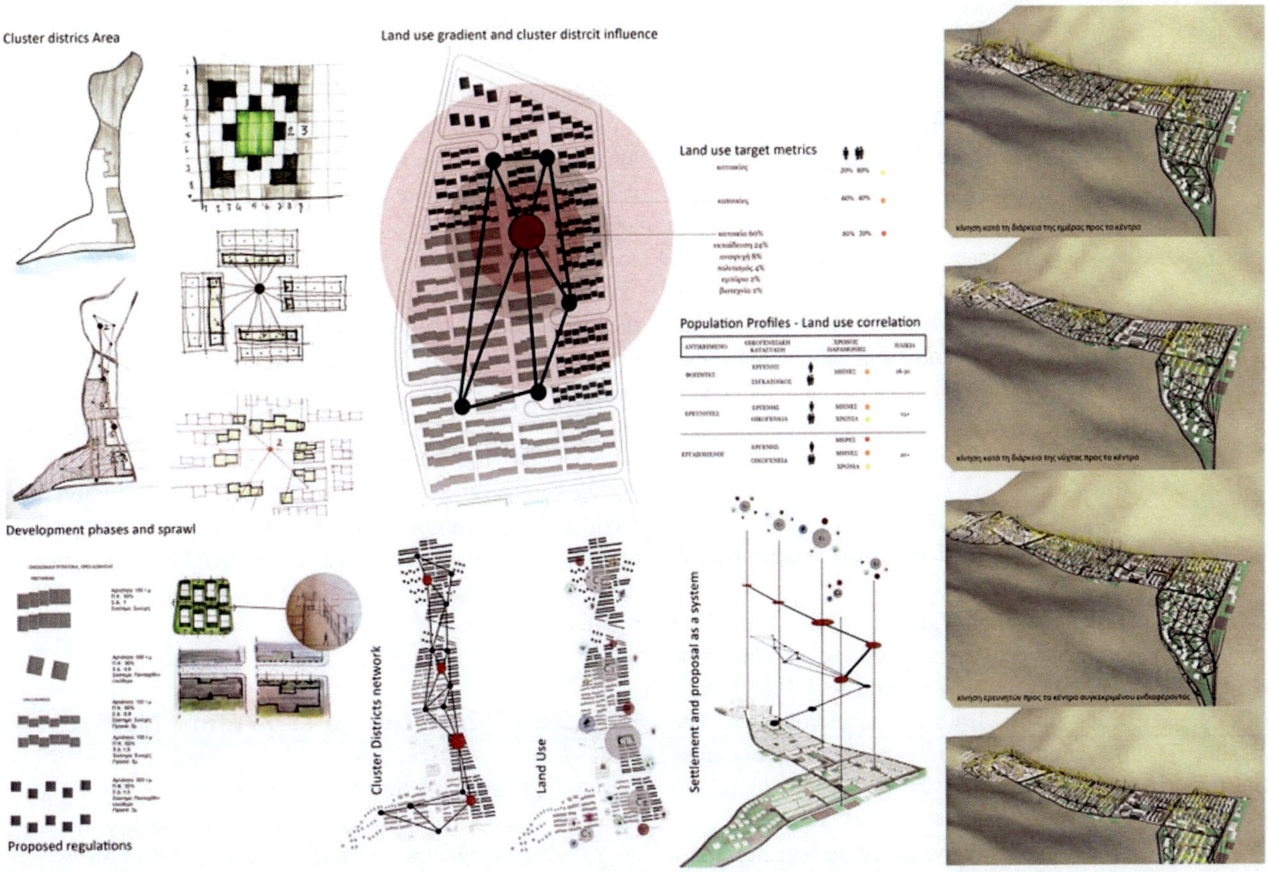

Fig. 10 Concept sketches, proposed land use and cover regulations, target metrics, cluster district network and land use diagrams, UNA evaluations

researchers community added will considerably alter this fundamental social characteristic of Aspra Spitia. The new hybrid society modeled will contain all three possible social protocols:

4. the traditional one, current workers that are second or even third generation, born in Aspra Spitia and working for the same employee as their parents;
5. the modern mechanized one;
6. the contemporary post-industrial one.

Preserving the original basic layout of the town, the two teams worked according to the following layers of analysis, synthesis and evaluation. On a functional level, existing commercial, social, educational and recreation infrastructure in the current public square are updated by means of the wider functional reorganization of the settlement. New, smaller scale, thematically targeted on culture and education spaces are added. Around them are organized residences, commerce, entertainment, etc., in various proportions of mixtures. Other secondary facilities are spread around the settlement. The above are interconnected axially as well as with the new set of local centers. An important relative parameter simulated is the degree of connectivity–isolation of central spaces which are being unified in wider public zone. The need to promote the feeling of community and the sense of place was expressed by a tactical survey of the existing and the re-planned space, in order to reveal possible nodes for the allocation of local centers with a more 'intimate' character.

PR02 and PR03 differentiated their approach to that of PR01, and in that they represent the second most important approach developed in the workshop, by addressing the issue of housing typologies and their organization in blocks and rows. Different types were further categorized beyond the logic already applied by DA. Apart from types corresponding to either seniority or family size, Teams B and C proposed new ones based on a variety of buildings and groups of houses generated according to the desired degree of openness to the public sphere and life. This variable was studied in parallel to that of the mixture of functions, as previously explained. Thus, PR02 proposed four different types of blocks, with their arrangement varying from being linear, to Γ and Π shapes, and even totally enclosed ones, whereas PR03 remained closer to the original DA linear syntax.

The last parameter put forward by Teams B and C was the functional and sociocultural correlation between permanent residents and researchers. At the first project the proposed research and communication center is part of the cluster infrastructure, and at the second part of the new open public space. These places have been simulated as an attractor of different people brought together by their knowledge, skills and wider interests and behavior. All these levels of analysis have been bound together with a layer of green spaces enriching the settlement's environmental footprint and the overall desired 'ambience' of the proposed space.

5 Discussion

5.1 Workshop Culture and Method Application

The first phase of the workshop allowed the time needed for the teams to build their cohesion, working on the same shared data and in line with the common toolkit and other references, i.e., the tutorials, lectures, discussions, etc. The workshop group started by immerging to the workshops subject matter, opened itself to the wider public in the middle of the procedure, then worked in smaller groups subcultures, i.e., the four teams, and ended up opening up by sharing their findings. Considered together, the three projects are varying in the ways they have used the method and design tools. Though, since the algorithm, the analytical data, the theoretical concepts entailed and the main strategic concept of the cluster districts were common, one would expect the teams to limit themselves in fewer aspects of the case study's attributes. This was an important issue discussed along the final presentation.

Although the participants recognized the limited time available, the intensity and complexity of the workshop's protocol and references, the workshop group agreed in the range of possibilities revealed in this attempt to parameterize the compositional logic and evolution of a historical twentieth-century urban heritage asset. A second issue raised, mostly by the tutors, was that although each team worked extensively on different concept variations and to a more limited extend, also redesigned their initial masterplans after evaluation processes, they did not share the overall analysis–synthesis–evaluation bidirectional process as clearly as it would be needed. A presentation including variations, both successive and rejected, leading to the final design proposal would have allowed to assess the overall behavior of each team's design microculture.

Examining further the ways participants have visualized their work, it has been noted that they have asked recourse to 'closed' representations, attempting to communicate their output as a finalized proposal, and not as work in progress piece of research. This has been explained mostly in terms of the need to further acknowledge and openly iterate the contribution of the adopted design method. In other words, the workshop has been matching up its goals in terms of knowledges, skills and design intentions. On the other hand, it may have contributed to the rise of an informed design awareness, but not to shaping a fully conscious new way of reasoning about architectural and urban design.

Last but not least, one final remark, put forward by one of the students, was that he would expect a parametric design seminar to open up an unimagined array of possible forms and structures. In this, the research team's position, one shared openly already since the first call, was that parametric design, following the brave experimentations of its first two decades, is now asked to contribute to the continuity of architectural evolution and not just propagate new species and architectural life forms. The authenticity of the definition is at least of equal importance to the complexity of the problems at hand.

5.2 Data-Driven Design and Formal Methods

Discussing further the results and the workshop experience, it can be argued that participants could early, easily identify with and apply geometric transformations, followed by an understanding, and in most cases application, of the associative capability of parametric models. Perspective obstacles became evident, when each team was challenged to combine design aspirations with qualitative–quantitative computational processes. Like Picon [29] and Zarzycki [24] argue, computer-aided design cannot be an endless exploration of potentialities, instead computational processes ought to augment the decision-making logic. Confronting such complex design agenda in an admittedly short timetable, as already discussed, participants could easily get entangled in a design labyrinth. On the contrary, examining the design outcomes shows that every team managed to envision design strategies and to deliver pertinent masterplans, which were able to defend through the presentation of a series of decisions, supported by the data design interplay. The latter raises the question of the background needed, for architects to be in position to apply formal methods mediated by data.

Regarding the educational background, methods and goals, the already referenced research aligns with our view and contributes to a critical part of the discussion. Our research agenda pursues to broaden the topic, in every opportunity given, by pointing out that in the current algorithmic culture, *new* design processes, software and their results are often idolized [30]. Also, like Greenfield [31] notes, whatever data is collected and measured by any means, consists of only a selected fraction of a large array available to study. Data selection and interpretation are already processes of curation and editing, thus designers cannot assume they are operating in a neutral and objective set. Hence, besides distancing from stringent formalism, architects need to build a critical awareness regarding data collection, analysis and application, in order to identify and understand the consequences of their design proposals. Especially in cases, where confronting architectural and urban design problems, the focus is in an all-inclusive social, technical and sustainable design.

6 Conclusion

In Aspra Spitia, DA have implemented a prototype industrial settlement which in its internal organization has the potential of an international node. A key asset of DA's plan was the care for a spatial organization, unfolding at various planning and design scales. In revisiting Doxiadis' pioneering analysis-based and data-driven anthropocentric ekistics, the research agenda discussed in this paper, in light of contemporary design tools and cluster development theory, supports that what will remain a small and geographically isolated community, may rise as a key regional agent. Possibly, defining the future of the market, and, thus, offering a new future meaning for the historical settlement. Face-to-face contact, personal informal relationships and, most importantly, a sense of common interest, complementarity, coordination and trust, are the pervasive social elements of the new cluster studied. Elements that defined the principles of the design and planning research were developed throughout the workshop. In this way, the original research agenda was implemented by means of the provided algorithm, the latter being the mediator between the settlement's past and its projected futures.

Acknowledgements The authors would like to thank Team A: Kalfaki M., Katrakis C., Lambrou K., Liofagos Y., Papamichelakis L. and Skouloudi I., Team B: Dimitriadou C., Kaproulia H., Karofilaki M., Marakakis M., Makrydaki E., Markou O. and Vomvolakis L., Team C: Dimopoulou E., Kordonouri T., Maragkakis P., Passos S., Vlatitsi N. and Zaharia V., Team D: Aintinides M., Mpagordaki M., Nikoloudaki D., Panoutsopoulou V., Rozani G. and Vlazaki A. for their participation and permission to publish the design outcomes.

References

1. Yiannoudes, S., Patsavos, N., & Tsesmetzis, V. (2015). 'Aspra Spitia 2015: Intentions and transformations. Constantinos A. Doxiadis' industrial settlement and its development. In A. Gospodini (Ed.), *Proceedings of International Conference on Changing Cities II: Spatial, Design, Landscape & Socio-economic Dimensions: Planning new cities* (pp. 1007–1018). Grafima, Thessaloniki.
2. Doxiadis, C. A. (1975). Specifications for the city of Anthropos (Man). In C. A. Doxiadis (Ed.), *Anthropopolis, city for human development* (pp. 101–190). New York: WW Norton & Company Inc.
3. Kyrtsis, A. (2006). *Constantinos A. Doxiadis: Texts, design drawings, settlements* (pp. 439–440). Ikaros, Athens.
4. Kostis, K. (2013). State and Enterprises in Greece: The history of 'Aluminium of greece', Κράτος και Επιχειρήσεις στην Ελλάδα: Η Ιστορία του "Αλουμινίου της Ελλάδος" (in Greek) (pp. 117). Polis, Athens.
5. Doxiadis Associates DA. (2020). ASPRA SPITIA a new "Greek" city' 1966 printed brochure in pdf, https://www.doxiadis.org/Downloads/aspra_spitia.pdf. Last accessed 22 April 2020.
6. Doxiadis Associates DA. (1965). Aspra spitia a new "Greek" city'. ARCHITEKTONIKA THEMATA/Architecture in Greece (pp. 53–57).
7. Tournikiotis, P. (2009). Aspra spitia or the legacy of C. A. Doxiadis. In G. Kazazi (Ed.) Constantinos Doxiadis and his Work (Vol. 2, pp. 209–228). Technical Chamber of Greece, Athens.
8. Kalafati, E. (2012). σπρα Σπίτια Βοιωτίας: Ζητήματα σχεδιασμού μιας νέας πόλης (in Greek), Aspra Spitia in Boeotia Issues of designing a new city. In V. Petridou, & N. Kyrkitsou, (Eds.), Εργάζομαι άρα κατοικώ: Η περίπτωση του συγκροτήματος κατοικιών των Μεταλλείων Μπάρλου στο Δίστομο Βοιωτίας, των Δ. και Σ. Αντωνακάκη (in Greek), I work therefore I live: The case of the complex of houses of the mines at Distomo of Boeotia, of D. and S. Antonakakis (pp. 17–24). University of Patras, Patras.
9. Porter, E. M. (1998). Clusters and the new economics of competition. *Harvard Business Review, 76*(6), 78–90.
10. Doxiadis, C. A. (1972). *The Two Headed eagle: From the past to the future of human settlements*. Athens: Lycabettus Press.
11. Wigley, M. (2015). Network Fever. In: Wigley, M. (Ed.) Buckminster fuller Inc. Architecture in the age of Radio, pp. 279–329. Lars Müller Publishers, Zürich.
12. Senske, N. (2011). A curriculum for integrating computational thinking. In: Cheon, J., Hardy, S., Hemsath, T. (Eds.), *Parametricism: (SPC) Proceedings of ACADIA regional 2011* (pp. 91–98). University of Nebraska-Lincoln College of Architecture, Nebraska.
13. Senske, N. (2014). Confronting the challenges of computational design instruction. In: N. Gu, S. Watanabe, H. Erhan, M. H. Haeusler, W. Huang, & R. Sosa (Eds.), *Rethinking Comprehensive Design: Speculative Counterculture, Proceedings of the 19th International CAADRIA Conference* (pp. 821–830). Kyoto Institute of Technology, Kyoto.
14. Psomopoulos, P. (Ed.) (2005). Special issue: A reader on ekistics, thirty years after C.A. doxiadis. EKISTICS/ΟΙΚΙΣΤΙΚΗ: The problems and science of human settlements *72*(430–435).
15. Faros, M. (dir.) Tzirtzilakis, Y. (Ed.), ERT S.A: Constantinos Doxiadis: A Romantic Pragmatist/Κωνσταντίνος Δοξιάδης: Ένας Ρομαντικός του Πραγματισμού (in Greek), Documentary, https://vimeo.com/84234212. Last accessed April 22, 2020.
16. Beirão, J., Nourian, P., & Mashhoodi, B. (2011). Parametric urban design: An interactive sketching system for shaping neighborhoods. In T. S. Zupancic, M. Juvancic, S. Spela Verovsek, & A. Jutraz (Eds.), *Respecting fragile places: 29th eCAADe conference proceedings* (pp. 225–234). Ljubljana: University of Ljubljana.
17. Beirão, J., Arrobas, P., Duarte, J. (2012). Parametric Urban design: Joining morphology and urban indicators in a single interactive model. In H. Achten, J. Pavlicek, J. Hulin, & D. Matějovska (Eds.), *Digital Physicality: Proceedings of the 30th eCAADe Conference* (pp. 167–175). Czech Technical University in Prague, Prague.
18. Pitts, G., Luther, M.: A parametric approach to 3D massing and density modelling. In H. Achten, J. Pavlicek, J. Hulin, D. Matějovska (Eds.) *Digital Physicality: Proceedings of the 30th*

eCAADe Conference (pp. 157–165). Prague: Czech Technical University in Prague.
19. Heumann, A., Syp, M., Holland, N., & Ringley, B. (2015). HumanUI for grasshopper, computer program. https://www.food4rhino.com/app/human-ui. Last accessed April 22, 2020.
20. Heumann, A. (2015). Human for grasshopper. *Computer Program*. https://www.food4rhino.com/app/human. Last accessed April 22, 2020.
21. City Form Labs. (2014). Urban network analysis toolbox for rhino (UNA), Computer program. https://cityform.gsd.harvard.edu/projects/una-rhino-toolbox. Last accessed April 22, 2020.
22. Sevtsuk, A., & Mekonnen, M. (2012). Urban network analysis toolbox. *International Journal of Geomatics and Spatial Analysis, 22*(2), 287–305.
23. Castro e Costa, E., Verniz, D., Varasteh, S., Miller, M., & Duarte, J. P. (2019) Implementing the Santa Marta Urban Grammar—A pedagogical tool for design computing in architecture. In: J. P. Sousa, J. P. Xavier, & Castro Henriques, G. (Eds.) *Architecture in the Age of the 4th Industrial Revolution—Proceedings of the 37th eCAADe and 23rd SIGraDi Conference* (pp. 349–358). University of Porto, Porto.
24. Zarzycki, A. (2011). Form-making Without Form Making. In J. Taron, V. Parlac, B. Kolarevic, J. Johnson, (Eds.), *Integration through Computation: Proceedings of the 31st Annual ACADIA Conference* (pp. 185–192). The University of Calgary, Calgary.
25. Doxiadis, C. A. (1963). Architecture in Transition (pp. 99–106). Hutchinson, London.
26. Doxiadis, C. A. (1956–1657). Gossip Square. In *General Reports, Constantinos Doxiadis Archive R-GA*.
27. Pyla, P. (2013). Gossip on the doxiadis 'gossip square': Unpacking the histories of an unglamorous public space. *Architectural Histories 1*(1), 28, 1–6.
28. Dorst, K. (2004). On the problem of design problems–problem solving and design expertise. *Journal of Design Research, 4*(2), 185–196.
29. Picon, A. (2004). Architecture and the virtual: Towards a new materiality. *New Technologies: New Architectures, 6,* 114–121.
30. Bogost, I. (2020). The cathedral of computation: We're not living in an algorithmic culture so much as a computational theocracy. The atlantic, https://www.theatlantic.com/technology/archive/2015/01/the-cathedral-of-computation/384300/. Last accessed April 22, 2020.
31. Greenfield, A. (2018). Machine learning: The algorithmic production of knowledge. In: *Radical technologies: The design of everyday life* (pp. 176–215). Verso, New York.

Alexandros Peteinarelis Architect Engineer Dipl. NTUA, M.Sc. Space, Design and Built Environment TUC. Working freelance and collaborating with architectural practices since 2006, involved in various type scale projects, responsible for architectural design and constructional details documentation. He has co-organized design workshops and has been a researcher at Technical University of Crete and University of Ioannina. As Adjunct Lecturer at School of Architecture TUC has taught digital media and special topics in architectural design. Appointed Adjunct Lecturer since 2018 at School of Architecture UOI, teaching architectural design and spatial representation courses.

Nikolaos Patsavos Dipl Arch AUTH, he is Associate Professor of Architecture at the University of Ioannina and the co-founding and managing partner of Ctrl_Space Lab architecture collaborative platform, where he has worked as project architect of research and applied projects, workshops, publications and exhibitions. Since 2001, he has been a researcher at the AA Graduate School, British School at Rome, NTUA, TU Crete and University of Thessaly. Since 2003, he has been teaching theory and design, at undergraduate and graduate level, at the AA, TU Crete and the Universities of Thessaly, Nicosia and Frederick.

L'Objet Invalide

Aniruddha Mukherjee

Abstract

Over the course of this paper, I will consider categories of thought within architectural discourse from the poststructuralist paradigm as a primary resource. The texts and authors referred to are specifically those that discuss and critically analyse matters of authorship and origination among the arts and architecture. The poststructuralist paradigm offers a body of commentary on the primacy of written text over speech and its problematic with respect to appropriation and translation of form as well as content. A particular discrepancy to be disentangled in my research is the nature of the theoretical basis established by architectural authors of the period with respect to an anticipation of digital modes of design. The architectural projects during the period can be considered early statements that also attempt to establish the limits of tools available to architectural production, specifically where this concerns architectural representation. And, thus, can be read as having embedded within them a tendency for or maybe an exposition of those limits. The paper aspires to uncover and elaborate the modes of thinking and the consequent problems of translation, into an aesthetic, i.e. considerations of form and content or the lack of clarity on any such legible distinctions.

Keywords

Deconstruction • Meaning • Poststructuralism • (Re)origination • Form • Content • Conceptualization • Spolia • Validity

A. Mukherjee (✉)
Universität Innsbruck, Institute of Urban Design (IOUD), Innsbruck, Austria
e-mail: aniruddha.mukherjee@student.uibk.ac.at

1 Introduction

Among its various motivations lies architecture's aim to produce meaning. As such, the relation between the architectural object and its meaning—irrespective of authorial intentions, can be generally observed as an unstable system of values. Such values are either inherent to architecture as they are appropriated across time or they are applied to architecture in pursuit of novelty, a new validity or valid by virtue of precedents.

In general, the theories of Jacques Derrida, Julia Kristeva, Michel Foucault and Gilles Deleuze are central to the poststructuralist period within philosophy and critical theory. We know that, Derrida, for instance, argued that a deconstructive method of analysis of textual works observes that the production of meaning can run counter to the sequence of the text and that the structures embedded in written language cannot be accepted as stable and finite systems [1], as advocated by the structuralist theorists of the 1950s and 60s. In a different proposition, the concept of Intertextuality—as initially conceptualized by Julia Kristeva, states that authors often establish meaning by alluding to other works. Here, interpretation is invoked primarily by referring to, through the use of linguistic or visual devices, the subjects' understanding and ability to access meanings embedded within the referred texts.

Poststructuralism and its impact on architecture, although not always concurrent to the emergence of philosophical texts, are now considered to have presented a critical set of questions to the modernist tradition and its various manifestations. The questions surrounding legibility and interpretation engendered various experiments within architecture that would attempt, not only to destabilize architectural dogmas of the post-war period but also aspired to a (re) origination of diversity, difference, formal memory and aesthetic meaning.

Although often analogous, poststructuralism seems to have gained traction among architects for its discursive

S. Eloy et al. (eds.), *Formal Methods in Architecture*, Advances in Science, Technology & Innovation,
https://doi.org/10.1007/978-3-030-57509-0_7

affordances as a framework for speculation on questions of form and its content. The works within the discipline of philosophy and linguistics became operative for their radical reconsiderations of this dichotomy.

In order for an inquiry into the poststructuralist modes of theorizing to be delineated, I will also discuss past examples that concern the symbolic ordering inherent to separation of arguments of/from architectural practice and abstract academic dissertations. Such distinctions, although seemingly oblique to the areas of concern, may help in the critical framing of specific problems of authorship that are under urgent investigation in present times. This method of seeing the discourse on the symbolic order helps make clear what a formal condition might mean or how it may be discussed in precisely intelligible terms. One which may allow access to different social capital for architecture but also the reverse where architecture's own formal capital can be reevaluated.

To think that here the author or the architecture may only be evaluated by its scientific reasoning may be an outdated and solitary mode of reading value itself. Any certainty afforded by science may also limit architecture's proclivity to limiting its own scientific material to only the latest. This, I submit, is the basis of my search for methods that may be capable of revealing the latent.

At first, what seems to be at stake, is the 'method of looking' into each architectural project that would require a recalculation or even in some way's a regression, of the project's conceptual bases. Methods that are at once conscious of two important aspects that can be considered common to most of these poststructuralist works—

1. The generative diagramming that was a part of the design procedures of these projects would have embedded in them certain biases of origins, for example, geometric primitives, trace of the Berlin wall, Mercator grid and so on.
2. The emphasis on the selection of modes of visualization in their resistance to specific problems, to name some, the fetishization of the author that comes with perspectival drawing or the plan as generator or even part to whole articulation and etc.

2 The Teleological Problem

Architectural types are widely regarded as emergent concepts that are most representative of a dominant culture during a particular time. As articulated by Anthony Vidler in The Third Typology [2], since the enlightenment era and into the modernist traditions, the material for classification of architectural forms has been the product of rationalistic and technological thinking. Pre-modern forms were understood to be, through an ontological characterization, natural primitives (primitive-hut) whereas modernist primitives were thought to be conveying an ontological system that was analogous to the processes of mass production (Industry). It is imperative here to note that the modernist characterization of architectural elements considered the machine or mass production to be a kind of natural system; a constructed nature that foregrounds processes of production as the bearer of meaning; industrial production as a kind of a stage at which meaning was being embedded into architectural production through its discreet material elements. These two typologies can be then understood as concerning themselves with the coexistence of meaning within the architectural element, where the architectural element always referred to another idea to be borne by it, as a place holder.

The third type, on the other hand, developed, in its critique of modernist values, a corollary to formal appropriation by means of teleological collapse—its purposefulness considered primarily for its formal condition without recourse to its socio-urban content. In other words, the continuity of architectural form was being proposed as an incoherence in relation to its traditional methods of constitution. A trajectory that required procedures of selection of architectural form, from within another nature (the city), to be reassembled across time and different contexts. In the third typology, the city was being considered, for all its differences, as a whole object that could be mined for figuration of novel forms. Although, conscious of the city being a collection of political and social institutions that performed specific functions, architects such as Aldo Rossi argued that the functioning of politics would just become re-originated as opposed to repeated in an architecture that curates its forms without any specific regard to traditional ways of reading the city. A kind of a continuity of codified architectural forms present within the eighteenth-century city. Here, it is also important to note that the eighteenth-century European city is a specific urban condition that is being read as having achieved a degree of formal maturity—a traditional object, by Rossi, Krier and other rationalists. That is taken as—informed then transformed to perform as form.

Before trying to establish the precise nature of the problem of tradition (tradition as determination in advance) in architectural work today, it is important to clarify what I mean by a traditional object within the discourse of the rationalists. The traditional object in this case refers to an object that claims to have the capacity of an architectural form and to bear a name on the basis of its accumulated cultural and ontological repetitions. This object clarifies its claims through formal associations to any repetitions of itself; before, after and in a moment of experience simultaneously. The object always remains significant across periods as a place holder of qualities and ideas that are repeated because of a precisely understood unchangeable origin. Such a sameness of architectural form affords continuous associations to sensibilities of space making, across time, by way of architecture's own conscious means of analyses and production. The rationalists found this sensibility of the traditional object as problematic but rich with expression and continuity, pro-figuration.

In their critique of the high modern practices, the rationalists insisted on the eighteenth-century city as a whole object for its characterization of public and private spaces as a differentiated continuity as opposed to the undifferentiated fragmentation of the modernist city.

In the case of Aldo Rossi, as has been discussed by Anthony Vidler, the City Hall project for Trieste bears the named reference to an idea of the eighteenth-century prison; a reference also then to Piranesi's demonstration of the prison as becoming an architectural expression of the disorientation of Western European society through the loss of religious faith and materialist reasoning. In the case of Piranesi's Imaginary Prisons, the works do not make references to a specific architectural assembly or a whole that would stand as a place holder for the prison but to formal effects that are exaggerated and staged within the picture frame through the use of technical and graphical tricks. The traditional object in this case would be fragments of recognizable architectural forms of the eighteenth century and before, considered worthy of another instance of repetition for its formal qualities but where its arche and telos are carefully disengaged and their coherent self-referentiality is left disturbed.

Aldo Rossi recognizes and chooses to reinforce the disturbances embodied within the picture frames of the Imaginary prisons as being of significant formal value. The disturbance that once stood for the attitudes of an epoch of disorientation and realignment of the human subject with respect to social collective organization, upon reintroduction by the rationalists, would allow for a critical re-projection of the socio-political concerns of another contemporary period. Such projections and/or readings were considered, by the rationalists as having a direct relation to the architecture that is to stage them. That is to say, the expectation was that the architectural form would inevitably force an engagement of a historical consciousness that increasingly lay dormant and even much maligned by the commentary of the modernists within contemporary Western European society.

The choice of the native city for its formal decomposition and a re-origination of the functioning of its forms through Rossi's techniques of abstraction, scaling, morphing etc., cannot be understood as a simple consequence of the failure of the modern movement. The analogy to the city has its basis in signification through concepts of Identity. Even within Vidler's schema of the three typologies, the first two refer to the identity that is derived from an external ordering while the third attempts to conceptualize an identity that looks inward to architectures own formal conditions. The faith that architecture and its collective memory are made up of identities that are often considered to be operative and essential is crucial to the motivation of the rational projects and the reading of the third typology as such. The subversion here lies not only in the foregrounding of architectures own formal and material history but more importantly the liberation of its subjugation to forms of validity by external associations. Architecture insisted on a restructuring of its self-consciousness and a reconstructed épistémè. The carried meanings of the elements of the city and their potential for confrontation with means of industrial production, for the rationalists during the period in question, is the primary mode of critique and maybe the only one.

Foucault, In the Order Of Things, makes a reference to the enlightenment period and the seventeenth century as an épistémè for its questioning of conventions and their specific relationship to categories of thought. In a subsequent interview with the anthropologist Paul Rainbow, Foucault tries to identify the problem of spatialization as one that is embedded in that which is consciously ignored.

> ...What is striking in the epistemological mutations and transformations of the seventeenth century is to see how the spatialization of knowledge was one of the factors in the constitution of this knowledge as a science...there was literally a spatialization of the very object of their analysis, since they gave themselves the rule of studying and classifying a plant only on the basis of that which was visible. They did not even want to use a microscope. All the traditional elements of knowledge, such as the medical functions of the plant, fell away. The object was spatialized. (Michel Foucault, The Order of Things, 1970)

Foucault, here, makes clear the significance of practices of postulation in the enlightenment period. He is not only suggesting that there is a shift in the approach as well as the method due to the governing rules of the sciences of the time but also that the rules are consciously made to govern unconscious assumptions about the object that a scientific discourse attempts to analyse. The new rationalists concerned themselves with clarity of postulation (abstraction) and free appropriation of architectural form through the use of a 'traditional architectural object'.

The subsequent proliferation of the analogous city, especially around North America and consequently around the globe, festered narrower motivations that were indifferent to any criticism. Its inability to conceptualize and operate consciously across points of liaison with other localized urban forms, especially within north America and around the globe, was responsible for the dilution of its critical potentials. This is also to say that 'The Eighteenth-Century European City'—The traditional object, itself can retrospectively be understood to have finite forms that may actually be responsible for the re-appropriation of these forms into means of mass production and the emergence of homogeneity within architectural processes; the very nature that the discourse meant to subvert.

The new rationalists' self-induced limits to selection of the material for articulation of value and meaning become clearly regressive when the discourse held 'becoming critical' as its only form of criticality. Its methodology of the negation of modernist functionalism remained limited to the act of selection of specific, and now seemingly arbitrary, set of primitive forms that relied on legibility through pictorial means of abstraction and assembly. These limits left a tense vacuum meant to be occupied by possible epistemological or even pedagogical infrastructure that a community of architects could access and operate with. In other words, the operative tenacity of individual authors willing to participate in the discourse remained the only measure for meaning. Left to authorial devices, architectural concerns of the new rationalists mutated into forms pro-theatricality and spectacles of mythology.

3 Linguistic Structure and Tradition

The problem of tradition has been presented by the post-structuralist theorists through different modes of analyses of language. Also, tradition as an architectural problem has been implicitly dealt with as a matter of concern at different periods and the traditional architectural object has been labelled as such and made legible through means of tactility and cognition at different moments throughout history. Architectural systems of ordering—Classical orders, Renaissance Hierarchy, Neo-Classical reordering and Modernist expansion, have been central to the unfolding of an epistemological clarity on some of these key moments.

During the decade following the end to WWII, the structure of language was beginning to gather credible scholarly interest for its self-referential system of notations and finite rules that could lend itself to infinite combinations. These conditions were being subjected to rigorous exploration along with their simultaneous introduction as a method of conceptualization in the western world. Architecture and its subsequent liaison with poststructuralist philosophies are rooted in the problem of legibility of a status-quo—a somewhat arbitrary signification yet precisely articulated pro forma and collectively agreed upon within the discipline. A set of conditions that became prone to prescription and presuppositions during the post-war era. Within architectural theory, this resulted in enthusiastic inquiry into structuralist theory based on Ferdinand Saussure's study of language and the conviction that the deep structure of language holds within it, possibilities of analogy that could be activated through architectural design.

Proponents of the structuralist paradigm within architecture, like Herman Hertzberger, Aldo Van Eyck and others from Team 10 argued that the self-referential system expressed by linguists was a valuable recourse to synthesizing modernist methods of architectural configuration. They argued that the system itself was sufficiently complex to afford infinite variations and that it could be discursive once conceptualized through architectural conventions of representation. This discursive value of the structuralists was to be in its attempts to allow for the daily eccentricities of human life to be housed within architectural form that does not attempt to articulate functional forms of living.

Having articulated the move away from functionalism of the modern movement in architecture, the structuralists insisted on appropriating forms of living from vernacular forms to be configured into assemblies and transplanted into urban or semi-urban Europe. The functionalism of the modern period was problematic for its rigidity and submission to capitalist means of production. The arguments, like the rationalists later on, were based on the fact that traditionally tested and mature forms of the Moroccan Kasbah were valuable for their spatial configurations. The architectural form was to submit to their configurational hierarchies and the assembly itself could have rules of self-referentiality that remain hidden as an under-drawing.

In the architectural discourse of the time formulated by the structuralists, quite literally the structuring of architectural form, explored means of fragmenting of architectural ideas through order and submission. The formal elements of architecture were to be applied to abstractions of traditional orders of organization. Part-to-whole relationships were to be established and made legible as fragments. Architectural space was to be read through its points of liaison with its own abstracted conditions and the rules governing the assembly were to be made explicit as concept. Architectural representation was to retain its role within production of works as a means for registering geometric relations that could be reduced to abstract essences of the meanings embedded within the architecture. Traditional means of representation remained medium specific in a way that suppressed any non-somatic reading of variation in reference to the objects of tradition that had been abstracted and reconfigured as novel forms. It is key here to recognize that treatment of the conventions of architectural representation, similar to that of the new rationalists, remained a continuous thorn in the flesh for their preference of an apolitical methodology. The architectural media at play continued to govern the treatment of the traditional object and its self-referentiality while rendering its form dormant and incapacitated with respect to its own political implications. The architectural medium, for the structuralists, was a means towards a conception of the under-drawing that would play no more of a formal game than to govern the abstracted; and where the degree of abstraction cannot be particularized within the rules of the game precisely because it inherits meaningful, discernible spatial qualities, like the example of the Moroccan Kasbah.

In her oblique approach to architecture as an art critic, but nonetheless very helpful to the understanding of possibilities of the medium, Rosalind Krauss, in her commentary on Clement Greenberg's idea of 'medium specificity' [3] attempts to clarify this confusion prevalent in criticism of what the role of the medium may be. The medium specificity of architectural work of the structuralists remained faithful to Clement Greenberg's analysis of Jackson Pollock, where he argues that Pollock's 'dripping' technique foregrounds the specific formal aspects within the medium of painting as a celebration of painting itself. Krauss, on the other hand insists that the techniques of Pollock engendered a new artistic medium that set a new proposition for all forms of artistic practices.

> ...in order to sustain artistic practice, a medium must be a supporting structure, generative of a set of conventions, some of which, in assuming the medium itself as their subject, will be wholly 'specific' to it, thus producing an experience of their own necessity (Krauss, 2000, p. 26).

Here, crucially, A certain experience of necessity becomes critical to architecture, especially since the introduction of computational tools within the discipline. As the digital medium began to assert its range of methodologies, the state of authorship, scholarship and intellection, approach to history and institutionally organized speculation, criticism and debate became symptoms of flimsy foundations that the discipline of architecture needed to address in urgency. Architecture then became aware, ironically only after Peter Eisenman's conceptualization and subsequent distancing from Derrida's archi-writing, of the disciplinary archaeology that had been insufficiently accessible to its own operatives. The speculative projects engendered by the digital, although the most visible among them remained perfunctory and overtly responsive to software instead of digital media as a means of archaeological intellection, had to have been felt as anticipations of a Foucauldian epistemological nature. To excavate, bury, remodel, re-originate, reorder, consolidate and fragment within the digital escaped all previous architectural and ontological premises. Most formally conceptual leaps would easily expose themselves to functions of software or relays of assembled coded definitions and thus lend themselves to practices of analogy to traditional architectural objects (e.g. ReFormed geometric primitives, Ellipsicoon, UN Studio 2015) or illustrated philosophical formulae (e.g. the fold, Yokohama Ferry Terminal, FOA).

Precisely why the thought, as Krauss suggests, that an artwork can, in its attempts at destabilizing its own formulation and institutionalization, reinforce the political charge of its media, irrespective of the reference to historical traditions, is a radical one even for times of densified conception like the present.

The work of Peter Eisenman found common motivation with that of Rosalind Krauss' project, as described by Jeffrey Kipnis [4], of attempting to exaugurate painting from Greenberg's claims to autonomy of the visual art forms [5]. Peter Eisenman's work became occupied with particular medium specificity(s) and the nature of autonomy within the architectural discipline with potent anticipation of digital techniques. It is imperative to clarify here that the digital media architecture made available to its own in the final decade of the twentieth century can be seen as a supplement to react to specific poststructuralist conceptual work being discussed among North American institutions. The post structuralist subject, newly confronted by Francis Fukuyama's End of History, became potent with possibilities of realism and realization of the need for re-origination. The turn to Deleuze's conception of the diagram came to be seen as the most accessible segue into the digital for three important reasons:

1. Architectural conceptualization based on semiotics had treated the diagram as a mode of 'presencing' instead of a static under-drawing pro forma.
2. The poststructuralists had sustainably maintained and demonstrated that architectural form could not be considered separated from its content insofar as processes were made explicit and discursive for the community to deliberate upon traces and origins of said form. (e.g. Mask of Medusa, John Hejduk, 1985 and Chora. L Works, Peter Eisenman/Jacques Derrida, 1985).
3. Authorship was no longer being considered as a direct consequence of medium specificity (Greenberg) but as the destabilization of conventions within architecture.

As suggested by Robert E. Somol in Dummy Text, or the diagrammatic basis of Contemporary architecture:

> ...Not only was history of form rewritten but Eisenman would subject 'Form' itself to perpetual revision through an exhaustive sequence of operations: transformation, decomposition, grafting, scaling rotation, inversion, superposition, shifting, folding, etc. And it is the catalogue of the procedures that becomes the subject matter of architecture, a disciplinary precondition to diagrammatic approach. (Diagram Diaries, 1999, p. 15).

In hindsight now, it seems as if the proponents of the digital and computational methods alike saw within these tools an ecosystem ripe for analogy to and illustration of philosophical analyses and hence chose to foreground works of Deleuze and Guattari for its tectonic possibilities and complimentary positioning to the unavoidable ubiquity of digital technology [6]. Specifically, the unique representation methods on offer simply by way of detailed NURBS geometrical data were considered and still is, as bearing direct meaningful relation to the idea of an architecture or a formal typology or real content that the geometry may be suggesting to innovate upon. Institutions of advanced architectural design and practices alike have been able to shape the ecosystem of semantic and analogous organization of styles, for the purposes of figuration over the last three decades either by considering a preceding architectural idea as a mature operative device simply by virtue of the discourse of an epoch (e.g. Figure-ground relationship of gestalt theories) or have used the precedent to the effect of consciously formulating games of rulemaking and codified protocol that render all of the meaning to an architectural work through its accomplishments on the finitude of medium. (Post-medium media and the return of Greenberg's autonomy for its own sake).

4 Les Invalides et Alia

On the other hand, within linguistic and philosophical circles, ideas of a post-deconstructive realism were being debated triggered by Derrida's commentary on Jean-Luc Nancy's work on the destabilized status within continental philosophy of phenomenological touch and embodiment. Nancy's work rejects the notion that the determination of objecthood is contingent to human correlation, i.e. the phenomenological recognition of objects as long as they are sensible through human experience [7]. Although not a poststructuralist idea in philosophy for its specific call to a form of realism, the treatment of the nature of difference, albeit within the bounds of the real, is a common trait of several of the authors discussed so far, in both architecture and philosophy.

Within the work of most of the poststructuralist architects like Peter Eisenman, Aldo Rossi, John Hejduk etc., the idea of a form becoming architecture through procedures of formal manipulation and most importantly a performing form seeking a state of presentness must be understood as a matter of origins or more precisely—constructed originals. The original device at play in different cases would either be a cultural or specifically architectural artefact that is chosen as a part of the characterization of the performance. The procedures subsequently begin to reveal and often draw attention to the characterization as an intellectual exercise. At the heart of these choices of origins remains a bias unexamined by contemporary discourse. The examination is not so much an investigation towards establishing a virtuous or even a common original artefact but its treatment within architectural operations.

As part of Eisenman's early investigations of semiotics, the House VI project stands out for its discursive clarity not only from the author himself but also multiple other sources that have contributed to the commentary. The commentary has always consisted of a confirmation of a poststructuralist ambiguity achieved through formal operations, but the nature of the sense left to a single discursive conclusion. For a brief description of the formal condition of House VI, it can be said that the house is a result of a performance involving a relief Ground, lateral Planes Framed within a three-dimensional grid as registers of the serial operations that help rationalize the volumetric and political conditions of the 9 square grid. The 9 Square-grid, arbitrarily chosen and introduced as a political object—'an original' because of its political charge, in this case refers to its assertion of a center that would govern the organization of the sensible experience of the house. The introduction of the 2 stairs—both operative, one as signifier separated from its signified meaning and the other as experiential marker of the diagonal axis within the square grid are left behind as the only available reading of non-central center. While the performance itself is characterized as a series of operations that authors deliberate as a game of formal architectural elements, subservient to a Virtual—Non-Physical system.

A fundamental category of measure for such ornamental work is that of a distinction between structure and merged surface. Any propositions for or against the use of such spolia in architecture must address an essential momentum towards or away from a structural thesis and towards a virtual system. This structural thesis manifests through varied means and media. The underdrawings (diagram), the spline control points, mesh nodes, diagram, grid as diagram, 9 square grid, monumental associations, registered cultural icons, historiographical continuity. Here, a Libidinal economy (Author's will to create) and political economy (reference pro stimuli) always go hand in hand. The old becomes real. Resists all manipulation through its register of time but remains in anticipation of a re-origination. A bit like an electric clock without batteries that remains accurate twice every day ends up referencing itself across time without reliance on any appropriation. Conventional activity is no longer relevant at this point. Reading the activity of the old for its presence is construction.

In the particular case of the House VI project, the drawings and diagrams all seem set to arrive at a singularity of presentness that Eisenman seems to be pointing to—The Diagonal Realized. However complex and differentiated the set of procedures, the discursive relevance is never brought into a wider array of its qualities. Even if the project aspires to set up an isolated model, void of scenography and situationist rhetoric, the discourse surrounding its discreet objects of formation can still be speculated upon and politicised as Flattened mass (lateral planes), ambiguous central plane (Alterity in an axonometric drawing), scaling (domesticity embedded in the 9 square-grid) and the performance regenerated for heuristic purposes. The possibilities within such speculation are not simply technological but also methodological in a poststructuralist sense where the author seeks constantly to read the interrelations through other such operations that are rooted firmly in the discourse surrounding the work; including the treatment, as a taxonomy of stable objects, of all its commentary on the problems of representation and the use of axonometric projections for all that has not been adequately discussed.

In retrospection, from today's vantage point, works of architecture can also be read as practices of spolia or originals —complex, differentiated, arbitrary yet architectural and of inherent paradigmatic significance. To clarify, I am considering spolia as a conceptual point of entry and a specific line of inquiry into poststructuralist works as important not only because of its bases in differentiation and self-referentiality as modes of criticism but also since its inadequate articulation as a formal commentary on the nature of abstraction within architecture. In its attempts at categorization, architectural discourse may have underestimated the disjunctive state of the original architectural object (a 3D 9-square grid to begin the design of a house in 1971, why?!) with respect to its paradoxical traditional cousins. That the original object—as memory, as volume, as mass, as gestalt experience or as representation may have an architectural precedent so remote —as site (situation) or as scenario (scene) to previously established modes of discourse, may lead to a wider system of values that the discipline could share with its other.

A possible entry into such questions could be to identify the real state of spolia (matter) and the formal qualities embedded within, that allow for a body to be one. This state does not depend on/is contingent simply on its part to whole problematics; instead, reading a body implies the attachment/application of a 'character' to a form.

Within the discipline of architecture, an example of such an application would be 'the diagram'. Where the logic of lines and shapes is used to represent a proposition but must always produce an excess to be read. Such diagrams reman in a state of becoming, through which a body may be forming. The movement of thought from the diagram to the body may not be one of physics but by means of slippage of specific meanings—articulated meanings through which multiple characters can occur.

The perception of such a 'Characteristic Body' then, depends on the legibility of the slippage for any meaning or thought to be made interpretative. A cultural attention to form will then become imperative through manipulations of form itself [8].

The origins of form continue to be a matter of concern to us today especially for this time that seeks all modes of flattening of its own diversity for purposes of inter-textual identities. Identity and meaning may remain emergent or rather in a state of tension with the original characters that play the game but the choices regarding the original, the act of choosing, must consider operations of spolia that recognize sites of conceptualization as antagonistic to scenarios of the sensible. The architecture of today must consider its operative capacities as a rethinking of the original not only for intellectual consumption but also as its basis for modes of evaluation of its own formation.

References

1. Kipnis, J., & Leeser, T. (1997). *Chora L Works: Jacques Derrida and Peter Eisenman*. New York: Monacelli Press.
2. Vidler, A. (1988). *The third typology, oppositions reader*. Princeton Architectural Press.
3. Krauss, R. (1994). *The optical unconscious*. MIT Press.
4. Kipnis, J. (2007). *Introduction to written into the void*. Yale University Press.
5. Krauss, R. (1987). *Death of a hermeneutic phantom: Materialization of the sign in the work of Peter Eisenman*. Peter Eisenman's Houses of Cards New York: Oxford University Press.
6. Deleuze, G. (1997). Postscript on societies of control. In *Neil Leach's rethinking architecture: A reader in cultural theory*. Routledge.
7. Derrida, J. (2005). *On touching-Jean-Luc Nancy*. Translated by Christine Irizarry. Stanford: Stanford University Press.
8. Mickey, S. Touching Without Touching. (2018). *Objects of post-deconstructive realism and object-oriented ontology*. Open Philosophy.

Aniruddha Mukherjee is an Indian architect and researcher, currently pursuing a PhD at the Universität Innsbruck. He has completed a Master of Arts in Architecture, specializing in Architecture and Urban Design at the Städelschule Architecture Class, Frankfurt. He lives and works as a freelance architectural designer and visual artist in Frankfurt.

Social and Ecological Concerns in the Built Environment

Upgrading Participation Through Computational Thinking in Architecture

Canan Albayrak de Brito Colaço and Zeynep Mennan

Abstract

Advances in socio-technologies have contributed to a cultural shift towards active participation in society and eventually gave rise to a participatory culture. In the design context, the distinction between pseudo- and genuine participation comes into prominence. This paper traces the intersection and convergence of design participation and design computation in architecture. The starting point of this research is an inquiry into the thread of user participation in architectural design grounded in the design methods movement as an alliance between participation and computation. For the purpose of studying relationships between participatory design and computational design, this paper revisits the premises of the 1971 Design Participation Conference together with early examples of architectural participation projects from the 70s (Lewisham Houses of Walter Segal, and Supports of John Habraken) and the computational realizations of these early participatory models (John Frazer's computational model of Segal Method, and MIT's Open Source Building Alliance's computational model of Supports). This exploration suggests that computational design thinking can be addressed to open and extend architectural design to a wider range of participants, upgrading the first participatory paradigm. The design of norm/objectile/genotype by the architect and that of form/object/phenotype by the user is studied to support this argument.

C. Albayrak de Brito Colaço (✉)
Instituto Universitário de Lisboa (ISCTE-IUL), ISTAR, Lisbon, Portugal
e-mail: canan.colaco@iscte-iul.pt

Z. Mennan
Department of Architecture, Middle East Technical University, Ankara, Turkey

Keywords

Design participation • Computational design • Norm versus form • Design research • Architectural design

1 Introduction

During the last decade, participation has become increasingly popular in many disciplines and contexts as a way to better connect users to designers. This cultural shift towards active participation in society eventually gave rise to a participatory culture as the dominant characteristic of contemporary society [11]. Participatory culture opposes the dichotomy between producers and consumers, extending to blurred boundaries between not only producers and customers, but also between author and receiver, designer and user, expert and layman. Increased interest and government policies making participation a requirement for approval and funding resulted in participation being abused and misused as a buzzword, as explained in the Architecture and Participation book by Jones, Petrescu and Till:

> Recently, government policy in Europe and the USA has made participation a necessary part of public work; it has thus effectively been institutionalized, another box among many to tick in order to get approval and funding. Participation becomes an organized (and potentially manipulated) part of any regeneration project, in which users are meant to be given a voice, but the process stifles the sounds coming out. [12, p. 25]

The contemporary participatory design agenda displays the distinction between genuine participation and pseudo-participation as an ongoing research topic with increasing significance [14]. Pseudo-participation is described as creating a feeling of participation and a focus on user education instead of power transfer and authorial distribution. This type of participation is used as a socially acceptable shield for legitimization and populism [21]. The fact that participatory design is practiced in different disciplines,

S. Eloy et al. (eds.), *Formal Methods in Architecture*, Advances in Science, Technology & Innovation,
https://doi.org/10.1007/978-3-030-57509-0_8

domains, and communities and makes it harder to define its de facto core concepts and characteristics. One of the most important issues in achieving genuine participation is the formulation of a shared language between users and designers. "The channels of communication between the expert and the non-expert are not transparent, and so participation remains dominated by the experts who initiate the communication on their own terms, circumscribing the process through professionally coded drawings and language." [21, p. 24]

To address this problem, our paper proposes to trace the intersection and convergence of design participation and design computation in architectural design. Two types of design participation models in architecture are studied within the context of this paper. The first group is defined by the architects who propose the use of computation for increasing user participation in design. The second group consists of architects less interested in the use of computational methods in addressing design participation issues, than computational thinking in order to share their design knowledge for guiding non-experts through complex design decision-making process.

2 The Beginnings of the Design Participation in Architecture

The Conference on Design Methods, held in 1962 in London, laid the origins of The Design Research Society, which was founded in the UK in 1966. Design Research Society gathered a group of people with shared interests in new approaches to the process of designing. The society held its first international conference in 1971 with the title "Design Participation." This conference brought architects, urban planners, architectural critics, artists, and scientists together, under the common aim of increasing participation in the design process and blurring the distinction between designer and user. These contributors covered a wide range of multidisciplinary topics, categorized by Nigel Cross in his edition of the proceedings as: social technology, participation in design and planning, adaptable environments, computer aids, and design methods [7].

The beginning of the participatory design in architecture is associated with this conference [15], where participation is introduced to architectural design as "the new wonder ingredient participation" [1]. The outcomes of the conference have brought forth discussions of do-it-yourself, democratization, adaptable environments, user-responsiveness, and personalization to the architectural discourse, with the common agreement on the fact that the traditional planning and design processes have become obsolete for their achievement. Accordingly, new design methods and models were proposed by contributors for a reorientation of knowledge during the design process between stakeholders, a real transfer of authority on design decisions, and a redefinition of the relationship between architecture and its user [7]. In his conference paper, Nicholas Negroponte mentioned that:

> Two apparently untreated movements have marked the development of theory of architecture in the past five years, that of participation and that of computation. In one case, we are talking about providing users of physical environments with a higher level of input to the criteria for design as well as to the design itself. In other case we are dealing with computers that aid the designer. One obvious convergence for these two growing lines of research occurs by making the user in participation be the designer in computation … I will call concurrence of participation and computation, responsive architecture; to be vigorously distinguished from flexible architecture, manipulative architecture or (even) adaptable architecture [19] p. 63.

In addition to Nicholas Negroponte, some other participants of the conference included Charles Eastman, William Mitchell and Christopher Jones, whose works have pioneered the combined research track of design participation and design computation. Jeremy Till highlights in his book Architecture and Participation that this conference formed "a strange alliance between systems theory, computer programmes and participatory rhetoric" and he continues "it is also interesting to note that many participation advocates such as Henry Sanoff came from a design methods background" [21] p. 24. He not only published one of the first books about architectural programming [20] but also developed important participatory design tools, method, and games [15].

This thread of user participation in architectural design originating in the design methods movement formed an alliance between participation and computation. Of course, not all architects working on user participation in architectural design during 70s and 80s were from the design methods movement. For instance, Walter Segal and John Habraken were renowned architects who embraced participatory design ideals [15].

John Habraken proposed the design method called "Supports," which is conceived to allow users to have meaningful participation in the design of their homes [10]. "Supports" addresses the use of social and technical developments with the aim of empowering users in the design process. In his book Habraken describes this structure as:

> A support structure is a construction which allows the provision of dwellings which can be built, altered and taken down independently of the others. On the one hand, support structures restore the natural relationship (between individual and dwelling), and on the other hand, they allow full exploitation of modern factory-based mass production. They overcome, therefore, the two principal shortcomings of mass housing [10].

The proposed supports system of Habraken should not be interpreted just as a skeleton construction of a large-scale building. Instead, it should be treated as a framework for an open system [10]. In "Supports," the constraints presented by the building are designed by the architects and the infill, which is the unconstrained part of the building, is designed by the inhabitant. In other words, "Supports" is a complete dynamism of structure. This structure is defined by the architect through a given field of relationships and the rules which govern them [10]. The infill is the parts chosen by the inhabitants within the rules determined by the architect. The inhabitant can decide on the size and the location of their unit within the defined framework.

In the late 70s, when Walter Segal was commissioned to direct the self-build scheme for Lewisham houses in South London, he interpreted Habraken's method where users were empowered to design their home and developed skills to adapt it further to their changing needs. The first project, Phase One, started in 1979, and its success brought the need for a second phase, which started in 1984. In total, twenty-seven families were given a piece of a plot to construct their own houses.

Walter Segal laid the set of principles for the self-builders. He developed an adaptable timber frame housing method that is easy to manipulate so that self-builders could be self-designers as well. This method, referred to as the Segal Method, had implications that reached beyond the simple construction of Lewisham council houses: It could produce many types of houses, one- and two-story buildings with flat or pitched roof, courtyards, split levels, and double-height space [2].

The design participation models presented above can be divided in two: on one hand in which researchers and designers use computation to increase user participation skills in design; and on the other hand they don't have specific interest in computational design. However, in later years, the computational realization of the works of Walter Segal and John Habraken proved that although they did not involve computerization themselves, their designs still inherit computational thinking. In the scope of this paper, only two computational realizations of these early design participation models will be discussed. Nevertheless, these examples could be further extended, for instance to Malagueira Houses of Álvaro Siza and José Duarte's shape grammar for Malagueira Houses.

3 Computational Realizations of Early Design Participation Models

In the architectural discipline, the design of systems that consist of elements combined with rules are is referred to as computational design [16]. The difference between computational thinking and computerization needs to be understood for apprehending computation and its importance in architectural design. It is not computerization, but computational thinking which has posed design as a procedure depending on defining relationships and considering the relationship-definition phase as an integral part of a broader design process. It requires the designer to take a step back from the activity of design and focus on the logic that binds the design together. This process of relationship creation requires a formal notation and introduces itself as an additional concept that had not previously been considered as part of design thinking [22]. Thus, computational design thinking comes into prominence by making the design process more open and explicit and opening the design for a wider range of participants to contribute to the process [21].

3.1 MIT Open Source Building Alliance's Computational Model of Supports

In 2003, Open Source Building Alliance (OSBA) has been established at MIT under the House_n research group. The OSBA proposes Open Source Building as a new design model alternative to impersonal mass housing and standardization [13]. Open Source Building is a design model that engages new tools (information technologies, low-cost electronics, inexpensive computation, affordable sensing technologies, numerically or robotically controlled fabrication technologies, high-performance materials, mass-customization strategies, and the integration of physical and logical infrastructure) with aspects of open source (where designers, engineers, users and other stakeholders at many different locations and organizations participate in design). Taking advantage of these new tools, the mission of OSBA is to make mass personalized designs affordable and accessible, while simultaneously addressing societal problems through the mass-customization of responsive buildings [13].

The proposed Open Source Building strategy can be seen to combine the use of modular components and the aspects of open building as developed by John Habraken. Habraken and his contemporaries had produced their models as technological imaginaries of the 60s and 70s. However, these news tools enable an actualization of these technological imaginaries. Similar to the support and fill system of Habraken, OSB also suggests the separation of the building into a chassis (providing structure, power, communication, etc.), and infill of mass customized modules (for interior fit-out, exterior facades, electronics, communication, etc.) configured by the user [13]. The Open Source building model offers a communication platform between users and designers, to offer personalized design solutions and unique environments to individuals. Their research presents a computational model and realization of ideas developed by Habraken. Their model enables home buyers to design their own house and allowing every home to be unique according to the needs of the consumers.

3.2 John Frazer's Computational Model of Segal Method

John Frazer and his team developed an interactive machine-readable modeling kit for the computation of the Segal method. The model enabled the self-designers to arrange the components in the physical model on an electronic panel. Every design element, such as 128 different types of panels, doors, windows, is labeled with an eight-bit code. This allowed a digital processor to automatically scan the model and identify all the parts and their locations so that it can automatically generate the plan of the building. In addition to these, the computational model could interpret the plan to produce area calculations, schedules, drawings, and three-dimensional views [9].

By using this computational physical model, a self-builder with no knowledge of architecture could not only create buildable designs according to Segal's method, but also design alternatives and quickly evaluate them in relation to some aspects, such as cost [9]. In 1982, the working computational model was completed. Even though Walter Segal, as the original designer of the system, was very excited about the model, his sudden death prevented the design model to be further experimented [9]. Frazer explains this aspect of the model as:

> We had some of Segal's ideas literally preserved in silicon and could have continued with the project after his death. On this occasion we decided not to, but it is an interesting notion that architects' ideas could be made permanently available in this way. Quite unlike a design 'in the style of', it would allow others to create projects using the actual generating programs used by deceased architects - a bizarre prospect given the strangeness of using these programs when the architect is alive. The Segal project showed us that there was real potential for client and user-involvement in design and a new potential for immortality for the architect. [9, p. 43]

In other words, the most important aspect of the computational model of the Segal Method is that it incorporated Walter Segal's design rules and much more of his expertise. As a result, people without any knowledge of architecture or computers could design a house by building a simple model [9].

The 27 Lewisham houses were self-designed and self-built using the Segal method. The fact that they were self-designed and self-built did not result in randomness. They all contributed to forming a meaningful whole. In this process, the role of the architect, Walter Segal, was not that of designing the final form, but the method that would enable everyone to design and build their own houses. With the expansion of the participatory paradigm through computation, the role of architects becomes to design the inner logic rather than the external form by means of defining design rules.

Frazer defines this new role of the architect, as designing a "genetic" code comprising the information of the design system with elements and rules for how to combine them. This "genotype" does not constitute the final product but supplies the instructions that describe the process of building the "phenotype." Rather than the architect, other actors in the design process such as the final user generate the final form by assigning fixed values to the "genotype."

This type of distribution of authorship between designer and user has posed design as a procedure which depends on defining relationships and the willingness and ability of the designer to consider the relationship-definition phase as an integral part of a broader design process. It requires the designer to take a step back from the activity of design and to focus on the logic that binds the design together. This process of relationship creation requires a formal notation and introduces itself as an additional concept that had not previously been considered as part of design thinking [22].

4 Design of Norm by Architects and Form by User

The discussion about the genotype-phenotype differentiation is further developed in the "Non-Standard Architectures" Exhibition [18] as that between norm and form, specifically in Bernard Cache's discussion of the condition of the object and his development of the concept of 'objectile' [3]. The pairs formed by genotype and phenotype, norm and form, objectile and object all refer to and contribute to the same idea of a new design model in which architectural authorship is split between more agents, "on one side, the designers of the general function; on the other, its final customisers, or interactors" [5].

The international architectural exhibition "Architectures Non-Standard" (Non-Standard Architectures) took place in 2003 at the Centre Pompidou, Paris. The exhibition presented the works of the architects through prototypes, physical models, and drawings along with a digital model presented by Objectile (Bernard Cache and Patrick Beaucé) for their Projectile Table Project. Visitors were allowed to experiment directly with the design tools to produce "non-standard" forms through interaction with the design models by assigning variables to their attributions. As a result, the visitor of the exhibition could also personalize and customize the norm and become the designer of the form. Mennan describes this as: "A new notion of form, gives rise to the notion of fluctuating norm, one which is in constant redefinition in an open-ended series formed by the non-determinacy of a formal catalogue." [17, p. 173].

The term "objectile," besides being the name of the architectural firm by Bernard Cache and Patrick Beaucé, defines a new type of object. This definition is created by Bernard Cache and further developed by Gilles Deleuze, in his book "The Fold: Leibniz and the Baroque" [8]. Deleuze lays foundations of the notion of objectile on the ideas of Leibniz related to the definition of families of curves depending upon one or several parameters. The final object is not defined by an essential form rather it is framed by parameters [8]. Deleuze notes that "The new status of the object no longer refers its condition to a spatial mold—in other words, to a relation of form-matter—but to a temporal modulation that implies as much the beginnings of a continuous variation of matter as a continuous development of form" [8].

According to Gilles Deleuze's and Bernard Cache's definitions, Objectile, a new kind of technical object, is not really an "object" but rather an "objectile" in Aristotelian terms: a generic, non-specific object that is no longer designed but calculated [3]. Carpo defines the objectile as:

> An objectile is a family or a class of objects that are individually different but 1 all similar, because they all share the same generative algorithm: the same code or mathematical function, or the same genotypic DNA, as it were. In technical terms, an objectile is an open-ended algorithm, or a generative, incomplete notation; and this necessarily means that every final incarnation of an objectile into an individual object requires the intervention of some additional agency that may be other than, and even unrelated to, the objectile's designer. [4]

In this matter, the difference between genotype and phenotype, or norm and form, is seen to be the same between objectile and object. Objectile comprises not the information of the final form but rather that of the design system with elements and rules as to how to combine them; and the object is the final product when fixed values are defined to the attributes of the objectile.

As a result, this type of duality between the design of norm (genotype, objectile) and form (phenotype, object) creates two classes of authors: the designer of the norm and the designer of the form. The role of the final designer can be carried by the end-user, customer, or client, who personalizes and customizes the general program to make it fit his desires [4]. These two authorial roles are separate and independent from one another. The author of the norm is the author of the modifiable source code. The end-user is the author of the usable artifact assigning values to parameters and turning the generic norm into a specific one as form.

The design models presented at the Design Participation Conference such as "Adaptive Conditional Architecture" by Charles Eastman, "Flatwriter" by Yona Friedman, and "Design Amplifiers" by Nicholas Negroponte; but also Supports of Habraken and the Segal Method are all instances of genotype/phenotype, norm/form and objectile/object in the way that they all contribute to the same idea of a design model in which architectural authorship is split between designer and user where designer is the author of the system and the user is the author of the final form [6].

5 Conclusions

The findings of this research show that computational design thinking has been used by architects to create systems capturing their design knowledge to guide non-experts through complex design decision-making processes. Two models of design participation in architecture have been discussed in this paper. The first model consists of researchers and designers who propose the use of computation to increase user participation in design. The second group comprises architects who did not use computational methods directly to address user participation. Even though Walter Segal and John Habraken did not involve computerization themselves, the computational realization of their models are explained in this paper to prove that their designs still inherited computational thinking. This type of computational design proves to be highly beneficial: It allows users to co-design with respect to formalize design rules, to explore various design options and to manage various design solutions.

The role of the architect shifts to that of the author of an open system (a norm, a generic notation or an objectile), while the user(s) become the author(s) of final form. This approach suggests a methodology that blurs the boundaries between expert and non-expert, designer and user. Moreover, by giving authorial rights, it empowers users during the design process and marks a clear demarcation from pseudo-participation.

Acknowledgements Fundação para a Ciência e Tecnologia through ISTAR Project FCT UIDB/04466/2020 is acknowledged for its support.

References

1. Banham, R. (1972). Alternative Networks for The Alternative Culture. In Cross, N. (ed.) *Design Participation Proceedings of the Design Research Society's Conference*. London: Academy Editions.
2. Broome, J. (1988). Segal method. *Architects' Journal*, 189(25).
3. Cache, B. (1995). *Earth moves the furnishing of territories*. Cambridge: The MIT Press.
4. Carpo, M. (2009). Authors, agents, agencies and the digital public. In *ARCH'IT*.
5. Carpo, M. (2013). Introduction. In *AD reader: The digital turn in architecture 1990–2010* (pp. 8–14). Hoboken: Wiley.
6. Colaço, C. A. (2018). *Framework for an architectural knowledge ecosystem through the distribution of authorship*. METU.
7. Cross, N. (1972). Here Comes Everyman. In *Design Participation Proceedings of the Design Research Society's Conference*, 11–13.

8. Deleuze, G. (1993). *The fold: Leibniz and the Baroque*. The Athlone Press Ltd.
9. Frazer, J. (1995). *An evolutionary architecture*. Architectural Association Publication.
10. Habraken, J. (1972) *Supports: an alternative to mass housing*. London: Architectural Press.
11. Jenkins, H. (2006). *Convergence culture: Where old and new media collide*. New York: New York University Press.
12. Jones, P. B., Petrescu, D., & Till, J. (2005). Introduction. In *Architecture and participation*. Spoon Press Taylor & Francis Group.
13. Larson, K., Intille, S., McLeish, T. J., Beaudin, J., & Williams, R. E. (2004). Open source building—Reinventing places of living. *BT Technology Journal, 22*(4), 187–200.
14. Luck, R. (2005). Learning to talk to users in participatory design situations. *Design Studies, 28*(3), 217–242.
15. Luck, R. (2018). Participatory design in architectural practice: Changing practices in future making in uncertain times. *Design Studies*. Elsevier Ltd, 1–19.
16. Menges, A., & Ahlquist, S. (2011). *Computational design thinking: Computation design thinking*. Hoboken: Wiley.
17. Mennan, Z. (2008). The question of non standard form. *METU JFA, 25*(2), 171–183.
18. Migayrou, F., & Mennan, Z. (2003). *Architectures non standard*. Centre Pompidou.
19. Negroponte, N. (1972) Aspects of Living in an Architecture Machine. In *Design Participation Proceedings of the Design Research Society's Conference*, 63–67.
20. Sanoff, H. (1977) *Methods of Architectural Programming*. Routledge Revivals.
21. Till, J. (2005). The negotiation of hope. In P. B. Jones, D. Petrescu, & J. Till (Eds.), *Architecture and participation*. Spoon Press Taylor & Francis Group.
22. Woodbury, R. (2010). *Elements of parametric design*. New York, London: Routledge.

Tools for the Co-designing of Housing Transformations: A Study on Interaction and Visualization Modes

Sara Eloy, Micaela Raposo, Fábio Costa, and Pieter E. Vermaas

Abstract

In this paper we present and assess tools for visualizing architectonic modifications of existing housing in co-design projects with inhabitants. These tools should enable inhabitants to explore and understand design variations of alterations of their houses. This contribution is part of ongoing research on the use of artificial realities for supporting the transformations of existing housing in architectonically responsible ways. Such transformations may be needed after the delivery of housing, say after five years or later, due to changed regulation, the need of updates or changed living conditions of inhabitants. For arriving at architectonically responsible transformations, we use shape grammar system for defining possible modifications of the housing. For empowering inhabitants to understand and explore these modifications to their housing, we develop a transformation grammar tool—MyChanges—to visualize the modifications by three visualization modes, from fully immersive to non-immersive. Interviews and tests with real inhabitants were performed, and preliminary conclusions show that a tool like the MyChanges would have a good acceptance among inhabitants.

S. Eloy (✉) · M. Raposo · F. Costa · P. E. Vermaas
Instituto Universitário de Lisboa (ISCTE-IUL), ISTAR, Av. Forças Armadas, 1649-026 Lisbon, Portugal
e-mail: sara.eloy@iscte-iul.pt

M. Raposo
e-mail: micaela.raposo@iscte-iul.pt

F. Costa
e-mail: fabio_dmc@hotmail.com

P. E. Vermaas
e-mail: p.e.vermaas@tudelft.nl

P. E. Vermaas
Philosophy Department, TU Delft, Delft, The Netherlands

Keywords

Participatory design • Generative design tool • Virtual reality • Users' feedback

1 Introduction

Since some decades, architecture promotes a participatory approach that gives inhabitants the power to co-determine architectural design. This approach leads theoretically to solutions that are closer to the inhabitants' desires and ensures an alliance between architectonic quality and inhabitant satisfaction. Personalization and appropriation is seen as an important property that creates a sense of belonging to a place and therefore improves the quality of life of inhabitants [1, 2]. But architecture also advances the "authorship" position that consists on maintaining the architects' concepts and decisions when refurbishing or altering houses. This can mean that no changes are allowed, or only those that architects sanction.

In an interview, Rem Koolhaas refers to the architect as someone who judges what is beautiful and what is not beautiful, what should stay and what should disappear, what should be shown and what should be hidden. Koolhaas continues saying that architects judge in relation to priorities in architecture and that those are personal judgments that aim at preserving the purity and freedom at stake [3].

The common strategy used by architects for both allowing personalization and maintaining the authorship position is offering several design solutions such that inhabitants can personalize their houses by picking the one of their liking. Examples of housing projects in which this strategy was followed are the Malagueira neighborhood in Évora, Portugal, and Silodam and the Borneo Sporenburg neighborhood in Amsterdam, the Netherlands. In Portugal, SAAL (*Serviço de Apoio Ambulatório Local*) was the largest social participation process in housing design that covered

S. Eloy et al. (eds.), *Formal Methods in Architecture*, Advances in Science, Technology & Innovation,
https://doi.org/10.1007/978-3-030-57509-0_9

territories from north to south of the country. The SAAL program was created to respond to a deficit in housing with good habitable conditions. Under this program, several housing complexes were built and some of them enabled citizens' participation in the design of their own homes. Although it was not developed during the SAAL period, the Malagueira neighborhood, designed by the architect Álvaro Siza Vieira, is a paradigmatic example of a participatory process in the spirit of the SAAL intention. The design of the Malagueira houses was developed in collaboration with the future inhabitants, and this process had impact on some design decisions, e.g., changes in the façades so that the houses had greater contact with the street than what was originally designed by the architect [4].

This first strategy seems, however, not to be enough since it covers only personalization during the design and construction of housing, whereas personalization will also occur at later stages. A further strategy adopted in architecture is therefore to also define possible future changes to houses. For instance, Siza Vieira laid down a set of rules that allowed the evolution of houses in Malagueira from one to five bedrooms in order to respond to the evolution of families through time. The set of rules Siza Vieira left also included other aspects of the design, e.g., limits to the height of the houses, the size of the openings, and the option to paint the paneling on the exterior walls and frames around windows and doors for respecting the vernacular language of the region [5, 6]. However, Malagueira was under huge criticism by the inhabitants and in the press. In fact, Malagueira was several times referred to as the "Arab quarter" looking monotonous and inhuman [4, 7]. While local authorities and some inhabitants who were directly involved in the process defended the good collaborative process and emphasized the qualities of the project, other inhabitants expressed discontent and made changes to the houses over time. Some of these changes respected the rules defined by the architect, and others were carried out illegally, such as the placement of stairs to the terrace of the upper floor.

In this paper, we present and assess tools that support this further strategy for architecturally responsible personalization. The tools are for defining and visualizing architectonic modifications of existing housing in co-design projects with inhabitants and should enable inhabitants to understand and explore modifications of their housing. This contribution is part of ongoing research on the use of artificial realities for supporting the transformation of existing housing in architectonically responsible ways. Such transformations may be needed after their delivery, say after five years or later, due to changed regulation (e.g., new regulation about safety or sustainability), the need of updates (e.g., new technical infrastructure), changed living conditions of inhabitants (e.g., the need for an extra room or working place) or other wishes by inhabitants (e.g., decoration). For arriving at architectonically responsible transformations, we use shape grammar system as developed by Eloy [8] and Eloy et al. [9] for defining possible modification of the housing. For empowering inhabitants to understand and explore these modifications to their housing, we developed tools to visualize the modifications. The longer-term goal of this research is to develop a transformation grammar tool to be applied to existing houses and to test it with real inhabitants to determine the acceptance by the inhabitants of the tool.

This paper is divided into three main sections. First, we present a brief state of the art regarding the view in architecture related to diversity, co-design and the use of digital tools to enable diversified solutions to be created automatically by a computer and used by inhabitants. Second, we present the development and test of a design system interface designed to enable inhabitants to obtain design solutions to alter their houses. We end with discussion and concluding remarks.

2 Diversity, Co-design and Generative Design Systems

2.1 Diversity and Co-design

Creating diversity is a goal that has been in the mind of architects for decades. This aim, when applied to housing design, would enable to satisfy inhabitants' by giving them solutions that would fit individual needs and aspirations. Pioneering work by John Habraken in the 1960s [10] on the Open Building approach was an important moment when architecture made an effort to address diversity in design responding to the individuality of the inhabitants. More recently, the incremental housing approach promoted by Reinhard Goethert [11] and the Elemental solutions for social housing [12] are successful examples of promoting diversity in architecture. Also the Dutch architectural office MVRDV is for long concerned with how architects can "handle the endless individual desires of prospective homebuyers" [13] and buildings like Silodam are attempts to respond to this question.

In the beginning of the twenty-first century, the introduction of sustainability issues and the reemergence of participatory and collaborative processes in architecture reintroduced the social debate in the architecture practice after a decade of star-architecture focuses on extraordinary and mediatic buildings [14]. Indeed, new thinking emerged about the relations between the place and its users instead of the focus being mainly on building as isolated unities that expresses subjectivity.

According to De Koning et al. [15] and Faliu et al. [16], co-design consists of the collaboration of all stakeholders in a project that reflects decisions taken together. These authors argue that in a participatory process, there are several steps needed from the provision of information, to the development of ideas and requirements within all stakeholders. Engaging people to participate in this process is easier if they are attached to the places to be designed. As Sanoff [17] says "citizens do not care about projects until they start affecting them."

2.2 Generative Design Tools

Generative design computer tools have been developed in the last 40 years to respond to the need to create alternatives of design. Shape grammar design systems [18] aim to explore large solution spaces in an iterative manner. As such, they can form an excellent means to support designers and architects in finding diversified design solutions.

Several authors have been researching how shape grammars can be applied to housing design and how they can include the needs of inhabitants. Some of the developed systems generate new designs based on the requirements defined by the user. These tools may start by inquiring the user for input data through questionnaires or image selection and icons, and then automatically generate solutions that best suit these requirements. Other options are to present to the user a step-by-step design generation which guides the user to the final design. MALAG [19] is a system that allows both the creation of new construction and the reproduction of existing houses using the Malagueira shape grammar [20]. Another system, the Layout Generation system [21], allows the user to define by stages the house layout while all the decisions taken are visualized. This system uses shape grammars to generate design solutions customized to the inhabitant. HouseMaker© (MVRDV and Axis.fm, 2012) is a vision developed by the architectural office MVRDV that all together hands over the design of houses to the future inhabitants. The aim was to develop a tool that categorizes all the elements of a house and offers a user-friendly interface with which inhabitants could customize their future houses regarding materials, shapes, rooms, sizes, layout, etc.

In the present research, we focus on transformation grammars which are grammars that enable the transformation of existing designs into adjusted ones [22]. Examples of these grammars applied to housing design are the work of Colakoglu [23] that aims at modifying traditional Hayat houses to fit contemporary wishes, and the work of Eloy [8] on the development of a transformation grammar to refurbish a specific housing type, Rabo-de-bacalhau in Lisbon, to meet the requirements of new life styles.

2.3 Representation and Visualization

Architects use several methods to describe and register their designs and most of them are used for centuries. Bi-dimensional floor plans, elevations and sections are very common although clients, non-designers, such as inhabitants, usually have difficulty in understanding these descriptions. In the last decades, computers have enabled new methods for representation and visualization that are more natural and closer to reality than ever before. Virtual reality (VR) and augmented reality (AR) are examples of such methods.

VR is referred as a completely virtual and immersive environment that follows or not the physical properties of real world [24]. To use VR, one needs to have a large set of screens (CAVE, Powerwall, etc.) or a head-mounted display (HMD). CAVE technology allows a relationship between the user and the virtual environment, where the user is totally immersed in a real-scale projection. HMD technology also allows great immersive although usually used on an individual basis contrary to the collective use of CAVE where several users can have a shared experience.

AR allows the interaction between real and virtual objects at the same time [25]. AR can be visualized through smartphones, tablets or HMD, like Microsoft HoloLens as well as large projections where people can observe and analyze the combination of the real environment with virtual elements.

Currently, there are several market software that enable interactive VR experiences; examples are MindeskVR[1] and FuzorVDC.[2] Both are software with interfaces for real-time editing projects making it possible to navigate, edit and modeling 3D CAD in a totally virtual environment without having to export the models. Enscape[3] is a plug-in that allows to walk-through a 3D model in real time (Revit, SketchUp, Rhino and ArchiCAD). Unity[4] and Unreal Engine[5] are two platforms for 3D creation for immersive and interactive virtual experiences, where models can be done in this software or imported from other 3D modeling software.

3 MyChanges

3.1 Goals and Methodology

In Eloy et al. [9], we presented a post-handover shape grammar for introducing inhabitants' wishes in the transformation of individual houses of the Malagueira housing complex by Álvaro Siza Vieira. This grammar was developed by a group of architects during a workshop and later refined, for capturing a balance between the visible

[1]https://www.umindeskvr.com.
[2]https://www.kalloctech.com.
[3]https://www.uenscape3d.com/.
[4]https://www.uuinty.com.
[5]https://www.unrealengine.com/en-US/.

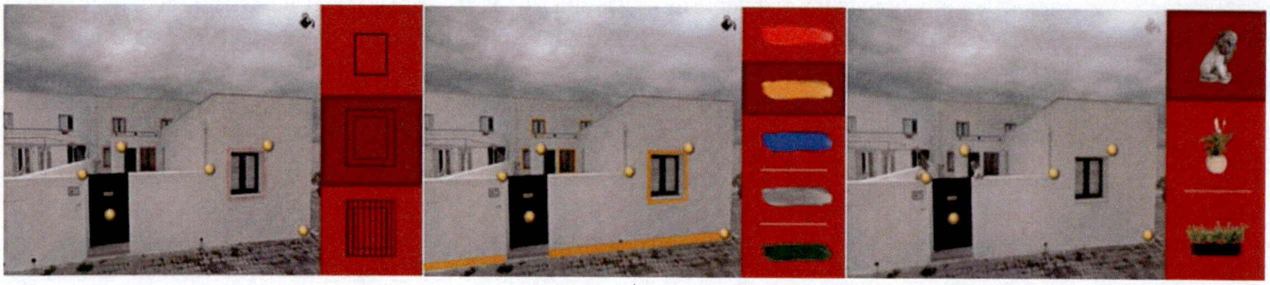

Fig. 1 Mock-up MyChanges, (left) window frame option selected, (middle), color menu and color selected (right) adornment menu and selected

transformations that Malagueira inhabitants did to their houses and the original architecture of Siza Vieira.

The goal of the research done for the present paper is to develop this Malagueira transformation grammar tool, from now on addressed as MyChanges, and test it with inhabitants of the Malagueira housing. The methodology used comprises four stages:

1. development of the Malagueira transformation grammar, done in [9];
2. development of a mock-up of the interface of MyChanges;
3. development of different modes of visualization from less to more immersive and from less to more visibility throughout the surrounding areas of the house; and
4. testing of MyChanges and the visualization modes with real inhabitants.

3.2 MyChanges Interaction and Visualization Development

The entire experience of MyChanges comprehends two main parts: the shape generation and the visualization.

The shape generation is currently not implemented in computer software but based on a simulation that reproduces some of the generation possibilities of the grammar. Although the grammar defined in [9] applies to all possible Malagueira facades, only one façade was used for the simulation.

The mock-up of the MyChanges interface is divided into two zones: first, a sidebar that corresponds to the menu, and, second, the middle area where a static image of the house façade to be customized is displayed. The customizable elements in the façade are identified with dots, on which the user can click so that the options appear in the side menu (Fig. 1). These elements are windows, doors, gate, wall, ornaments and railings. There is also an icon in the upper right corner of the viewing area that allows access to the color options of the wall paintings (Fig. 1). This icon is disactivated when the option selected does not allow painting (Fig. 1, right). The goal is that users can combine the customization of several of these elements as they prefer. Due to software limitations, in this mock-up it is only possible to reach predefined solutions.

To reach the result, the user clicks on the customizable element tag on the façade element he/she wants to customize and select the customization option he/she wants from the menu. This step is repeated until the user is satisfied with the drawing and by clicking on the color icon, which appears as soon as a paintable personalization is activated, he/she can select the color to be applied to all paintable elements. The result is visualized in a non-immersive screen-based static image superimposed on a photograph of one original façade of the house.

For visualizing the outcomes, we developed three different possibilities:

1. a non-immersive screen-based visualization using a tablet and where the final design is visualized as a static image;
2. a semi-immersive visualization using a smartphone and where the final design can be seen by a 360° render of the site; and
3. a fully immersive visualization using an HMD and where the final design can be freely navigated by the user.

The non-immersive screen-based visualization (A) uses a photo of a house in the neighborhood and superimposes digital elements as wall, window and door frames, and decoration (Fig. 1).

For the semi-immersive visualization using a smartphone (B), we modeled in Autodesk Revit[6] a representation of the Street Malagueirinha. With Enscape's library we add environment, sky, vegetation, cars and people, that were placed in the Revit model. After detailing the 3D model, we rendered a 2D panorama to be used in a smartphone screen (Fig. 2). For seeing the image using a smartphone, we used the app VR Media Center.[7]

[6]https://www.autodesk.pt/products/revit/overview.
[7]https://play.google.com/store/apps/details?id = com.xojot.vrplayer&hl = en_US.

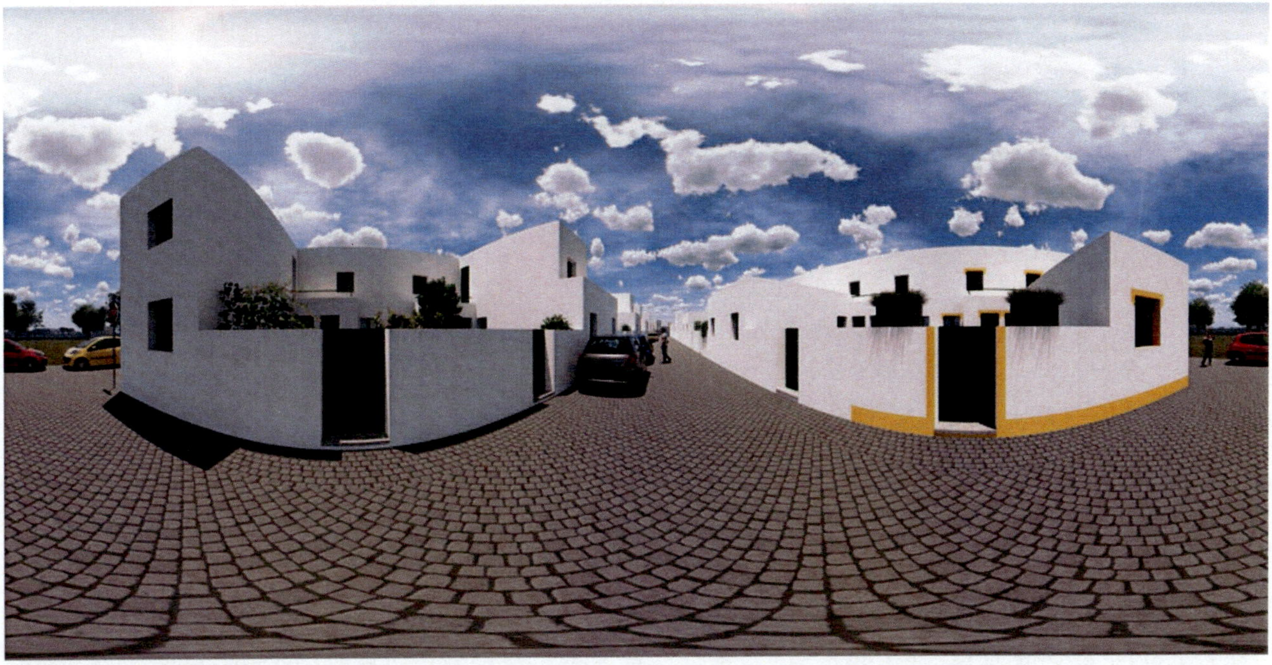

Fig. 2 Non-3D 360° image

Fig. 3 Navigation in VR

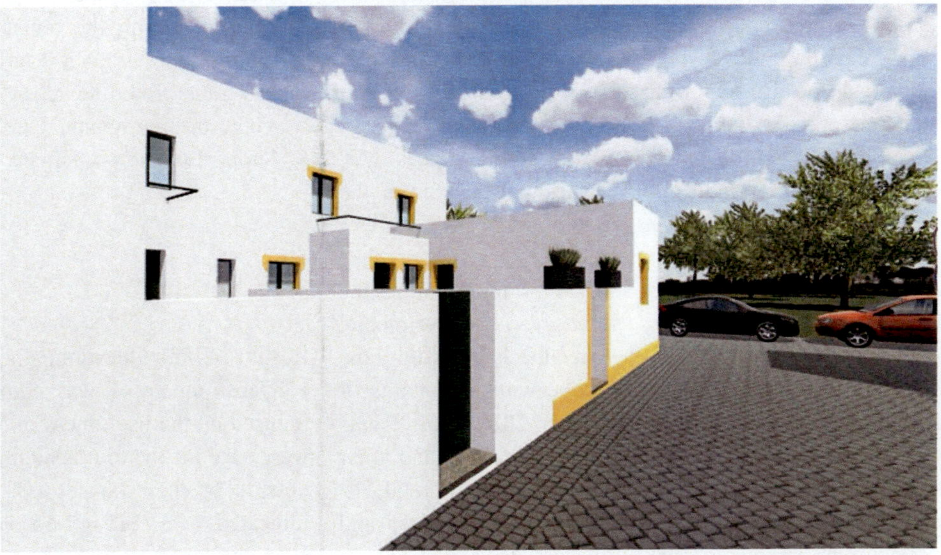

For the fully immersive visualization using the HMD Oculus Rift[8] (C), we used the same Revit model previously made. In order to create a simulation of the neighborhood of Malagueira, we grouped three streets with three different customized façades so that users, if wanted, could navigate through the model. To place this model in VR mode and see it in a fully immersive way, we imported an FBX file exported from Revit into Unity 2019.2.12f1 software. Through asset store of Unity, we add all environment elements, as well as street sounds and people (non-player character—NPC) with movement (Figs. 3 and 4). The built model from Unity was then used in Oculus Rift.

3.3 Experiment

The following protocol refers to the experiments that were performed with real users (inhabitants of the Malagueira neighborhood) to assess the MyChanges tool.

[8]https://www.oculus.com/rift/.

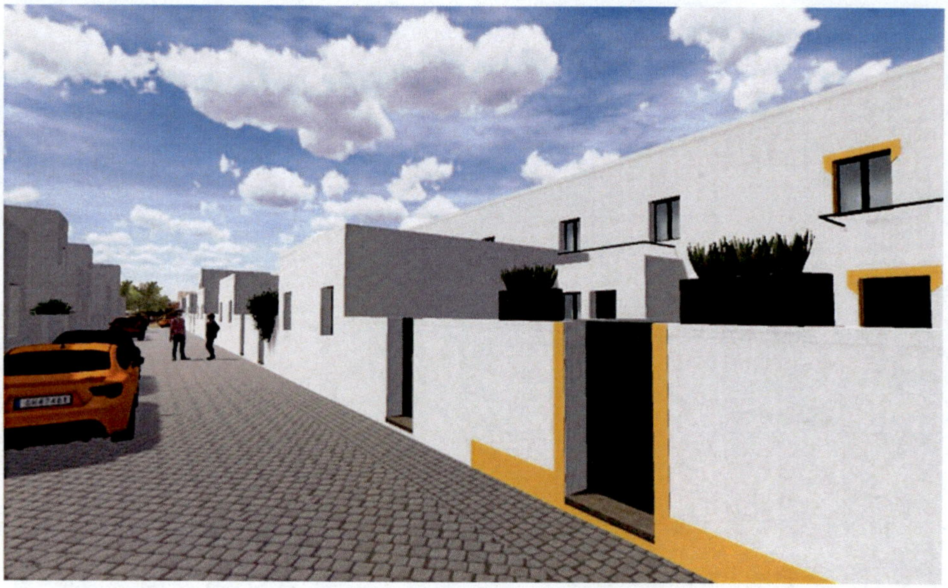

Fig. 4 Customized house in VR with animated NPC

The goal of the experiment was to assess:

- The satisfaction of users with such a tool regarding the customization possibility that it offers to the design of the façade of their house. This satisfaction was measured not just regarding the customization possibilities presently encoded but also about the idea of customizing through such an app.
- The satisfaction of users regarding three modes (A), (B) and (C) of visualization of the final designs of the façade, as described above.

The different ways (A, B and C) of representing the transformed façade of the house were given to experimental subjects in order to obtain their feedback regarding the potential of the tools. Tests were done with inhabitants of Malagueira and took place in February 2020. Participants were recruited from the neighborhood in cafes and their offices from 10 a.m. to 6 p.m. for one day. In total, 12 inhabitants took part of the experiment, seven women and five man with an average age of 59 years (SD = 12.2).

The protocol used to perform the experiment is as follows. Participants were approached by three researchers who:

1. Explained the goal of the study and ask for their permission to participate and gave the informed consent form to sign.
2. Shown the tablet with the generative design mock-up *MyChanges* and explained the generation possibilities, guiding the participant when using the system.
3. Gave the participant Questionnaire 1 to fill and offered support to the participant with filling it.
4. Shown the participant three modes of visualization (these modes of visualization were presented to different participants altering the order): (A) A static image of the transformed façade using a tablet; (B) A 360º render of the street with the transformed façade, visible using a smartphone; (C) A VR navigable model of the street with the transformed façade, visible using a HDM (Fig. 5).
5. Gave the participant Questionnaire 2 to fill and offered support to the participant with filling it.

3.4 Results

Results of both Questionnaires 1 and 2 are shown in Table 1.

Questionnaire 1 was composed of three questions that focused on the usefulness of MyChanges. When participants were asked if the tool is helpful for inhabitants to change the outside of their houses (Q1), the average of participants indicated 4.33 (SD = 1.15), which confirms that they consider the tool helpful.

The average of participants indicated 4.50 (SD = 1) when asked if they would use the tool in case they wanted to change the outside of their house (Q2), which indicates a high level of acceptance and intention to use.

When asked if they would be happy to use this tool alone, or with their family or friends (Q3), the average of participants indicated 4.42 (SD = 1.16), which says that they would be pleased to use the tool accompanied by friends and family.

Questionnaire 2 was composed of seven questions, five close and two open questions, that focused the use of the interface of MyChanges.

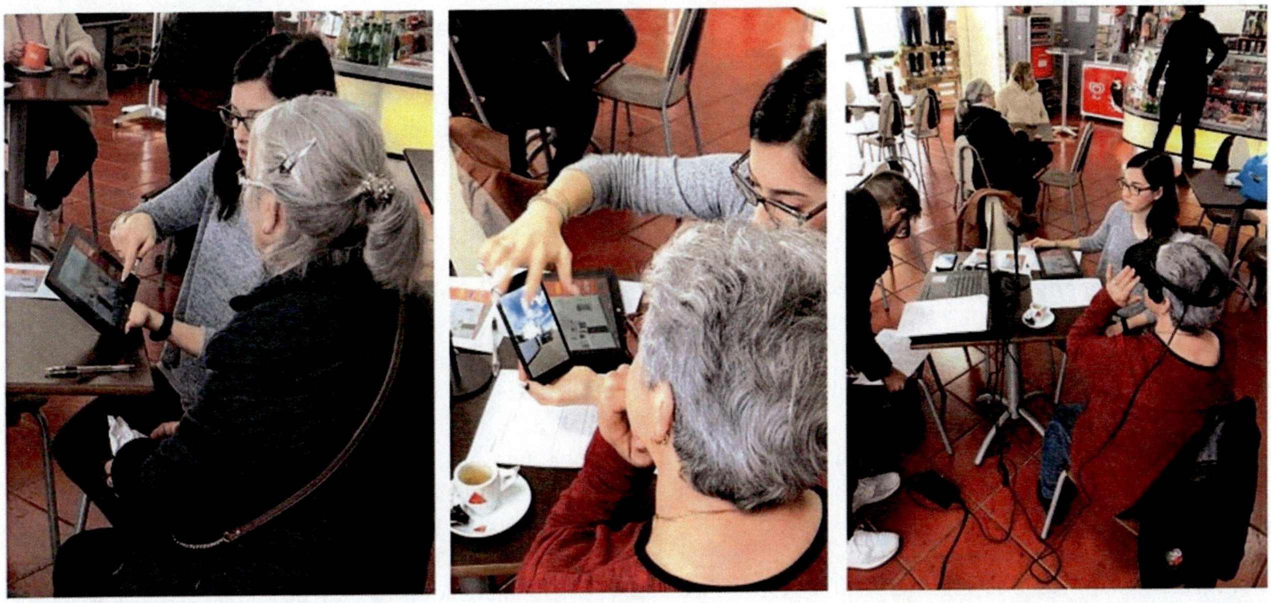

Fig. 5 Participants during the experiments, (left) using the MyChanges tool, (middle) using the smartphone to visualize and (right) visualizing with the HMD

Table 1 Responses of Questionnaires 1 and 2

			Questionnaire 1			Questionnaire 2				
Gender		Age	Q1	Q2	Q3	Q4	Q6	Q7	Q8	Q9
F		44	4	5	5	*A*	*A*	1	5	NA
F		44	5	5	5	*A*	*A*	3	5	5
F		60	5	5	5	*C*	*C*	5	1	5
F		67	5	5	5	*C*	*C*	4	2	5
F		67	1	3	1	*A*	*A*	5	1	1
M		67	5	5	5	*B*	*C*	5	3	5
M		85	4	2	4	*C*	*C*	1	4	3
M		47	5	5	5	*A*	*A*	5	2	5
F		68	4	5	5	*A*	*B*	5	2	5
M		56	5	5	5	*C*	*C*	5	2	5
F		52	5	5	4	*C*	*C*	5	4	5
M		55	4	4	4	*A*	*A*	4	5	5
F = 7	*M*	59.33	4.33	4.50	4.42	*A* = 6	*A* = 5	4.00	3.00	4.45
M = 5	SD	12.12	1.15	1.00	1.16	*B* = 1	*B* = 1	1.54	1.54	1.29
						C = 5	*C* = 6			

NA—the participant did not answer. Questions 1, 2, 3, 7, 8 and 9 where answered using a scale from 1 to 5 being 1 the lowest score and 5 the highest score

In the beginning of this questionnaire, participants were asked "which way of looking at the changes do you prefer to use?" (Q4). Six participants chose A (tablet), one chose the B (smartphone) and five chose C (HMD). When asked to explain their choice (Q5), most participants that chose A said that their preference was related to the fact they could simulate changes in the table and to the fact it is simpler to see. Participants that chose C justified by saying the "it looks real", "you fell in the place" and the "image is bigger". These results let us think that, if the transformation tool was implemented in VR, there was a chance that people would prefer an HMD to both simulate changes and visualize. There was only one participant that chose B, and the reason was that this medium gives him a global view of the surrounding areas.

Participants where then asked which way of looking at the changes made them better understand how their house will look like? (Q6). Five participants chose A (tablet), one chose the B (smartphone) and six chose C (HMD). Also, in this question responses were, as in question 1, divided between A (tablet) and C (HMD). There was a correspondence between choosing the same device in both questions 4 and 6. There were two participants who chose differently from question 4 to question 6, and it is interesting to observe that both chose a more immersive and complete view in the second question (B → C and A → B).

Regarding the importance of observing the whole street when exploring changes to their house (Q7), the average of participants indicated 4 (SD = 1.54), which says that they think it is a bit important to see the entire street when exploring the changes to their house façade.

One important aspect for us to inquire was the level of wish to use this tool alone or together with a professional designer (Q8). Regarding this question, participants were divided. The average of participants indicated 3 (SD = 1.54), from which five indicated they would prefer or maybe prefer to use it alone and six indicated they would prefer or maybe preferred to use it with a professional.

We were also intrigued by what the inhabitants think concerning the level of control that should exist, or not, when altering houses in Malagueira. For that we asked if participants think that such a tool should control the changes to the houses (and therefore maintain the special character of the neighborhood) (Q9). The average of participants indicated 4.45 (SD = 1.29), which says that they think that the tool should exist and some of them highlighted that different features could be integrated. There was an outlier in the group of participants who responded with a 1 (meaning—"I think it should not exist"). This person was very concerned with what is going on with the "savage" customization in Malagueira.

Regarding what this tool could offer more (Q10), four participants said that the tool is good as it is now. Four other participants suggested that more options could be integrated in the customization options such as "the interior layout, furniture and color" and "options to paint the bottom line in the house when the street in not flat." The participant that indicated that the tool should not exist (in Q9) said the tool is good but fails since it presents solutions to inhabitants that are forbidden.

This last comment from one participant regarding the aspect of "forbidden" was often present in the conversations with inhabitants during the interviews. The tension within inhabitants regarding the possibility to alter the houses (mentioned by Mota [5] and Rodrigues [6]) was felt during the stage of attracting participants. Indeed, several inhabitants refused to talk when alterations in architecture were mentioned.

4 Discussion and Concluding Remarks

In this paper, we presented and assessed the MyChanges tool for visualizing architectonic modifications of existing housing in co-design projects with inhabitants that should enable the inhabitants to understand and explore alternatives for modifications of their housing. The paper focused on inhabitants of the Malagueira neighborhood in Évora, Portugal, and concerned their satisfaction with the tool and three possible interfaces to simulate alternatives of customization of their houses' façades. Interviews and tests were performed with inhabitants, and results show that a tool like MyChanges would have a good acceptance among inhabitants. We, however, note that the interviews and tests were done with 12 participants, which is a too small number for a quantitative and statistically significant analysis. Moreover, participants were chosen among the ones available in cafes and by appointment on the day of testing, which may have created bias since, e.g., inhabitants working outside the neighborhood were not part of the test. Hence, our results can at best allow a qualitative analysis of the acceptance of MyChanges.

4.1 Future Work

For future work, we envision enabling a direct customization while in the immersive virtual environment, not just viewing but also editing in VR with Oculus Rift and its controllers. Future work will also include the definition of a grammar that enables inhabitants in Malagueira to simulate the customization not only of the original façades, but also façades that are already customized and that the inhabitants wants to change. Such a tool involves a different approach in ICT development since it implies new features: (i) the recognition of the façade and its elements; (ii) the recognition of the already added elements to the façade; (iii) and diminished reality to hide what exists and is suggested to be removed.

Further work will also include interviews with architects about whether the modification design tool MyChanges can solve the task of finding a balance between customization modifications and saving the original architecture of housing, and about what improvements can be made to this tool for better solving this task.

4.2 Concluding Remarks

The MyChanges tool is meant as a way to respond to the issue that customization poses to architecture. It is aimed at supporting housing customization that accommodates the evolving needs of inhabitants. And it is aimed at supporting customization that complies with the original architectural language of the housing by letting architects define the tool (the original architect, or others). Such tool can be used as a way of mediating the communication between architects and users in participatory design processes as advocated in the Open Building approach [10] and aiming at the users' acceptance. Also, the tool can be used as a way for users to express their wills by directly designing them in co-design processes as the ones explored in the U_CODE project.[9]

The research presented in this paper builds from Koolhaas' thoughts [3] related to the judgmental aspect of architecture regarding beauty by giving tools by which architects analyze and judge inhabitants' decisions on customizing their houses. This issue is then addressed by giving the inhabitants the power to co-determine changes to their housing in line with the "authorship" of architects.

By this solution, architects need to define identity-preserving refurbishment options for "context." For that the use of a shape grammar encoding transformation rules is a solution. Shape grammars are generative design tools that deliver customized design solutions based on the same principles of design for every single design problem. Since shape grammars uses languages of design, they force customization solutions to be in the language of design and therefore keeping the identity of the buildings.

Acknowledgements This project was partially funded by grant EEA Grants FBR_OC1_020 - ISCTE "Artificial Realities: Virtual as an aesthetic medium in architectural ideation." Also, we had support from projects FCT UID/Multi/04466/2019 and UIDB/04466/2020. Micaela Raposo has a Ph.D. research grant by Fundação para a Ciência e Tecnologia through grant SFRH/BD/146044/2019.

References

1. Coelho, A. B. (2010). *Habitação Humanizada*. Lisboa: Laboratório Nacional de Engenharia Civil.
2. Gehl, J. (2010) *Cities for people*. Island Press.
3. Zaera-Polo, A., & Koolhaas, R. (1992). Rem Koolhaas/OMA, 1987–1992. El Croquis, n°53.
4. Mota, N. (2017). *A progressive attachment* (pp. 89–107). Consuming architecture: On the occupation, appropriation and interpretation of buildings.
5. Mota, N. (2014b). Design ed Self-help. Producing closed forms for open buildings (pp. 1–13). *ETH Zürich*.
6. Rodrigues, A. L. (2015). A experiência da Quinta da Malagueira (1977): Ensinamentos e cautelas na concretização de habitação de custos-controlados, para uma arquitetura avisada. In *3° CIHEL - congresso Internacional de Habitação no Espaço Lusófono*. São Paulo (Brasil), pp. 1–14.
7. Mota, N. (2014a). A progressive attachment: Accommodating growth and change in Álvaro Siza's Malagueira neighhood. In D. Maudlin, & M. Vellinga (Eds.) *Consuming Architecture. On the occupation, appropriation and interpretation of buildings* (pp. 89–107). London: Routledge.
8. Eloy, S. (2012). *A transformation grammar-based methodology for housing rehabilitation*. Universidade Técnica de Lisboa.
9. Eloy, S., Dias, M. Â., & Vermaas, P. E. (2018). User-centered shape grammars for housing transformations: Towards post-handover grammars. In *Proceedings of Sigradi 2018 Technopoliticas*. São Paulo (Brasil): IAU USP.
10. Habraken, N. J. (1972). *Supports: An alternative to mass housing*. London (UK): The Architectural Press.
11. Goethert, R. (2010). Incremental housing: A proactive urban strategy. *Monday Developments*, (September), pp. 23–25.
12. Aravena, A., & Iacobelli, A. (2016). *Elemental: Incremental housing and participatory design manual*. Ostfildern: Hatje Cantz Verlag.
13. MVRDV and Axis.fm. (2012). Architecture in mind. The House-Maker. In MVRDV and The Why Factory (Eds.), The Vertical Village. Individual, Informal, Intense. NAi Publishers, pp 169–194.
14. Corullon, M. (2016). Apresentação. *Arquitetura em Diálogo—Alejandro Zaera-Polo* (pp. 7–12). UBU: São Paulo.
15. De Koning, J. I. J. C., Crul, M. R. M., & Renee, W. (2016). Models of co-creation. *Service Design, 2016*, 266–278.
16. Faliu, B., et al. (2018). Design and Prototyping of an Interactive Virtual Environment to Foster Citizen Participation and Creativity in Urban Design. In *27th International Conference on Information Systems Developmen*, (Sweden), pp. 1–13. https://doi.org/10.1007/978-3-030-22993-1_4.
17. Sanoff, H. (2008). Multiple views of participatory design. *International Journal of Architectural Research, 23*(1), 131–143. https://doi.org/10.15368/focus.2011v8n1.1.
18. Stiny, G., & Gips, J. (1972). Shape grammars and the generative specification of painting and sculpture. In O. Petrocelli (Ed.), *The best computer papers of 1971*. New Jersey: Auerbach.
19. Correia, R. C., Duarte, J. P., & Leitão, A. M. (2010). MALAG: A discursive grammar interpreter for the online generation of mass customized housing. In *Proceedings of the Workshop in 4th Conference Design Computing and Cognition*.
20. Duarte, J. P. (2005). A discursive grammar for customizing mass housing: The case of Siza's houses at Malagueira', *Automation in Construction, 14*(2 SPEC. ISS.), pp. 265–275. https://doi.org/10.1016/j.autcon.2004.07.013.
21. Veloso, P., Celani, G., & Scheeren, R. (2018). From the generation of layouts to the production of construction documents: An application in the customization of apartment plans. *Automation in Construction, 96*, 224–235. https://doi.org/10.1016/j.autcon.2018.09.013. (Elsevier).
22. Knight, T. (1989). Transformations of De Stijl art: The paitings of georges vantongerloo and fritz glarner. *Environment and Planning B: Planning and Design, 16*(1), 168–217. https://doi.org/10.1068/b160051.
23. Colakoglu, B. (2005). Design by grammar: An interpretation and generation of vernacular hayat houses in contemporary context. *Environment and Planning B: Planning and Design, 32*(1), 141–149. https://doi.org/10.1068/b3096.
24. Milgram, P., et al. (1994). Augmented reality: A class of displays on the reality-virtuality continuum. *Systems Research*, 2351 (Telemanipulator and Telepresence Technologies), pp. 282–292. http://doi.org/10.1.1.83.6861.
25. Tori, R., & Kirner, C. (2006). Fundamentos da Realidade Aumentada (pp. 22–38). *Fundamentos e Tecnologia de Realidade Virtual e Aumentada*. In R. Tori, C. Kirner, & R. Siscouto, p. 422.

[9]See about the U_CODE (Urban Collective Design Environment) project here http://www.u-code.eu/.

Sara Eloy Eloy is an architect, Ph.D. in architecture. Eloy's main areas of research include digital technologies applied to architecture, shape grammars, virtual and augmented reality, analysis of the building space, namely considering space perception, and housing rehabilitation. She is Director of the Information Sciences Technologies and Architecture Research Center (ISTAR-Iscte) where she is a fellow researcher. She has curated several exhibitions being the latest "Artificial Realities: Virtual as an Aesthetic Medium for Architectural Ideation," in the scope of Lisbon Architecture Triennale 2019. She has participated in national and international research projects and published her work in several journals.

Micaela Raposo Raposo is an architect, M.Sc. in architecture and Ph.D. student in Iscte. Raposo's Ph.D. topic Housing Co-design: A Framework Definition based on Generative Design Systems aims at synthetizing the potentials of generative design to address the customization of houses in a co-design process. In her master thesis Ver a Arquitetura através das tecnologias digitais, Raposo designed and tested an augmented reality app SeeARch for city information on architecture.

Fábio Costa Costa is MSc in architecture and Assistant Researcher at ISTAR-Iscte in the project Artificial Realities: Virtual as an aesthetic medium in architectural ideation. The project explores how artificial realities, namely virtual and augmented reality, are used as an aesthetic medium for architecture ideation stages, namely during participatory design processes. In his master thesis O Virtual sobre a Realidade: Estudo de ferramenta para aumentar maquetas reais, Costa researched processes for augmenting information on physical model buildings with augmented reality. He designed, supported the development and tested the tool Arch4models.

Pieter E. Vermaas Vermaas is a philosopher of technology and Ph.D. in philosophy. Vermaas does research at the Philosophy Department of TU Delft, on the structure of design processes and their applications in engineering, architecture and society at large. He edited the volumes "Philosophy and Design: From Engineering to Architecture" and "Advancements in the Philosophy of Design." Vermaas is the editor-in-chief of the book series "Philosophy of Engineering and Technology" and "Design Research Foundations."

Nature Pleated Surfaces

Maria João de Oliveira and Filipa Crespo Osório

Abstract

This paper presents a concept design methodology that intends to enable users to design a self-supporting structure based on the terrestrial plants' biological properties and on origami geometric principles that were tested on an eight-hour workshop at eCAADe/SiGraDi'2019 in Oporto. Focusing on rigid and flat-foldable origami surfaces, the methodology invites its users to deal with several architectural aims, constraints and analysis based on the geometric rules of this origami type. These surfaces behave in a particular way through folding, they depart from a planar state and after the folding process they arrive to a new planar state (flat-foldable Origami). If the faces remain rigid and flat during the folding process, the only change happens at the creases that behave as revolute joints. This way the folding of the surface is directly related to rigid kinematics allowing for a geometry-based digital simulation. After the aims, constraints and analysis definition, the methodology users are conducted through an abstraction process that explores the three fundamental forms of nature adaptation strategies—morphology, behavior and physiology, giving rise to a hybrid biological principle that will lead to the conceptual design process. The design process is conducted through analysis, experimentation with paper prototypes and the use of parametric design, enabling this way the emulation of the design process, where the functionality, geometry, robustness and aesthetics are tested and discussed.

Keywords

Biomimetics • Rigid origami • Methodology • Emulation • Parametric processes

1 Introduction

> At this point in our global ecological crisis, the survival of humanity will require a fundamental shift in our attitude toward nature: from finding out how we can dominate and manipulate nature to how we can learn from her. Fritjof Capra [1]

It could not be wrong to say that Nature is probably the most efficient, reliable and sustainable architect of the planet. Across the centuries, the survey of Nature and the use of basic and local resources has inspired humans, leading to experiment and exceed formal, structural and material boundaries, and also to the creation of new tools and methodologies [2].

The present research intends to contribute to a sound relationship between humans and our planet flora by developing an architectural design methodology based on vascular plant events formalized through foldable origami surfaces. Architects are the great thinkers and builders of space. If, for hypothesis, architects could shape their constructions based on the "minimum inventory, maximum performance" motto, human constructions could not only improve human health, spatial conditions and functional adaptation, but also improve a building's life cycle by relating and integrating its environmental resources and the inhabitants' needs while decreasing greenhouse gas emissions, energy consumption, resource usage, waste generation, among others.

The last two decades present several prominent experiences that cross nature and origami themes: on one hand, the (re)rising of nature into the architectural panorama through the use of minimal resources and maximum performance from the utilization of the new digital tools, on the other

M. J. de Oliveira (✉) · F. Crespo Osório
Instituto Universitário de Lisboa (ISCTE-IUL), ISTAR, Lisboa, Portugal
e-mail: Mjoao.oliveira@iscte-iul.pt

S. Eloy et al. (eds.), *Formal Methods in Architecture*, Advances in Science, Technology & Innovation,
https://doi.org/10.1007/978-3-030-57509-0_10

hand the use of origami surfaces to potentiate the creation of exploratory geometries, material research and structural behavior. The combination of both vocabularies may generate a new and innovative architectural semantic.

In a very obvious way, biomimetic design using origami surfaces could be referred the series of Oribotic flowers designed by Gardiner from 2004 to 2011 [3] or Blumen Lumen flowers developed by Foldhaus in 2014 [4]. In both cases, the designers used origami surfaces to create kinetic flowers that open and close in response to diverse stimuli.

In a non-so-obvious way, there are examples of the utilization of origami surfaces to materialize concepts and buildings such as Vega, Gonzalez e Gutierrez that in 2009 developed an experimental shading canopy with their students at the Universidad Técnica Federico Santa Maria in Valparaíso, Chile [5]. The developed self-supporting structure explored the deployment of catenary vaults created through the aggregation of modules with variable aperture which were designed and fabricated through techniques of folding and pinning—Botterfold. The parametric structure was designed based on the local solar trajectory in order to generate the most possible shadow at the hottest hours of the day. The canopy was composed of 1200 triangular aluminum modules, disposed in order to provide shading comfort to its potential inhabitants. Bathed in the late oceanic sunsets, the optical characteristics of the structure create an ephemeral effect of sparkles which evokes its neighbor, the sea.

Later in 2016, at the Detmold University of Applied Sciences, Tal Friedman developed the Foldfinding-Origami Pavilion [6]. Linking the self-supportive mechanism of the thin shell of a flower to a self-supporting thin-shell folded structure, Friedman created a full-scale pavilion folded out of aluminum boards which were parametrically designed to fold into the flower-like shape inspired by the techniques of Origami. The final form is comprised of two "flower" modules connected together to form a gate structure resembling the outline of a Japanese Torii. Each flower is made of four sections which acquire a precise position when folded. Each folded piece contains around 12–20 interconnected flat faces. This technique saves fabrication time and building tolerance. This project is a proof-of-concept model to conclude that folded structures can be fabricated to full scale while maintaining their stiffness and self-supporting stability. By reducing the need for additional supporting structural systems, a new kind of thin-shell lightweight structures made from just one material emerges.

In 2017, architects Cavada, Barcala and García drew inspiration from folded paper in Origami, to create a temporary installation for the architecture and design festival Concéntrico 03 in Logroño, Spain [7]. The fundamental idea started with the statement that a paper sheet alone does not sustain itself, but when joined in a series of precisely folded sheets, they are capable of sustaining not only themselves but other forces. The installation was built from 39 wooden panels joined together with hinges and assembled without any other supporting structure or sub-structure. The gaps between the timber panels allow light to seep through, giving the structure the appearance of a glowing lantern at night.

Also, in 2017, Heather and Ivan Morison developed the Look! Look! Look! sculptural pavilion for the National Trust's historic Berrington Hall in Herefordshire [8]. Inspired by the eye-catchers buildings of the Georgian era and origami, the structure resembles a pineapple. The pavilion has a centralized plan and the folds depart from a radial crease pattern crossed all around twice by a v-line, which in the folded position leads to a progression in an accordion-like manner on the horizontal direction and two inversions of sense on the vertical direction. The structure is composed by 90 faces, erected through timber panels and coated with a translucent pink fabric, anchored to the ground through a metallic foundation frame. When seen from the outside the installation appears strong and sculptural and from the inside, it has a translucent and delicate appearance allowing light to pass freely from the entrances and the open ceiling.

In the presence of such examples, the authors of this paper have decided to make an experiment by joining two distinct Ph.D. investigations, one on the field of the biomimetics and the creation of a biomimetic architectural process [9], and the other on the field of kinetic origami surfaces to be used in architecture and supported by the parameterization of origami crease patterns and their folding path [10].

The fundamental purpose of this project was to test if a biomimetic design process, that departs from the deep knowledge of terrestrial plants adaptation strategies, could use origami geometry and parametric design as a formalization of the goal of the biomimetic architectural process and produce foldable self-supporting structures.

Rigid origami surfaces can be achieved through tessellations that follow particular rules which lead to the generation of polygons constrained to rotate around their shared edges. These surfaces may follow rigid kinematics rules and may be able to fold from a planar state into another, collapsed planar state. This way the surfaces may undertake a range of configurations before collapsing into a bundle, thus occupying the least possible space and the use of only rigid faces simplifies the parametric simulation by allowing it to be purely geometric.

To test the feasibility of the combination, two Ph.D. researches were conducted the Nature Pleated Surfaces workshop for the eCAADe/SiGraDi'2019 in Oporto.

2 Nature Pleated Surfaces: The Workshop

2.1 Workshop Outline

This section presents the summary of the work developed at "Nature Pleated Surfaces Workshop," held in September 2019 during the eCAADe/SiGraDi'2019 in Oporto and the retrieved results.

The workshop had the participation of students and professionals from different fields, such as parametric design, biomimetic architecture and origami design research, and also from different ages and countries—Chile, Bolivia, Brazil, Iran, Portugal and Turkey.

The workshop was conducted through eight hours and was divided into four distinct phases. The first was a theoretical phase where the students were introduced to the geometry of rigid and flat-foldable origami principles, its types, fundamental axioms and foldability rules. The second phase was also theoretical and concerned the presentation of plants' biological principles, focusing on the specific case of the terrestrial vascular plants and its adaptation strategies as well as ways to interpret those strategies and bring them into the architectural practice.

The third phase was dedicated to the practical application of the theories. The students were divided into groups and started the development of their projects supported by online research, construction of diagrams and concept definition. From the definition of the conceptual intention, the groups started to experiment with paper folding and parametric design to achieve rigid and flat-foldable origami crease patterns. The fourth and final phase was entirely devoted to the design studio and to the materialization of the project of each group.

The main goal of this eight-hour experience was to develop a self-supporting structure, based on the biological adaptation principles of the terrestrial plants and on the geometry of rigid and flat-foldable origami. As final product, the groups should achieve: (1) an architectural challenge definition, by determining the function of the project and a corresponding degree-4 origami pattern for the self-supporting structure; (2) a meme event construction based on plants morphological adaptation (the conceptualization of an abstract idea of design); and (3) a parametric and physical model of the designed solution.

2.2 Rigid Origami Geometry

In what regards the origami geometry, the students were first introduced to some elementary concepts such as the basic type of folds and how to parameterize them using Grasshopper for Rhinoceros. The basic folds in origami are the mountain and valley folds, and the parameterization of such folds corresponds to a simple rotation of each face of + or −90° around the shared edge between the two faces (Fig. 1).

Then were explained the seven Huzita–Justin Axioms for folding operations, which are very similar to the Euclidean axioms for constructions with straightedge and compass. These axioms allow to find creases and angles through folding. From these basic concepts, the students were then introduced to the rules for rigid and flat foldability, such as the Maekawa–Justin and Kawasaki–Justin theorems and concepts such as the degree of a vertex.

In origami, the degree of any vertex corresponds to the number of creases that depart from the vertex. It is the identity of the creases (Mountain or Valley), their assignment around the vertex and the angles between each crease that determines if the faces around a certain vertex can be rigid and flat folded. An origami tessellation can be described as a degree-n tessellation if all the interior vertices of the crease pattern have degree-n. So, the students were then introduced to some rigidly and flat folding tessellations of degree-2, degree-4 and degree-6 crease patterns (Fig. 2), as well as to the resulting folding of the tessellations at different stages.

Finally, were presented, and provided to the students, several parameterizations of rigidly folding patterns developed in Grasshopper for Rhinoceros, as demonstrated on Fig. 3. These parameterizations allowed the students to test and fold some crease patterns of degree 2, 4 and 6. Through the definitions, the students were able to manipulate the geometry of the base crease pattern by changing the dimensions of the faces through sliders for the coordinates of the points that defined the faces. The definitions allowed also to determine the size of the tessellation to fold, that is the number of faces that composed it and were constructed in a way that obliges for the following of the rules for rigid and flat foldability. This way the students could manipulate the tessellations and make them fold through a folding slider which helped them to understand deeper the geometry relations of the folding of multiple faces.

2.3 Biomimetic-Based Methodology

The second phase of the workshop concerned the predefined biomimetic design path matrix (Fig. 4) composed by three main domains: the architectural domain, nature domain and the construction of the artifact domain.

Architecture Domain—Identification, Analysis and Diagnose

The architectural domain aims to define the architectural main function.

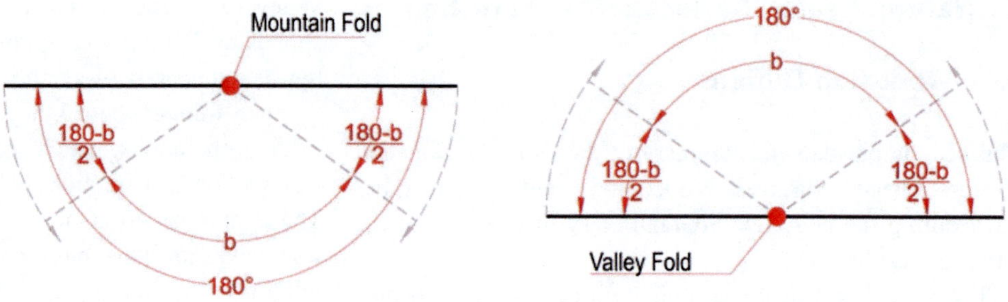

Fig. 1 Front view of the parameterization of mountain and valley folds

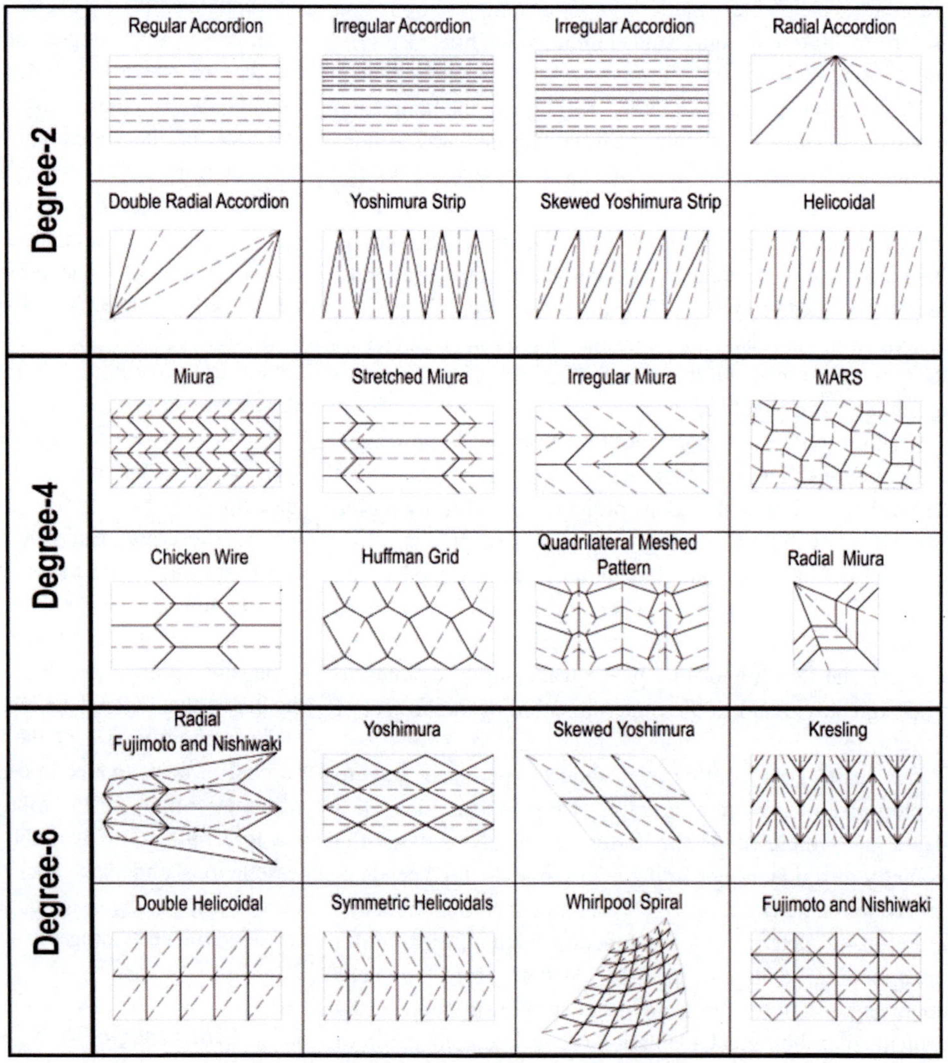

Fig. 2 Examples of rigidly folding origami tessellations of degree 2, 4 and 6

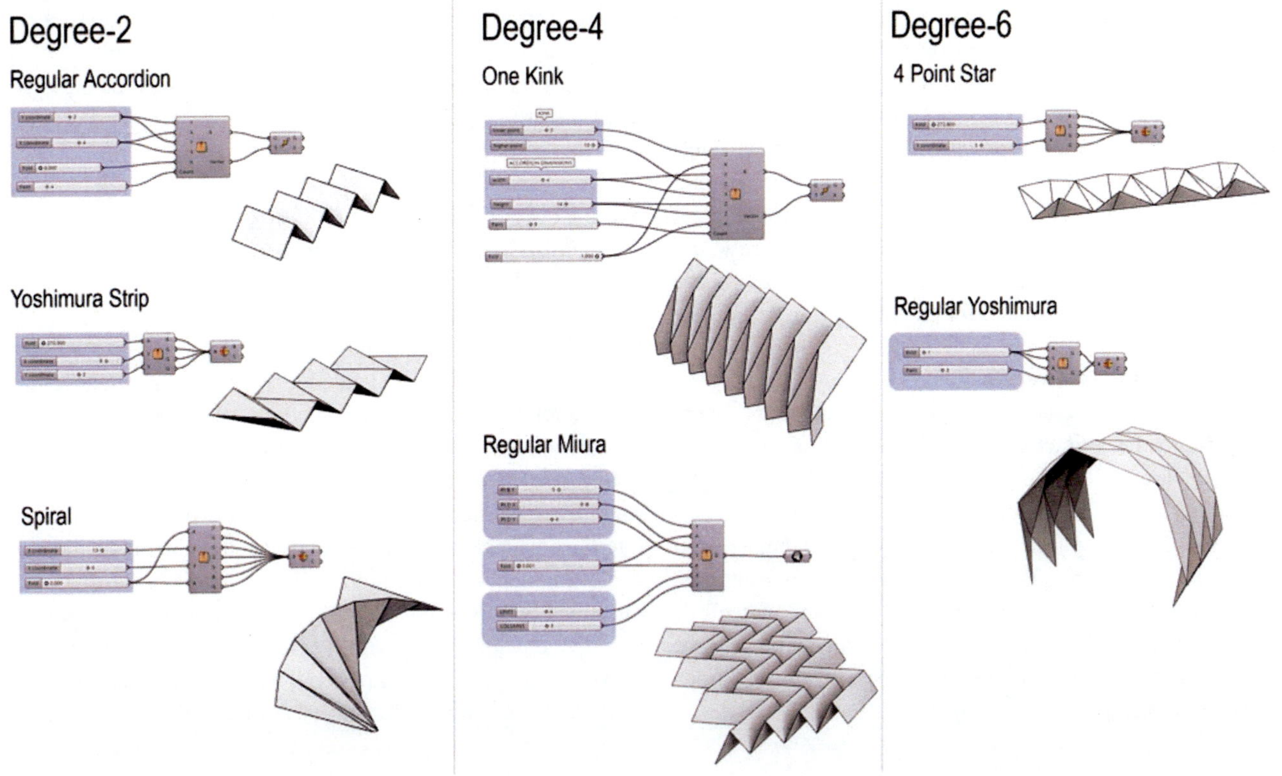

Fig. 3 Provided folding parameterizations

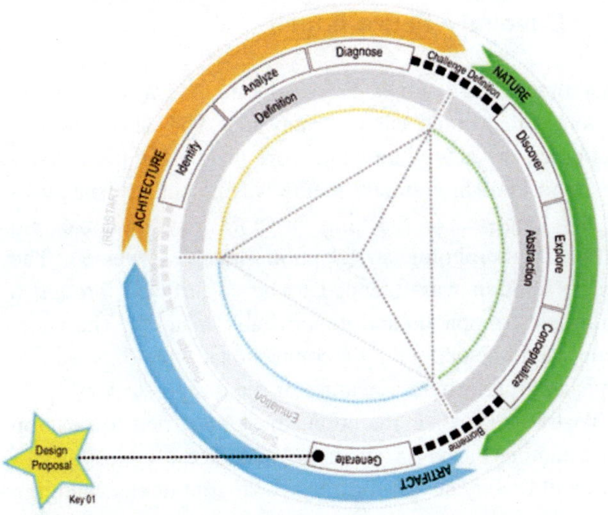

Fig. 4 Biomimetic design path matrix

Participants were free to choose their structure function as well as to choose their own rigid and flat-foldable origami crease pattern. These factors were a determinant mojo factor to engage participants into a design analysis and diagnose process essential to conceive an abstract idea of the structure that could be related with terrestrial vascular plants adaptation strategies events.

Nature Domain—Discover, Exploration and Conceptualization

The nature domain departs from the previous design solution intention focusing into the morphological adaptation strategies of terrestrial vascular plants discovery, by exploring its events and potential features translating it into a design conceptualization.

To the participants were given several online databases, and a synthetic list of the most well-known and documented morphological adaptation strategies of terrestrial vascular plants. Departing from the previous domain, where participants had diagnosed their self-supporting structure main function and defined possible origami rules and potential geometric solutions, at nature domain, participants were then invited to enrich their structure functions and performative behavior by linking morphological adaptation strategies to their design solutions. This is an abstract process that begins with the identification of determine morphological strategies with which participants identify some of their initial potential ideas that could enable and potential their design structures. We called this process as the biomimetic meme construction. The biomimetic meme construction is composed by the extraction of strategies, patterns, material and performative features from the original morphological strategies that combined give rise to a completely new and artificial Biomimetic meme. This biomimetic meme becomes the strategical model for the design project.

Artifact Domain—Generate

From the design conceptualization, participants were engaged into the artifact domain, by generating parametrically their design solutions, and confronting them with the inevitability of its materialization.

At this domain, participants were ready to engage the parametric design task. To generate their design solutions, participants used Rhino + Grasshopper software, as well as 80gr sheet paper. Origami theorems and axioms were determinant for the definition of the design form and foldability principle, as well as performative features.

3 Projects

From the first two theoretical phases of the workshop, the students developed five different design solutions. Each project was based on different natural systems and organisms, and each arose different issues during the design process and the prototypes production. Regarding origami geometry, one group used the Yoshimura pattern and the others used variations of degree-4 tessellations.

3.1 Project 01: Urban Furniture (Bárbara Zandavali + Yağmur Yeni)

The first project, developed by Bárbara Zandavali and Yagmur Yeni, was called Urban Furniture, and the aimed function was to create an urban shelter with shading properties. Their biomimetic meme was essentially based on symmetry, fractal and under-story plants as a way of radial spread in several directions. These morphological adaptation strategies conducted the group to a radial solution, with dynamic properties (Fig. 5).

The designed solution was a foldable self-supporting structure with three essential anchor points. They used a degree-4 crease pattern with several kinks to generate an ellipsoidal section shading canopy. The intention of the group was to use the same design to generate different structures by changing the positions of the anchor points. By having the same surface and placing differently the anchor points was possible to create structures with different radius, apertures and covered area, due to the geometry of the origami pattern (Fig. 6).

3.2 Project 02: Foldable Hangar (Guilherme Giantini + Mateus Pimentel + Mauro Couceiro)

Foldable hangar was the most ambitious project in terms of scale since the group intended to generate a structure to cover a long area. Their biomimetic meme was sustained on succulent and fractal morphological adaptation strategies, giving rise to a fractal and lighting pattern that could be static or dynamic, materialized through a monolayer structure that sustained translucent and opaque materials and that could be deployed through a linear movement (Fig. 7). Sustained by this initial premise, the origami selected pattern was the chicken wire with vertices so close that almost arrived to the Yoshimura pattern. The designed pattern was well achieved since its fractal type of repetition allows for good structural resistance, essential for the covering of large areas (Fig. 8).

3.3 Project 03: Mini Orchid Yard (Débora Oliveira + Pedro Engel)

Mini orchid yard aimed at developing a water recycling flowerpot that could capture rainwater. The design project was based on canopy and carnivore plant adaptation strategies, and their biomimetic meme was essentially sustained by the hygroscopic, packing structures, hydrophobic and symmetry morphological adaption strategies (Fig. 9). The design solution was intended to be constructed through a multi-layered, opaque and impermeable material. The water cycle would be defined by a convectional three-dimensional movement performative feature of the design strategy.

To formalize their intention, the group tried to develop the container based on an irregular degree-4 crease pattern that could have an organic form with continuous changing curvatures (Fig. 10). Unfortunately, they were not able to reach their intention neither in paper nor through the Grasshopper definitions.

3.4 Project 04: Ibiscus Water Collector (Juliana Jaramillo + David Diaz)

Working at the installation scale, the Ibiscus water collector main function was to collect water for dry public spaces, at

Architecture Constraits	Symmetry	Assymetry	Fractal Plants	Under-story plants	BIOMIMETIC MEME
Shelter		X		X	X
Shell	X	X			X
Radial	X		X		X
Meme Strategies					
Dynamic		X	X	X	X
Static	X				
Meme Pattern Features					
Density			X	X	X
Monolayer				X	
Horizontal				X	
Continuous	X			X	X
Meme Performative Features					
Radial Movement	X		X		X
Generative			X		

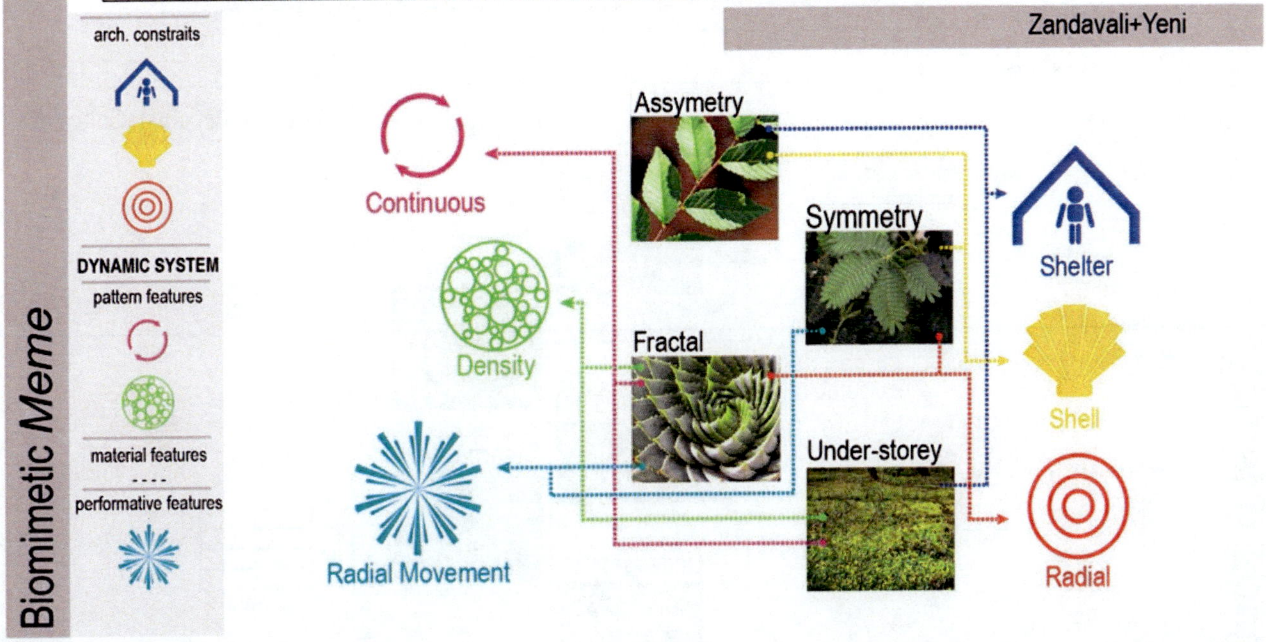

Fig. 5 Urban furniture—biomimetic meme construction

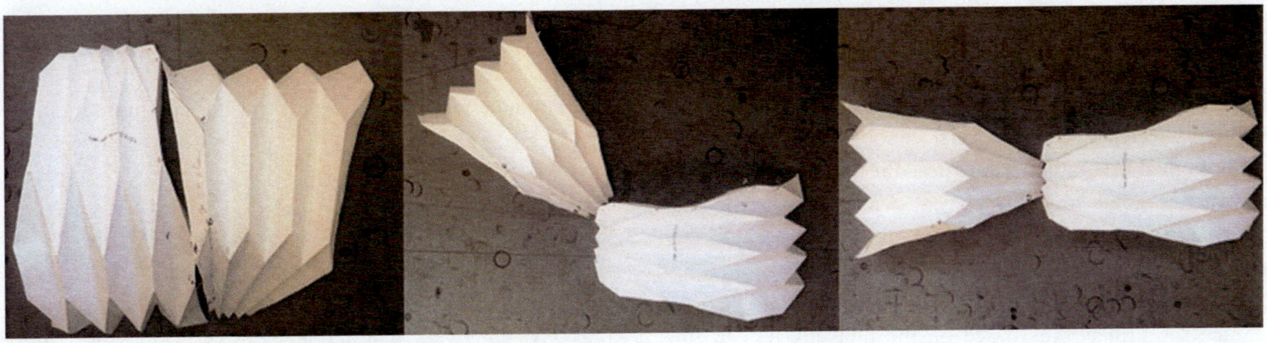

Fig. 6 Urban furniture—exploratory mockups

Architecture Constraits	Fractal	Succulent	Stoma	Symmetry	Under-storey plants	BIOMIMETIC MEME
Lighting	X	X			X	X
Self-Supporting Shelter		X	X	X		X
Meme Strategies						
Dynamic		X	X			
Static	X			X	X	X
Meme Pattern Features						
Loose		X	X			
Monolayer			X	X	X	X
Multilayer	X	X				
(re)Scalable	X	X	X			X
Meme Material Features						
Translucent			X		X	
Opaque	X	X			X	X
Flexible	X		X		X	X
Meme Performative Features						
Radial Movement	X				X	
Linear Movement	X	X	X		X	X

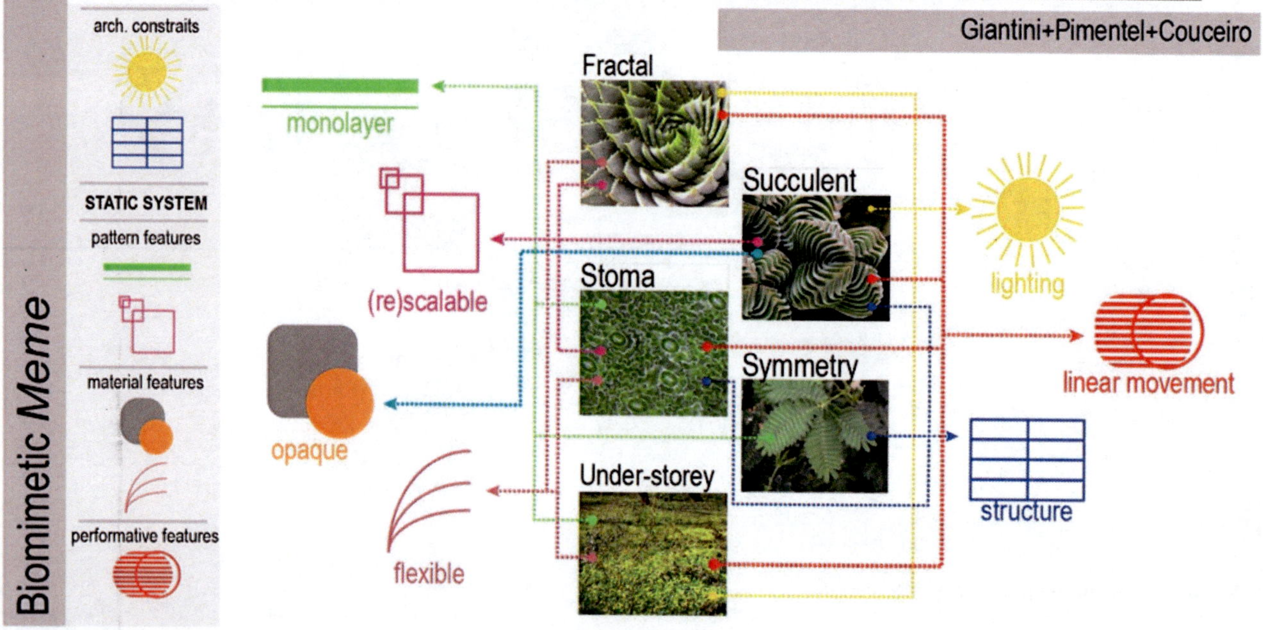

Fig. 7 Foldable hangar—biomimetic meme construction

the same time that it could provide a shading shelter for the inhabitants. Sustained by vessels, canopy and hydrotropic morphological adaptation strategies, the biomimetic meme was a static, radial, monolayer event, with impermeable and rigid material features, that had the ability to storage and distribute water, in order to provide water to the surrounding living species (Fig. 11).

Since the project was based on the hibiscus flower, the origami crease pattern was quite straight forward. It departed from a radial progression of folds with one kink that allowed for the closure of the structure and to the slight inflexion that would conduct water into the central part of the structure (Fig. 12).

3.5 Project 05: FourToSix Walkway (Mona Ghandi + Victor Silva)

FourToSix walkway was conceived as a movable and mutable structure. The primary main function for this self-supporting structure was to serve as an art gallery. Departing from the stoma and inflorescence as fundamental

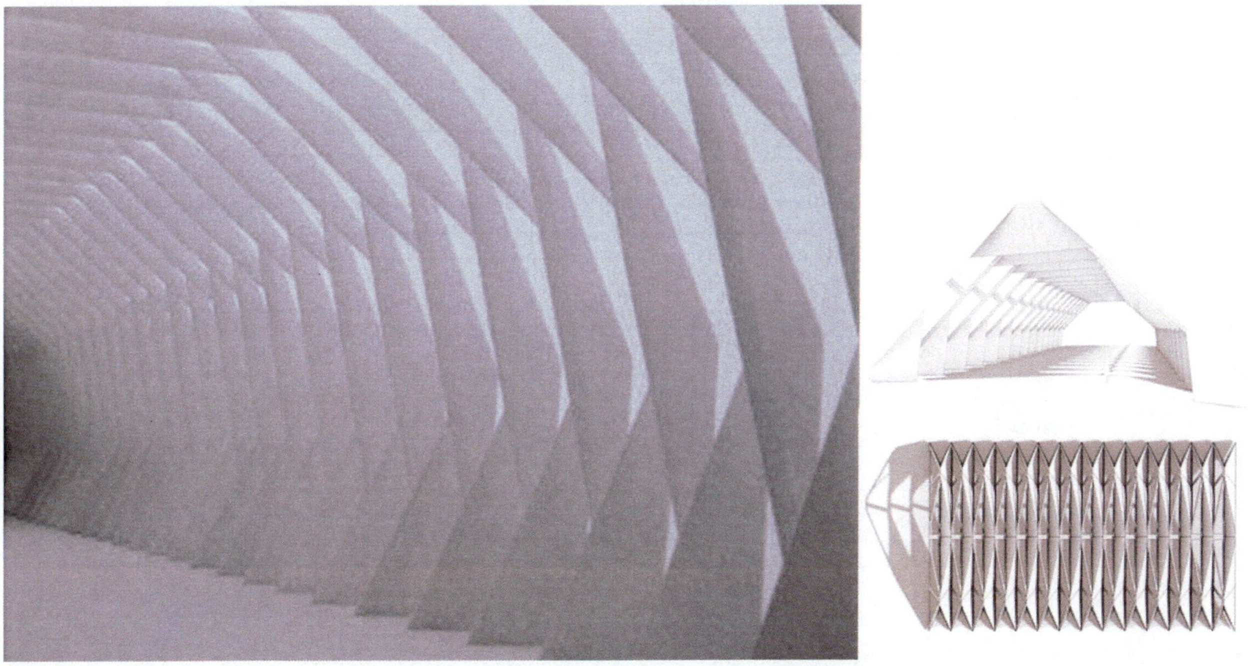

Fig. 8 Foldable hangar—3D simulation

morphological adaptation strategies, the biomimetic meme gave rise to a linear and dynamic structure, made of translucent and light materials in order to be (re)assembled and storage whenever needed (Fig. 13).

To materialize the linear and dynamic structure, the group chose the Skewed Yoshimura pattern that could be locked when touching the ground through the use of triangular faces (Fig. 14).

The five groups were able to present design proposals for structures based on rigid origami while exploring terrestrial vascular plants morphological strategies. From the study and abstraction of the meme pattern, material and performative features, all reached a global intention and some even considered the motion hypothesis and/or the mechanical implementation. Nevertheless, projects 02 and 03 are considered the less positive, based on the group's performance and work presentation.

Project 02 was not able to achieve the intended level of abstraction that could allow for the creation of a reliable biomimetic meme. And so, their design solution its technical and geometrical efficient, however lacks from biomimetic inspiration. In the case of project 03, the biomimetic meme was very strong but also very complex, and the students did not have enough time to conceive their parametric design solution.

On the overall, the final evaluation of the projects reflects a successful experience and a positive answer for the hypothesis of combining a biomimetic design process with a formalization through rigid origami geometry. At the end of the workshop, the students were invited to share their ideas about the workshop, the most positive and negative points on their view, which will be extremely useful for the implementation of future similar workshops.

Architecture Constraits	Cannopy Plants	Carnivorous	Hygroscopic	Packing Structures	Symmetry	BIOMIMETIC MEME
Expandable	X		X	X		X
Storage		X		X	X	X
Meme Strategies						
Dynamic	X	X	X			X
Static				X	X	
Meme Pattern Features						
Loose	X					
Monolayer					X	
Multilayer	X	X	X	X		X
Compactability		X	X	X		
Meme Material Features						
Translucent	X				X	
Opaque		X	X	X	X	X
Impermeable	X	X		X	X	X
Permeable			X		X	
Meme Performative Features						
Convection		X	X			X
3D Movement		X		X		X
Radial Movement	X					
Linear Movement			X			

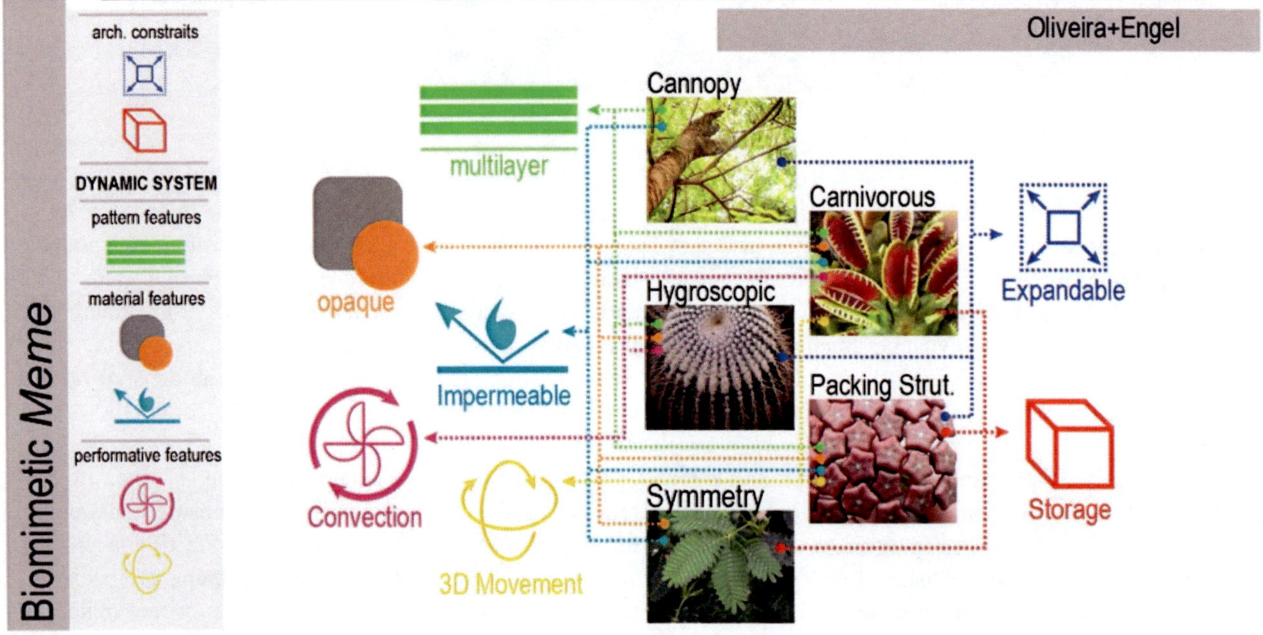

Fig. 9 Mini orchid yard—biomimetic meme construction

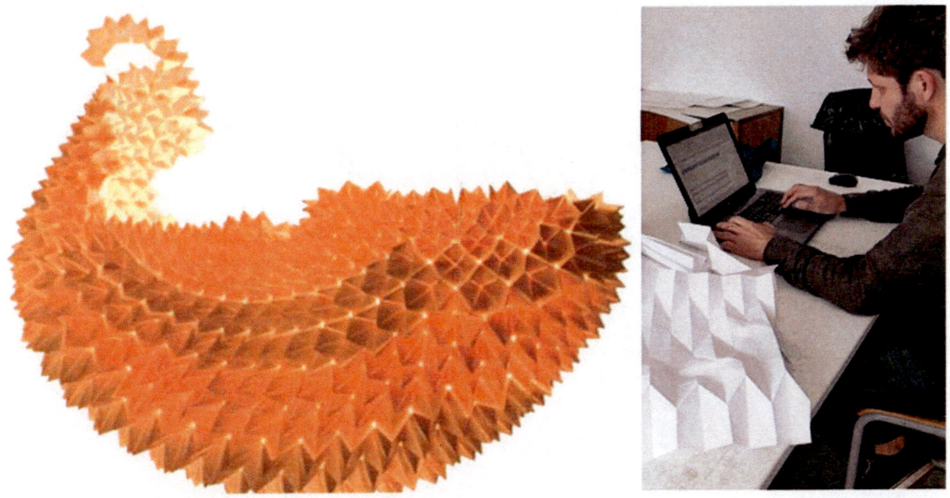

Fig. 10 Mini orchid yard—3D simulation and exploratory mockups

Architecture Constraits	Cannopy Plants	Hydrotropism	Tracheid	Phyllotaxis	BIOMIMETIC MEME
Collect Water		X	X		X
Provide shading	X			X	X
Meme Strategies					
Dynamic			X		
Static	X	X		X	X
Meme Pattern Features					
Radial	X	X			X
Monolayer				X	
Multilayer	X				
Symmetry	X		X		X
Meme Material Features					
Translucent	X				
Opaque	X	X	X		X
Flexible		X	X	X	X
Meme Performative Features					
Radial Movement	X	X			X
Unidirectional Movement			X	X	X
Tracking		X			

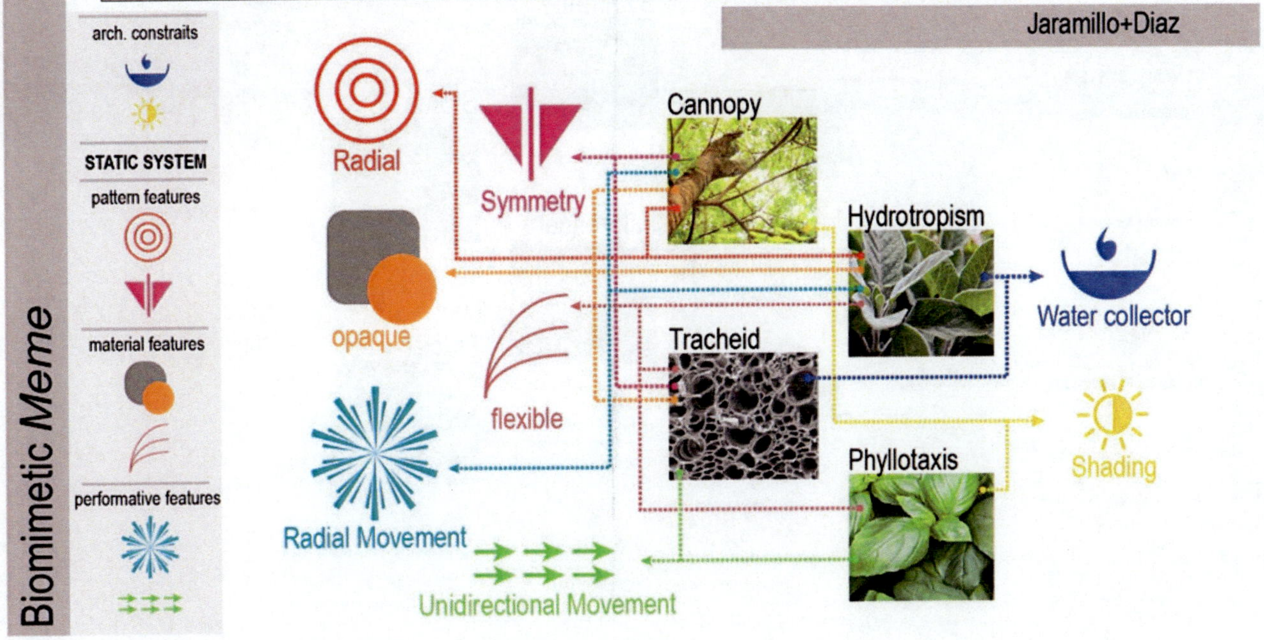

Fig. 11 Ibiscus water collector—biomimetic meme construction

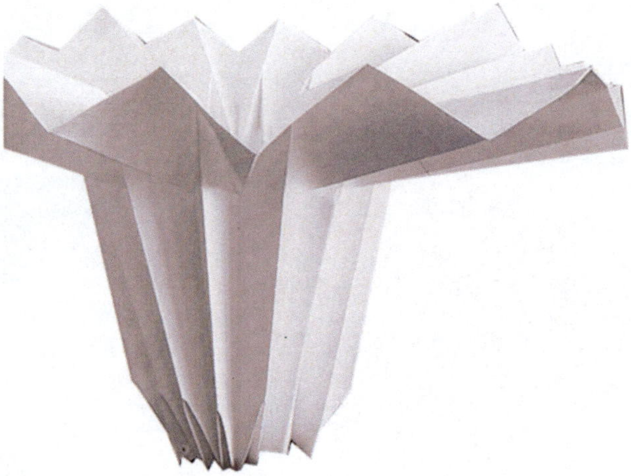

Fig. 12 Ibiscus water collector—exploratory mockup

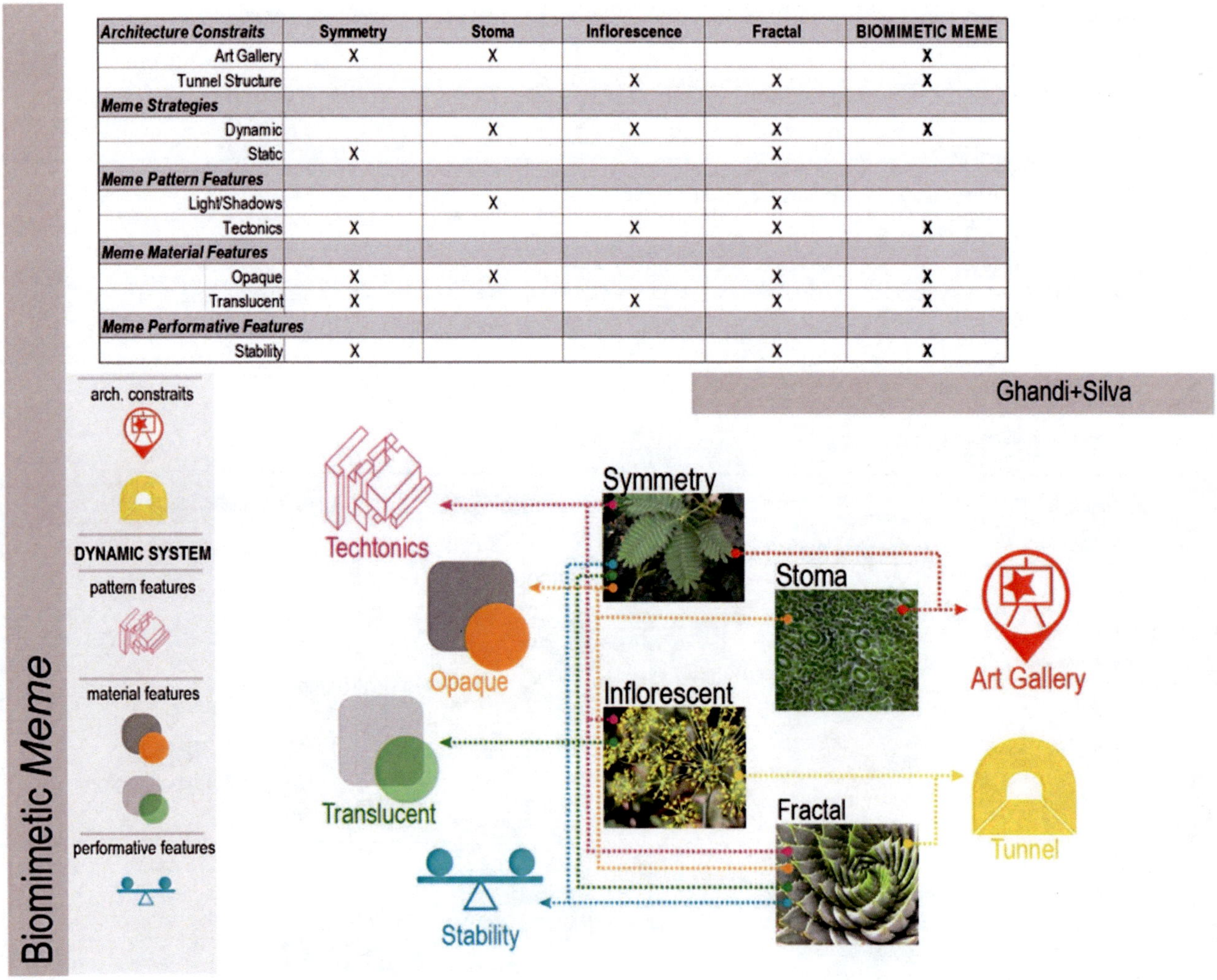

Architecture Constraits	Symmetry	Stoma	Inflorescence	Fractal	BIOMIMETIC MEME
Art Gallery	X	X			X
Tunnel Structure			X	X	X
Meme Strategies					
Dynamic		X	X	X	X
Static	X			X	
Meme Pattern Features					
Light/Shadows		X		X	
Tectonics	X		X	X	X
Meme Material Features					
Opaque	X	X		X	X
Translucent	X		X	X	X
Meme Performative Features					
Stability	X			X	X

Fig. 13 FourToSix walkway—biomimetic meme construction

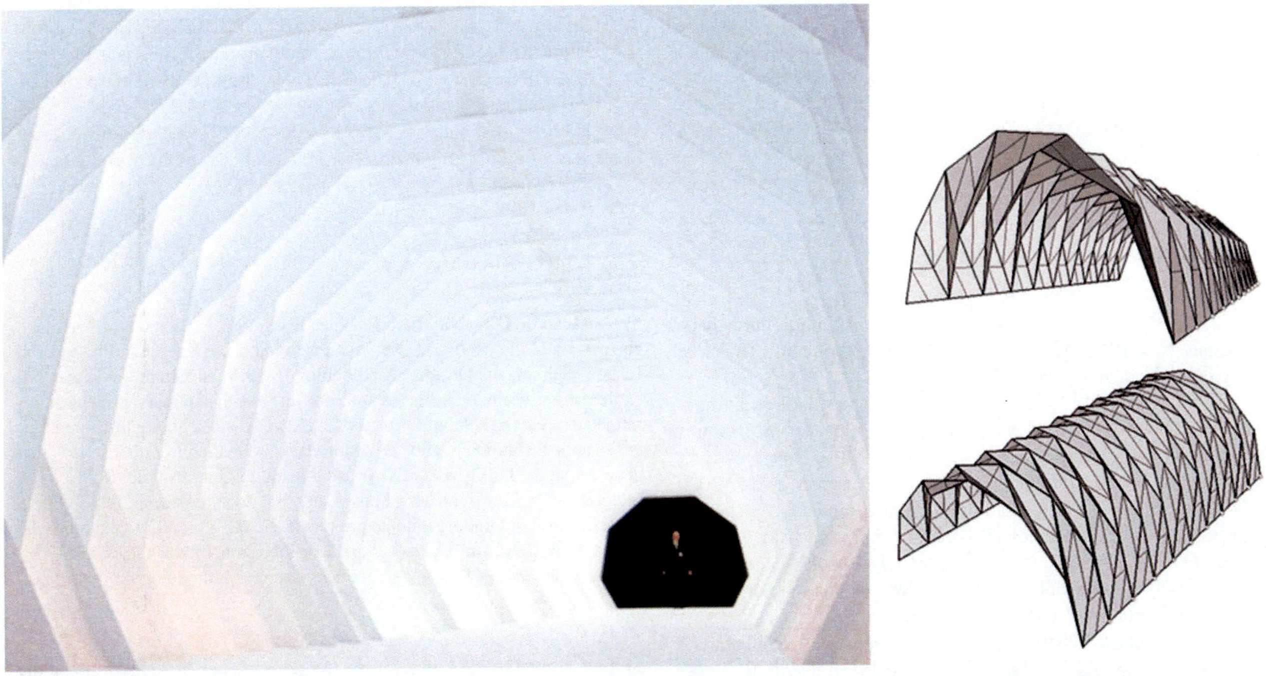

Fig. 14 FourToSix walkway—3D simulation

4 Conclusions

From the Nature Pleated Surfaces workshop was possible to retrieve valuable information, regarding the implemented biomimetic method, the ability of students to learn about origami geometry, but most of all regarding its application procedure.

During the *Architectural* domain, at the origami phase could be important to produce some practical exercises with the participants using paper and then implementing those same exercises on Grasshopper. Origami geometry is quite complex, and the exercises could help an origami beginner to engage easier on this type of geometry and understand in a tangible way the implications of the folds and their design. At the end, almost every group was able to understand the basic principles of origami and how to manipulate the pattern in order to achieve a determined geometric concept, especially with the paper models, but this was only possible with direct aid from the tutors.

During *Nature* domain, the link elements between the origami geometry and the terrestrial vascular plants that enable the search and exploration through the terrestrial plant's adaptation strategies, should be rooted and extracted from the architectural function definition, and not from the origami geometry. In fact, all the five groups rooted they *Biomeme* on the architectural functions. This was one of the most relevant discoveries during this process.

At the final *Artifact* domain, when conceiving the parametric design solution, some of the participants anchored their design solutions on the biomimetic material features. This revealed to be an essential factor for the parametric design construction, elevating the projects to a more realistic and concrete state.

However, the generation of the parametric grasshopper definitions was not so well achieved. The groups were not able to create their definitions from scratch, most used the given definitions and altered them to reach their intended geometry. It was clearly too ambitious to expect that the students would be able to learn and produce so much work on their own in only eight hours.

The workshop could be much more successful if it had at least three days (twenty-four hours) for the development of the four phases. For forward experiences, the first day should concern the first phase, that is origami geometry experimentation with paper and parameterization, engaging the participants into origami theorems and geometric rules. The second day should concern phase two, getting familiar with the biomimetic process, exploring terrestrial vascular plants morphological and even other adaptations strategies. And finally, the third day would be used for phases three and four, by devoting their time to the parametric and simulation design studio. This way is believed that the students could go deeper on every presented subject and develop the ability of being autonomous while using the presented concepts and thus reach more interesting and complete project proposals.

Acknowledgements The authors would like to thank the enthusiasm and passion with which all the participants have devoted to this experimental workshop. Thank you very much Bárbara Zandavali,

David Diaz, Débora Oliveira, Guilherme Giantini, Juliana Jaramillo, Mateus Pimentel, Mauro Couceiro, Mona Ghandi, Pedro Engel Victor Silva and Yağmur Yeni.The second author would also like to thank Fundação para a Ciência e Tecnologia (FCT) that funded the research through the scholarship with reference number SFRH/BD/100818/2014.

References

1. Capra, F. (1997). *Web of life*. London: HarperCollins Publishers.
2. Oxman, N. (2010). Material-based design computation (Doctoral Dissertation), Cambridge: MIT.
3. Gardiner, M. (2004). Oribotics [origin]. (Online) matthewgardiner.net. Accessed May 15, 2020. Available https://www.oribotics.com/art/Oribotics_origin.
4. Lumen, B. (2014). (Online) Foldhaus. Accessed May 15, 2020. Available https://www.foldhaus.com/blumen-lumen.
5. Britto, F. (2012). Botterfold/Taller Precise Parametrics Permutations, October 29, 2012. (Online) ArchDaily Brasil. Accessed May 15, 2020. Available https://www.archdaily.com.br/78228/botterfold-taller-precise-parametrics-permutations.
6. Origami Pavilion. (2016). (Online) Tal Friedman Architecture & Design. Accessed May 15, 2020. Available https://talfriedman.com/origami-pavilion.
7. Wang, L. (2017). Beautiful timber pavilion unfolds like origami, July 31, 2017. (Online) INHABITAT. Accessed May 15, 2020. Available https://inhabitat.com/beautiful-timber-pavilion-unfolds-like-origami/.
8. Look! Look! Look!/Studio Morison, 11 July 2017. (Online) ArchDaily. Accessed May 15, 2020. Available https://www.archdaily.com/875341/look-look-look-studio-morison/.
9. Oliveira, M. J., & Osório, F. (2017). KINE[SIS]TEM'17 From Nature to Architectural Matter International Conference Proceedings. DINÂMIA'CET-IUL, ISCTE-Instituto Universitário de Lisboa, Lisbon, Portugal.
10. Osório, F. C., Paio, A., & Oliveira, S. (2018). Origami tessellations: Folding algorithms from local to global. *Boletim da Aproged, 34,* 73–78.

Maria João de Oliveira Maria João de Oliveira is a Lisbon based Architect, co-founder of the MOFO, LDA Architecture and Design, and a Professor at Lisbon School of Design. Methodology, form, structure and material explorer, her main interests are focused in transdisciplinary projects, biomimetic research, shading structures, CAD/CAM Technologies, open science and technology and environmental sustainability, activities that develops at the DINÂMIA´CET at Iscte-Instituto Universitário de Lisboa as Integrated Ph.D. researcher. Master in Architecture since 2008, with a specialization in Digital Architecture (Iscte + FAUP), in 2020 she aquire the Ph.D. in Architecture degree with the conclusion of the thesis in Architecture "Bioshading Systems a Concept Design Methodology".

Filipa Crespo Osório Filipa Crespo Osório graduated in Architecture at the Department of Architecture of the University of Coimbra (DARQ-FCTUC) in 2006. After graduation she worked in architecture offices in Barcelona and Lisbon and in 2012/13 returned to the academic world to attend the Course on Advanced Studies in Digital Architecture at Iscte. In 2020 she concluded her Ph.D. thesis at Iscte, ISTAR with title "Origami Surfaces for Kinetic Architecture", with a grant funded by the Foundation for Science and Technology (Ref. SFRH/BD/100818/2014).

Building Circulation from a Syntactic Context in Relation to Indoor Thermal Environment

Leen Hiasat and Pedro Januário

Abstract

Building circulation layouts channel materials, energy, information, and people from one location to another inside a building and reflect the overall building spatial organization which is sought to affect indoor thermal conditions. This paper is a theoretical review that performs a cross-thinking of researches in the fields of building circulation design and indoor thermal conditions. The methodology consisted of three steps. First, the general theories in indoor circulation design are reviewed, and a syntactic analysis approach is chosen to understand the movement patterns. Second, the adaptive approach of naturally conditioned spaces along with its affecting factors is addressed as a primary evaluator of indoor thermal conditions. Third, previous studies of both fields are presented, to confirm the relation between the syntactic measures to social and economic measures, and the geometric measures to indoor environmental measures. Based on the discussed concepts in both fields, a new thinking approach is proposed, relating syntactic measures to indoor thermal measures. This approach offers an early indoor thermal feedback on primary design decisions of the building layout. Further experimental work with quantitative measures is needed to confirm this proposition.

Keywords

Building circulation • Syntactic approach • Indoor thermal environment • Space syntax • Configuration networks

L. Hiasat (✉) · P. Januário
Lisbon School of Architecture, CIAUD Research Center, Universidade de Lisboa, Lisbon, Portugal
e-mail: lhiasat@yahoo.com

1 Introduction

> Just as blood vessels of the vascular network distribute energy and materials to cells in an organism, road networks distribute energy, materials, and people in an urban area
>
> Samaniego and Moses [41]

Buildings are constructed to provide people with a space that protects them from extreme outdoor conditions; a good building should provide a comfortable environment for its occupants. This includes thermal, visual, and acoustic comfort as well as indoor air quality ([40], p. 183).

Focusing on thermal and lighting needs, with the nowadays concerns of limits of non-renewable energy sources and their environmental impact, research has been directed toward optimizing design parameters along with construction techniques and material selection to reduce energy consumption.

However, the primary design decisions taken in early design stages which define buildings' spatial structures should also be given more attention as they cannot be changed later on in the life cycle of the building.

These decisions are mainly related to building circulation including corridors, ramps, elevators, stairways, and other physical paths of movement. As these spaces are found to be a key organizing mechanism of layouts, they connect interior spaces, establish the building's exterior connection, and accordingly reflect the overall spatial organization of the building. They also make up a great amount of the overall building's area, and they are the spaces assigned mainly for the movement of people through, around, and between the parts of the built environment.

Therefore, movement corridors have always been optimized according to several criteria with considerations for privacy, safety, the efficiency of use, environmental aspects, and a lot more [25, 26, 32].

Even with the huge amount of researches done to confirm the effect of improved building fabric over the energy consumption [2, 27, 35], the amount of energy required for

S. Eloy et al. (eds.), *Formal Methods in Architecture*, Advances in Science, Technology & Innovation,
https://doi.org/10.1007/978-3-030-57509-0_11

heating and cooling of buildings is still significant. Thus, we need to start new approaches of thinking that would produce ideal conditions of indoor thermal comfort leading to further reductions of used energy.

The present research will discuss a new approach that suggests the existence of a connection between syntactic measures and indoor thermal environment and to discuss whether energy sources represented in heat, air, and humidity move and accumulate inside buildings just like human beings do.

This approach will consider design decisions related to the building's spatial structure formed out through building circulation layouts. These aspects are believed to be valuable issues to be evaluated in the early design stages to ensure a successful and optimized building design.

2 Definitions and Scope

2.1 Building Circulation

Spaces used for movement form an integral part of any building organization and occupy a significant amount of the volume of a building.

The circulation system is often referred to as a "skeleton" that forms the supporting structure of the building [23]. It is also seen as a dedicated area for a movement that connects various parts of a building or settlement into a coherent network of spaces, creating flows, linking activities together, and making the space continuous [39].

The idea of movement through space is one of the central themes in architectural theory, design, and practice. A fundamental issue in building circulation is the specific paths that people take when moving from one space to another. Therefore, the sequence in which the units are connected constitutes an important design aspect.

Circulation Design. Designing floor plans and circulation systems for efficient navigation and orientation is considered a complex problem in architectural design. Finding a functional arrangement of elements in a plan is considered a complex activity because it considers multiple criteria [47].

Contemporary existing methods for circulation planning are based on analyses of the human body dimensions and different user groups, which set architectural standards for general building dimensions including corridors, stairs, landings, and others [32].

Good circulation inside buildings and urban environment requires also many design criteria to be met; such as handling anticipated flows, providing access to desired spaces, and sometimes optimizing the length of journeys between connected spaces for occupational and construction costs [39].

Building circulation is used in spatial programs as well to present the designer's intents and to foresee the user's circulation through the building. They also have certain codes regarding security, safety, and performance that they need to comply with [25]. Consequently, many methods were developed to measure distances of circulation areas in buildings. And some literature suggested codes and guidelines for designing circulation systems [26].

Nevertheless, Ireland [21] explained that circulation design is a neglected part of the design process for it is generally considered as a percentage value of the total floor area; a standard value depending on the building's type. He argues that circulation is not merely an optimization, where it is thought that the most efficient is to organize rooms with minimal circulation space. Rather, they should be thought of as positive elements that affect our perception of the building's form and space. That is why, this approach here will shed light on another aspect of circulation design related to its syntactic context to combine a semantic and behavioral perception of these spaces.

Circulation from a Syntactic Context. Many definitions of circulation have contained the concept of accessibility as a base characteristic for determining circulation spaces; that circulation is what offers accessibility and connection to other spaces of an organization [20, 24, 29, 39, 42].

Hence, many researchers have talked about the role of architectural characteristics in determining circulation patterns; how movement can be interpreted as form and form as a structure of the movement, and how the patterns of space affect users' movement [12, 31, 43]; Cassirer, E., 1955 *apud* [36]. Peponis has also explained circulation as "a pattern of discrete transitions from one space to another" [36]; he implied that the "disposition and arrangement of boundaries" is what structures space and organizes potential movement.

Then, upon these patterns of accessibility, a certain amount of control is often required [3]; this control differentiates spaces as integrated and segregated according to the amount of connectivity they acquire, which leads into a hierarchy in organizations [18]. For example, Willoughby had defined authority, workflow, and social contact as determinants of control in categorizing different hierarchies for office building layouts [45].

In buildings, where a large amount of circulation control is applied (e.g., hospitals and schools) or in buildings where circulation constitutes the major part of the venue (e.g., airports and train stations), the design can be seen to reflect significantly on the building's organization. In these buildings, the circulation patterns constitute the main factor in producing its spatial organization.

Marshall had distinguished between three types of system representations: the constitutional which is only hierarchal; the configurational which is topological; and the compositional which is absolute geometric with distances [28]. The configurational form is used to show topological aspects that relate to the element's abstract geometry and relationships between them, more than other properties that relate to exact locations and metric distances. They represent layouts conceptually as a pattern of interconnecting lines and points, in which points represent locations and lines represent movement corridors.

This configurational approach was considered as the basis for modeling the network of spaces in the space syntax theory. According to the geometric representations, algorithms are used to produce abstract diagrams based on spatial adjacencies. Then, through mathematical calculations, syntactic analyses can be performed on these graphs to produce syntactic measures [1].

The calculated syntactic measures can include many; however, according to their definitions, they might have interesting relations to a whole different set of measures:

- Connectivity; counts the directly connected spaces; which define the spaces that have direct effects on each other.
- Integration; defines segregated from integrated spaces; which indicates the amount of movement in these spaces.
- Control; it counts the amount of control a space has over others; which determines to what extent space can change the amount of movement happening in others.

This paper focuses on architectural configuration networks represented through the building's circulation layouts, as they are associated with conveying materials, energy, information, and people from one location to another inside a building.

2.2 Building Indoor Environment

Nowadays, the indoor environment is an important factor in building design, as it affects human health, comfort, and performance. It is also directly related to minimizing energy and consumption of natural resources.

Buildings are constructed to provide people with a space that protects them from (extreme) outdoor conditions. The successful indoor environment is dependent on an understanding of the multiple parameters that can be included in building design.

Indoor Environmental Definitions. Many researches were studied to differentiate the terms used to define the indoor environmental conditions based on diverse factors affecting them, among these definitions are: **environmental performance** with factors of costs of components, energy and water [16]; **environmental comfort** with factors of thermal, acoustic, natural lighting, and functionality [10, 14]; **indoor environmental quality** with factors of air quality, air temperature, relative humidity, and acoustics [4, 40]; **thermal performance** concerned with the thermal insulation criteria and energy use [37]; and finally, **thermal environment** or **thermal comfort**, which are mainly concerned with factors of air temperature, mean radiant temperatures, air relative humidity, air velocity along with people clothing and metabolic rates [11, 13, 33, 38, 48].

In the research of Da Graça et al. [14], several design solutions were suggested to have an important role in improving the different building's environmental conditions. These solutions included form, orientation, location of functions, and openings as well as the appropriate choice of site. While in Ali et al. [4] study, different strategies of passive design techniques were proposed for the same purpose, among these were orientation, building shape, and ventilation strategies. Some of these factors imply the configuration aspect.

Indoor Environment from a Thermal Approach. When talking about the indoor environmental quality in the ASHRAE standards, many factors are included, such as thermal comfort, air quality, contaminants, odors, acoustics, and illumination [6, 7]. All of these factors affect the acceptable indoor environment. However, each of these has been tackled in separate publications and for that they have their specific standards, guidelines, and measuring factors.

ASHRAE standards [8] define the thermal environment as "*The conditions that affect a person's heat loss.*" While ISO 7730 [22] defines it as "*A condition obtained when the internal heat production in the body is equal to the loss of heat to the environment.*"

These definitions imply dependability on thermal interactions between the human body and its surrounding environment [11], which affects occupant's thermal satisfaction. This satisfaction has been defined in these two standards through a range of comfort zone defined by ASHRAE as "*Those combinations of air temperature, mean radiant temperature, and humidity that are predicted to be an acceptable thermal environment at particular values of airspeed, metabolic rate, and clothing insulation*" [8].

The definition indicates that what helps people in determining their comfort or discomfort status comes directly from temperature and moisture sensations of the skin, as well as behaviors engaged to regulate this status [6].

On the other hand, the thermal environment approach is considered the basis for evaluating the indoor thermal conditions of free-running buildings. These basic buildings were defined as naturally ventilated buildings that do not have artificial heating or cooling and those which their indoor conditions will follow those outdoors, but with a modified extent [33, 38]. For these buildings, the different worldwide

Table 1 Comparison of thermal comfort factors according to different determining methods ([7]—edited by authors)

Comfort zone method	Airspeed (m/s)	Humidity ratio (kgH_2O/kg dry air)	Metabolic rate (met)	Clothing level (clo)	Temperature ranges (°C)
Graphical method	<0.2	<0.012	1.0–1.3	0.5–1.0	22–28 °C
Analytical method	<0.2	Any	1.0–2.0	0–1.5	Calculated using the ASHRAE thermal comfort tool
Elevated airspeed method	$0.2 < x < 0.8$	Any	1.0–2.0	0–1.5	Up to 3° increase on graphical
Adaptive method	Any	Any	1.0–1.3	Free to adjust	20–31 °C

standards have produced a specified method called the adaptive approach [8, 22]. This approach applies to naturally conditioned spaces as the increased air velocities can help occupants in adapting to warmer environments which widens the comfort zone including temperature, airspeed, and humidity levels. Table 1 shows a comparison of methods.

Finally, the cooling and heating load calculations established by the standards require the use of these basic measures to calculate heat balance and energy loads accordingly [6].

Consequently, the factors that will be addressed to evaluate indoor thermal conditions are six: air temperature, radiant temperature, airspeed, humidity, metabolic rate, and clothing insulation. The first four are determined through indoor environmental conditions, while the last two depend on the occupants.

3 Previous Studies

3.1 Syntactic Circulation Studies

Syntactic measures have been correlated with different other measures in many studies to analyze different projects and reflect upon the design–decision process.

In a study done by Narvaez et al. [30], the syntactic approach was used to decode accessibility of the city's morphology and relate it to different economic variables of taxes and housing typology prices. The purpose was to anticipate where the economic developments are headed and hence direct the investments in the right direction. Another study, in Amman–Jordan, attempted to understand the social, cultural, and economic effects on changing the spatial structure of public spaces [5]. Through measuring retail income, it was concluded that higher physical connectivity relates to the higher social and economic growth of the area, therefore affecting the decisions taken for future urban developments of public spaces.

Furthermore, various studies had evolved in generating circulation typologies within buildings. In a study done by Natapov et al. [31], the syntactic analysis was conducted to redesign a circulation pattern, while preserving the spatial organization of an existing building, for the aim of creating easier wayfinding. While Nourian et al. [34] have developed a computational syntactic design process, using the syntactic analyses in a parametric environment to investigate the possibility of reaching concrete plan layouts from an abstract connectivity pattern, this parametric approach helped in analyzing the effect of preliminary decisions made by designers on the likely spatial performance of buildings.

Another study that basically attempted to analyze the effect of a dual transportation system of a small city on social exclusion and transportation measures had also managed to relate "ringy" spatial structures to reduced CO_2 emissions in that city [43]. Interestingly, this study was unique in relating syntactic measures to urban environmental ones.

3.2 Circulation and Environmental Studies

A lot of studies have addressed the effects of building geometry over energy consumption and thermal performance. Depecker et al. [15] had tried to relate the heating consumption of the buildings with their shape. They characterized the shape of the building by the shape coefficient which is defined as the ratio between the external skin surfaces and the inner volume of the building. The results showed that the energetic consumption is inversely proportional to the compactness of the building (lower coefficient value) in cold climates but is not applicable in mild climates.

Using the same concept of compactness on different morphologies along with parameters of glazing percentage and distribution, and orientation, a study was performed to evaluate the degree to which a building design meets the relevant thermal insulation criteria [37]. And with the same factors, two other studies managed to produce a calculation method that can predict the building's annual cooling and total energy use based on the shape of the building [2, 35].

Roulet [40] in his definition of a comfortable living environment discussed many factors including thermal insulation, ventilation rate, temperature control, daylighting, and spatial arrangements inside a building, while Zomorodian and Nasrollahi [48] had analyzed the energy demands and investigated the architectural design parameters of orientation, window-to-wall ratio, sun shading, building shape, and space organization.

Another interesting study was done by Dobbelsteen et al. [16], where the focus was on the different shapes and stacking typologies to assess the environmental life cycle of office buildings with equal spatial demands. It had led to interesting relationships between building geometry and its environmental performance.

While the most relevant study to the scope of this approach is a research by Dogan et al. [17] that had managed to prove the significance of interior subdivisions over the energy use intensity of a building. The results showed a 15% difference between a paradigm that considers a core perimeter scheme and another that considers the different interior subdivisions in a scheme. Which confirms that configuration networks constituted through building circulation impact indoor environments.

4 The New Approach

4.1 Discussion

When Chris Webster had suggested evaluating an urban form through the concept of accessibility, it became easier to relate form to specific scientific parameters, because accessibility is easier to be measured [44]. His way of measuring accessibility was merely geometrically represented through distance and time between locations.

However, in the Natural movement theory as presented by Hillier; the configuration of space is argued to have a factor in the choice of destinations, which is thought to be independent of other metric properties [20]. Therefore, accessibility was measured from a topological aspect.

Accordingly, in the space syntax theory, it was argued that movement is the fundamental part over which buildings are generated and configured. Any layout was defined with two essential components; movement and occupation of space, in which permeability between spaces is what makes a functional layout [18, 19]. Thus, the syntactic approach suggested analyzing spatial settings from a topological nature instead of a geometrical nature. It is sought that density and volume of movement in configuration networks are intrinsically associated with space's locations, rather than the geometric length between them [9]. Hence cost, time, and length of movement will not be as important as locations and connections themselves.

Kanyar had argued that the physical accessibility pattern of a building implies how potential movement is distributed and which local parts of a configuration may have a stronger effect on potential movement [24].

And since building circulation is the spaces assigned mainly for the movement of people through, around, and between the parts of the built environment, they are the main reflector of a building's spatial configuration and the movement shaper. These spaces make up a great amount of the overall building's area and have a greater connection to the outer environment than other interior spaces [38]. They are also the spaces for moving matter and energy around the building, and hence, they are assumed to have a great effect over the indoor environment.

Bernardi and Kowaltowski [10] have mentioned that factors of indoor thermal conditions are usually affected by the organization of urban elements, architectural design, and user behavior.

The previously mentioned researches in building's indoor environments have confirmed a relationship between the building's shape and circulation spaces measured in a geometrical aspect to the indoor thermal environment.

Space syntax theory had also investigated relationships between spatial layouts and a range of social, economic, and environmental aspects such as density, land use, land value, crime, and other environmental aspects such as CO_2 emissions, which affect and are affected by the pattern of flow for people and objects [43]. Later on, some of these investigations have evolved into tools based on the theories to generate models; e.g., urban movement model, land use model, and crime model.

Consequently, if the human movement was spatially guided by topological rather than metric factors and were related to different social, economic, and environmental measures, then why not the indoor thermal environment as well.

4.2 Concepts

The new approach suggests that applying certain syntactic measures in building circulation layouts can help in predicting its indoor thermal environment in the early design stages. Since circulation measures represent matter movement through the building, it might as well relate to energy movement. Just like circulation layouts determine how likely it is for a person to reach a space inside a building, it might as well determine how likely it is for solar heat, cold, or hot breezes, and humidity to reach a space inside the building.

Spatial structures are one of the primary design decisions taken in the early design stages of a project. They are usually expressed in a form of abstract graphs that reflect the buildings discrete spaces and the different movement options according to the specified program but ignore other architectural articulations. These structures cannot be changed later on in the life cycle of the building.

The space syntax approach allows performing syntactic analyses on these graphs and producing syntactic measures that might be correlated with measures of the indoor thermal environment. And performing this analysis can help the designers in understanding the pattern consequences of their design moves, the relation of new and existing patterns, and in general the relation of parts and wholes in buildings and cities [18].

To formulate the syntactic-thermal relation, different cases of spatial structures should be simulated and spatially analyzed against indoor thermal environment measures. A general comparison should be made between the results, reading the measures, understanding them, and finding similarities and differences. Then, establish a cause–effect relationship through correlation and regression analyses. And a multi-objective optimization method is suggested, as the one utilized in a research by Zhang et al. [46]. Using this method in a parametric environment offers the ability to control the two sets of variables to be able to find the optimal trade-offs between the options and to understand the effects the variables have on each other.

If these abstract graphs with minimal design details were able to relate to indoor thermal conditions in a design tool that would offer real-time feedback of every design move done by the architects, it would improve greatly the architectural design practice along with the indoor thermal environment.

5 Conclusions and Future Works

In this paper, studies and followed global procedures that related building circulation to the indoor environment were explored. Their approaches were analyzed and compared to establish a ground base for the new approach proposed in this paper.

The syntactic approach of analyzing spatial structures has proved to relate to different social, economic, and environmental aspects of human life. It also showed more significance regarding movement aspects than geometric properties.

The literature review confirmed a relationship between circulation measures and indoor environmental conditions from a geometric approach. Nevertheless, it might as well still exist from a syntactic approach.

The proposed approach suggests a method on how to realize and formulate the syntactic-thermal relation. Further empirical and experimental work following the method explained will be done in future works to validate the new approach. Out of this, a mathematical relationship can be calculated, which allows the spatial measures to be used predictively of the thermal environment.

This approach can also be developed in the future into a thermal model that can be applied on different layouts of buildings under variable climatic conditions.

References

1. Al_Sayed, K. T. (2014). *Space syntax methodology* (4th ed.). London: Bartlett School of Architecture, UCL.
2. AlAnzi, A., Seo, D., & Krarti, M. (2009). Impact of building shape on thermal performance of office buildings in Kuwait. *Energy Conversion and Management, 50,* 822–828.
3. Alexander, C. (1966). A city is not a tree. In 206, ed. *Design.* London: Council of Industrial Design, pp. 1–17.
4. Ali, H., Almomani, H., & Hindiyeh, M. (2009). Evaluating indoor environmental quality of public school buildings in Jordan. *Indoor and Built Environment, 18*(1), 66–76.
5. Aljafari, M. (2014). *3.4.3 Emerging public spaces in the city of Amman, Jordan: An analysis of everyday life practices.* Dortmund, Germany: Thesis submitted to the Faculty of Spatial Planning at Dortmund University for the award of the degree of Dr. Ing..
6. ASHRAE. (2009). *Handbook fundamentals.* Atlanta: American Society of Heating, Refrigerating and Air-Conditioning Engineers Inc.
7. ASHRAE. (2016). *Standard 55 user's manual,* Atlanta: American Society for Heating, Refrigerating and Air-Conditioning Engineers, Inc.
8. ASHRAE. (2017). *ANSI/ASHRAE standard 55-2017—Thermal environment conditions for human occupancy.* Atlanta: American Society for Heating, Refrigerating and Air-Conditioning Engineers, Inc.
9. Batty, M., & Rana, S. (2004). The automatic definition and generation of axial lines and axial maps. *Environment and Planning B: Planning and Design, 31,* 615–640.
10. Bernardi, N., & Kowaltowski, D. (2006). Environmental comfort in school buildings: A case study of awareness and participation of users. *Environment and Behavior, 38*(2), 155–172.
11. Bessoudo, M., Tzempelikos, A., Athienitis, A., & Zmeureanu, R. (2010). Indoor thermal environmental conditions near glazed facades with shading devices—Part I: experiments and building thermal model. *Building and Environment, 45,* 2506–2516.
12. Bitgood, S. (1989). Chapter 17: Problems in visitor orientation and circulation. In *Visitor Studies,* 1(1), pp. pp. 155–170.
13. Corgnati, S. P., Filippi, M., & Viazzo, S. (2007). Perception of the thermal environment in high school and university classrooms: Subjective preferences and thermal comfort. *Building and Environment, 42,* 951–959.
14. Da Graça, V., Kowaltowski, D., & Petreche, J. (2007). An evaluation method for school building design at the preliminary phase with optimisation of aspects of environmental comfort for the school system of the State São Paulo in Brazil. *Building and Environment, 42,* 984–999.
15. Depecker, P., Menezo, C., Virgone, J., & Lepers, S. (2001). Design of buildings shape and energetic consumption. *Building and Environment, 36,* 627–635.
16. Dobbelsteen, A., Thijissen, S., Colaleo, V., & Metz, T. (2007). Ecology of the building geometry—Environmental performance of different building shapes. In *Proceedings of the CIB World Building Congress.* Cape Town, South Africa.
17. Dogan, T., Saratsis, E., & Reinhart, C. (2015). *The optimization potential of floor-plan typologies in early design energy modeling.* Hyderabad: International Building Performance Simulation Association (IBPSA).
18. Hillier, B. (1996). *Space is the machine.* London: University of Cambridge.
19. Hillier, B., & Hanson, J. (1984). *The social logic of space.* Cambridge: Cambridge University Press.
20. Hillier, B., et al. (1993). Natural movement: or configuration and attraction in urban pedestrian movement. *Environment & Planning B: Planning & Design,* 29–66.

21. Ireland, T. (2008). *Sniffing space*. Milan, 11th Generative Art Conference.
22. ISO 7730. (2005). *ISO 7730 international standard ergonomics of the thermal environment—Analytical determination and interpretation of thermal comfort using calculation of the PMV and PPD indices and local thermal comfort criteria*. Geneva: International Organization for Standardization.
23. Jiang, B., & Liu, X. (2010). Automatic generation of the axial lines of urban environments to capture what we perceive. *International Journal of Geographical Information Science, 24*(4), 545–558.
24. Kaynar, I. (2005). *Visibility, movement paths and preferences in open plan museums: An observational and descriptive study of the Ann Arbor hands-on Museum*. Delft: Techne Press.
25. Lee, J-k, et al. (2010). Computing walking distances within buildings using the universal circulation network. *Environment and Planning B: Planning and Design, 37*, 628–645.
26. Llewelyn-Davies, (2000). *Urban Design compendium 1: Urban design principles*. London: English Partnerships, The Housing Corporation.
27. Marks, W. (1997). Multicriteria optimisation of shape of energy-saving buildings. *Building and Environment, 32*(4), 331–339.
28. Marshall, S. (2005). *Streets and patterns*. London: Spon Press Taylor & Francis Group.
29. Mitchell, W. J. (2005). E-topia: Information and communication technologies and the transformation of urban life. In G. C. Manuel Castells (Ed.), *The network society from knowledge to policy* (pp. 325–333). Washington DC: Center for Transatlantic Relations.
30. Narvaez, L., Penn, A. & Griffiths, S. (2012). Space syntax economics: Decoding accessibility using property value and housing price in Cardiff, Wales. In *Proceedings of the Eighth International Space Syntax Symposium*. Santiago de Chile.
31. Natapov, A., Kuliga, S., Dalton, R. C., & Hölscher, C. (2015). Building circulation typology and space syntax predictive measures. In *Proceedings of the 10th Space Syntax Symposium (SSS10)*. London.
32. Neufert, E., & Neufert, P. (2000). *Architects' data* (3rd ed.). Hoboken: Blackwell Sciences.
33. Nicol, F., & Pagliano, L. (2007). Allowing for thermal comfort in free-running buildings in the new European Standard EN15251. In *Proceedings of the 2nd PALENC Conference and 28th AIVC Conference on Building Low Energy Cooling and Advanced Ventilation Technologies in the 21st Century*. Crete Island, Greece.
34. Nourian, P., Rezvani, S., & Sariyildiz, S. (2013). A syntactic architectural design methodology: Integrating real-time space syntax analysis in a configurative architectural design process. In *Proceedings of the Ninth International Space Syntax Symposium*. Seoul.
35. Ourghi, R., Al-Anzi, A., & Krarti, M. (2007). A simplified analysis method to predict the impact of shape on annual energy use for office buildings. *Energy Conversion and Management, 48*, 300–305.
36. Peponis, J., et al. (1997). *On the description of shape and spatial configuration inside buildings: Convex partitions and their local properties (1)*. London: Space Syntax First International Symposium.
37. Pessenlehner, W., & Mahdavi, A. (2003). *Building morphology, transparence, and energy performance*. In *Proceedings of the Eighth International IBPSA Conference*. Eindhoven, Netherlands.
38. Pitts, A. (2013). Thermal comfort in transition spaces. *Buildings, 3*, 122–142.
39. Puusepp, R. (2011). *Generating circulation diagrams for architecture and urban design using multi-agent systems*. University of East London ed. London: A thesis submitted in partial fulfillment of the requirements of the School of Architecture and Visual Arts, University of East London for the degree of Doctor of Philosophy.
40. Roulet, C.-A. (2001). Indoor environment quality in buildings and its impact on outdoor environment. *Energy and Buildings, 33*, 183–191.
41. Samaniego, H., & Moses, M. E. (2008). Cities as organisms: Allometric scaling of urban road networks. *Journal of Transport and Land Use, 1*(1), 21–39.
42. Twombly, R. (2003). *Louis Kahn, essential texts* (1st ed.). New York: W. W. Norton & Company.
43. Vaughan, L. (2007). The spatial syntax of urban segregation. *Progress in Planning, 67*, 205–294.
44. Webster, C. (2009). *Pricing accessibility: Urban form, missing markets and the efficient allocation of property rights over shared urban resources*. Stockholm, Sweden, Urban Buzz Position Paper.
45. Willoughby, T. M. (1975). Building forms and circulation patterns. *Environment and Planning B, 2*, 59–87.
46. Zhang, A., et al. (2017). Optimization of thermal and daylight performance of school buildings based on a multi-objective genetic algorithm in the cold climate of China. *Energy and Buildings, 139*, 371–384.
47. Zimring, C., & Craing, D. L. (2001). Chapter 7: Defining design between domains: An argument for design research a la Carte. In C. Eastman, W. Newstetter, M. McCracken (Eds.), *Design knowing and learning: Cognition in design education* (pp. 125–146). Oxford: Elsevier Science Ltd.
48. Zomorodian, Z., & Nasrollahi, F. (2013). Architectural design optimization of school buildings for reduction of energy demand in hot and dry climates of Iran. *International Journal of Architectural Engineering & Urban Planning, 23*, 41–50.

Leen Hiasat Ph.D. candidate from the Faculty of Architecture at the University of Lisbon, searching in the fields of circulation design from a syntactic approach and the indoor environment, a former lecturer at Al-Zaytoonah University of Jordan, tutoring in architectural fields, an architect who has worked for consulting offices providing complete architectural services for different projects throughout Jordan, obtained a Master's degree from the University of Jordan for the thesis entitled Syntactic Analysis of Urban Spatial Networks' Influence on Land Values—Bridges and Tunnels of West Amman, and obtained a Bachelor's degree in Architecture from the University of Jordan.

Pedro Januário Pedro Miguel Gomes Januário, graduated in Architecture (1994) from the Faculty of Architecture of the Technical University of Lisbon. In 1994, he was invited to teach at FAUTL in the field of Visual Communication, having taught CAD disciplines at the same time that he developed his activity as a freelance architect (Formosinho Sánchez; Arqui III; Sua Kay; CESUR; World Monument Found; etc.). As a specialist in the CAD area, he was Computer Consultant at Caixade Imagens and Luso Cuanza, being certified by Autodesk. In 2008, he completed his PhD at Escuela Técnica Superior de Arquitectura at the Polytechnic University of Madrid with a theme related to the Tagus Opera. At FAUTL, he held several positions, namely as Coordinator of Pedagogical Internships at FAUTL 2006–2007, Vice-President of the Pedagogical Council between 2009–2012, and currently as President of this body since February 2012; he belongs to the Senate of the University of Lisbon and collaborated in the creation of the Scenography Degree. He is currently co-coordinator of the specialization course in Paper Architectures and Parametric Design and responsible for three R&D projects at CIAUD, collaborating on two others. He has several articles published in international and national magazines and has presented communications at international congresses. He has several articles published in international and national magazines and has presented communications at international congresses and at CIAUD.

What About if Buildings Respond to My Mood?

Luísa Almeida, Sara Eloy, and Ana de Almeida

Abstract

This work analyzes the possibilities of interaction between the built environment and its users, focused on the responsiveness of the first to the emotions of the latter. Transforming the built environment according to the mood, feelings, and emotions of users, moment by moment, is discussed and analyzed. The main goal of this research is to define a responsive model by which the built environment can respond in a personalized way to the users' emotions. For such, computational technical issues, building construction elements and users' interaction are identified and analyzed. Case studies where occurs an interaction between the physical space and users are presented. We define a model for an architecture that is responsive to the user's emotions assuming the individual at one end and the space at the other. The interaction between both ends takes place according to intermediate steps: the collection of data, the recognition of emotion, and the execution of the action that responds to the detected emotion. As this work focuses on an innovative and disruptive aspect of the built environment, the recognition of the new difficulties and related ethical issues are discussed.

Keywords

Responsive architecture • Interaction • Users' responses • User's emotions • Ethics

L. Almeida · S. Eloy (✉) · A. de Almeida
Instituto Universitário de Lisboa (ISCTE-IUL), ISTAR, Av. Forças Armadas, 1649-026 Lisbon, Portugal
e-mail: sara.eloy@iscte-iul.pt

L. Almeida
e-mail: lbarreiros01@gmail.com

A. de Almeida
e-mail: ana.almeida@iscte-iul.pt

A. de Almeida
Centre for Informatics and Systems, University of Coimbra, Coimbra, Portugal

1 Introduction

Current technology allows us to build in such a way that buildings can be increasingly adaptable and interactive with their users and assume innovative forms of response. In this line of thought, the conception of the built space has been modified along with the technological evolution that allows the space to be morphologically transformed and reactive to its occupants.

The flexibility of the built limits is no longer virtual but is now physical. When, in the twentieth century, the walls were freed from their supporting function and could be replaced by glass panels, we witnessed the dilution between interior and exterior space and between different interior spaces. Now, we want to take the next step, in which "the boxes" that surround spaces can be transformed.

For a building to be truly able to adapt and respond effectively to the occupants, it is necessary to incorporate technological and spatial systems that allow an effective interaction between the building and the occupant. These systems must include a group of sensors that enable the interpretations of events to which an automatic system may respond, coupled with a set of actuators able to act upon the environment. A system that responds to environmental stimuli will have, e.g., temperature sensors that, when reaching a certain level, act by increasing or decreasing the air conditioning. On the other hand, a system that responds to the users' stimuli will have possibilities like facial recognition that allows to identify that the user is tired and to suggest a suitable lighting level for that state. These responsive systems use artificial intelligence that offers intelligent systems to analyze and act toward a user's state. Even though these systems allow high degrees of interactivity, its usage should be discussed since its use will imply that people give up personal control and their emotional states will be invaded.

Responsive architecture is currently focused mainly on changes regarding light, thermal, and sound aspects of the

S. Eloy et al. (eds.), *Formal Methods in Architecture*, Advances in Science, Technology & Innovation,
https://doi.org/10.1007/978-3-030-57509-0_12

built environment. Thus, it is not the user who prompt the changes but the environment who dictates them without his/her approval. This work proposes to analyze the mechanisms and discuss the consequences of having the user and his/her emotions dictating changes in space. Therefore, this work analyzes the possibilities of an interaction between the built environments and their users, focused on the responsiveness of the first to the emotions of the latter. Transforming the built environment according to our mood, feelings, and emotions, moment by moment, is discussed and analyzed in the scope of this paper.

The main goal of this research is to define a responsive model by which the built environment can respond in a personalized way to the user's emotions. For such, the emotions in stake, the technical and the constructive elements of interest to be incorporated in the spaces are identified.

The paper has seven main parts. In section two, we define responsive architecture. In the following two sections, we explore two types of responsive architecture: first the architecture that is responsive to environmental aspects (as light, temperature, wind, sound, etc.), and next a section targeting architecture that is responsive to the users' behavior and emotions. In the fifth section, we propose a model to be adapted when responsive architecture to user's stimuli is to be adopted. As this work focuses on an innovative and disruptive aspect of the conception and use of space, both from the architectural and technological point of view, the recognition of the difficulties that arise as well as their effects is discussed at the next section. The work ends with conclusions.

2 Responsive Architecture

The use of responsive architecture enables to have the control of the building comfort conditions, so that a building can act more efficiently (e.g., energy management or noise attenuation). The concept of "responsive architecture" was first introduced in the 1960s by the architect Nicholas Negroponte [19]. The model proposed by Negroponte, inspired by cybernetics theories, derives from the incorporation of computing in architecture [27].

Responsive architecture is often associated with kinetic systems. Although kinetic systems have existed for several centuries, it was only in the twentieth century that architects began to discuss extensively the integration of this type of systems in architecture. However, the possibility of implementing these systems was only possible by the end of the twentieth century, due to technological advances that were taking place in mechanics, electronics, and robotics.

One of the first names to materialize a responsive structure in terms of architecture was Buckminster Fuller, who created the US pavilion Montreal Biosphere for Expo 67 in Montreal. This project consisted of a geodesic dome, in which a complex shading system inspired by the biological processes of the human body was created to maintain the internal temperature [3]. Although Fuller's first experiments have not been successfully achieved, they open the way for responsive architecture to become more and more a reality beyond theory.

Another name that stands out in this field of study is Tristan-d'Estree Sterk. Sterk criticized that the focus was at that time too centered on esthetic aspects rather than the functional aspects that responsive architecture allowed [36]. As a starting point, Sterk uses the premises created by the precursors of responsive architecture, and creates a model considering the latest technological developments [16]. One of the most important premises to keep in mind for this author is the way that the user's needs and wants are considered in the solutions.

The Fun Palace project by Cédric Price was another important moment since it created a unique synthesis of the theories and technologies that were under development at the time, adopting concepts such as cybernetics and artificial intelligence [24]. In fact, in this project, Price believed that through the correct use of technology the occupants of a space would be able to have total control over the environment in such a way that the building would be responsive to their needs and the countless activities that would take place there [13]. "The Fun Palace wasn't about technology. It was about the people." [23]. The Fun Palace was a revolutionary project, in which the main aim was to create a space for people to interact directly with it.

2.1 Kinetic Systems

In 1970, William Zuk and Roger Clark published the book Kinetic Architecture [39], which inspired a new generation of architects to design buildings with kinetic components. Since then, and due to constant technological developments, kinetic solutions have been developed and applied. The type of kinetic movement applicable to a building element allows its shape to vary in four ways: translation, rotation, scale, and deformation. The ability to control the various transformations and movements is an important feature of kinetic systems.

Thanks to technological developments, it is now possible to explore various types of computerized control systems that can be directly activated (activated on the object) or indirectly activated (activated outside the object and through information coming from sensors). Indirect systems can be isolated, when the information comes from only one sensor, or they can be networked, when the information comes from a network sensor system [8].

The fundamental components of a responsive system are sensors and actuators. Sensors are responsible for measuring space properties, such as detecting movement, light, temperature, and humidity. In current buildings, the most used sensors are the ones for detecting presence, thermal, humidity, and gas. Sensors broadcast messages to the actuators, which then execute the appropriate commands. Actuators thus have the responsibility to execute actions. These elements control devices such as dimmers, heating and blinds controllers, valves, among others [19].

3 Architecture Responding to Environmental Stimuli

The use of a responsive architecture gives the possibility of improving the occupant's comfort by controlling the building conditions and enabling a more efficient usage regarding, e.g., energy saving and noise reduction.

Frazer [10] in the 90s also experimented with environmental feedback and interaction with the users and external factors. In his experiences, the "environmental information [whether sound, light, color or movement] was fed to a simplified environmental model which formulated responses based on an evolutionary algorithm" ([10], p. 75).

Other interesting examples of responsive environments are the BIQ (Bio Intelligent Quotient) House by ARUP in Hamburg (2013), the One Ocean Pavilion by SOMA Lima in South Korea (2012), and the Homeostatic Façade System by Decker Yeaon in New York.

In the BIQ House, the shading system (SolarLeaf) consists of a layer of microalgae placed between two layers of laminated glass, whose position adjusts according to the movement of the sun. The controlled growth of these microalgae allows to create shading. In the One Ocean Pavilion, the façade shading system mimics the movement and shape of a fish's gills. The lamellas that make up the system are both resistant and flexible, allowing elastic deformations, so that light enters the building. The movement of these lamellae is achieved through actuators located in the upper and lower limits that apply compression forces, and that are individually controlled, producing a dynamic result [22]. The Homeostatic Façade System is a self-regulating shading system which automatically adjusts to the outside weather conditions. It is composed of two layers of glass and a layer containing a structure of an intelligent material called dielectric elastomer. This inner layer transforms electrical energy into mechanical work, which enables the alteration of its shape. The expansion of its shape occurs when the structure receives an excessive amount of sunlight. One of the greatest advantages of this prototype is its low energy consumption [4].

4 Architecture Responding to Users' Stimuli

In 1969, Andrew Rabeneck said that the basic purpose of design is to avoid uncertainty through forecasting. For Rabeneck, the way to increase the useful life of a building was to design it in such a way that, anticipating the future needs of its users, it could continue to respond to them properly [33]. For that, architects would have to take advantage of all technological advances in cybernetics, such as process automation.

Several authors have been struggling with issues related to responsive architecture and the impact that such solutions can have on spaces and their occupants [5]. Archigram in 1964[1] defined the concept of the Control and Choice Dwelling in which inhabitants can control their environment, and the dwelling is populated with robotized elements. Along this concept, Archigram is divided between the "hard" elements that stay in the same place and the "soft" ones that move and are transformable by the interaction of the inhabitant with the building.

In 1972, Charles Eastman created the concept of adaptive-conditional architecture, related to an automatic space adjustment typology, based on spaces and users' feedback. This concept is focused on automation systems. The author describes how these systems could control the building and compares them using the analogy of a thermostat controlling the temperature in a room. Eastman's system is divided into four elements: sensors distributed to monitor temperature changes; a control system or an algorithm that together with the sensors interprets the results and compares them with the predefined instructions; actuators; and a control mechanism that the inhabitants can regulate according to their preferences [5].

Another interesting example is that of the architect Michael Mozer, who in the late 90s of the twentieth century, implemented a system in an old school. Mozer's system, by observing the behaviors and patterns of its inhabitants, programs itself and learns to anticipate and therefore accommodate the needs of its users. Mozer and his students installed sensors and actuators to control light, ventilation, and heating. The sensors monitored the temperature, ambient light, sound, and movement in each room, and the actuators, e.g., acted on opening of doors and windows, among other. By observing the behavior of the inhabitants, the system could predict what actions would be needed to accommodate the users' needs. For example, the system predicts the period in which the rooms are occupied, initiating a process in which the system controls the correct temperature and light for the occupants, thus providing the greatest possible comfort [38].

[1]See http://archigram.westminster.ac.uk/project.php?id=109.

The model that comes closest to the theme of an architecture responsive to the occupant's emotions is presented by Kas Oosterhuis—"a house with a character of its own, sometimes unyielding, sometimes flexible, at one time sexy, at another unpredictable, stiff and unfeeling." [29]. This author presents a vision of a "time-based architecture," in which buildings can change their shape, configuration, appearance, and environment in response to occupation patterns. Therefore, buildings will be adaptive, interactive, and responsive. Later Oosterhuis with the Hyperbody Reach Group tries to materialize this model through the prototype "NSA Muscle" [18]. NSA Muscle is an inflatable interactive structure that reacts to the touch and presence of occupants, with the ability to physically reconfigure itself. Pneumatic actuators, known as 'muscles,' orchestrate movement through changes in length, height, and width through the pressure associated with each of these. The various actuators work together to change the space in all directions, making the prototype constantly twist, bend, and rotate. The prototype reinforces the idea that architecture becomes a game where its occupants become the players, and architects are the programmers of that game. "We are becoming programmers of behaviors instead of makers of dead objects." [29].

The cerebral hut artistic installation explores the relationship between architecture, interaction, movement, and human thought. Tradition indicates that space influences occupants. However, this artistic installation is intended to reverse this relationship by creating the possibility for the occupant to reconfigure the physical limits of space through thought. In order to create a space that responds to the occupant's thoughts, a device that measures brain frequencies is used. By measuring the level of concentration and the blinking of the occupant, and therefore, interpreting this data the system would create movement. The movement is created by modules inspired in Japanese origami art, which are inserted in hexagonal modules repeated throughout the structure. Thus, the installation allows its occupants to control its physical limits with their thinking [28].

5 A Model for User's Responding Architecture

5.1 Recognition of Emotions

Although there is no consensus from the scientific community about the definition of emotions, they still play a fundamental role in human social life [12]. Paul Young "characterizes emotion as a special mental event involving integrated changes in feeling, behavior, and physiology" ([12], p. 334). Gendron and Barrett refer other authors' definitions of emotion, as William James in the end of the nineteenth century: "bodily changes follow directly the perception of the exciting fact, and that our feeling of the same changes as they occur is the emotion." For Shu et al. ([35], p. 1), "emotion often refers to a mental state that arises spontaneously rather than through conscious effort and is often accompanied by physical and physiological changes that are relevant to the human organs and tissues such as brain, heart, skin, blood flow, muscle, facial expressions, voice, etc." Mulligan and Scherer [26] say that x is an emotion if: x is an affective episode; x has the property of intentionality (i.e., of being directed); x contains bodily changes (arousal, expression, etc.) that are felt; x contains a perceptual or intellectual episode, y, which has the property of intentionality; x is inherited from the intentionality of y; x is triggered by at least one appraisal; and x is guided by at least one appraisal.

Emotions imply a relationship between an experience and a particular event, an object or an environment [11]. Ekman refers that, in his studies with colleagues, he had found that every investigator had obtained evidence for six emotions (happiness, surprise, fear, sadness, anger, and disgust combined with contempt) [6]. Another relevant fact brought by Ekman's research is that he states that observers in Western cultures judge the emotion shown in facial expressions and that there are evidences of universality in those facial expressions. The ability to recognize emotions is one of the characteristics of emotional intelligence [7], an aspect of human intelligence that has been discussed as being more important than mathematical and verbal intelligence.

5.2 Affective Computing

Affective computing is the area of computing that enables computers to feel, recognize, respond, and adapt appropriately to the user's emotions. This field intersects with cognitive sciences, psychology, and HCI (Human–Computer Interaction). Affective systems use emotions as input and, using machine learning, learn to recognize emotions, and to adapt to the user through positive or negative feedback.

MIT Media Lab has created a research department for affective computing, in which software and hardware tools are developed. The developed research targets the analysis of how the user can interact with the computer. There are two types of interactions: self-report and current expression. Self-report systems give the users the freedom to communicate their emotions. Usually these systems present different possibilities, where the user selects the icon that best represents an emotion. The advantage of this type of systems is that the user has full control of the message he/she wants to send. However, the user must stop whatever he/she is doing to communicate with the system. Current expression systems can recognize the user's emotion without interrupting a task. Nevertheless, this type of system takes away the user's control over the message sent, and therefore, the user loses his/her privacy [31].

In order to perform emotional recognition, a technological system needs to collect and combine different user's signals, information about the user's context, and preferences. Of the best-known strategies for emotional recognition, we find is facial identification that might be combined with other physiological manifestations, such as breathing, blood pressure, skin conductivity, muscle tension, among others [35]. Khalili and Moradi [17] study on emotions detection used several biometric sensors as electroencephalogram, galvanic skin resistance, temperature, blood pressure, and breathing. Authors conclude that, for emotion assessment, EEG signals perform better than other physiological signals.

One of the biggest technological challenges in this field is how will a computer recognize emotions correctly, when they are not even always recognized by humans themselves. In fact, one of the biggest disadvantages of facial recognition is that individuals can manipulate it by altering their facial expression to disguise their real emotions. If a more accurate recognition could be accomplished, e.g., through brain waves, this would be avoided.

5.3 Relation Between Space and Emotions

Another aspect that needs to be addressed is how buildings will react to human emotions, so to say, how to identify the appropriate correspondence between different human emotions and the spatial characteristics that will best respond to them. Human perception of space is a subtle act of the human body and mind. Space is self-evident and mathematically described, but the human perception of space is not. Spatial data spans spatial characteristics that need to be feed to a responsive system. For a space to effectively respond to emotions and moods, there is also the need to correlate those spatial characteristics with the emotions believed to be common for all human beings.

Although studies have already been conducted and some results have been published, still correlating spatial properties with human perception and consequent emotions is still at its infancy. Leite et al. [20] show how safe and unsafe spaces, triggered by using or not handrails in ramps and stairs, influence human perception. Sun et al. [37] investigate the influence of surface slope on human gait characteristics, finding that pedestrians' step length was decreased during ramp descent. Franz et al. [9] studied the quantitative relations between the experience of architectural spaces and physical properties and found that participants prefer spaces whose proportions are close to ratios near to the golden ratio. Grobman and Shemesh [15] also studied shape properties of space and concluded that non-experts showed no preference toward symmetrical spaces and were significantly more interested by a curvy space than the experts, who showed a tendency to prefer a sharp space. A recent study related to emotions by Quesnel and Riecke [32] shows how virtual reality simulating space can enhance awe and goose bumps.

Architects like Pallasma [30] have focused on studying on how architecture is designed for the human senses. In these studies, Pallasma refers to the task of architecture to "create embodied and lived existential metaphors that concretize and structure our being in the world" ([30], p. 71). Although architecture is mainly concentrated in the eye, there are a multitude of sensory experiences that can be incorporated in architecture by haptic, olfactory, and auditory stimuli.

The previous mentioned studies, as well as those from other authors, focus on the following architectural elements: form, material, color, texture, scale and proportion, light (and shadow), temperature, and sound. All these elements and combinations between them may have a correlation with emotions felt by occupants. These elements can be materialized in floors, walls, ceilings, windows, and doors, by "hard" or "soft" elements as referred by Archigram.[2] Fixed or "hard" solutions, e.g., the fix covering of a floor with a carpet, may trigger feelings of comfort in some users and may also trigger feelings of disgust associated with dirtiness in others. Using soft and mutable elements opens the possibility of adapting the responses to different users at different moments of the day.

These studies show us that, by manipulating the environmental conditions or qualities of a space, predetermined emotions can be induced or provoked in its occupants. It is not a question of creating an emotional architecture, but one of increasing the transformability of the spaces themselves. Psychology and perception of emotions can then be used as a theoretical basis for understanding the personal and subjective nature of the experience of spaces. Observing users experiencing materials, shapes and proportions may enable the mapping of space according to the emotions provoked. Regarding the mutation of space this can be achieved in several ways, from a more physical one, by an elastic and transformable spatial envelope, to a more visual based one by using optical illusions, with the projection of images combined with light, sound, and temperature, providing scenic effects.

[2]See http://archigram.westminster.ac.uk/project.php?id=109.

5.4 The Responsive Model

The model presented here for an architecture that is responsive to the user's emotions assumes the individual at one end and the space at the other, as represented by Fig. 1. The interaction between both ends takes place according to intermediate steps: the collection of data, the recognition of emotion, and the execution of the action that responds to the detected emotion. This last step can be accomplished by using an evolutionary algorithm that searches for the solution that best responds to the detected emotions and current environment sensing, which corresponds to an issuing of the necessary orders for the mutation of the space.

As the occupant goes through the space, different sensors start collecting data. This data is analyzed by an intelligent system, in which machine learning models identify the current emotion felt by the occupant. Then, a genetic algorithm is going to search for the proper response, that is, the best state of spatial elements for the space mutation, and using installed actuators, will rearrange the space accordingly.

Emotional recognition can be carried out through various combined instruments, like biometric sensors and video, that analyze the user's physiological and behavioral variables. Studies like [1, 20, 25] use biometric data from electrodermal activation (EDA), electromyography (EMG), electroencephalogram (EEG), and electrocardiogram (ECG) while facial emotion recognition is the focus in [21, 34].

As a scenario where precise data for the identification of the user's emotion is to be collected, we propose that data could be acquired directly from the user's brain where emotions are generated by using adequate instruments. In this case, an helmet is able to perform an electroencephalogram (EEG), that records the bioelectric activity of the brain as in Balan et al. [1], where the authors use it to recognize fear. The next phase will perform data analysis and pattern recognition for emotion identification. In face of a recognized emotion, together with ambience smart data and physical contextualization, a genetic algorithm will return the best spatial elements combination for the space to respond to the users feeling. The end of this cycle is the physical space mutation itself.

The system uses an adaptative machine learning algorithm and includes parameters that change over time, to adapt to the data being received and to learn the user's reaction to the changes in the space. The analysis of this data is continuous. The main data source is the brain, and the data are brain waves signals that must be related to emotions, attention, and cognitive processes. The evolutionary algorithm has an adjustment mutation factor which, in addition to responding to human emotion by giving indications to the space to change, also analyzes the individual's reaction to this change, incorporating positive or negative feedback into the current spatial state, adjusting it for a more customized change in the future, thus creating increasingly knowledgeable and stimulating systems.

5.5 Discussion

Imagining a space that complements our existence and reacts to our moods is an ambition that is closer to be fulfilled.

The first question is how this model can be materialized. Although in an early stage, the advancement of technologies leads us to believe that, with new and more flexible materials, it will be possible to provide space mutation. The customization trend reaches its maximum expression with this proposal. Here, we overcome the issue of adornment in architecture, special coverings, furniture, and instead we conceive an almost second skin of an architectural space that acts, moment by moment, as an extension of ourselves. As said before, we propose that the mutation of space may be done either through transformable spatial envelope or by using optical illusions that provide scenic effects. The connection between what is desirable to happen in the space and the user's emotion at a given moment is the most difficult issue to establish. We can define associations of spatial configurations with emotions, but the real purpose will only be fulfilled when the space will be able to adapt to a given individual, designing itself in a customized fit.

A second question is related to the way that a space should react to an emotion, suppose that the user is sad, should the space react to cheer him/her up or should it embrace the user and comfort him/her in their sadness. Such decisions would need to be customized to the personality of users. Although we are all different, it is still possible to find common denominators. Thus, a set of basic spatial parameters that respond to generic emotions for any human being could be created. For that, we use architectural spaces and elements, and we associate them with the feelings and emotions they generically induce. Kate Gordon refers to "horizontal lines as "peaceful," vertical as "ambitious," diagonal as "action," and curved lines "voluptuous" [14] Gordon continues saying that curves are graceful and avoid the harshness of some straight lines." Therefore, when the user shows to be stressed, the response of the space could be to emphasize the referred predominance of horizontal lines, for instance using lights. If, by contrast, the intended response was to react to a lethargic state, e.g., if the user was in his/her working space, vertical lines, or even diagonal lines for greater dynamism would be highlighted. When the emotion recognized by the building indicates that the user is tired, the space's response could be related to emphasizing more organic shapes (like rounding off the edges). In fact,

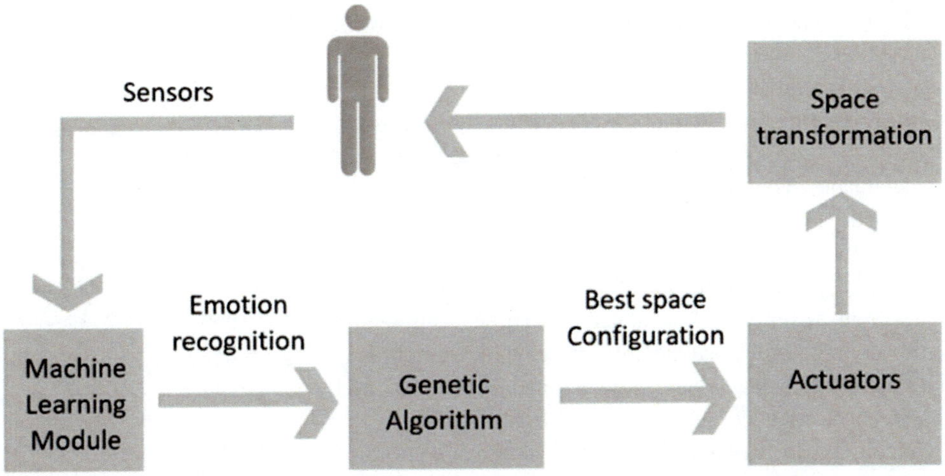

Fig. 1 Model for architecture design that is responsive to the user's emotions

Gordon [14] refers that "curve suggests smooth and easy movement." As such, we envision not the traditional masonry walls and flat concrete ceiling but more flexible elements like membranes, fabrics, and lights associated with simple kinetic structures.

A third question to be posed is the personal and collective nature of the proposal. As it is presented, this proposal targets mainly one user in a space that can be customized for him/her. When in a public area, as a museum or a restaurant, several people are together such model needs to be extended, and a smarter approach for the dialog between building and user must be taken into consideration. In such situations, one side of the problem is the technological solution adopted which needs to collect data from a group of people and analyze it based on a criterion of fitness, e.g., most predominant or most effusive "wins." Another side of the problem is the social issue, since the solution is for the majority, it might be not adapted to a small but relevant group of users.

6 Do We Want Responsive Architecture?

This theme raises ethical questions regarding the use of a responsive architecture: that of the user's privacy and behavioral manipulation.

On the one hand, we are witnessing a rampant data collection by the digital devices and services we use, much of this was, and still is, done opportunistically and without the user's informed consent. To use technologies, users by default give permissions for applications to receive personal data that is often sold to third parties for unknown purposes.

On the other hand, we accept some collection of data for our own convenience, e.g., for optimizing Internet searches to get the best possible response to our requests. The information that we provide to strangers about our preferences and behavioral patterns helps to manipulate and condition our behavior very effectively. Suggestions for shopping and places to visit are a first step in this manipulation that "leads" us to let go of our own options and accept others' as better for us.

In a model where constant monitoring of the occupant is required in order to optimize spatial qualities, the biggest questions are: Who else will access the data collected? and, Will technology manipulate our behavior in a way we do not want?

Even if the data acquisition is performed anonymously, intelligent systems already have the possibility to create profiles and easily discover the user's identity. In the end, what will be the degree of privacy that the user has and of the security of personal data?

7 Concluding Remarks

The introduction of technology into the built environment is a paradigm shift that allows architects to abandon the static characteristic of buildings and to enable them to acquire a kinetic character. With the evolution of computing, artificial intelligence, and affective computing, it becomes possible to further customize space, as well as the creation of a time-changing architecture that fits the occupant's emotions. The latent paradigm shift in this type of architecture is the search for a model in which the occupant, through his/her emotions, directly influences the space.

For Frazer, the new architecture will be unpredictable and based on behaviors. The emphasis will be on the evolutionary and descriptive processes which will enable the "expression of an equilibrium between the endogenous development of the architectural concept and the exogenous influences exerted by the environment" [10]—Fraser estimates that the future will not be a "static picture of being, but a dynamic picture of becoming and unfolding—a direct analogy with a description of the natural world."

The theoretical basis that was developed in this work went through the identification of the possibilities of responsive architecture in two different ways. Firstly, the building reacting to external stimuli such as temperature; secondly, as the focus of this paper, the building reacting to the occupants. The analysis of current responsive architecture practices allowed to identify innovative examples and the available technologies that can enhance a time-based architecture. In this paper, we defined a model for responsive architecture based on its occupants' emotions and which could act on both physical flexible elements like membranes and fabrics associated with simple kinetic structures and non-physical ones like projections and lights.

According to Beesley et al. "the next generation of architecture will be able to sense, change, and transform itself" [2]. In an architecture that is responsive to the emotions of the occupants, it is important to combine the different areas of knowledge, such as computing, psychology, among others, and therefore, the architect should include other players in the design process.

Acknowledgements This publication was partially funded by Fundação para a Ciência e Tecnologia under project UIDB/04466/2020 and UIDP/04466/2020.

References

1. Bălan, O., et al. (2019). 'Fear level classification based on emotional dimensions and machine learning techniques'. Sensors. *Multidisciplinary Digital Publishing Institute, 19*(7), 1738. https://doi.org/10.3390/s19071738.
2. Beesley, P., Hirosue, S., & Ruxton, J. (2006). Toward responsive architectures. In *Responsive architectures: Subtle technologies*. Riverside Architectural Press.
3. Bullivant, L. (2006). *Responsive environments : Architecture, art and design, V & A contemporary series*.
4. Decker, M. (2013). Emergent futures: Nanotechology and emergent materials in architecture. In *Conference: Tectonics of Teaching—Building Technology Educators Society (BTES)* (February).
5. Eastman, C. (1972). Adaptive-conditional architecture. In *Design participation, proceedings of the design research society's conference Manchester* (pp. 51–57). London: Academy Editions.
6. Ekman, P. (1992). Are there basic emotions? *Psychological Review, 99*(3), 550–553.
7. Elfenbein, H. A., Marsh, A. A., & Ambady, W. I. N. (1997). Emotional intelligence and the recognition of emotion from facial expressions. In L. F. Barrett & P. Salovey (Eds.), *Emotions and social behavior. The wisdom in feeling: Psychological processes in emotional intelligence* (pp. 37–59). New York: The Guilford Press.
8. Fox, M., & Kemp, M. (2008). Interactive Architecture. In *Interactive architecture*.
9. Franz, G., Von Der Heyde, M., & Bülthoff, H. H. (2005). An empirical approach to the experience of architectural space in virtual reality-exploring relations between features and affective appraisals of rectangular indoor spaces. *Automation in Construction, 14*(2 SPEC. ISS.), 165–172. https://doi.org/10.1016/j.autcon.2004.07.009.
10. Frazer, J. (1995). *An evolutionary architecture*. London: Architectural Association. Available at: https://opasquet.fr/dl/texts/Frazer_An_Evolutionary_Architecture_1995.pdf.
11. Frijda, N. H. (1994). Varieties of affect: Emotions and episodes, moods, and sentiments. *The Nature of Emotion: Fundamental Questions*.
12. Gendron, M., & Barrett, L. F. (2009). Reconstructing the past: A century of ideas about emotion in psychology. *Emotion Review, 1* (4), 316–339. https://doi.org/10.1177/1754073909338877.
13. Glynn, R. (2005). *Fun palace—Cédric price*. Available at: http://www.interactivearchitecture.org/fun-palace-cedric-price.html.
14. Gordon, K. (1906). *Esthetics*. New York: Henry Holt and Company.
15. Grobman, J. Y., & Shemesh, A. (2015). Space and human perception—Exploring our reaction to different geometries of spaces. In *20th International Conference of the Association for Computer-Aided Architectural Design Research in Asia CAADRIA*.
16. Henriques, G. C. (2015). *Sistemas responsivos: relevância, estado da arte e desenvolvimentos* (pp. 200–206). https://doi.org/10.5151/despro-sigradi2015-60042.
17. Khalili, Z., & Moradi, M. H. (2009). Emotion recognition system using brain and peripheral signals: Using correlation dimension to improve the results of EEG. In *2009 International Joint Conference on Neural Networks* (pp. 1571–1575). IEEE. https://doi.org/10.1109/ijcnn.2009.5178854.
18. Kievid, C., & Oosterhuis, K. (2003). *Muscle NSA, a basis for a true paradigm shift in architecture*. Available at: http://www.hyperbody.nl/research/projects/muscle-nsa/index.html.
19. Kolarevic, B., & Parlac, V. (2015). Adaptive, responsive building skins. In *Building dynamics: Exploring architecture of change*. https://doi.org/10.4324/9781315763279.
20. Leite, S., et al. (2019). Physiological arousal quantifying perception of safe and unsafe virtual environments by older and younger adults. *Sensors, 19*(2447), 1–19.
21. Liu, C., et al. (2018). Multi-feature based emotion recognition for video clips. In *Proceedings of the 20th ACM international conference on multimodal interaction (ICMI'18)* (pp. 630–634). New York, USA: Association for Computing Machinery.
22. Maier, F. (2012). *One ocean—Thematic pavilion for EXPO 2012*. Available at: https://www.detail-online.com/article/one-ocean-thematic-pavilion-for-expo-2012-16339/.
23. Mathews, S. (2005). The fun palace: Cedric price's experiment in architecture and technology. *Technoetic Arts, 3*(2), 73–92. https://doi.org/10.1386/tear.3.2.73/1.
24. Meagher, M. (2015). Designing for change: The poetic potential of responsive architecture. *Frontiers of Architectural Research. Elsevier, 4*(2), 159–165. https://doi.org/10.1016/j.foar.2015.03.002.
25. Ménard, M., et al. (2015) *Emotion recognition based on heart rate and skin conductance*. https://doi.org/10.5220/0005241100260032.
26. Mulligan, K., & Scherer, K. R. (2012). Toward a working definition of emotion. *Emotion Review, 4*(4), 345–357. https://doi.org/10.1177/1754073912445818.
27. Negroponte, N. (1970). *The architectural machine: Towards a more human environment*. Cambridge: MIT Press.
28. Office Ozelo. (2015). *Cerebral hut*. Available at: http://www.ozeloffice.com/new-page.
29. Oosterhuis, K. (2003) *Hyperbodies : Toward an E-motive architecture*. Birkhäuser.
30. Pallasma, J. (2005). *The eyes of the skin. Architecture and the senses*. Hoboken: Wiley.
31. Picard, R. W. (2000). Toward computers that recognize and respond to user emotion. *IBM Systems Journal*.
32. Quesnel, D., & Riecke, B. E. (2018). Are you awed yet? How virtual reality gives us awe and goose bumps. *Frontiers in Psychology*, 1–22. https://doi.org/10.3389/fpsyg.2018.02158.

33. Rabeneck, A. (1969). Cybermation: A useful dream. *Architectural Design*, 497–500.
34. Samadiani, N., et al. (2020). A multiple feature fusion framework for video emotion recognition in the wild. *Concurrency Computation Practice and Experience*. https://doi.org/10.1002/cpe.5764.
35. Shu, L., et al. (2018). A review of emotion recognition using physiological signals. *Sensors, 18*(2074). https://doi.org/10.3390/s18072074.
36. Sterk, T. D. (2003). Building upon negroponte: A hybridized model of control suitable for responsive architecture. *Automation in Construction, 14*(2), 225–232.
37. Sun, J., et al. (1996). The influence of surface slope on human gait characteristics: A study of urban pedestrians walking on a inclined surface. *Ergonomics, 39*(4), 677–692.
38. World's "Smartest" House Created By CU-Boulder Team. (1998). Available at: https://www.oecd-ilibrary.org/docserver/518742334033.pdf?expires=1580242037&id=id&accname=oid015503&checksum=AF990ADD908741CBBB25670E25137FB4.
39. Zuk, W., & Clark, R. H. (1970). *Kinetic architecture*. Van Nostrand Reinhold.

Luísa Almeida is M.Sc. in Architecture with the title Architecture Becomes Alive—Arquitetura Responsiva às Emoções do Ocupante. Almeida is interested in researching on topics like responsive architecture, digital, and emerging technologies. During the last years, she collaborated in the co-organization of several editions of the FISTA (Forum of Iscte School of Technology and Architecture), and participated in several workshops like KINE[SIS]TEM'17 (Summer School) at Iscte and a Summer Studio at Bartlett School of Architecture, UCL.

Sara Eloy Eloy is an architect, Ph.D. in Architecture. Eloy's main areas of research include digital technologies applied to architecture, shape grammars, virtual and augmented reality, analysis of the building space namely considering space perception, and housing rehabilitation. She is the Director of the Information Sciences Technologies and Architecture Research Center (ISTAR) where she is a fellow researcher. She has curated several exhibitions being the latest "Artificial Realities: Virtual as an Aesthetic Medium for Architectural Ideation," in the scope of Lisbon Architecture Triennale 2019. She has participated in national and international research projects and published her work in several journals.

Ana de Almeida Almeida has a Ph.D. in Applied Mathematics, with a specialization in Computer Science. Her research interests include Algorithmics, Complexity Theory, Machine Learning, Pattern Recognition, and Evolutionary Computation. She is particularly interested in evolutionary strategies for multi-criteria problems, as well as the construction of self-adjusting predictive models for real applications. She has chaired or co-chaired important international conferences in Complexity Sciences and Computer Science, participated in fundamental and in applied research projects and in research and innovation projects between academia and industry. She is an IEEE Senior Member, IEEE IMS Portuguese Chapter founding member and an ACM member.

Borealis Sound an Interactive Wall for Situational Awareness: The Impact of Responsive Architecture on Users

Beatriz Couto and Sara Eloy

Abstract

This paper refers to the presence of visual communication elements in public spaces using interactive multimedia surfaces. The state of the art, the development of a proposal, and its evaluation are presented. The interactive surface applies to school of music "Escola de Música do Conservatório Nacional," to which a design project was done in an academic context. The proposal aims to explore the incorporation of digital multimedia surfaces enhancing new dynamics while users walk in the interior of the building. Visual content will be displayed, and the interaction between users and the multimedia surfaces is done by the sounds produce by people and instruments. To assess the intuitive nature and relevance of the proposal, satisfaction and usability tests were conducted with the potential users. It was concluded that users appreciated the proposal and was able to perceive the effect people's presence have in the interaction with the multimedia surface.

Keywords

Responsive architecture • Interaction • Users' responses • Multimedia • Interactive wall • Sound • User's testing

1 Introduction

This paper concerns the presence of visual communication elements in public spaces using interactive multimedia surfaces. The state of the art, the development of a proposal, and its evaluation are presented. The interactive surface applies to Escola de Música do Conservatório Nacional—EMCN (School of Music of the National Conservatory of Lisbon), to which a rehabilitation and expansion design project was developed in an academic context (Fig. 1). Since the task involved a music school, the aim was to develop a proposal that would explore interactive multimedia surfaces that react to what characterizes this space—the sound. The problem identified in EMCN to which this research address is the noisy corridors of the school. The solution presented aims to explore the incorporation of digital multimedia surfaces enhancing new social dynamics while users walk in the interior of the building. Visual content will be displayed, and it will react according to the interaction with the sounds produced by people and instruments. This proposal has therefore a dual goal since on one side it acts as an entertainment and on the other as a device to control the sound level of the public areas of the school creating a better learning environment.

To achieve the proposed objectives, a methodology structured in three stages was adopted. The first stage corresponded to the literature review and the research of different examples of multimedia surfaces in order to analyze the current systems and the relevance of their use. Theoretical and technical bases related to the creation of these surfaces were also investigated. The methodologies for technology development that were studied and were instrumental to the development of the proposal and allowed the definition of the interactive proposed system design as well as its later evaluation based on user experience. The second stage included two moments. First, the methodology to be applied to the practical proposal was defined. This methodology incorporates the development of the proposal and addresses the architectural and technological aspects. After defining the proposal, a prototype was created to simulate and test the interaction performance. At this moment, usability and satisfaction test of the proposed system were carried out in order to assess the intuitive nature

B. Couto
CGI, Av. José Malhoa 16, 1070-159 Lisbon, Portugal
e-mail: beatrizcouto0@gmail.com

S. Eloy (✉)
Instituto Universitário de Lisboa (ISCTE-IUL), ISTAR, Av. Forças Armadas, 1649-026 Lisbon, Portugal
e-mail: sara.eloy@iscte-iul.pt

S. Eloy et al. (eds.), *Formal Methods in Architecture*, Advances in Science, Technology & Innovation,
https://doi.org/10.1007/978-3-030-57509-0_13

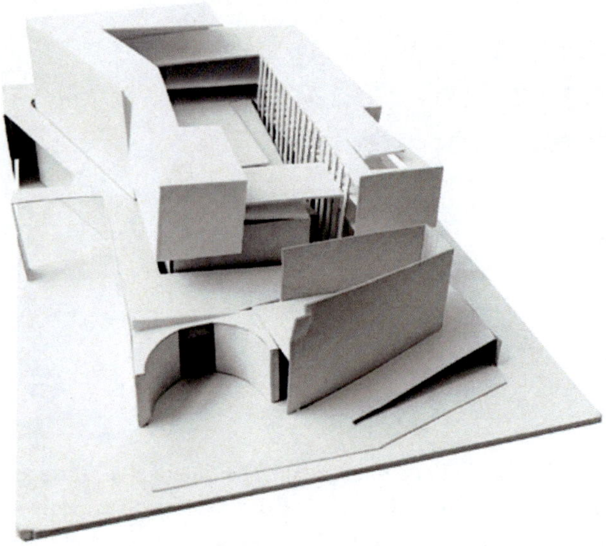

Fig. 1 Model of the proposal

and relevance of the proposal. The last stage was related to the analysis of the results obtained with the prototype tests. This analysis aimed at assessing whether the defined proposal corresponds to the intended one.

This paper is divided into three main parts. Section 2 describes interactive multimedia surfaces and the state of the art. Section 3 describes all the steps of the development of the proposal here presented for an interactive multimedia surface from the definition of the initial conditions to the performance and analysis of usability tests. The paper ends with a section of discussion and concluding remarks.

2 Interactive Multimedia Surfaces

According to Achten [1], interactive architecture is a special case of "responsive buildings" which are the ones that can anticipate and change their behavior or content according to a schedule. The particularity of an interactive surface is to create a user internal representation. This means that activity (ies) that potentially could be carried by the user(s) were anticipated and introduced in the building system.

Based on their own design installations, Fischer and Hornecker [6] analyzed and compared different spatial configurations in relation to their interaction system. According to the authors, interactive Media Façade interventions, "takes account of the social use of space and of the existing structure of the built environment." In the SMSlingshot installation designed by the authors for a square, they identified seven types of spaces related to the experience of an interactive surface. These are: Display spaces, Interaction space, Potential Interaction Space (PIS), Gap spaces, Social Interaction Spaces (SIS), Comfort spaces, and Activation spaces. According to the authors, "depending on installation type, context, and environment, these spaces can differ in size and position" ([6], p. 4).

Indaverea [9] considers three approaches to make "media surfaces." The first consists of creating from scratch digital content to be projected onto a building. An example of this approach is the spatial augmented reality "Perdi o meu coração em Lisboa" (I lost my heart in Lisbon), made by Oskar and Gaspar at Terreiro do Paço, Lisbon, in 2014. The second approach consists of capturing data from the environment and its subsequent interpretation, using digital means to project the content on a building. According to Indaverea [9], the Tower of the Winds (1986), designed by Toyo Ito, is an example of this second approach. The third approach integrates an interaction with the public, which will thus have a direct influence on the content on display. The difference between the second and the third approach is that while the second reacts to environmental stimulation, the third reacts to human stimulation. As for the content, the author distinguishes the following functions: artistic, playful, community media, public service, and news.

This way of classifying interactive surfaces is nevertheless not consensual. In fact, Gehring and Wiethoff [7] present another way of classifying interactive surfaces, focusing on the communication process between people and buildings. These authors classify the types of interactive surfaces, enumerating what distinguishes them from each other, based on their most striking characteristics. Water Light Graffiti[1] produced by Antonin Fourneau and Art2m, presented at the third edition of LUMINA[2] festival in Cascais 2014, is a good example of an installation with different types of interaction. This installation was considered a breakthrough in terms of interaction in urban architecture by mixing a natural element—water—with technology, LEDs, and humidity sensors. The installation allowed an interaction between the public and natural phenomena such as rain. Another example is Perspective Lyrique, interactive video mapping, designed by the team of architects 1024 and held at the Celestins Theater in 2010, under the Fête des Lumières Festival in Lyon.[3] The artists have integrated an interactive device that causes a succession of visual distortions which were activated by voice. Finally, the SMSlingshot installation consists of a slingshot that "shoots" messages to a facade. It allows the public to momentarily leave a mark on the facade and sends signs in the form of color ink balloons [6].

Achten and Kopriva [2] proposed a methodological framework for interactive architectural design developed

[1]Water light Graffiti in: http://www.waterlightgraffiti.com/about-wlg/.
[2]Lumina (2014) in http://lumina.pt/2014/pt/lumina/.
[3]1024 blog in: https://1024d.wordpress.com/category/event-project/prespective-lyrique/.

into four basic stages: Analysis; Concept generation; Simulation; Evaluation. Later, Achten [1] states that to develop interactive systems there are no design methods that ensure an appropriate solution since part of the development can be addressed in a methodological way, but another substantial part has to be seized through iterative processes and prototypes to test ideas. In order to form a theoretical basis for the design of interactive architecture. Achten [1] also defines the principles of agent theory, a key concept originated in the field of artificial intelligence. This concept considers the analysis, the design of complex systems and includes two complementary moments which are the communication directly with other objects, and the changes to which it is subjected autonomously without an entry/order. The term "building as agent" describes the interaction, or responsiveness, as if it was a person who is being described. According to that principle, the author suggests five points that support the global design of interactive systems: system response type; type of interaction for the user; type of interaction for the system; style of interaction between the user and the system; and type of interaction activity.

Besides these functional and methodological aspects of developing an interactive wall, it is also relevant to be aware of its potential social impact. The presence of digital devices in the places we inhabit are enabling us to access a large set of information in a totally new way. The addition of interactive features enables these devices to be customized to their users which opens a new dimension of space appropriation. Informative panels exist already everywhere we go, with information on outside weather conditions, indoor temperature and humidity, available parking places, and the level of CO_2 in, e.g., an underground parking area, among other examples. This situation awareness and the information given are helpful and enable users to make informed decisions. Situation awareness has been recognized as a critical foundation for successful decision-making across a broad range of situations.[4] This concept has roots in military action as most of the technology development nowadays. Endsley [5] refers that for complex situations people are having trouble coping with data explosion and that the "answer lies in understanding how people process the vast amount of data around them to arrive at effective performance." Situation awareness comprises three steps: perception—gathering information about the situation; understanding—comprehending the meaning of the information acquired; prediction—anticipating the outcome of future events.[5] These steps are fundamental for good decision-making.

There are several studies on the development and impact of situation awareness information displays in several fields as in [4, 8]. Most of these studies assess the developed systems by usability tests and do not focus on discussions on how interactive spaces are changing the way we inhabit, appropriate, and use indoor spaces. On this matter, Snibbe and Raffle have a very interesting research based on several years of own experience developing interactive systems for public exhibits. The authors use the concept of social immersive media built on simulated narrative models that "create strong emotional responses and social engagement through visceral interaction" ([14], p. 1447).

3 Borealis Sound System Development

The proposal was prepared using a methodological system present in Achten and Kopriva [2] and Achten [1]. The methodology used begins stepwise in order to specify the main criteria that allow the future implementation. At the end, a simulation is carried out through a proof of concept prototype. The proposal development was done throughout the following five stages: (i) definition of the initial conditions; (ii) analysis of the context; (iii) system specification—requirements; (iv) usability testing; (v) analysis of the results.

3.1 The Creative Process

Initially, three different proposals were considered for the current project. Although all the proposal defines distinct interactions, all have the common principle of promoting a connection between sound and image, since they would be inserted in a music school. We briefly describe each of the three proposals (A, B, and C) since "to get a good idea, get lots of ideas." (Fudd's first law of creativity, [13]).

Proposal A (Fig. 2)—The Memory of the Path—consists of the installation of six LED panels, aligned with walls along the north corridor of floor 0 of the EMCN. In this proposal, pressure sensors placed on the floor sense people passing and sends information to an actuator to control images and color of lights projected on the walls, varying according to the users' path.

Proposal B (Fig. 3)—Musical Fragments—also consists of the application of LED panels in the north corridor of floor 0 of the EMCN. This corridor is a crossing area for students, and the panels aim to promote a playful activity. For this proposal, the participation of more than one student is necessary, so that, when touching on a panel, it emits a short piece of music. The purpose of the interaction is to challenge participants to identify the music playing. In order to decipher the music, it would be necessary to interact with

[4]For more information on Situation Awareness see https://en.wikipedia.org/wiki/Situation_awareness.

[5]Situational Awareness matters, see https://www.samatters.com/.

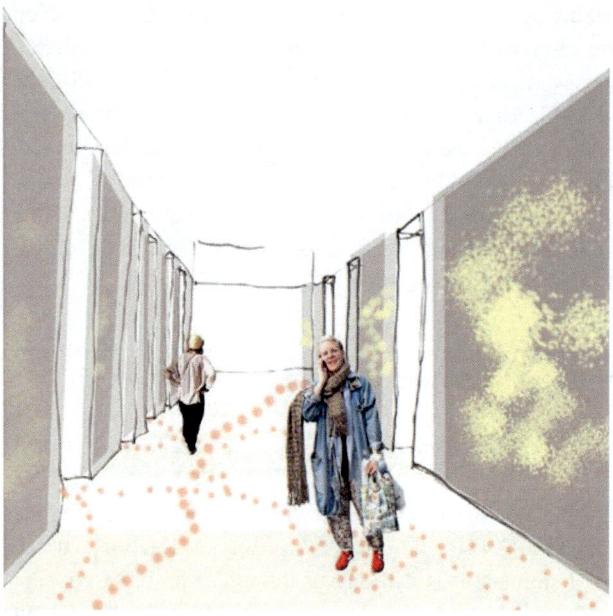

Fig. 2 Proposal A—memory of the path

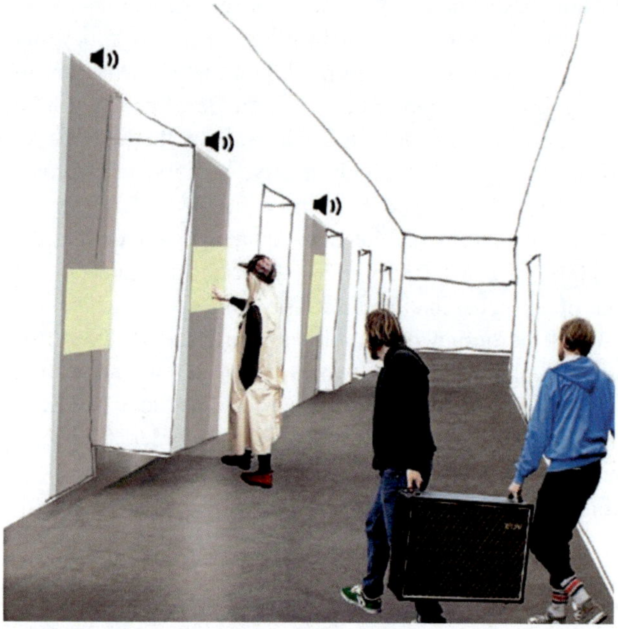

Fig. 3 Proposal B—musical fragments

more than one panel, with each panel emitting only a certain part of the music to be deciphered.

In proposal C (Fig. 4)—Propagation, Superposition—the panels would be placed perpendicular to the wall. Consequently, these would have a greater impact as they delimit the students' path along the south corridor of floor 1. Both sides of the panel are to be used with one screen on each side. The interaction is triggered by touch, which causes the light in a certain panel to turn on. This light spreads to the following panels, each time with less intensity—until the light goes out. The intensity of the light corresponds to the intensity of the touch; the more force is applied, the greater the light emission, making it possible to propagate it to the last panel. The corridor would thus be exposed to a constant rhythm of lights with different intensities.

Besides the aim of entertainment, the second and more relevant aim related to the proposals is to provide a better learning environment.

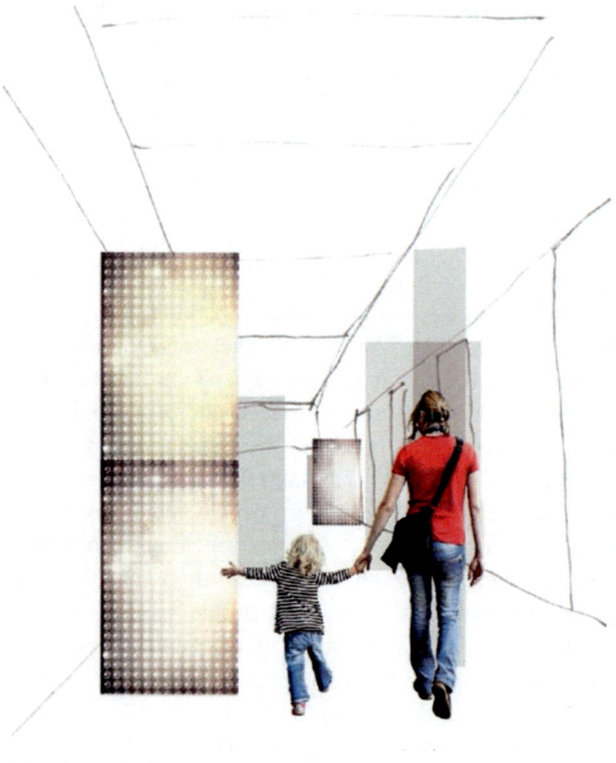

Fig. 4 Proposal C—propagation, superposition

3.2 Stage 1—Definition of the Initial Conditions

The interactive wall is proposed to the ground floor of the north corridor in EMCN. The design presupposes that all visual content displayed on the screens are triggered by the sound/noise produced by people, devices, and musical instruments. According to Bullivant [3], "the interaction with a multimedia installation allows people to participate in a different world by adding a transformative effect."

The performance criteria for the visual effects were defined regarding the amplitude and frequency of the sound. These criteria were chosen in order to provide users with a sensory experience that entangles them with the combination of both sound and visual effects. A second aim was to

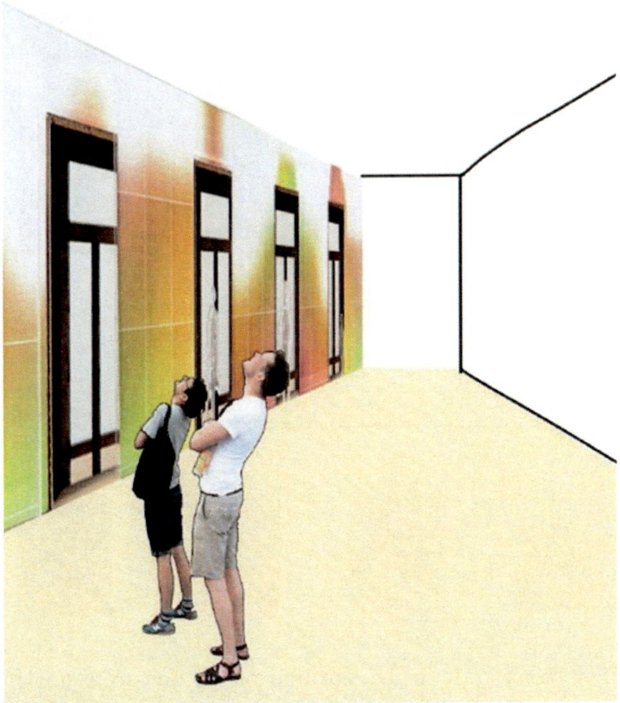

Fig. 5 Interactive multimedia wall for EMCN

control the volume of the sound in a place where silence, sound clarity needed, and noise reduction during the classes is mandatory.

The current proposal consists of a set of LED matrices (Fig. 5). In order to adapt the visual content to the chosen real situation, the performance of the installation is divided into two distinct periods, a quieter in a low-amplitude sound environment and a second one in a higher amplitude sound environment. Each one works within a stipulated time frame, the period (I) occurs during the classes, and the period (II) occurs in the absence of classes. At the period (I), the content reacts depending on the amplitude, and during the period (II), the content reacts depending both on the amplitude and frequency. These two periods are distinguished in terms of both the visual content displayed and the volume picked up not in terms of interaction. For the period (I), two scenarios were designed: (1) a single person in the corridor, (2) a group of people in the corridor. For the period (II), only one extra scenario was designed: (3) student rehearsing in the corridor. Scenarios (1) and (2) may also occur at the period (II).

We then applied the five points that support the creation of a global design of interactive systems following Achten [1]. The points are type of system response; type of interaction for the user; type of interaction for the system; style of interaction between the user and the system; and type of interaction activity.

The designed system is reactive regarding the type of interaction since it includes as a condition that any sound produced in the corridor triggers an action in the displays. In such a design, the user is active since he/she is the main sound source of interaction. The type of interaction by the user is designed as unintentional since the system does not require a voluntary user participation. Thus, the system is said to be indirectly associated with the user. When the user is active and the system is indirect, the resulting interaction style is called assistant [1]. In assistant interaction to operate the system does not imply the user to have the knowledge of how the system works.

Controlling the sound produced in the corridor and acting as a mean of entertainment are the main objectives of the proposed interactive wall. When the sound produced in the corridor exceeds a preset limit (85 dB, equivalent to a human scream) an alert message is transmitted (red stain). To avoid disturbance in the school, it was intended that this visual signal should serve as a limit indicator for the volume and not has a disturbing alert.

Exploration was considered as the interaction activity. The combination of this exploration activity with the defined objectives (entertainment and sound regulation in the corridor) describes the type of user experience which is based on improvising. This type of experience allows the user to explore various options and configurations of the content, without a pre-defined objective or performance criterion.

To set out principles for the performance, an analysis of the properties of sound was done. To explore the amplitude property in both periods (I) and (II), types of sounds were associated with a low, medium, and high amplitude (sounds considered weak, medium, and strong). The frequency was introduced as a performance criterion in the period (II) by analyzing the frequency that the three types of instruments more likely to be tested in the hallway such as wind, strings, and percussion. The proposed three scenarios can help distinguishing the dynamics occurring in each period. For them, seven different criteria were proposed which in turn cause seven variations of visual content, two for Scenario (1), two for Scenario (2), and three for Scenario (3).

The Northern Lights were chosen to be the visual content of the interactive wall because of their beauty. In nature, these lights are the result of weather events. In the proposed application, the images of the Northern Lights will, in analogy, result of the different sounds captured by the sensors (microphones). The goal was not to reproduce their beauty, but rather to take advantage of the visual impact that is embodied in the movement of the lights in all directions.

In the first two scenarios, the performance criterion is the amplitude. The visual content (color stain) varies in size—when the amplitude is low, the stain is small, and when the amplitude is medium, the spot is bigger. If the amplitude reaches high, the visual effect, "freezes" and a red stain will

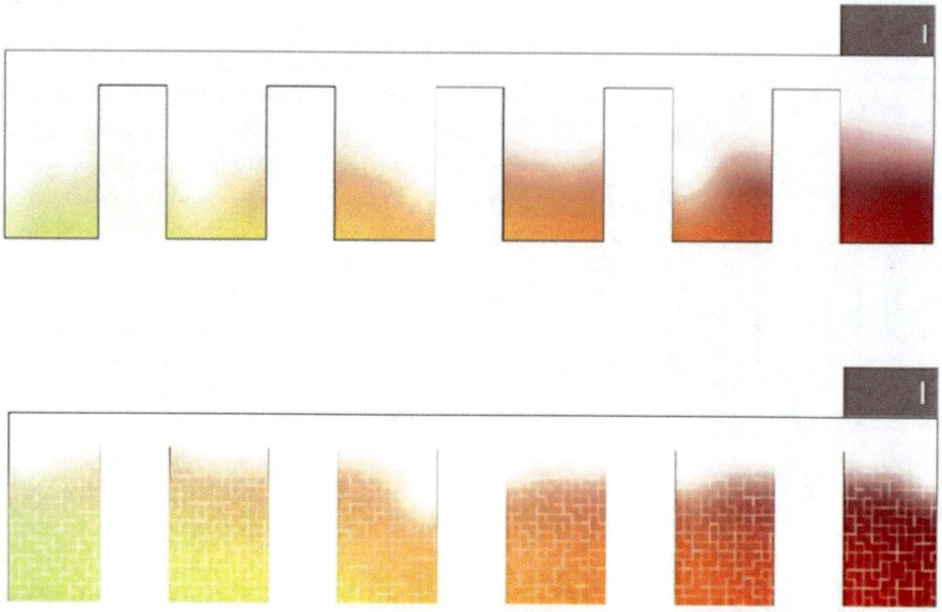

Fig. 6 (Top) first visual content, (bottom) second visual content

flood the corridor. This red light intends to act as an alarm calling for the users' attention to reduce the volume (too loud) that they are producing. This reaction of the interactive system condemns the "noise" considered negative by all the users of the school. To the third and final scenario created, a color scale was associated with a range of frequencies: In the presence of low frequencies the colors are warm; for mid-range intermediate colors will be used; for high-frequencies cold colors.

3.3 Stage 2—Analysis of the Context

A total of six audio samples collected online and a sample recorded for this purpose were used. These samples are associated with the designed scenarios with the goal of matching each sound to a type of situation. Sounds samples included a person walking, a phone ringing, people talking out loud. The samples for the instruments were bass, guitar, and violin.

The samples were reproduced in the Abbleton Live 9[6] software in order to export the spectrum graphics for each sound. As support for the definition of the specific performance criteria proposed spectrum graphs were produced. This allows an analysis of the amplitude and respective frequencies, making it possible to distinguish the loud from the low sounds as well as the sounds in which the low, medium, or high frequencies predominate.

For scenario 1, in the situation of reaction to a single person, we defined that the amplitude corresponding to the sound of the steps in the corridor reaches 36 dB, whereas the amplitude corresponding to the sound of the cell phone corresponds to 65 dB. The first performance criterion is defined by a maximum amplitude of 36 dB—when this value is reached, the installation generates the first visual content. The second performance criterion is defined by an amplitude greater than 65 dB, thus generating the second visual content. The two visual contents created for the present scenario can be found in Fig. 6.

In scenario 2, a group of people is in the corridor. The two sounds present in this scenario correspond to a conversation; the first sound is of a medium amplitude with 65 dB and the second sound a high amplitude with 72 dB. Each amplitude gives rise to two more visual contents (Fig. 7).

Scenario 3 corresponds to a moment when one or more students are rehearsing in the corridor during class hours. The sounds present in this scenario correspond to a bass, an acoustic guitar, and a violin. In this scenario, it was defined that the visual content appears when the amplitude of the sound produced is above 10 dB. It was concluded that the bass sound is predominantly low, the guitar sound is predominantly medium, and the violin is predominantly high. Thus, each type of frequency generates distinct visual content. Considering the three frequency variations, these represent the fifth, sixth, and seventh performance criteria. It was decided to associate a color scale to each of these frequencies; in the presence of a low frequency, the fifth content occurs in red tones; for medium frequency, sixth content, yellow tones and for a high frequency, seventh and last content, blue tones are shown (Fig. 8).

[6]See https://www.ableton.com.

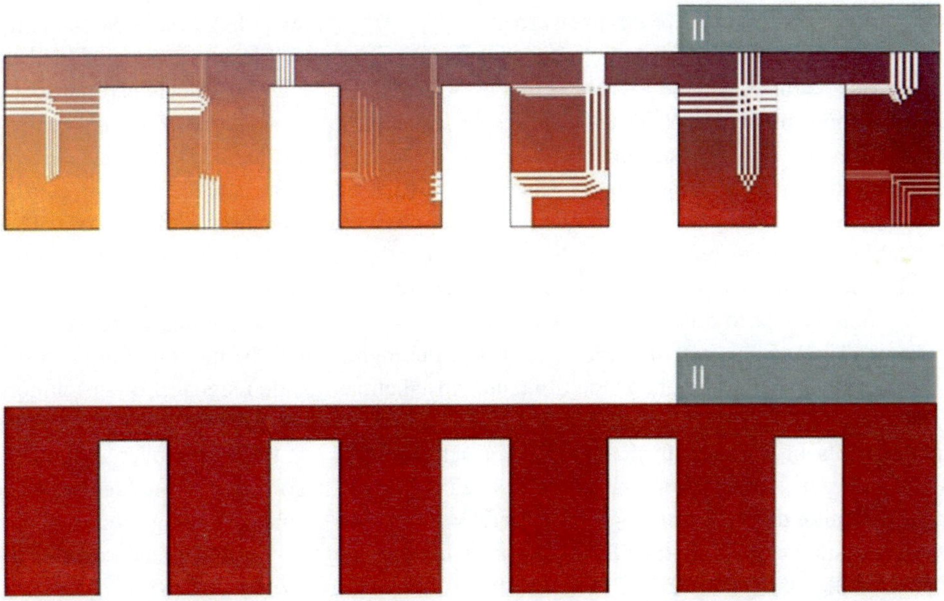

Fig. 7 (Top) third visual content, (bottom) fourth visual content

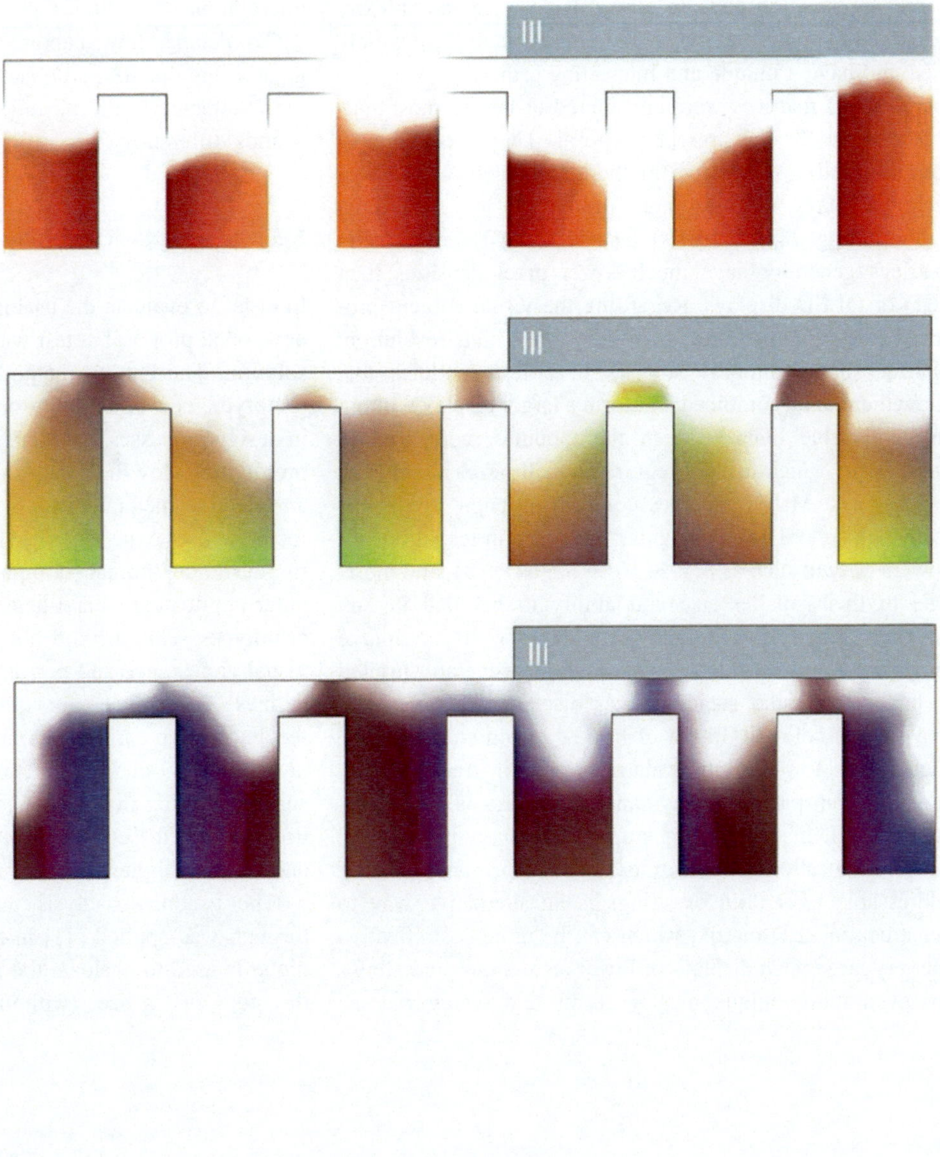

Fig. 8 (Top) *fifth visual content*, (middle) sixth visual content, (bottom) seventh visual content

3.4 Stage 3—System Requirements and Specifications

The development of this proposal required research about digital signal, display controllers, and sound capture regarding the type of approach to achieve the expected results. Even though the proposal was designed for the EMCN, it was considered the possibility of implementing it in other public spaces, and therefore, the technical requirements should adapt to different space characteristics.

Digital signage can refer to various display technologies. "Video wall" generally refers to multi-monitor installations for large format video presentation. These consist of arrangements of standard format monitors (such as 16:9) with thin bezels, with common monitor sizes being 47 or 55 in. The monitor display technologies vary between: thin-film transistor liquid crystal display (TFT-LCD or LCD); organic light-emitting diode (OLED); E-Paper (non-illuminated display technology that uses charged white and black particles in microcapsules to reproduce grayscale images); LED matrices for large scale displays; and Flip-disk displays which are now obsolete in terms of cost and complexity compared to modern display technologies but are still widely used and have a unique and interesting esthetic.

The LED matrices were considered to be the most suitable display for the present proposal. They are the most common and cost-effective method of implementing large scale displays. Pixels are produced by individual light-emitting diodes (LEDs) instead of TFT arrays. LED matrices generally have much lower pixel densities than LCD or OLED displays. Regarding the visual content proposed, the monitor does not need a very high resolution. Matrices are commonly sold as bezel-less modules that allow them to be combined and form a larger display without any noticeable lines between the modules, compared to video walls which usually require a small bezel in-between the displays. Matrix modules come in different sizes, normally square (1:1) or rectangular (1:2), and in resolutions of powers of two: such as 8×8, 16×32, 64×64, and more. Due to their smaller size and ability to be tiled in any direction, LED matrices allows to cover walls of unique dimensions, unlike video walls which are generally limited to large rectangular standard sizes displays. LED matrices may be full color (RGB) or of limited or individual colors, such as the low-resolution matrices used in transportation signage. Transparent displays and LED curtains are a variation of LED matrices where a transparent base material and wide pitch allow for a translucent appearance. Using addressable LED strip or strings is an alternative way to construct an LED matrix, which can be more cost-effective for very large pitch displays of lower resolutions, and allows for even more unique display forms and configurations, including textile based "soft" displays. Extra low-resolution displays can be combined with diffuser materials for novel lighting effects and unusual pixel formation.

In terms of display controllers, the hardware requirements for digital signage can be reduced to a small computer with a very capable graphics card, or several computers networked together; some video walls have a dedicated computer for each monitor. LED matrices have lower computational requirements given their low resolution. A single controller can drive many matrix modules, but a small computer may still be necessary to process the sound inputs and produce a video signal. Low-resolution matrices or LED strips/strings have even lower requirements and are likely easily driven by a microcontroller, possibly even without the need for a computer to produce the video signal, depending on how the sound is being processed.

For the sound capture, the omnidirectional microphones would be used as sound sensors hanged along the corridor (omnidirectional condenser low profile for AT4022413 recording). These microphones capture the sound independently of the sound source in relation to the direction of the microphone.

To decide, how capture sound in the space depends greatly on the desired visualization and how the viewer reacts/interacts with it. Indeed, microphones can capture sounds differently from different angles.

3.5 Stage 4—Usability Testing

In order to evaluate the usability and user satisfaction of the developed proposal, a test was held using a prototype of the solution. The chosen method for the test, known as paper prototype, is a low-fidelity prototyping test commonly used in design processes that focus on user experience [13]. Rapid prototyping (low-fidelity prototyping or paper prototype) is an effective method to be used in design processes with a focus on user experience [13]. It is a method generally used for designing human–computer interfaces, websites, computer applications, hand-held devices, and proof-of-concept prototypes. This method not only allows to show users the visual appearance and performance of an interface, but also defines a technique for quickly obtaining criticism and feedback in an early development phase. This would not be possible if another method of design conception was used, such as, coding in the application language or by a simulation made with software. These methods require more time and involve higher costs.

The usability analysis method called cognitive walkthrough was applied [11] in the tests. This evaluation technique is used to evaluate the interface design, by simulating the steps that a user (without any prior knowledge of the

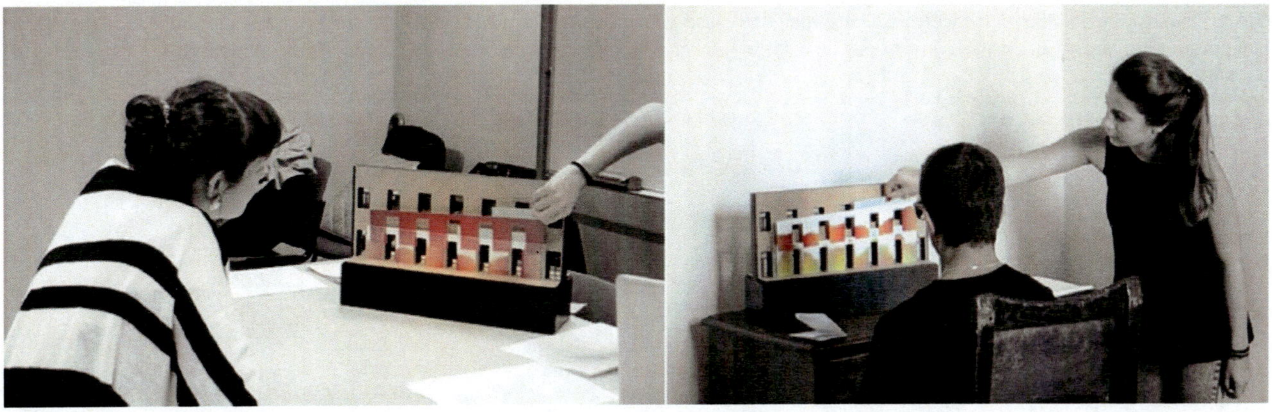

Fig. 9 (Left) a test performed at Iscte, (right) a test performed at EMCN

system) would do to reach a goal [11]. In this method, the user is being guided throughout the experiment by a "facilitator" who simulated the performance of the installation. The demonstration of the performance of the proposal starts in a simple scenario and leads the participant to the most complete scenario. It was intended to verify if the visual contents used in the three scenarios were effectively assimilated as messages and if not, proceed with the improvement of the project.

20 individual tests were carried out on a sample of Iscte architecture students and EMCN students. These tests were carried out to assess the relevance of the installation in the school context and the degree of satisfaction of the participants as well as usability issues.

The study was divided into four stages:

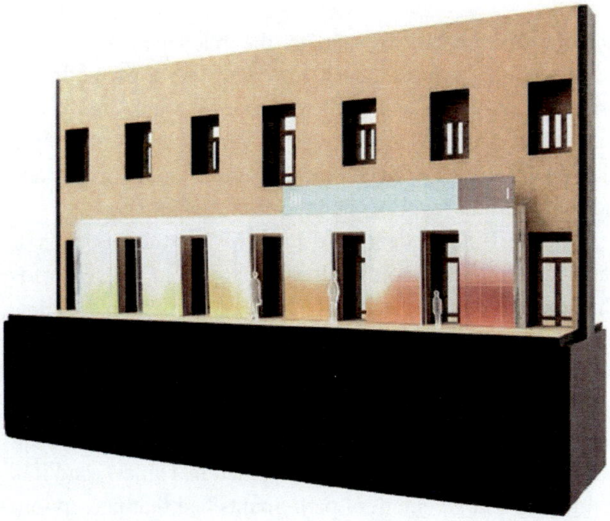

Fig. 10 Prototype for tests

- Stage I: The first stage consisted of creating a 1:50 scale model of the location where the interactive surface is to be proposed and the interactive wall itself (Fig. 10). Medium-Density Fiberboard (MDF), acrylic, black spray paint, and glue were used for the model. The visual contents for the three scenarios were printed on paper. One of the fundamental aspects of rapid prototyping is to avoid using complex technologies to simulate the interactive behavior of the proposal. In this way, all alteration of visual content was handled "manually" by the researcher, ensuring that the illusion that the prototype created simulated the performance of the interface in a similar way to the real scenario.
- Stage II: The second stage consisted of receiving the participants, presenting the goals of the simulation and asking the participants to sign an informed consent form.
- Stage III: In the third stage, tests were carried out (Fig. 9). The registration of observations and comments of the participants during and after the tests, allowed to carry out a qualitative analysis of the proposal presented. Two researchers were present during the tests to ensure that no interruption was needed.
- Stage IV: After performing the tests satisfaction and usability questionnaires were given to participants who allowed to obtain a quantitative assessment of the proposal.

3.6 Stage 5—Results Analysis

A qualitative assessment and a quantitative evaluation were performed with the data from the previous stage. The results of the observations, the written suggestions, and videos recorded during the simulations constituted the data to do the qualitative assessment. The quantitative evaluation was done with the posttest questionnaire, which focused on

Table 1 Number of participants that were asked regarding the cause–effect relation between the action done (production of sound) and the visual content displayed. Participants answered questions on a scale of level 1 to level 7, where level 1 corresponds to "not noticeable" and level 7 to "very noticeable"

	Scenario 1				Scenario 2				Scenario 3			
	Iscte	EM.	Total	%	Iscte	EM.	Total	%	Iscte	EM.	Total	%
Level 1												
Level 2												
Level 3												
Level 4	1		1	5	1		1	5	1		1	5
Level 5	4	1	5	25	5	1	6	30	3	3	6	30
Level 6	4	3	7	35	3	2	8	25	6		6	30
Level 7	6	1	7	35	6	2	6	40	5	2	7	35

fundamental aspects to validate the relevance and effectiveness of the proposal, such as: being aware of the transformation of the visual content; understanding the visual content present on each scenario; the awareness of the relationship between visual content and sound properties; and the intuition level of the interactive experience.

Both the Iscte students and the EMCN students considered that the visual content have changed in all scenarios which allowed us to conclude that all participants understood that the proposed panel will be in constant transformation.

Participants were asked in what extent they perceived that there was an alteration of the visual content related to an action in the space. Results are presented in Table 1, and it is visible that a large extent of participants had that perception. Participants were also asked how intuitive the interactive effect in the visual content caused by the sound was 45% of the participants found the experience intuitive (level 6) and 40% considered it very intuitive (level 7). The remaining 15% considered it in level 5.

EMCN students had no difficulty in locating the model in the context of the school. The students made several suggestions to the visual content presented, namely a greater variation of visual content, in addition to the association of frequencies with a color scale and variation in the size of the stain associated with the amplitude. It was also suggested to use timber as a performance criterion. EMCN students inquired if two instruments were playing with different frequencies, the installation would recognize the two instruments and react differently to each one. Finally, a positive observation was made regarding the implementation of the proposed installation for the EMCN. One student comment that this approach would be a much more interesting way of alerting students for the noise they cause in the corridors. Currently, the way to prevent this situation is by using papers posted on the walls of the corridors with the message "Silence, there are classes taking place." In general, the participants showed interest in the proposal.

4 Discussion and Concluding Remarks

The goal of the proposal was to create a surface for a school of music that would act as an informative piece with the aim to change users' behaviors to promote a better study environment as well as to be entertaining while users pass by in the public areas of the school.

The development of the proposal for an interactive wall for EMCN focused on the configuration of the interactive installation, the relation with the place where it will be inserted, the definition of the performance criteria of the interaction and the creation of visual content. Tests were carried out with a low-fidelity prototype. The used methodology constitutes an inexpensive and effective method to assess the impact of a design a priori its execution and therefore prevents wrong decisions to be made. Since the proposal implies a direct relation with users (person–machine interaction), it was essential to carry out tests that simulated the way the relation is established in order to assess its effectiveness. The usability and satisfaction tests performed to assess the proposed system reveled that users understand the message of the wall regarding the noise control and they consider this method as an interesting way of doing such a control.

The result of this study allows a reflection on how media façades can add value and solve identified problems in society, in the present case a problem identified by the end users of EMCN—students and employees. We envision that such an interactive wall as the one proposed could be used in other environments where an automatic triggered soft alert (rather than, e.g., an audio alarm or an alert given by a person) would potentiate a more conscious behavior by users. Places like libraries, working areas, hospitals and clinics, and schools (when classes are occurring) are good examples for the use of such devices.

Another reflection that emerges from this study is how will multimedia surfaces reshape our relationship with the built environment, and in what extend will it change the way we appropriate those spaces. Such surfaces introduce a type of mutable element with digital contents that does not exist in traditional static physical buildings. Such a new intruder in buildings will have an impact on the organization of space and on its appropriation by users. For Oddey and White [12], large digital displays can be appropriated as windows that connect the interior space to a particular visual field extending the physical space and playing with the notion of the presence. Digital contents will no longer be confined in our pockets and bags but are already assuming positions on the rigid surfaces that we are accustomed to be static and with which we will start to interact with. Snibbe and Raffle [14] also show how media walls promote social engagement and how we can control user behavior by choosing different interaction techniques. Kuikkaniemi et al. [10] say that such interactive walls have the potential to motivate individual and group interaction and will transform passive viewing into an involved performance. These authors also highlight that digital displays augment spatial information and make it more flexible and engaging compared to when analog signs are used.

Acknowledgements Authors acknowledgement the advisory of Professor Pedro Faria Lopes. This publication was partially funded by Fundação para a Ciência e Tecnologia under project UIDB/04466/2020 and UIDP/04466/2020.

References

1. Achten, H. (2014). One and many: An agent perspective on interactive architecture'. In *ACADIA 2014 Design Agency: Proceedings of the 34th Annual Conference of the Association for Computer Aided Design in Architecture* (pp. 479–480). Toronto: Riverside Architectural Press.
2. Achten, H., & Kopriva, M. (2010). A design methodological framework for interactive architecture. *Future Cities* (pp. 169–177). ETH Zurich: Zurich.
3. Bullivant, L. (2007). '4dsocial', *AD 4dsocial Interactive Design Environments, 77*(188).
4. Chen, Y. V., Qian, Z. C., & Lei, W. T. (2016). Designing a situational awareness information display: adopting an affordance-based framework interaction design. *Informatics, 3*(6). https://doi.org/10.3390/informatics3020006.
5. Endsley, M. E. (2001). Designing for situation awareness in complex system the challenge of the information age. In *Proceedings of the Second International Workshop on Symbiosis of Humans, Artifacts and Environment.*
6. Fischer, P. T., & Hornecker, E. (2012). Urban HCI: Spatial Aspects in the Design of Shared Encounters for Media Façades. In *Proceedings of the 2012 ACM Annual Conference on Human Factors in Computing Systems* (pp. 307–316). Available at: http://www.ehornecker.de/Papers/UrbanHCISpatial.pdf.
7. Gehring, S., & Wiethoff, A. (2014). Interaction with Media Façades. *Informatik-Spektrum, 37*(5), 474–482. https://doi.org/10.1007/s00287-014-0818-0.
8. Hoferlin, B., et al. (2011). Interactive auditory display to support situational awareness in video surveillance. In *The 17th International Conference on Auditory Display (ICAD-2011).* Budapest, Hungary.
9. Indaverea, M. (2010). Fachadas Tecnológicas: de como la arquitectura se convierte en un media. In *Actas del Segundo Congreso Internacional sobre Imagen, Cultura y Tecnología* (pp. 59–70). Universidad Carlos III.
10. Kuikkaniemi, K., et al. (2011). Interactive screens will life. *Computer*, 40–47.
11. Mano, A., & Campos, J. C. (2004). Aplicação de um cognitive walkthrough: estudo de caso. In *Interação 2004—1a. Conferência Nacional em Interação Pessoa-Máquina* (pp. 256–258).
12. Oddey, A., & White, C. (2006). Introduction: The potentials of space. In A. Oddey & C. White (Eds.), *The potentials of spaces: The theory and practice of scenography & performance* (pp. 11–24).
13. Rettig, M. (1994). Prototyping for tiny fingers. *Communications of the ACM, 37*(4), 21–27.
14. Snibbe, S. S., & Raffle, H. S. (2009). Social immersive media. In *CHI 2009 ~ New Media Experiences 2*, 1447–1456.

Beatriz Couto Couto is an architect, M.Sc. in Architecture with the theme Boeralis Sound: Parede Interactiva em Projeto de Arquitetura. Since her graduation, Couto has earned experience as an architect by working at Cidades Invisiveis, CHP Arquitetos, and CF Arquitetos. She kept learning real-time video manipulation and the applicability to architecture by taking a course at DubLab, and at AntiFrame, she learned how to plan and organize art exhibitions. After a degree in UX and UI Design at EDIT, she started working as a UX/UI Consultant at CGI where she has been working on different projects related with Energy and Utilities. She is also a Scrum Master Certified (CSM) from Scrum Alliance.

Sara Eloy Eloy is an architect, Ph.D. in Architecture. Eloy main areas of research include digital technologies applied to architecture, shape grammars, virtual and augmented reality, analysis of the building space, namely considering space perception, and housing rehabilitation. She is the Director of the Information Sciences Technologies and Architecture Research Center (ISTAR) where she is a fellow researcher. She has curated several exhibitions being the latest "Artificial Realities: Virtual as an Esthetic Medium for Architectural Ideation," in the scope of Lisbon Architecture Triennale 2019. She has participated in national and international research projects and published her work in several journals.

Adaptability and Resilience of the Built Environment

Check for updates

Space, Adaptability and Use: Triangulating Research Criteria for Adaptability Retrieval

Carolina Coelho

Abstract

This paper presents a methodology to assess spatial adaptability by means of a broad range of criteria from intertwined research cultures, from a more analytical to a more cultural analysis of space use. In order to thoroughly assess all these stages of the building's lifecycle that impact on its adaptability, it focuses on several procedures undertaken for collecting data on this regard, but foremost, on the formal methods in architecture and spatial visualisation of the results. Thus, it proposes a multi-criteria methodology composed by three sequential milestones that describe specific building conditions on this behalf, towards a final retrieval of the building's adaptability. Milestone I describes the building's physical features and overall configuration, Milestone II describes all the possible activities it can allocate and Milestone III describes its effective appropriation. Finally, the retrieval of spatial adaptability is undertaken by a correlation of the most representative variables prior obtained in each stage. It is proposed that the three milestones enable the collection of a wide array of complementary data, whose representations also vary, but that overall provide a robust outlook on spatial adaptability.

Keywords

Spatial adaptability • Use • Triangulated research • Multi-criteria analysis • Entropy

C. Coelho (✉)
Centre of 20th Century Interdisciplinary Studies, FCTUC
Department of Architecture, University of Coimbra, Coimbra, Portugal
e-mail: carolina.coelho@uc.pt

1 Introduction—Gathering Research Criteria

This paper discusses a methodological approach to spatial adaptability by means of a broad range of criteria from intertwined research cultures. It argues that the use of diverse procedures, from a more analytical to a more cultural analysis of space use, provides an in-depth outlook on the building performance overall and, specifically, a more robust conclusion on its spatial adaptability potential.

For that purpose, the paper departs from the premise that adaptability ranges from the original design conceived by the architect, in regard to the intrinsic attributes provided during the design; to the spatial fruition the building enables, through appropriation by its users over time. It is also based on the definition of adaptability as placed by Mário Krüger as "the ability of the built form to maintain compatibility between activities and spaces, as those vary"[1] ([21], p. 1169) in line with the work of his peer in Cambridge, William Fawcett [8].

In order to thoroughly assess all these stages of the building's lifecycle that impact on its adaptability, from a wide-ranging research[2] [4], this paper focuses on the several procedures undertaken for collecting data on spatial adaptability, but foremost, on the formal methods in architecture and spatial visualisation of the results. Thus, it proposes a multi-criteria methodology composed by three sequential milestones that describe specific building conditions on this behalf, towards a final retrieval of the spatial adaptability of the building under assessment, resorting to the "description retrieval" mechanism as defined by Hillier and Hanson [17]

[1]Translation from the original quote in Portuguese.

[2]See Coelho, C.: *Life within architecture from design process to space use. Adaptability in school buildings today—A methodological approach.* Ph.D. Thesis in Architecture, Departamento de Arquitetura da Faculdade de Ciências e Tecnologia, Universidade de Coimbra, Coimbra [4].

S. Eloy et al. (eds.), *Formal Methods in Architecture*, Advances in Science, Technology & Innovation,
https://doi.org/10.1007/978-3-030-57509-0_14

and subsequently interpreted by Peponis [25] and Psarra [26].

This three-stage approach claims that assessing adaptability comprises: (I) the space and its features designed by the architect, (II) the activities that can be allocated to each space throughout its use, (III) the appropriation and behaviour in space that is site-specific by each community of users. Hence, it argues that adaptability is a spatial feature that is outlined by the architect in the design, which is accomplished during its effective use, and which can change according to the users' spatial needs through time. Lastly, it also proposes that adaptability is assessed not only by analytical procedures but also by the people's overall behaviour in space, as well as their individual sense of belonging, by means of a more cultural approach to evidence-based design.

In order to demonstrate how this methodology is developed, which criteria it involves, and how it visually displays that data, it has been applied to a contemporary Portuguese school in Coimbra, that is both a regular curriculum school and also a Music Conservatory. This choice on the case study is justified by the current need for spatial adaptability in schools, and even more so, in schools with several student profiles. This is due to the evolving changes in the pedagogical curriculum, faculty and student body and technological devices and strategies for knowledge transmission. Thus, schools hold a pressing urge for spatial adaptability, for coping with the current and future teaching and learning requirements and for continuing to support the upcoming formal and informal pedagogical activities in both formal and informal active learning environments throughout all the building. Quoting the Organisation for Economic Co-Operation and Development:

> Educational facilities need to accommodate both the known and identifiable needs of today, and the uncertain demands of the future. (OECD [24], p. vii)

2 A Triangulated Methodology for Assessing Spatial Adaptability

2.1 Milestone I—Description of the Space

The initial stage of this methodology focuses on the space designed by the architect and the attributes conceived at an earlier phase that physically enable it to shelter a larger or narrower, specialised or more general, site-specific or wide-ranging array of uses.

Proceeding from Brand's [2] "shearing layers of change", in which six spatial components are defined—specifically: "site", "structure", "skin", "services", "space plan" and "stuff"—these are associated with a distinctive lifespan and, hence, a higher or lower frequency on the need for change through time. Thus, the design choice on either of these elements may constrain the building's overall performance and, ultimately, enable or hamper the pursuit of a certain type of uses on that particular space. Likewise, Habraken's [12] concepts of "supports" and "infill" also relate the building's elements to their adaptative potential, which is also developed on his further work *The structure of the ordinary: form and control in the built environment* [13], assuming the building's intrinsic change through time and its embedded relation with the inhabitants:

> We must learn to look afresh at the intricate ongoing symbiosis between people and built matter. (Habraken [13], p. 7)

Consequently, Milestone I associates spatial adaptability with the feasibility capacity enabled by the building's architectural attributes and, so, it describes the building according to its physical features and overall configuration. This can be assessed by a detailed description of the space regarding: dimension—height, length and weight; configuration—openness or partitioning; coating—on floors and ceilings; networks—power access and connectivity; environmental conditions—temperature and acoustics; and accessibility—direct or by lift, staircase or ramp. So, this methodology departs from the architectural drawings: plans, sections and elevations, at a general scale, and then proceeds to a subsequent detailed examination on its materialisation, in order to perceive the building's physical provision and, thus, its capacity constraints towards its intended function. This is followed by a study of the activities related to each space, providing an initial outlook on the spaces' general uses.

For such purposes, the data is collected in coloured schemes edited on the building's plans and exploring the location of the diverse activities enabled on each space: the specialised and the general rooms, the areas for larger or smaller audiences or groups, for formal and informal uses, or even for each pedagogical curriculum (Fig. 1). This will depict the possible hybridisation or segregation of uses throughout the spatial sample.

Specifically, from the analysis of the proposed case study as a means to materialise and validate this methodology, it is possible to conclude on a mixed-use building, where different pedagogical curricula are assembled, and whose learning environment is enriched by spaces for a variety of learning situations and users, which adequately depicts the contemporary learning paradigm and that already conveys a high adaptability potential:

> Schools, like people, should be one-of-a-kind organisms. They need space that is adaptable to changing needs. They need flexible facilities that can, at a minimal cost, be converted from classrooms to seminars, to individual study space, to study rooms for five students. Architecture for education should facilitate the evolution of future schools. (Brubaker [3], p. 192)

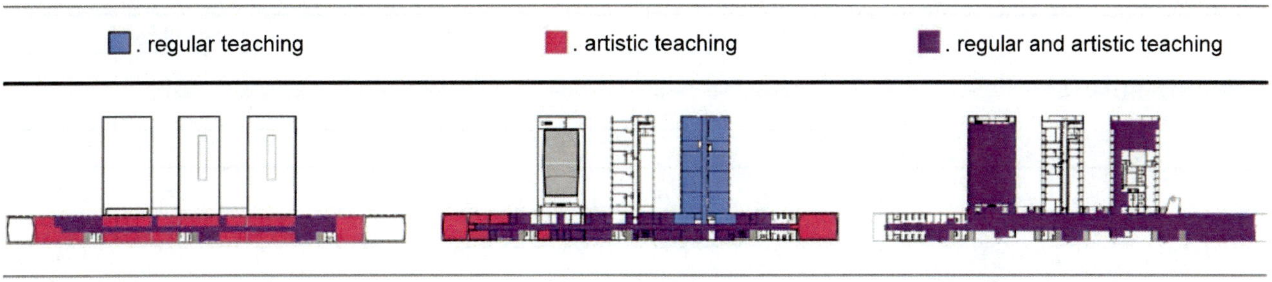

Fig. 1 Milestone I—Plans of the case study identifying the types of learning spaces for each curricular profile by colour

Subsequently, and still under this milestone, the application of the methodology proceeds with a space syntax analysis that describes the morphology of the building in regard to the integration of the spaces that compose the case study. This is based on the assumption that social dynamics is deeply rooted on the spatial system, as sustained by the original studies at the Bartlett Unit for Architectural Studies that developed this approach:

> The idea is that, with an objective and precise method of description, we can investigate how well environments work, rigorously relating social variables to architectural forms. (Hillier et al. [16], p. 49)

For a more insightful examination, the building is assessed by means of convex spaces, axial lines and isovists [5, 6] that will provide maps with the warmer and colder colours, as traditionally used by space syntax, enabling a clear understanding of the shallow and deep spaces and their overall representativeness. This can be complemented both by the accurate figures provided by Depthmap for integration, but also by other attributes, such as connectivity and mean depth, whose maps can add to the interpretation of the results attained. In addition, the intelligibility analysis also bears significance for this study, explaining how the system is perceived by its users and, ultimately, relating spatial configuration with behaviour, movement and standing in space.

Furthermore, and within the visibility analysis, the study on isovists also explains the patterns of co-presence and encounters in space, and hence, the ones more prone for multi-use, namely by social interaction and informal gatherings. This assumes that spaces with high visual integration and isovists with a broader visual scope enable a more intense use and potentially high spatial adaptability, as described by Benedikt: "Potentially this may allow a number of spatial behaviors to be explained and predicted as field-dependent behaviors." (Benedikt [1], pp. 48–49).

All in all, the visual outputs from the space syntax analysis consist of the several maps for integration as the global measure that "takes into account the relations of a space to every other space in the system" (Hillier and Hanson [17], p. 109), but also for the remaining local attributes. In addition, this can be added to the detailed scrutiny of the figures provided by Depthmap, and all this data can be compared by means of charts or scattergrams for a more immediate and visual output (Figs. 2 and 3).

This has been applied to the case study, and for a more detailed study, it has been done by floorplans for a specific approach to each respective configuration. The study also focused on the maximum, average and minimum values and the reasons why the results demonstrated a broader or narrower range on the figures encountered. Moreover, as some figures highly diverged from the remaining, the outliers have also been dismissed at a final stage of comparison, for a more fine-grained analysis.

From the application of this milestone to the case study, it has been concluded that the corridors are the most integrated spaces and have the highest connectivity, but as the three floorplans hold different configurations, it was also found that it was the open space on the ground floor, considered both as a standing space and a circulation, that held the highest integration and also the highest visibility. This is developed by the axial line analysis, which we argue to be the most appropriate space syntax approach to adaptability, since it gathers a set of convex spaces in an overall entity:

> Therefore, when analysing this school by means of axial lines rather than convex spaces, the outcomes are more representative of the spaces' actual layout and foremost of its pedagogical enriched curricula. Corridors with movement, associated with socialisation and communication, can be considered active learning environments, having in this school a very integrated position, metaphorically denoting its pedagogical significance. (Coelho and Heitor [5], p. 10)

2.2 Milestone II—Description of All Possible Activity Allocations

Milestone II of the methodology proposes to assess spatial adaptability by describing all the possible activities that the building can allocate in each space. This is based on the data collected on the prior stage and associates the building's

ATTRIBUTE SUMMARY		Ground Floor			First Floor			Second Floor		
		convex	axial	vga	convex	axial	vga	convex	axial	vga
		rn								
INTEGRATION	minimum	0,76	1,39	1,78	0,83	1,39	1,69	1,13	1,86	2,08
	average	1,34	2,9	5,15	2,08	2,26	3,32	2,7	6,05	4,24
	maximum	3	7,79	9,24	7,35	5,33	6,76	14,17	113,29	8,07

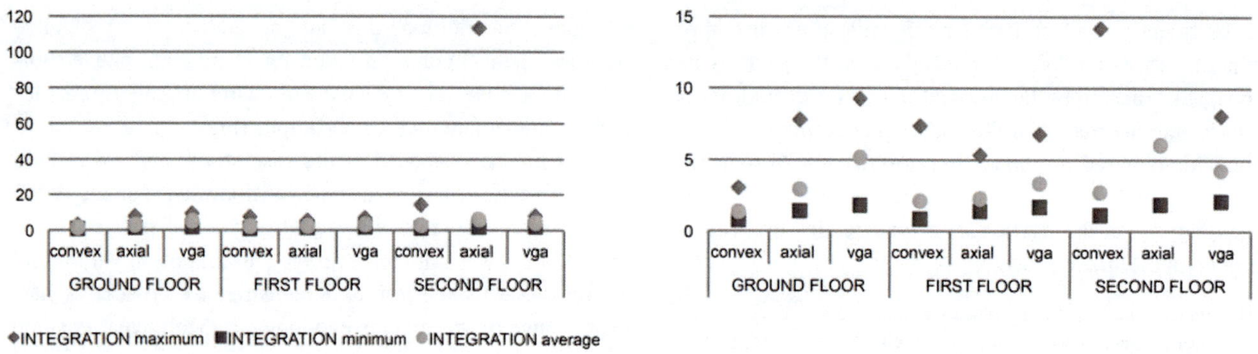

Fig. 2 Milestone I—Integration r_n comparison between convex, axial and VGA analysis (table and scattergram with and without outliers)

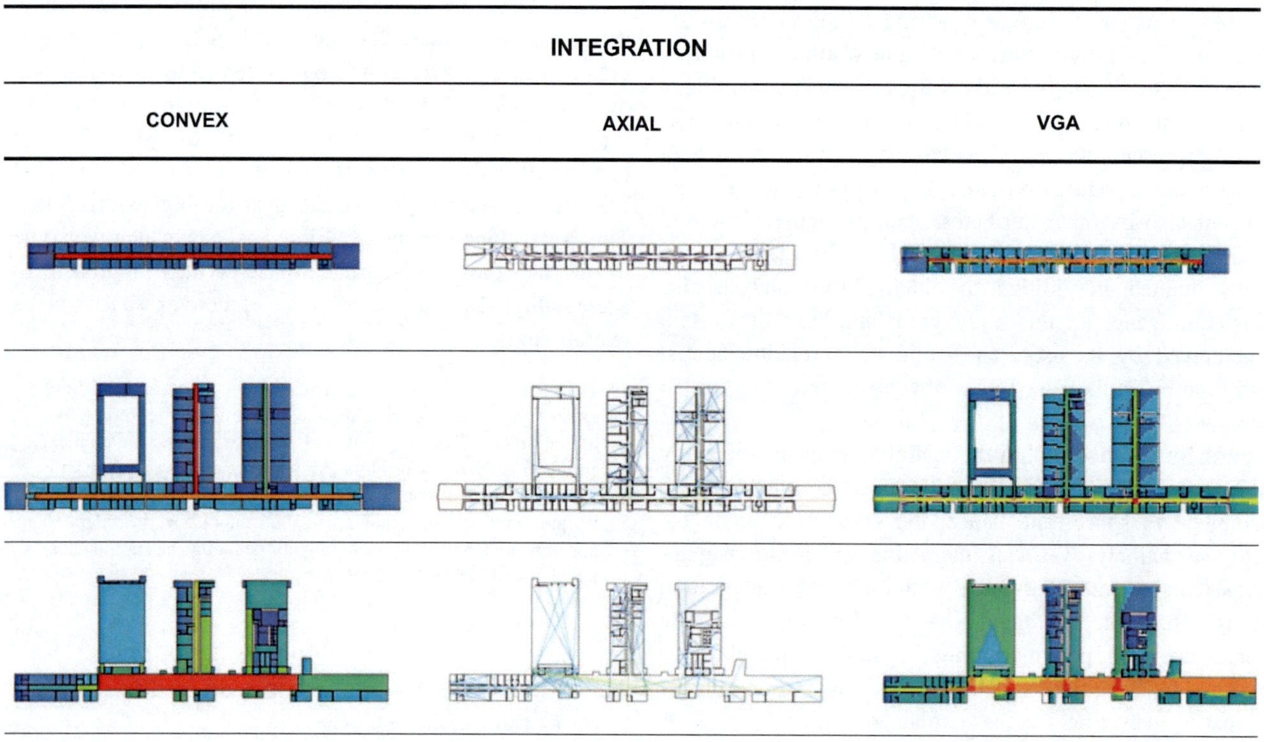

Fig. 3 Milestone I—Integration map for convex, axial and VGA analysis

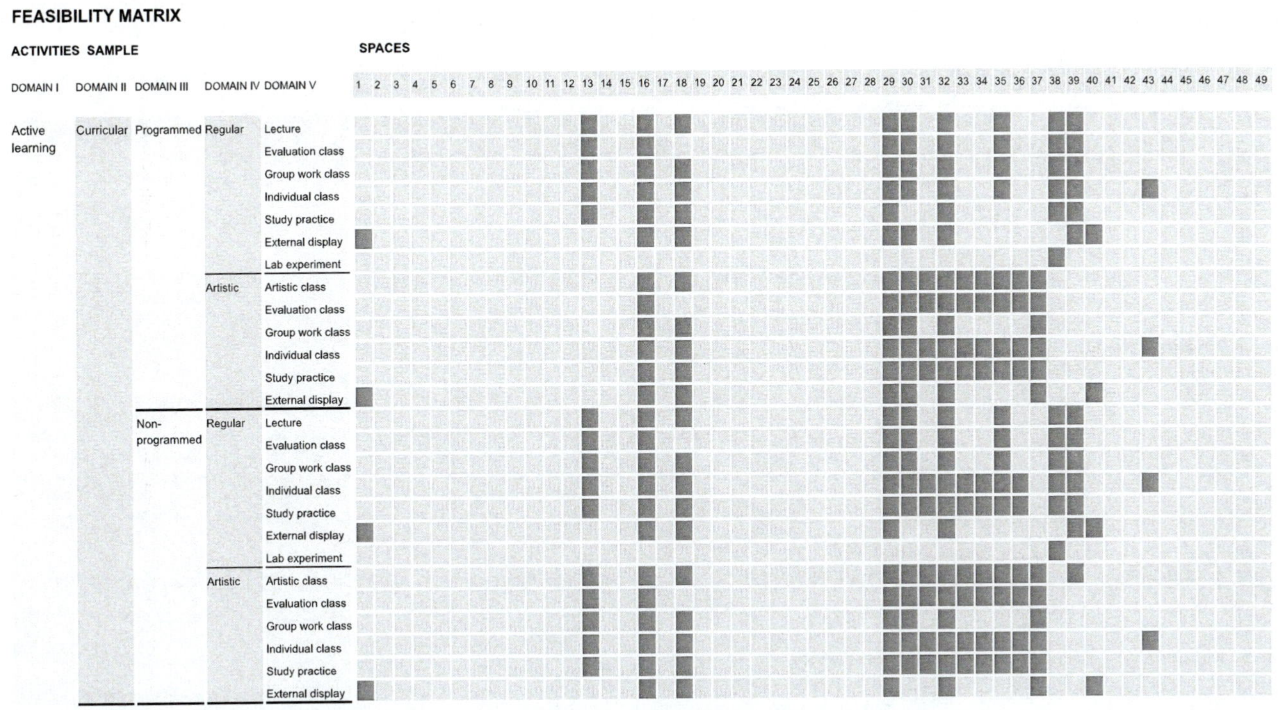

Fig. 4 Milestone II—Feasibility matrix (partial)

physical attributes with the activities that it can admit, defining its overall allocational potential.

If the initial milestone is represented by a more graphical outcome by means of the coloured schemes on the plans and the maps, this milestone is much more analytical and relies on a set of matrices to match the activities to the spaces according to their thresholds of feasibility. This is achieved by listing the case study's existing spaces, the main activities defined for that building and the spatial features that may condition or support its use. Then, it is possible to match all the possible activities that each identified space can hold by means of a feasibility matrix that conveys the spaces' "loose-fit" potential (Fawcett [7], p. 5).

This has been applied to the case study as perceived on Fig. 4, which displays each binary match between a space and an activity by a black square. Hence, by examining this chart, it is possible to establish another immediate conclusion on the allocational potential of each space according to the density and scattering of the black squares. Firstly, if the matrix presents a wide array of black squares, this implies that the spaces can cater for a large scope of activities, which conveys a higher spatial adaptability. Furthermore, if the activities are assorted by categories or domains—specifically curricular and supportive, programmed and non-programmed activities, for different users—the analysis of the matrix also enables additional conclusions, not only on the number of possible activities for each space, but also on their type. Ultimately, the wider in range and the more scattered the activities on the matrix, the more adaptable the space for different uses and users. According to Krüger [22]:

> Adaptability is the result of the reorganisation of the operational patterns of activities with the minimum environmental and spatial change. In this case, it should be considered the locational tolerance of activities, aiming at minimising the costs of possible changes in spatial configuration. (Krüger [22], p. 17)[3]

This occurs in certain spaces from this particular case study. Specifically, the main hall can be the set for both external and internal performances and exhibitions, for coping with the different students' profiles and for standing and moving. Similarly, spaces such as the library also demonstrate several allocations, as well as medium size classrooms, which can cater for different pedagogical curricula. On the contrary, supportive spaces such as the reception office, storage or the locker rooms, often hold a single use.

Additionally, and also under this analytical approach to adaptability, Milestone II parallels the concept of adaptability to entropy, as claimed by William Fawcett [8] in Cambridge, on his Doctoral Thesis *A Mathematical Approach to Adaptability in Buildings*. So, it resorts to the concept of entropy to depict the degree of uncertainty on the allocation of an activity to a space [18] and it weighs the

[3]Translation from the original quote in Portuguese.

ALLOCATION MATRIX _ DOMAIN III

ACTIVITIES SAMPLE SPACES

I	II	III	1	2	3	4	5	6	7	8	9	10	11	12	13	14	15	16	17	18	19	20	21	22	23	24	25
I a)	II a)	III a)	2	0	0	0	0	0	0	0	0	0	0	0	5	0	0	12	0	10	0	0	0	0	0	0	0
		III b)	2	0	0	0	0	0	0	0	0	0	0	0	10	0	0	12	0	10	0	0	0	0	0	0	0
	II b)	III c)	15	0	0	0	0	0	0	0	0	0	0	0	12	0	0	15	12	15	9	12	0	0	0	2	4
		III d)	15	3	3	3	3	0	0	0	0	0	0	0	12	0	0	15	12	15	9	12	0	0	0	2	4
I b)	II c)	III e)	9	0	0	0	0	0	0	0	2	0	0	0	6	0	0	6	9	6	9	9	0	0	2	4	4
		III f)	12	6	6	6	6	3	3	3	0	2	2	2	6	2	2	6	12	9	12	9	2	2	2	4	4
	II d)																										
allocations per space			55	9	9	9	9	3	3	3	2	2	2	2	51	2	2	66	45	65	39	42	2	2	4	12	16

I	II	III	26	27	28	29	30	31	32	33	34	35	36	37	38	39	40	41	42	43	44	45	46	47	48	49
I a)	II a)	III a)	0	0	0	12	11	4	12	4	4	8	4	6	6	6	2	0	0	2	0	0	0	0	0	0
		III b)	0	0	0	12	10	6	12	6	6	9	6	6	6	7	2	0	0	2	0	0	0	0	0	0
	II b)	III c)	0	0	0	15	13	6	15	6	6	7	6	11	7	11	12	2	2	6	0	0	0	0	2	0
		III d)	0	0	0	15	13	6	15	6	6	7	6	11	7	11	12	2	2	6	0	0	0	0	2	0
I b)	II c)	III e)	6	2	2	6	6	0	6	0	0	0	0	2	2	2	6	4	4	4	2	2	2	0	1	0
		III f)	7	2	2	6	6	0	6	0	0	0	0	2	2	2	9	4	4	4	2	2	0	3	0	2
	II d)																									
allocations per space			13	4	4	66	59	22	66	22	22	31	22	38	30	39	43	12	12	24	4	4	2	3	5	2

DISTRIBUTION MATRIX_DOMAIN III

ACTIVITIES SAMPLE SPACES

I	II	III	1	2	3	4	5	6	7	8	9	10	11	12	13	14	15	16	17	18	19	20	21	22	23	24	25
I a)	II a)	III a)	2/55	0	0	0	0	0	0	0	0	0	0	0	5/51	0	0	2/11	0	2/13	0	0	0	0	0	0	0
		III b)	2/55	0	0	0	0	0	0	0	0	0	0	0	10/51	0	0	2/11	0	2/13	0	0	0	0	0	0	0
	II b)	III c)	3/11	0	0	0	0	0	0	0	0	0	0	0	4/17	0	0	5/22	4/15	3/13	3/13	2/7	0	0	0	1/6	1/4
		III d)	3/11	1/3	1/3	1/3	1/3	0	0	0	0	0	0	0	4/17	0	0	5/22	4/15	3/13	3/13	2/7	0	0	0	1/6	1/4
I b)	II c)	III e)	9/55	0	0	0	0	0	0	0	1	0	0	0	2/17	0	0	1/11	1/5	6/65	3/13	3/14	0	0	1/2	1/3	1/4
		III f)	12/55	2/3	2/3	2/3	2/3	1	1	1	0	1	1	1	2/17	1	1	1/11	4/15	9/65	4/13	3/14	1	1	1/2	1/3	1/4
	II d)																										
allocations per space			55	9	9	9	9	3	3	3	2	2	2	2	51	2	2	66	45	65	39	42	2	2	4	12	16
S = ENTROPY			1,578	0,637	0,637	0,637	0,637	0,000	0,000	0,000	0,000	0,000	0,000	0,000	1,732	0,000	0,000	1,729	1,379	1,746	1,378	1,376	0,000	0,000	0,693	1,330	1,386

I	II	III	26	27	28	29	30	31	32	33	34	35	36	37	38	39	40	41	42	43	44	45	46	47	48	49
I a)	II a)	III a)	0	0	0	2/11	11/59	2/11	2/11	2/11	2/11	8/31	2/11	3/19	1/5	2/13	2/43	0	0	1/12	0	0	0	0	0	0
		III b)	0	0	0	2/11	10/59	3/11	2/11	3/11	3/11	9/31	3/11	3/19	1/5	7/39	2/43	0	0	1/12	0	0	0	0	0	0
	II b)	III c)	0	0	0	5/22	13/59	3/11	5/22	3/11	3/11	7/31	3/11	11/38	7/30	11/39	12/43	1/6	1/6	1/4	0	0	0	0	2/5	0
		III d)	0	0	0	5/22	13/59	3/11	5/22	3/11	3/11	7/31	3/11	11/38	7/30	11/39	12/43	1/6	1/6	1/4	0	0	0	0	2/5	0
I b)	II c)	III e)	6/13	1/2	1/2	1/11	6/59	0	1/11	0	0	0	0	1/19	1/15	2/39	6/43	1/3	1/3	1/6	1/2	1/2	1	0	1/5	0
		III f)	7/13	1/2	1/2	1/11	6/59	0	1/11	0	0	0	0	1/19	1/15	2/39	9/43	1/3	1/3	1/6	1/2	1/2	0	1	0	1
	II d)																									
allocations per space			13	4	4	66	59	22	66	22	22	31	22	38	30	39	43	12	12	24	4	4	2	3	5	2
S = ENTROPY			0,690	0,693	0,693	1,729	1,745	1,373	1,729	1,373	1,373	1,381	1,373	1,611	1,684	1,615	1,600	1,330	1,330	1,705	0,693	0,693	0,000	0,000	1,055	0,000

Fig. 5 Milestone II—Allocation and distribution matrices

entropy of each space according to Shannon and Weaver's formulation [27]:

$$S = -\sum_{i} p_i \ln p_i$$

For calculating the entropy of a space, a set of matrices are needed, which are paralleled with mathematical calculations [6]. Specifically, this methodology departs from the feasibility matrix and then proceeds to an allocation matrix and to a subsequent distribution matrix, and lastly, it estimates the final results by using the formulation (Fig. 5). This enables to associate a figure to a space, and to rank them according to their adaptability potential.

When applying this to the case study of the music school, this methodology determined that the spaces with higher entropy were the library, music studios, the main hall and large collective spaces, which can cater for larger or smaller audiences.

2.3 Milestone III—Description of Effective Events

After assessing all the possible activities that can be allocated to each space, Milestone III recalls the assumption that spatial adaptability is validated throughout the building's actual use, and so, it describes its effective appropriation. This approach is based on literature references such as Whyte's [28] studies in American cities and Gehl's [11] in Northern Europe, mostly focused on public spaces and collective uses.

Ultimately, these outputs may either limit the scope of all the possible activities, or may also enlarge it by pointing out different uses that occur in each space and which may differ from the envisioned ones on the design process, quoting Worthington:

> Users are continuously making design decisions as they move scenery around on a day-to-day basis to match changing demands. (Worthington [29], p. 17)

For assessing effective use, Milestone III resorts to procedures such as observations and focus groups, which depict the actual building's fruition throughout and, on a more fine-grained analysis, how is it perceived by different groups of users and what are the reasons that support the remarked behaviour.

First, this stage is based on observation matrices that report the users' dynamics on the building's plans. These recognise their overall density in each space and can also specify both the schedules in which these occur and the type of activities undertaken there.

The visual outputs from this procedure comprise drawings in the plans, using lines to represent movement, arrows to inform on the direction of that movement, dots for locating the activities in space and circles to pinpoint the location of the neutral observers. The thickness of the lines drawn by neutral observers also indicates the density of that movement and a colour scheme can be used to indicate the distinctive activities. Furthermore, the observation matrices may also be completed by field notes taken by the observers for a more detailed report.

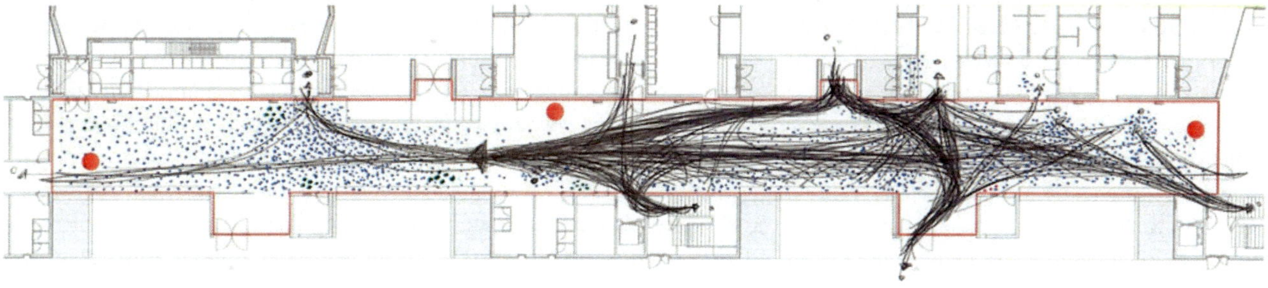

Fig. 6 Milestone III—Observation matrix

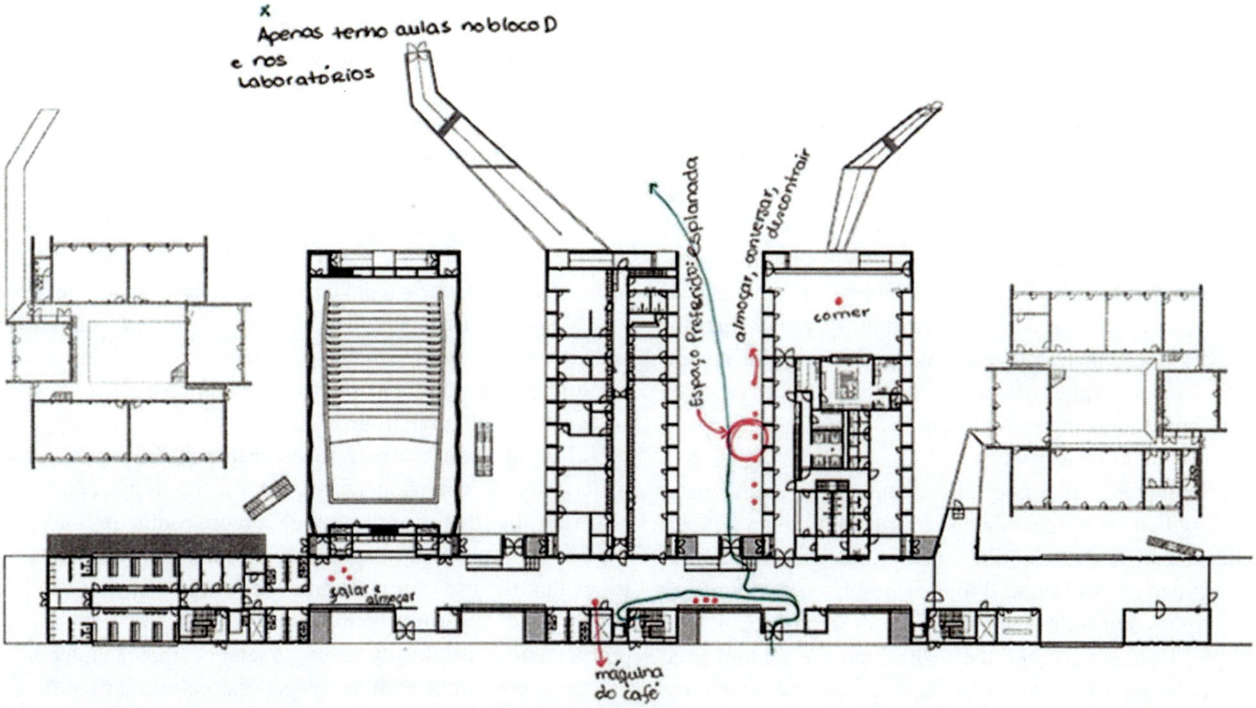

Fig. 7 Milestone III—Drawing made during a focus group

In the case study of this school, observations have identified whether the spaces are more prone to formal or informal activities, programmed or non-programmed events, or even as a set for learning or only as supportive of the learning process. In a compared analysis, it is possible to conclude that the spaces with higher densities are the circulations and the ones that act as common spaces for the general school community (Fig. 6).

Moreover, this procedure also enables a time comparison, by paralleling the same space's fruition in different time schedules, concluding on its use on the morning arrival, the afternoon exit, the mid-day break or the regular working hours. This is particular evident on a school, whose space use is defined by a weekly schedule that delineates the spaces' formal activities.

After observations of collective activities, users are set into groups and asked to comment on the results, adding further information on their appropriation of space, which may potentially differ amongst the several groups. This is reported by photos, sound and video recording, drawings and notes on the plans from each participant, regarding personal appropriation of space and their most frequent individual routes (Fig. 7).

In a school, the focus groups identified for this milestone are teachers and students from each pedagogical profile, which determined different uses from each one, and foremost, a more formal functional scheme by the teachers, as opposed to the students, and a more spontaneous and wide-ranging appropriation from the artistic school students rather than the regular curricular students.

3 Conclusion—Criteria Correlations for Retrieval of Spatial Adaptability

The retrieval of spatial adaptability can be undertaken by a correlation of the most representative variables obtained from each milestone, by means of scattergrams that enable to examine how the independent variable explains the dependent variable and how statistically valid that correlation is according to the probability values (*p* values).

This can be translated by the correlation between the data collected on Milestones I and II, resulting on integration and entropy scattergrams, whose determination coefficients and regression lines inform on how the variables are related and how meaningful the correlations are (Fig. 8).

When applied to the case study, it has been concluded that the highest entropy spaces are also spaces with high morpho-syntactic integration, mostly when this approach correlates axial line integration to axial line entropy [6], rather than convex spaces per se. On a closer look and resorting to scattergrams that depict the values by floor plan, it is concluded that the highest correlation is for the ground floor due to its scattered spatial distribution, rather than for the upper floors that have a double-loaded corridor layout. This is also a floorplan with a high intelligibility, as reported by the space syntax analysis and its correlation between integration and connectivity.

Furthermore, Milestone III complements this data, informing that some spaces are used differently than initially intended, and it circumscribes all possible activity allocations defined on Milestone II to the ones that actually occur. This implies constraints on the use of some spaces. For example, large spaces like the auditorium are seldom used by smaller groups, due to environmental and economic reasons: first, the users explained that larger spaces did not conveyed a sense of comfort and belonging in space; and secondly, is was not cost-effective to hold those activities on over-provided spaces. On the contrary, medium-size classrooms enable a wide scope of learning activities, and the niches on the corridors or on larger spaces are frequently occupied by groups of students, particularly when these are well-integrated spaces and with high connectivity to the main corridors and formal classrooms. In addition, it was also concluded that some furniture acts as a gathering device, whose location determines a meeting spot, or whose features enable a more comfortable space for standing with the peers. This sustains that other criteria impact spatial adaptability, namely: informal fruition, the sense of belonging in space and management decisions.

Besides, light, heating and acoustics are also determining factors for informal and spontaneous social gathering, rather than for formal events that are previously determined by the school management on the weekly schedules. Thus, this methodology recognises that the definition and design of spatial features–such as: layout, configuration, dimension, coating, environmental conditions, accessibility and furniture—are relevant for adaptability.

Lastly, after gathering the data from different research provinces and critically assessing their correlations, it has been concluded on a high adaptability from spaces where formal and informal activities may occur, as well as moving and standing, from which the main hall is the most paradigmatic example, acting as a "learning street" [14] and an active learning environment.

All in all, the three milestones enable the collection of a wide array of complementary data, whose representations also vary, but that overall provide a more robust outlook on the building's spatial adaptability. Hence, this paper establishes that the use of a wide array of formal methods can be congregated towards a more robust outlook on a common problematic. Specifically, it is concluded that adaptability gathers inputs from both analytical and cultural criteria:

- Milestone I—aims to perceive the space's ability to allocate the uses. Thus, it resorts to methods for detailing space and its features, first, by architectural drawings: plans, elevation and sections at different scales; and

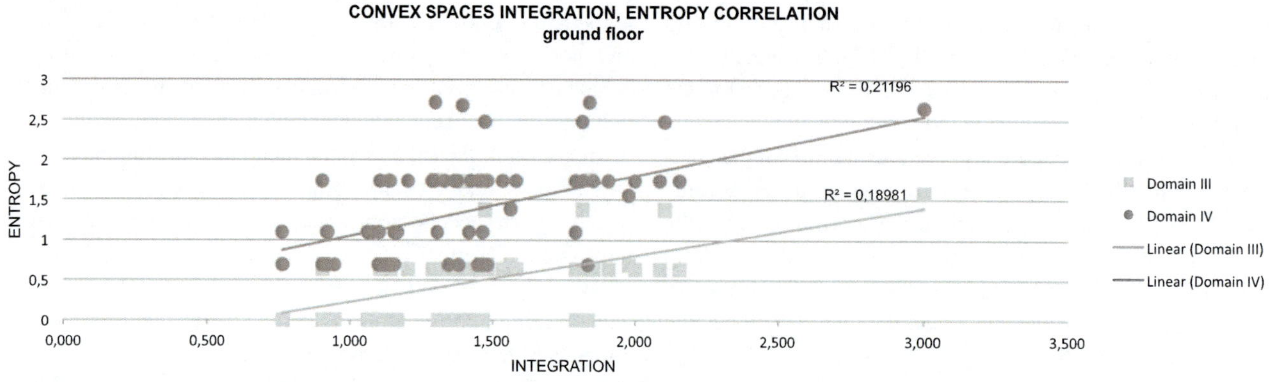

Fig. 8 Correlations—Convex spaces correlation scattergram—Integration, entropy for the ground floor

secondly, by a space syntax analysis using graphs and maps and, eventually, tables and scattergrams for further analysis.

- Milestone II—focuses on all possible uses. So, it lists the set of activities that can be allocated to each space and then uses matrices to describe the biunivocal matches between spaces and activities. This is the most analytical procedure towards adaptability, which also engages mathematical formulations, namely for weighing the entropy of a space and for pairing it to its adaptability potential, which can also be visually explained by charts and matrices.
- Milestone III—embraces appropriation. Therefore, it relates to the activities that effectively take place in a real-life scenario. This is necessarily related to the users and, logically, comprises social research methods. Observations and focus groups use drawings on plans and field notes that can be written both by the participants and also by neutral observers, depicting routes, densities and types of events in space. Likewise, the records from these meetings and from the observations, by means of photos, sound and video recording also display detailed information on individual appropriation. Finally, the spatial features that enable these observed or described behaviours in space can be examined and improved.
- Description retrieval of adaptability—is achieved by a critical assessment of each procedure, and by determining quantitative and qualitative correlations between them all, specifically between syntactic integration, spatial feasibility and informal appropriation. This is visually established by scattergrams that correlate the most significant values from each of the prior milestones, calculating their determination coefficients, the trendlines and the p values of overall correlations, or of specific ones for further detail.

Overall, the proposed methodology consists in a multi-criteria analysis, that is also diverse in the research areas that it engages and on the outcomes that it provides, aiming towards a triangulated conclusion that entangles different architectural approaches to space and, ultimately, a more detailed and enlarged understanding on the concept of adaptability. For this research this is very representative, acknowledging that providing for adaptability involves introducing it in the brief, affording it in the design and enabling its fulfilment along the building's lifecycle. This identifies adaptability not just as an abstract measure, but rather a closer one to the context, that potentiates encounters, movements and a broader spatial fruition, subject to quantitative and qualitative interpretation.

Lionel March [23], in the *Environment and Planning B Anniversary Issue*, reviews the contributors and the contents of the previously published papers, whose inputs to knowledge advancement have been extensive in distinct areas of research. Inherently, Krüger [19, 20] and his summarised remarks from his Doctoral Thesis in Cambridge, Fawcett [9, 10] and his papers on activity schedules, Hillier's widespread works on space syntax [15] and Benedikt's [1] studies on isovists, are mentioned for their work, along with different research subjects and approaches such as those of Stiny and Gips, Steadman, Cooke, Knight, among many others. March even admits his "bias" towards some, to finally underline the relevance of all valid contributions:

> From a scholarly viewpoint that imbalance is to be regretted. There is too little serious work being undertaken in architectural and design scholarship to be partisan. (March [23], p. 12)

This assists to demonstrate that knowledge may advance by means of diverse inputs and study fields and, ultimately, that different research cultures may be reconcilable towards a common problem. The gathering of different procedures, perspectives, formal methods and outputs to a shared problematic is here argued to be vital for a more comprehensive and robust conclusion, rather than a single one that may not provide it per se and, furthermore, spatial adaptability is considered a common denominator that triangulates these approaches towards knowledge advancement in architectural research.

References

1. Benedikt, M. (1979). To take hold of space: Isovists and isovist fields. *Environment and Planning B, 6,* 47–65.
2. Brand, S. (1994). *How buildings learn: What happens after they're built*. New York: Viking Adult.
3. Brubaker, W. (1998). *Planning and designing schools*. New York: McGraw-Hill.
4. Coelho, C. (2017). *Life within architecture from design process to space use. Adaptability in school buildings today—A methodological approach*. Ph.D. Thesis in Architecture, Departamento de Arquitetura da Faculdade de Ciências e Tecnologia, Universidade de Coimbra, Coimbra.
5. Coelho, C., & Heitor, T. (2017). Adaptability retrieval in artistic learning environments. In T. Heitor, M. Serra, J. Silva, M. Bacharel & L. Silva (Eds.), *Proceedings of the 11th International Space Syntax Symposium* (pp. 7: 1–19). Lisbon: Instituto Superior Técnico, Departamento de Engenharia Civil, Arquitetura e Georrecursos.
6. Coelho, C., & Krüger, M. (2015). Towards a methodology to assess adaptability in educational spaces: An entropy approach to space syntax. In K. Karimi, L. Vaughan, K. Sailer, G. Palaiologou, & T. Bolton (Eds.), *Proceedings of the 10th International Space Syntax Symposium* (pp. 17: 1–20). London: Space Syntax Laboratory, The Bartlett School of Architecture, University College London.
7. Fawcett, W. (1976). School management and school design. *Journal of Architectural Research, 5,* 10–21.
8. Fawcett, W. (1978). *A mathematical approach to adaptability in buildings*. Doctoral Thesis. S.l., University of Cambridge, Cambridge.

9. Fawcett, W. (1979a). All possible and most probable schedules in organizations. *Environment & Planning B, 6*(2), 123–154.
10. Fawcett, W. (1979b). Catalogue of activity schedules for small organisations. *Environment & Planning B, 6*(3), 293–300.
11. Gehl, J. (2011). *Life between buildings: Using public space*. Washington, DC: Island Press (Original edition from 1987).
12. Habraken, N. (1972). *Supports: An alternative to mass housing*. London: The Architectural Press (Original edition in Dutch from 1961).
13. Habraken, N. (1998). *The structure of the ordinary: form and control in the built environment*. Cambridge: The MIT Press.
14. Hertzberger, H. (2008). *Space and learning: Lessons in architecture 3*. Rotterdam: 010 Publishers.
15. Hillier, B, Leaman, A., Stansall, P., & Bedford, M. (1976). Space Syntax. *Environment and Planning B, 3*, 147–185.
16. Hillier, B., Hanson, J, Peponis, J., Hudson, J., & Burdett, R. (1983, November 30). Space syntax: a different urban perspective. *The Architects' Journal, 178*(48), 47–54, 59–63.
17. Hillier, B., & Hanson, J. (1984). *The social logic of space*. Cambridge: Cambridge University Press.
18. Jaynes, E. (1957). Information theory and statistical mechanics. *Physical Review, Series II, 106*(4), 620–630.
19. Krüger, M. (1979a). An approach to built-form connectivity at an urban scale: System description and its representation. *Environment and Planning B, 6*, 67–88.
20. Krüger, M. (1979b). An approach to built-form connectivity at an urban scale: Variations of connectivity and adjacency measures amongst zones and other related topics. *Environment and Planning B, 6*, 305–320.
21. Krüger, M. (1981). Maximização da adaptabilidade. *Ciência e Cultura, 33*(9), 1169–1182.
22. Krüger, M. (1992). *Caracterização e Programação de Edifícios Complexos*. Departamento de Engenharia Civil. Instituto Superior Técnico: Universidade Técnica de Lisboa.
23. March, L. (1998). [8+(6)+11] = 25+x. *Environment and Planning B: Planning and Design*, Anniversary Issue, 10–19.
24. OECD—Organisation for Economic Co-Operation and Development. (2001). *Designs for learning: 55 exemplary educational facilities*. Paris: OECD Publishing.
25. Peponis, J. (2001). Interacting questions and descriptions. How do they look from here? In J. Peponis, J. Wineman, & S. Bafna (Eds.), *Proceedings of the 3rd International Space Syntax Symposium* (pp. xiii–xxvi). Atlanta: Georgia Institute of Technology.
26. Psarra, S. (2003). A bird's eye view from the ground. In J. Hanson (Ed.), *Proceedings of the 4th International Space Syntax Symposium* (pp. 1–8). London: University College London.
27. Shannon, C. E., & Weaver, W. (1949). *The mathematical theory of communication*. Urbana: University of Illinois Press.
28. Whyte, W. H. (1980). *The social life of small urban spaces*. Washington, D.C.: Conservation Foundation.
29. Worthington, J. (2007). Enabling place-making: The role of the school in supporting the community. In *Building excellence: Exploring the implications of the curriculum for Excellence for school buildings*. Edinburgh: The Scottish Government.

Carolina Coelho is an architect from the Department of Architecture, University of Coimbra (Darq FCTUC). She has concluded her Doctoral Thesis *Life within architecture from design process to space use. Adaptability in school buildings today—A methodological approach*, at the Centre for Social Studies and Darq FCTUC. Her Thesis has been awarded with The Glen Earthman Outstanding Dissertation Award. Her findings have been presented in peer review publications and international conferences. She is Assistant Professor at Darq FCTUC for Theory and History of Architecture and supervises Master Theses. Currently she is an integrated researcher from the Centre of 20th Century Interdisciplinary Studies (CEIS20) and incorporates the research projects (Eu)ropa. Rise of Portuguese Architecture, and CoReD. Collaborative Re-design with Schools.

Adapting Analysis Workflows to Humanitarian Needs: Different Road Network Models and Tools

Camilla Pezzica, Valerio Cutini, Clarice Bleil de Souza, and Alessandro Santucci

Abstract

Urban planners and decision-makers need to provide, with increasing frequency, a rapid response to the many natural hazards looming over cities today, while relying on limited information and computational resources. This research explores viable ways to support them in developing performance-driven proposals via urban modelling by: (i) exploring potential synergies between configurational analysis methods, open computing platforms and collaborative geo-data; (ii) testing the interchangeability of different road network models in a set of suitable case studies. Getting fast, free and ubiquitous access to updated spatial information should allow decision-makers and experts to gauge relevant centrality and resilience properties of urban grids and bears opportunities for quantitative scenario testing beyond institutional boundaries. Therefore, this paper suggests linking knowledge and metrics from different street network analysis models and adapting analysis workflows to humanitarian needs to foster deliberation in strategic decision-making and tactical planning in urban disaster contexts. The results obtained in a set of suitable case studies indicate that, under certain conditions, time and computational resources can be saved, and interdisciplinarity boosted, by suitably interchanging primal and dual analysis models and tools. Nonetheless, further tests are needed to establish thresholds and generalize results.

Keywords

Street network analysis • Space Syntax • Volunteered Geographic Information • GIS • Python scripting

1 Introduction

According to the data in the Emergency Events Database [1], which is collected by the Centre for Research on the Epidemiology of Disasters of Université Catholique de Louvain (UCL) in Belgium, in the past few decades, urban planners and decision-makers have been dealing with increasingly frequent occurrences of natural hazards, turning into disasters. Additionally, cities are becoming increasingly complex [2], bringing new challenges to the provision of a rapid technical response to urban perturbations. Therefore, the international research community has targeted disaster management as a growing priority [3], highlighting the limitations of reductionist approaches in disaster recovery planning [4, 5]. It follows that professionals tasked with setting up recovery plans after urban disasters need to obtain a good understanding of the events, as they unfold in such dynamic contexts, to prevent unexpected and long-lasting impacts on future development processes [6]. To this end, they require access to up-to-date spatial information and a rich toolkit of flexible analytical methods.

A new research stream linking novel planning paradigms (e.g. performance-based planning) and urban modelling has highlighted the importance of models and simulation tools in achieving urban performance [7], with some authors suggesting to promote sustainability by addressing urban resilience [8]. Several studies (Table 1) have shown that street network analysis models offer many opportunities to support deliberation in strategic and tactical hazard-related planning,

C. Pezzica (✉) · V. Cutini · A. Santucci
Department of Energy, Systems Territory and Construction Engineering, University of Pisa, Largo Lucio Lazzarino, 56122 Pisa, Italy
e-mail: Pezzicac@cardiff.ac.uk

V. Cutini
e-mail: valerio.cutini@ing.unipi.it

A. Santucci
e-mail: santu.ale@gmail.com

C. Pezzica · C. Bleil de Souza
Welsh School of Architecture, Cardiff University, Bute building, King Edward VII Avenue, Cardiff, CF10 3NB, UK
e-mail: Bleildesouzac@cardiff.ac.uk

S. Eloy et al. (eds.), *Formal Methods in Architecture*, Advances in Science, Technology & Innovation,
https://doi.org/10.1007/978-3-030-57509-0_15

Table 1 Applications of street network analysis models in urban disaster scenarios

Disaster type	Analysis outcome	Disaster phase	References
Earthquake	Highlight abrupt changes inflicted to the urban morphology, quantifying dynamic changes in levels of centrality/attractiveness	Emergency	[9, 10]
Earthquake, generic disaster	Refine evacuation models and emergency/relief plans	Emergency/relief	[11–13]
Flooding	Help to prioritize post-disaster interventions, while minimizing indirect economic losses	Relief/recovery	[14]
Earthquake + tsunami	Clarify the long-term effects of post-disaster plans with respect to reconstruction claims	Relief/recovery	[9, 15]
Earthquake, hurricane	Study and improve community resilience	Relief/recovery	[9, 16]
Earthquake, flooding	Quantify the capacity of an urban grid to absorb and recover from physical shocks	Preparedness	[9, 10, 17–20]

opening possibilities for scenario testing through iterated simulations of 'what if' situations to become part of the technicians' workflow.

However, there is still a lack of methods, suitable to support decision-makers operating under extreme pressure of time and resources in developing performance-driven planning proposals after urban disasters [21]. In particular, despite their potential in disaster scenarios, street network analyses often present the following workflow's limitations: time requirements, reproducibility issues and sample size constraints; besides problems of input data consistency, accessibility and update frequency, which affect the scalability, generalizability and interpretation of the configurational analyses' results.

Therefore, this research exploits user-generated data and state-of-the-art computational tools as well as the richness of available configurational analysis models and methods (illustrated in Sect. 2) to allow humanitarian planners performing street networks analyses flexibly and at a low cost. At the same time, it seeks to enhance workflows' reproducibility and streamline the analysis process, for instance by limiting the constant switching between different applications and the need of intermediate files via a better integration of various analysis steps (data retrieval, import, elaboration and export). The study directly challenges the idea that there is one correct way to translate reality into a model and suggests adopting a pragmatic approach when there is limited availability of time, data and resources, to generate evidence useful to guide decision-making of experts under uncertainty [22, 23].

The paper expands the scope of a previous study by the same authors, involving a sensitivity assessment of rapid Space Syntax analysis workflows' outputs to modelling variations [24]. The main conclusions of [24] were the following: (i) analysis requirements can be reduced by filtering, generalizing and re-modelling open street networks' data as appropriate before performing the calculations; (ii) analysis outputs are reliable so long as the steps involved in model construction are sound and traceable, as well as semi-automated.

Building on these results, this paper further explores synergies between the use of volunteered geographic information (VGI), open computing platforms and different configurational analysis methods. Furthermore, it qualitatively assesses the interchangeability of two different road network models, as this would not only improve computational costs but also models' integration across disciplines. To this end, the study performs a comparative configurational centrality and resilience analysis of three types of mobility networks, namely the drivable, the cyclable and the walkable, in two case studies with different urban grid morphologies. This test is complemented by a demonstration of the method in a real post-disaster scenario and a critical discussion with suggestions for further research.

2 Configurational Analysis

In line with Space Syntax theory [25], this paper assumes the urban grid, i.e. the system of freely accessible connections in a settlement corresponding to the road network (see Fig. 1), to be the physical driver of natural movement in cities, supporting all dependent urban sub-systems, among which commerce and services. Urban grids can be studied in terms of their configurational properties as certain characteristics of an urban grid when modelled as a graph (i.e. a topological structure made of nodes and edges) are good indicators of the level of complexity, connectedness and resilience of a city [19]. Additionally, adopting a network approach to the study of the urban grid offers a unique opportunity for

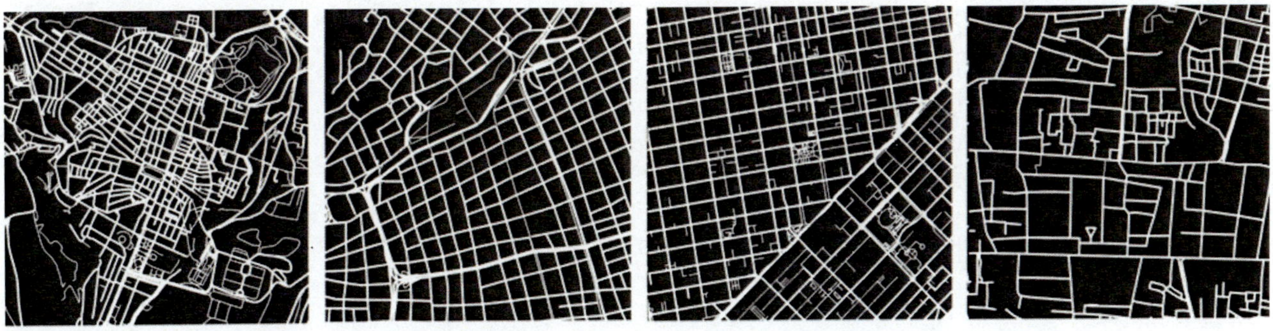

Fig. 1 Samples of 1 square mile urban grids of L'Aquila, Cali, San Francisco, Yogyakarta extracted with OSMnx

transferring knowledge from numerous scientific disciplines, such as transport engineering, social studies and economics, which rely on network models to study reality [26].

2.1 Primal and Dual Approaches

Beyond a common graph theory background, classic network analysis and Space Syntax mainly differ in the way they translate the urban network into a graph. In the former case (primal approach), streets are represented as graph edges and crossroads as nodes. In the latter case (dual approach), streets are translated into nodes and edges represent their topological connections. Figure 2 illustrates how the same fragment of urban grid is modelled in these two cases. Past research [27, 28] has shown that both modelling techniques can be successfully deployed to study properties of urban street networks, despite one approach can be preferred to the other depending on the objective of the analysis. Due to primal graph models of urban street networks being metric and dual ones being often topological, these two models have been used for different purposes in various domains, with little methodological cross-fertilization. For instance, since in a dual graph, streets can intersect a virtually illimited number of other streets because, differently from a primal graph, the number of connections per node is not constrained by the physical size of road intersections; the dual graph is more suitable to study scale-free behaviours in urban road networks [29]. Moreover, the dual approach is generally preferred by planners who work with roads as design units.

The Space Syntax Angular Segment Analysis (ASA) developed by Turner in 2007 [30] as an alternative to the axial analysis has, however, integrated the use of metric distances in the dual graph analysis shortening the distance between the primal and the dual approaches. Despite some differences between the two models remain, this opened possibilities for Space Syntax to share with classic street network analysis the same data source, namely Road Centre Lines (RCL).

More recently, Batty [31] made a first attempt to mathematically link the two approaches, but the process involves non-trivial matrix computations and, for the moment, seems far from being fully operational.

Nevertheless, this paper will show that for certain tasks such as network centrality assessment (i.e. measuring the distribution of attractiveness or spatial accessibility among the segments of a road network system) and for certain grid structures, the two analysis models can be conveniently interchanged, allowing professionals to spare precious time and computational resources in the initial planning stages.

2.2 Centrality Indices

Some properties of urban street networks can be effectively described using the multifaceted concept of centrality and related metrics (Table 2). In space syntax, *Integration* and

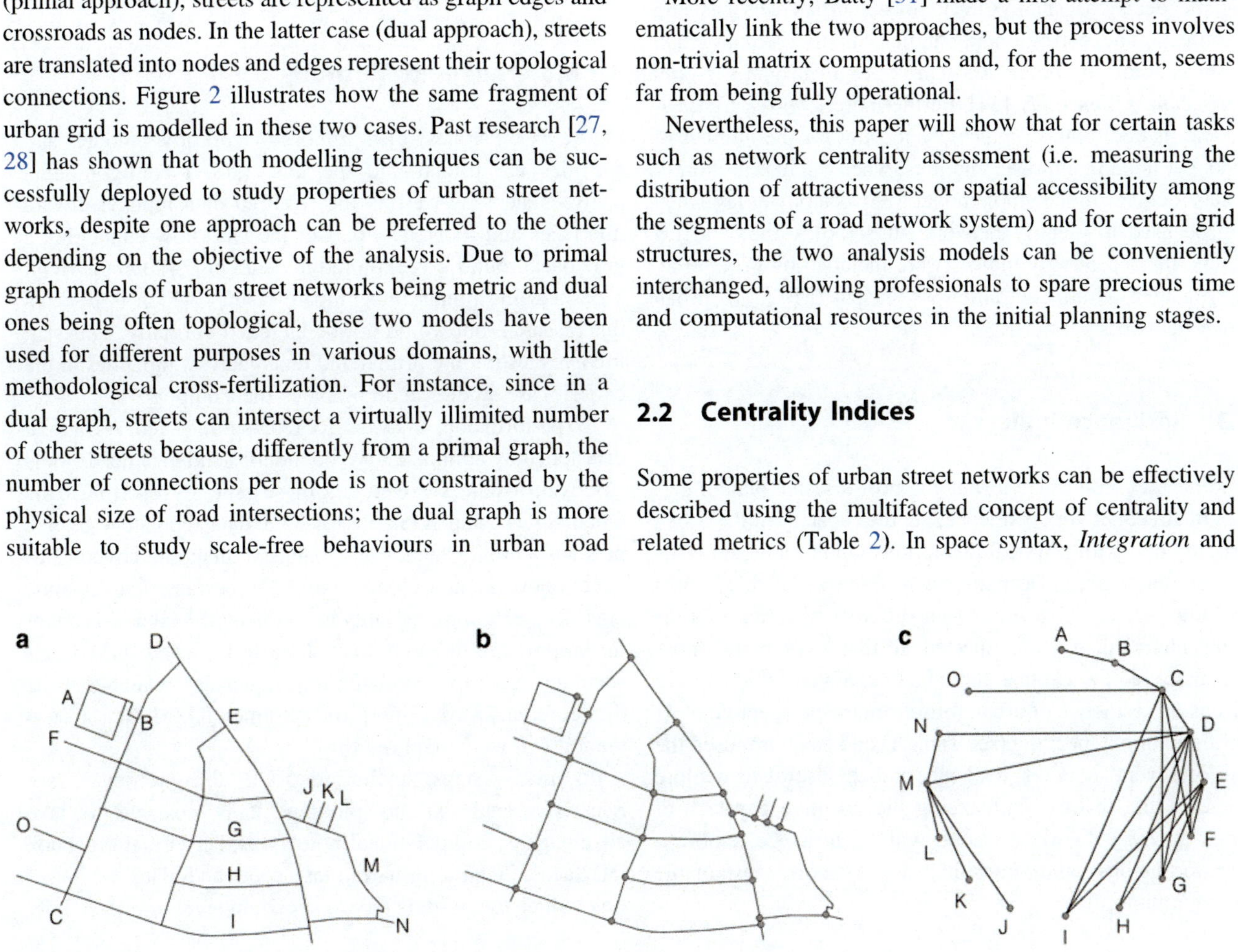

Fig. 2 Modelling of a street network (**a**) into a primary (**b**) and into a dual (**c**) graph

Table 2 Centrality indices

Centrality index	Concept	Primal	Dual
Integration or NAIN	Being near		X
Closeness centrality	Being near	X	
Choice or NACH	Being between		X
Betweenness centrality	Being between	X	
Connectivity	Being linked		X
Degree centrality	Being linked	X	

Choice are used to identify road networks' to-movement and through-movement patterns, respectively. Hillier and Iida [32] showed that a higher level of *Integration* correlates with a higher density of commercial activities and a greater pedestrian flow. Thus, *Integration* and *Normalised Angular Integration* (*NAIN*) depict the level of global/local accessibility and attractiveness of a street and are higher when a segment is configurationally closer to all the others. *Choice* and *Normalised Angular Choice* (*NACH*) measure how often a road segment falls into the shortest route connecting every origin–destination pair and therefore identify traffic bottlenecks as crowding usually happens where all shortest paths overlap. Consequently, *NACH* is generally used to inform logistic decisions and *NAIN* to locate urban services and resources. The corresponding indices in the primal graph analysis are in order *Closeness Centrality* [33] and *Betweenness Centrality* [34]. Furthermore, *Degree Centrality* and its dual twin *Connectivity* account for the number of links per node in a model, and thus, when the distribution of nodes' degrees in the graph model follows a power law, they can be used to identify network hubs [26]. Among a few others, these centrality indices were included by Porta et al. [27] in their Multi-Centrality Assessment (MCA) of urban street networks.

2.3 Resilience Indices

Another key concept relevant for post-disaster planning is that of street network resilience, i.e. the capacity of the urban grid to cope with physical disruptions while maintaining its global functioning. Designing a resilient network requires striking a good balance between different qualities such as being hierarchical and connected, so that it resembles more the structure of a leaf than that of a tree. Several studies have discussed resilience of urban form from the perspective of its configurational meaning (see Table 1) and have proposed the use of several network resilience indices useful to explore some of the features influencing the adaptive capacity of urban street networks to cope with disturbance, embrace transformations and transition to a more sustainable configuration.

The calculation of two of these, namely the *Average Number of Connections per Node* (primal graph) and *Mean Connectivity* (dual graph), immediately follows that of the centrality indices in Sect. 2.2, and thus, their quantitative comparison can be done for all the analysed networks without further computations. These two indices are a coarse-grained measure of the ability of a grid to maintain the road system linked despite randomized network disruptions [35]. Further metrics and resilience concepts are discussed in Sect. 4.4 (Table 4) but will not be analysed as formal correspondences between the primal and dual graph indices are less well established and more research would be needed to clarify mutual similarities and differences.

3 Materials and Methods

Figure 3 illustrates the proposed workflow model and specifies the primal and the dual (ASA) configurational analyses steps as well as the required data input, software interfaces and tools. The part of the workflow containing a grey background was explored in detail in [24]. On the right, a dashed line points to 'compare result', which is done in this research, but would not be included normally. This step involves testing the primal and dual analysis pipelines in the chosen case studies and compares their outputs.

To perform both analyses, RCL data is required. As natural disasters defy administrative boundaries and data needs to be strictly up to date, we overcome these issues by resorting to the Open Street Map (OSM) [36] VGI data repository, which presents a homogeneous and extensive mapping coverage of street networks at a global level [37], besides rich semantic tags. As performing a quality assessment of VGI data is out of the scope of this research, we assume the retrieved OSM data is accurate. A set of automated techniques are available in the literature, anyway, to perform automatic OSM data assessment if and when required [38].

In many steps, including RCL data retrieval (see Sect. 3.2) analysis and plotting, it is possible to take advantage of computational notebooks, i.e. virtual notebook interfaces, integrating narratives, code and data visualization, which allow users having a certain level of interactivity

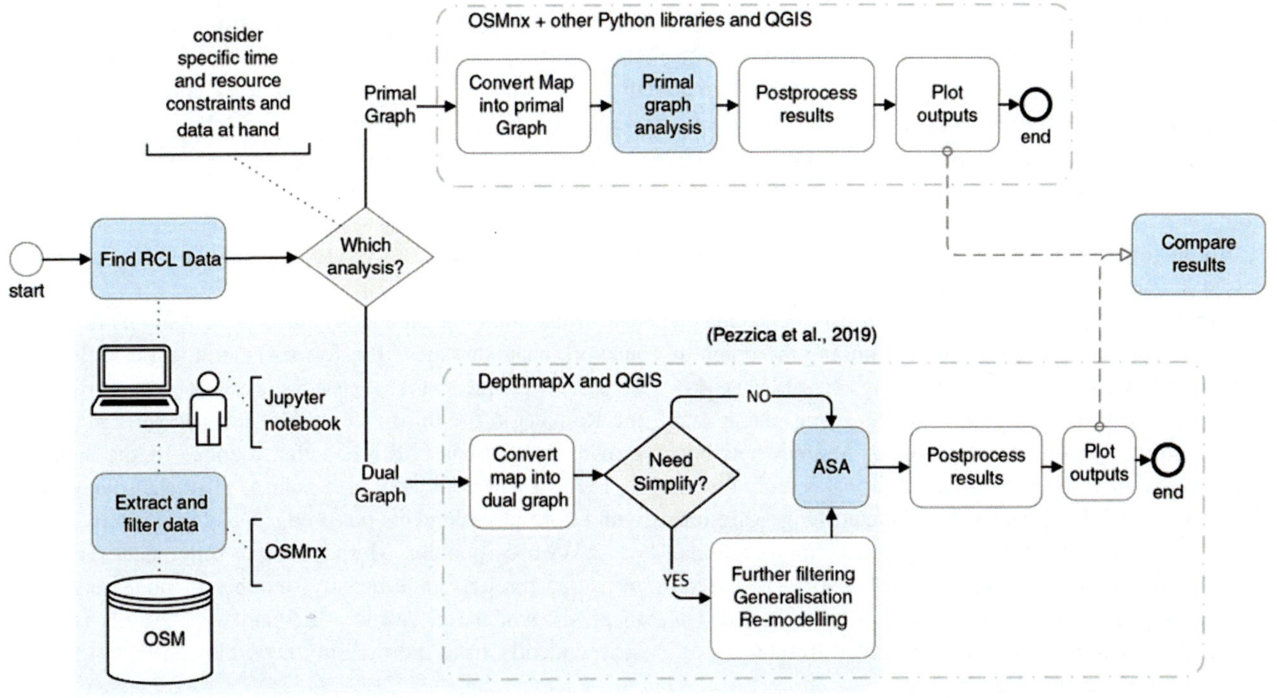

Fig. 3 Dual and primal configurational analysis workflows

with analysis results when running computations and sharing replicable workflows with other professionals and researchers [39]. Additionally, computational notebooks can be run without the need of installing many specialized software. This way, advanced analysis methods can be easily tested by users, including those who are new to urban data science and geo-computation, to verify the possibility to integrate them in the existing urban planning processes. Specifically, this research makes use of Jupyter computational notebooks [40] as the interface to run some Python [41] libraries useful for OSM data retrieval, analysis and visualization. Some of the advantages of using Python for programming are that it is a high-level language diffused among non-developers, currently integrated in several GIS and CAD applications to customize professionals' workflows via scripting. Additionally, the community of Python contributors is very active, and related resources and libraries are freely accessible.

The OSM primal graph analysis is conducted using the OSMnx [42] and Networkx [43] Python libraries for the core calculations and Matplotlib [44] (a data visualization library) for plotting results. The ASA (dual graph analysis) is performed in DepthmapX [45], which is one of the leading Space Syntax analysis software. OSMnx and QGIS [46] (an open GIS software) were used for data preparation and post-processing, respectively, as explained in [24]. The latter is required to enable an effective visual comparison of the colour-coded maps generated by both models, by applying the same data classification method (i.e. natural breaks or quantile equal count) to the primal and dual analysis outputs. This step requires a preliminary export of the analysis results in Shapefile (from Jupyter) and MapInfo formats (from DepthmapX) for import into QGIS.

3.1 Case Studies

Two cities were selected for the study, namely L'Aquila in Italy (location of the devastating 2009 earthquake) and Cali in Colombia (hit by the 1995 earthquake). They seem suitable to demonstrate the application of the method as they are located in disaster-prone areas and have a sensibly different dimension, scale and spatial morphology. The authors' familiarity with both cities—one of which was analysed in [9] using a traditional configurational analysis workflow—allows overcoming the need for extensive empirical validation of analysis outputs. Moreover, choosing L'Aquila as a case study enables the demonstration and critical discussion of an application possibility of the proposed method in a recent disaster scenario.

3.2 Data Preparation

Querying the OSM API using OSMnx is the easiest way to retrieve the required custom extracts of OSM RCL data, while avoiding the use of intermediary platforms. This requires only a couple of lines of Python code and requests can be done by simply inputting one of the following: (a) name of a city/province/region/country; (b) coordinates of a point and a radius; (c) coordinates of a point and the length of the side of a square; (d) a shapefile containing a custom boundary. For Cali, a 'type a' query was made, while for L'Aquila it was used a custom boundary equivalent to the one used in [9]. Then, the library creates a buffer around the area of interest of a given radius (here 500 m) to associate to each node a correct street count before clipping the graph to the requested extension. As shown in Fig. 4c, OSMnx presents an option to automatically retrieve a non-planar and corrected primal graph representation of a street network, and the possibility to perform the analysis either on an undirected (always both ways) or a directed (edges are one or two ways) graph; the latter taking into account directional movement. Additionally, by filtering of OSM roads' labels, OSMnx allows the automatic extraction of different types of mobility networks, including but not limited to the:

1. 'Walk' network, which contains all streets pedestrians have access to, including paths through parks, links between buildings and residential streets.
2. 'Drive' network, which contains all streets that are drivable and public apart from service roads.
3. 'Bike' network, which contains only the roads and paths used by cyclists.

Noticeably, the quick automatic generation of multiple graphs from one base map holds a great potential for post-disaster applications, because it can make different mobility analysis workflows more fit-for-purpose, streamlined and easy to replicate through automation [24]. Section 4 presents the results obtained in a Centrality (4.2) and Resilience (4.3) assessments of these three mobility networks in the two cities analysed.

4 Results and Discussion

The urban grids of Cali and L'Aquila present highly different network characteristics. The former is an historic settlement of 69,393 people, which existed since before the arrival of the Romans in the third century BC and has an organically grown, dense urban grid. The latter, founded by the Spanish in 1536, is characterized by a coarse, planned, square grid with informal settlements on its edges, and is currently home of 2,319,655 people. Morphological differences can be measured through node density (number of nodes per area mapped), which for Cali is ~2.5 and for L'Aquila is ~5, independently from the mobility network.

4.1 Processing Time

Table 3 shows that primal graphs present a sensibly lower magnitude of nodes compared to dual graphs. This is relevant when it comes to processing time as it was observed that increasing the number nodes increases computational speed in a nonlinear way, meaning the primal graph analysis is significantly faster than the dual graph one for larger network models. A rigorous assessment of computational performance between the two approaches, however, was not

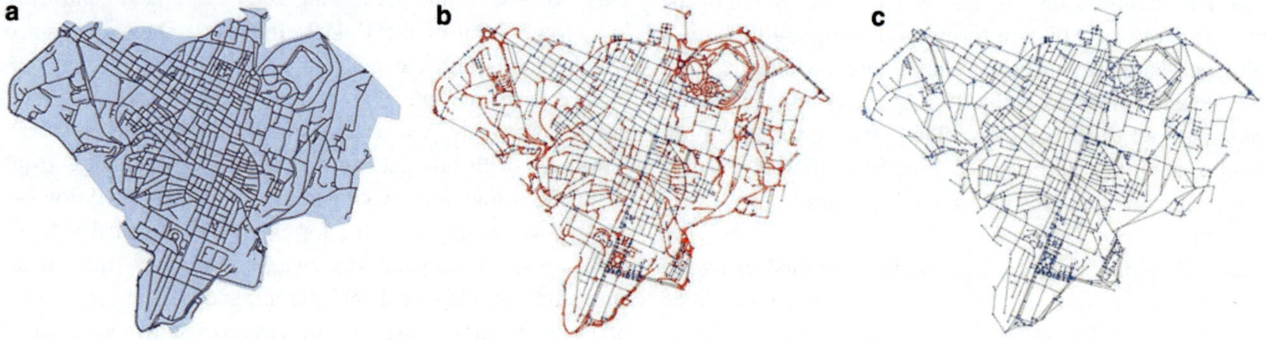

Fig. 4 L'Aquila network. **a** Official data (CTR); **b** OSM original and **c** OSM simplified data

Table 3 Number of nodes in primal and dual models in the Walk, Bike and Drive networks

City	Walk primal	Walk dual	Bike primal	Bike dual	Drive primal	Drive dual
L'Aquila	1,837	8,564	1,318	6,730	923	4,864
Cali	29,110	72,436	25,317	65,964	22,812	58,924

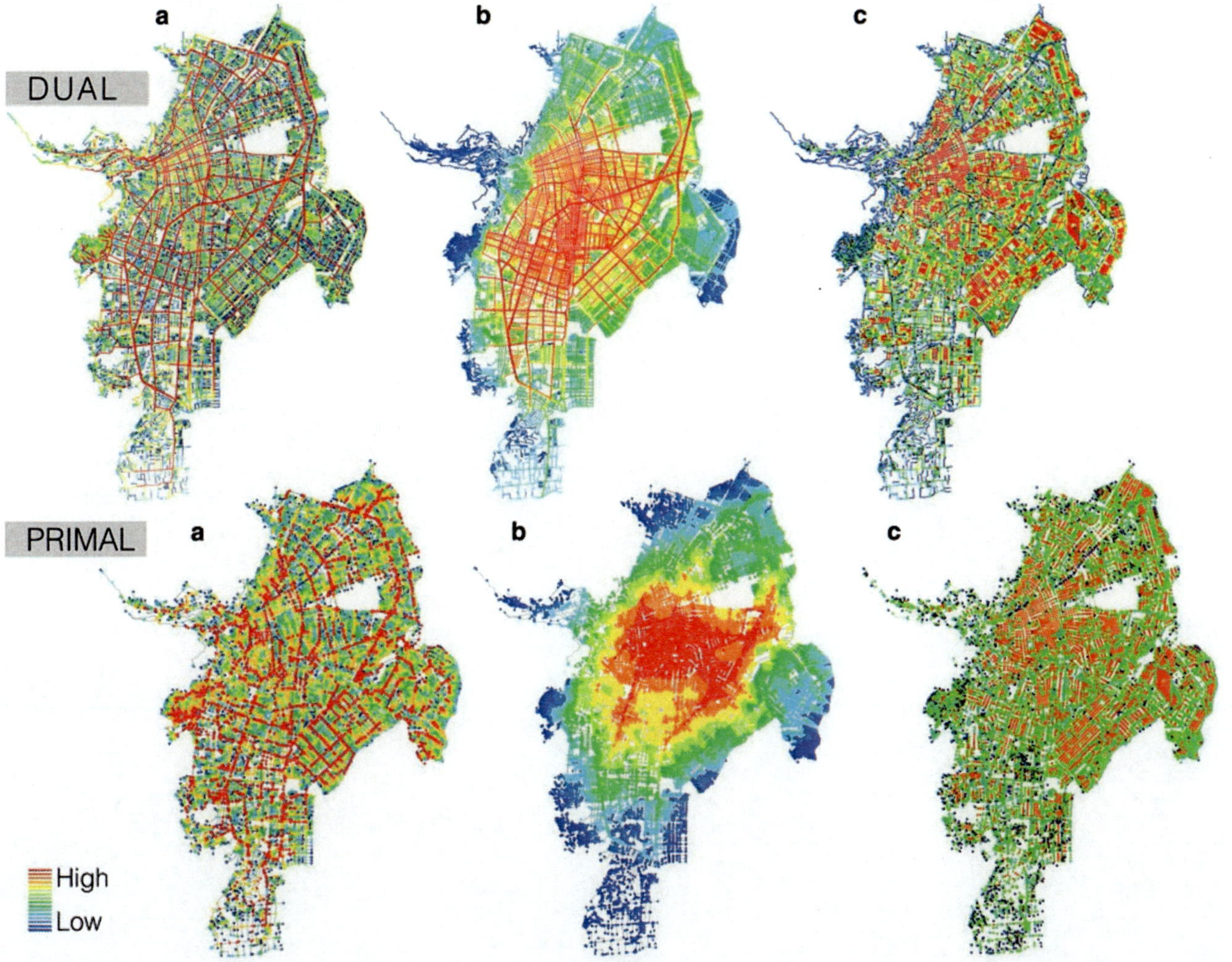

Fig. 5 Cali Walk network global centrality analysis: **a** NACH and Betweenness Centrality; **b** NAIN and Closeness Centrality; **c** Connectivity and Degree Centrality

undertaken as this would have required comparing the efficiency of the two analysis algorithms, which was out of the scope of this research. Nonetheless, by looking at the number of nodes alone, it was possible to preliminary assess the potential incremental benefits in opting for the lightest model, i.e. the primal graph, for performing the analysis.

Yet, an important aspect to consider is how much informative primal models' results are for urban design and planning tasks, i.e. how close their outputs are to those of dual graph model, as the former may result harder to read to urban designers and planners who are used to work with streets as intervention units.

4.2 Centrality Assessment

Figure 5 shows an acceptable correspondence between the outputs of the two global analyses for the Walk network of Cali, which can be highly advantageous to speed up the centrality assessment of this fairly large network system. In particular, it is observed a high correlation between the distribution of *Betweenness Centrality* and *NACH* (a) and between that of *Connectivity* and *Degree Centrality* (c). This could be due to the combined effect of the similar mathematical definition of the indices and of the grid-like structure of Cali's Walk network, which presents little angular variations. A weaker correspondence, although arguably still acceptable for a preliminary planning evaluation, can be observed between the distribution of *NAIN* and *Closeness Centrality* (b). This could be primarily due to a more significant difference in how these two indices are calculated requiring more in-depth investigations if a correspondence is to be established.

In the Walk network of L'Aquila (Fig. 6) we observe a less evident correspondence, despite overall considerations align with those made for Cali when it comes to indices' similarities. This may imply that the two methods are not

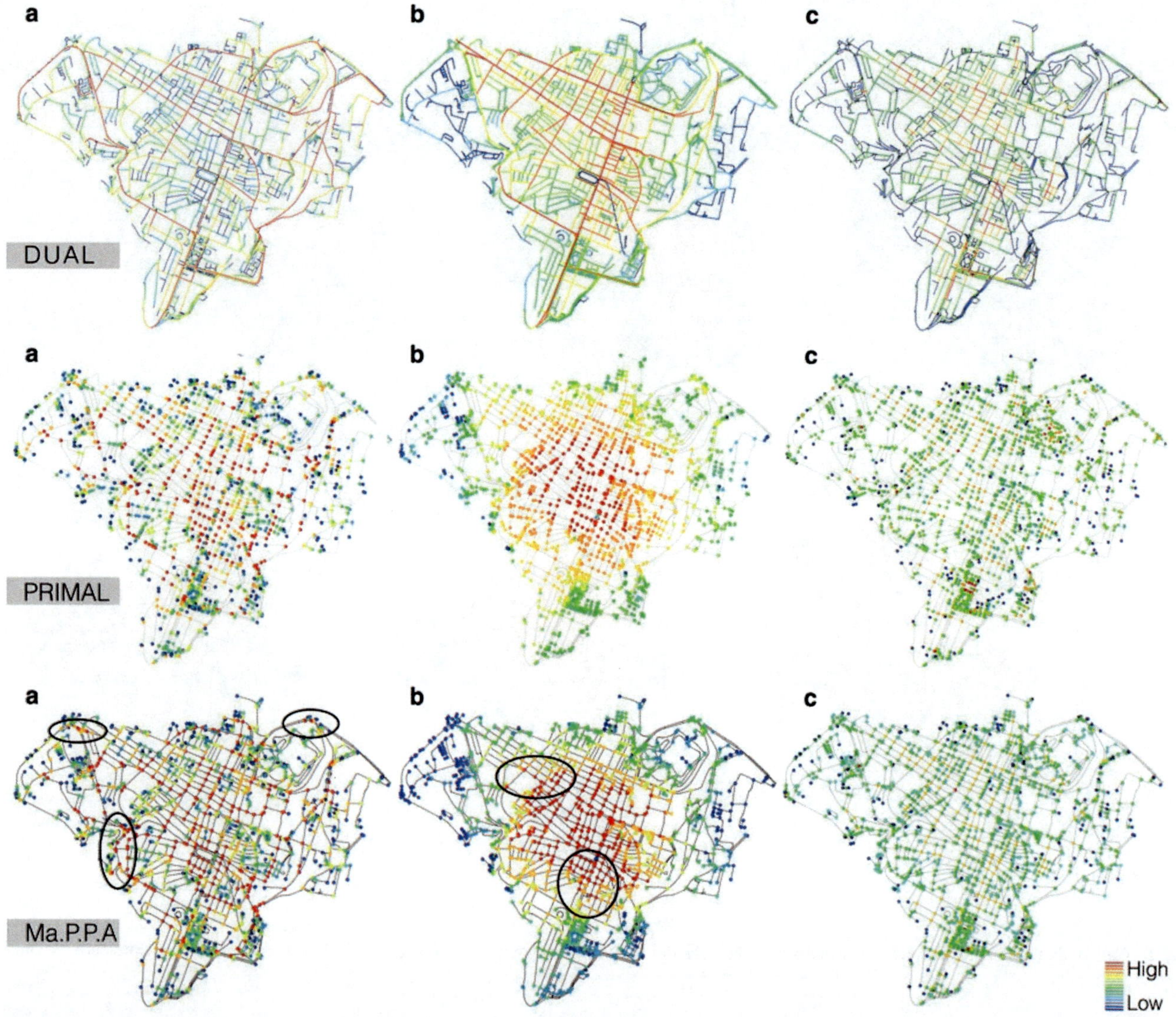

Fig. 6 L'Aquila Walk network global Centrality analysis: **a** NACH and Betweenness Centrality; **b** NAIN and Closeness Centrality; **c** Connectivity and Degree Centrality

interchangeable in smaller, organic street network models. The lower correspondence observed in L'Aquila Walk network is likely to depend on the different way nodes are distributed in the primal and dual graphs. Improvements in establishing comparable node distribution between the two graphs in terms of their spatial logic may be achieved by repositioning primal nodes in visually strategic locations, according to angular variations, via the automatic generation of new points in GIS as proposed by Cutini et al. [47] in their Mark Point Parameter Analysis (Ma.P.P.A.).

Figure 6 shows that, a better correspondence in terms of *Betweenness Centrality* (a) as well as *Closeness Centrality* (b) can be achieved by comparing dual graphs (row 1) with primal Ma.P.P.A. graphs (row 3). This hypothesis is confirmed by comparing the result obtained in all cases for *Degree Centrality* (c), as this index is not sensible to angular variations, and thus, the only differences registered in the Ma.P.P.A. graph are relative to the location of additional nodes. Nevertheless, since, in cases like that of L'Aquila (i.e. reduced network sizes), performing a dual graph analysis does not require more than a few minutes, it is probably easier to opt for it directly.

The primal and the dual graphs could still be analysed in parallel whenever, in addition to street segments, configurational information about intersections points is needed (e.g. if the focus is on certain pieces of infrastructure) or if the use of certain primal graph indices is well established to tackle the specific task at hand.

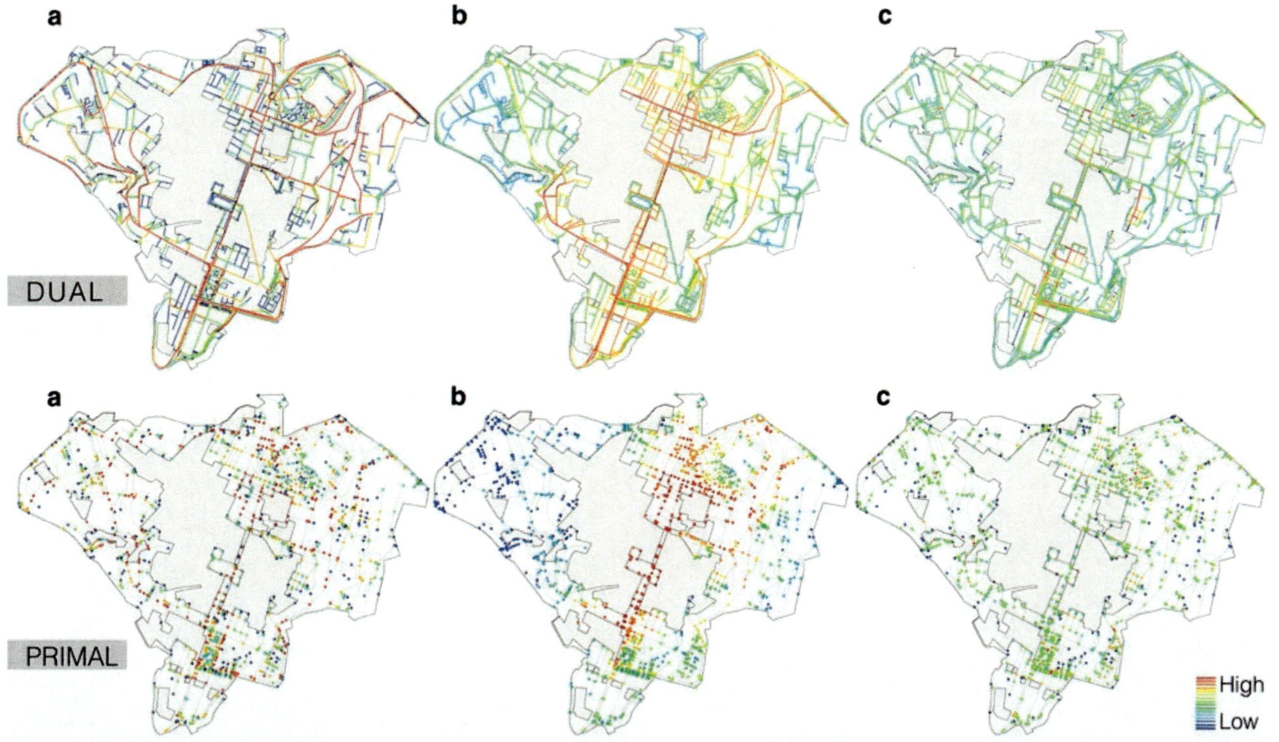

Fig. 7 L'Aquila 2009 Walk network global centrality analysis: **a** NACH and Betweenness Centrality; **b** NAIN and Closeness Centrality; **c** Connectivity and Degree Centrality

4.2.1 Disaster Application

Moving from the study of an unaffected urban network system to that of a city hit by the actual perturbation of a strong disaster event, the grid of L'Aquila soon after the 2009 earthquake (6.3 Richter scale) is analysed to show how this method can be exploited to highlight the impact of such an event on the morphology of a city, observing the corresponding changes in levels of centrality distribution. Figure 7 illustrates the results of the primal and dual graph analyses after cutting out from the original Walk map the so-called red zone, which defines the perimeter of the urban area ($\sim 1{,}600{,}000$ m^2) declared off limits for transit in the aftermath of the disaster, due to safety reasons. Noticeably, by eliminating a big chunk of the organic urban fabric, both the *NACH-Betweenness* and the *NAIN-Closeness Centrality* pairs correlate more closely than in Fig. 6, while the correspondence between *Connectivity* and *Degree Centrality* is maintained. In [9], a similar study was made to gauge the properties of L'Aquila's network immediately after the 2009 disaster, using a classical axial analysis [25]; which offers an opportunity to cross-compare—to an extent—the two dual analysis outputs. The maps in Fig. 8 of [8] p. 102:12, depicting the distribution of *Choice* (not-normalized version of *NACH*), *Integration* (not-normalized version of *NAIN*) and *Connectivity* in the same area, present a good semantic correspondence with those in row 1 of Fig. 7.

Figure 7 highlights the deep gaps that earthquakes can cause in urban settings, which may endure in the long term. To date, more than 11 years after the 2009 disaster, a part of this area (mainly in the West) remains precluded to public use due to difficulties in the reconstruction process. Arguably, in such a context, having had access to configurational analysis outputs could have supported a better recovery planning. For instance, areas with insufficient levels of accessibility could have been clearly identified and some of the spatial issues, which resulted from the subsequent realization of L'Aquila's 19 temporary housing sites [5], could have been addressed early on.

4.2.2 Other Mobility Networks

Taking Figs. 6, 8 and 9 as an illustrative example, it is noted that a high morphological similarity is maintained in the three mobility network models of L'Aquila. The general considerations made for the Walk seem in fact transferable to its Bike and Drive networks, which present a similar distribution of centrality values. Similar observations, although results are not visually displayed in this paper, were made for Cali.

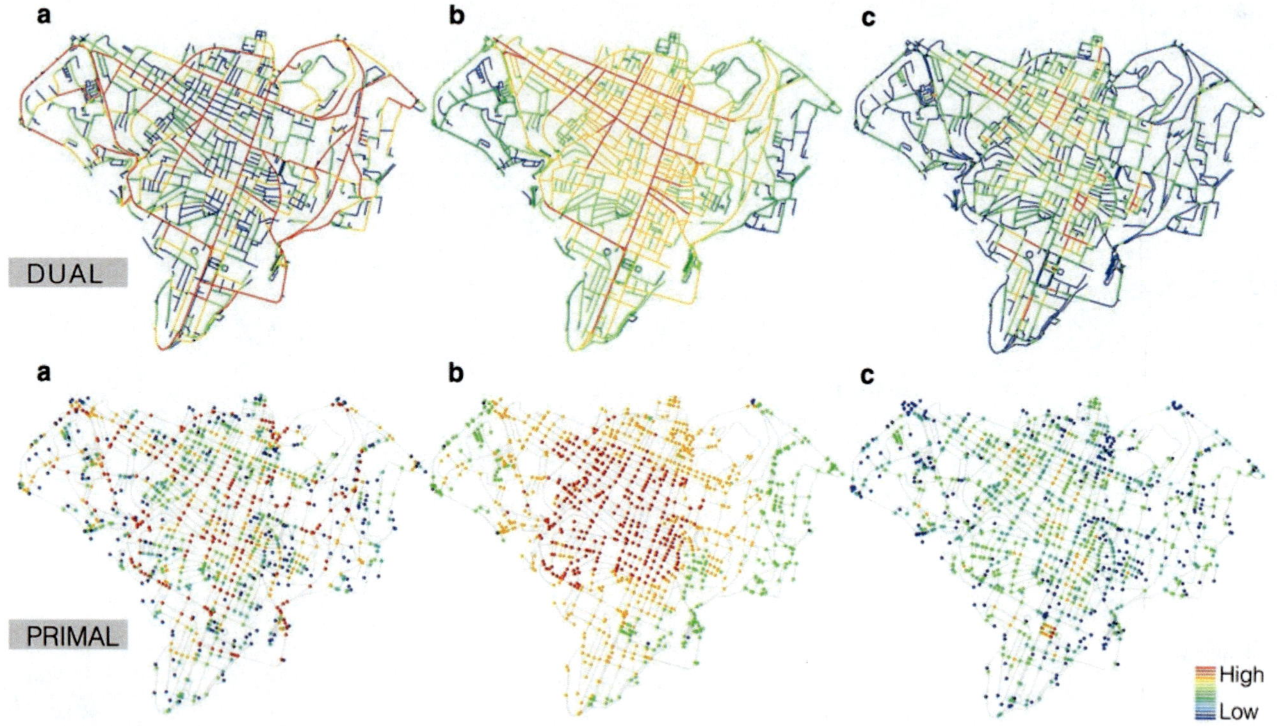

Fig. 8 L'Aquila Bike network global centrality analysis: **a** NACH and Betweenness Centrality; **b** NAIN and Closeness Centrality; **c** Connectivity and Degree Centrality

4.3 Resilience Assessment

Figure 10 illustrates the results obtained in the two case studies and across the three mobility networks for the following road network resilience indices: *Mean Connectivity* (Cali d. and L'Aquila d.) and *Average Number of Connections per Node* (Cali p. and L'Aquila p.). The plot illustrates that Cali performs better than L'Aquila with respect to both indices in all mobility networks, because of its higher redundancy of connections. Additionally, the Walk networks of both cities result on an average more interconnected while the Bike networks result the least so.

Despite some differences can be observed among the absolute values of the indices, the analysis results for the two cities and across the three mobility networks are largely comparable in both network models, with L'Aquila having on average 2–3 connections per node and Cali having on average 3–4.

According to Nilsson and Gil [29], who conducted a synchronic analysis of several urban street networks, the similarity between primal and dual analysis results should have been expected because the ASA (dual graph analysis) was performed on a traditional RCL segment map, where urban streets are split into several segments according to angular variations despite being in principle a unique entity. In fact, the RCL segment map is inherently a disaggregated network model that does not account for the principle of 'good continuation' of streets, which is somehow embedded in the classical axial map [25] (used in [9]) and in other street network models based on 'natural roads' [48]. This also explains why Cali's dual graph model, due to its regular grid-shaped structure, results on an average more interconnected than its primal twin, whereas in L'Aquila it is the other way around.

4.4 Further Considerations on Network Resilience

Values lower than those of the pre-disaster situation were found for L'Aquila 2009 Walk network, being 2.6 the *Average Number of Connections per Node* and 2.42 the *Mean Connectivity*. Moreover, the indices plotted in Fig. 10 indicate that the denser and organic grid of L'Aquila was (and still is) likely to be more configurationally vulnerable to future urban disasters than the coarser and grid-like grid of Cali. This confirms the previous findings by Sharifi [20], who identifies planned grids as more configurationally resilient. Despite this holds potentially important implications for humanitarian planning, more research on road network resilience and further testing with additional indices would be needed to explore this point in detail. As a preliminary step in this direction, Table 4 suggests some links between relevant primal and dual network indices, based on

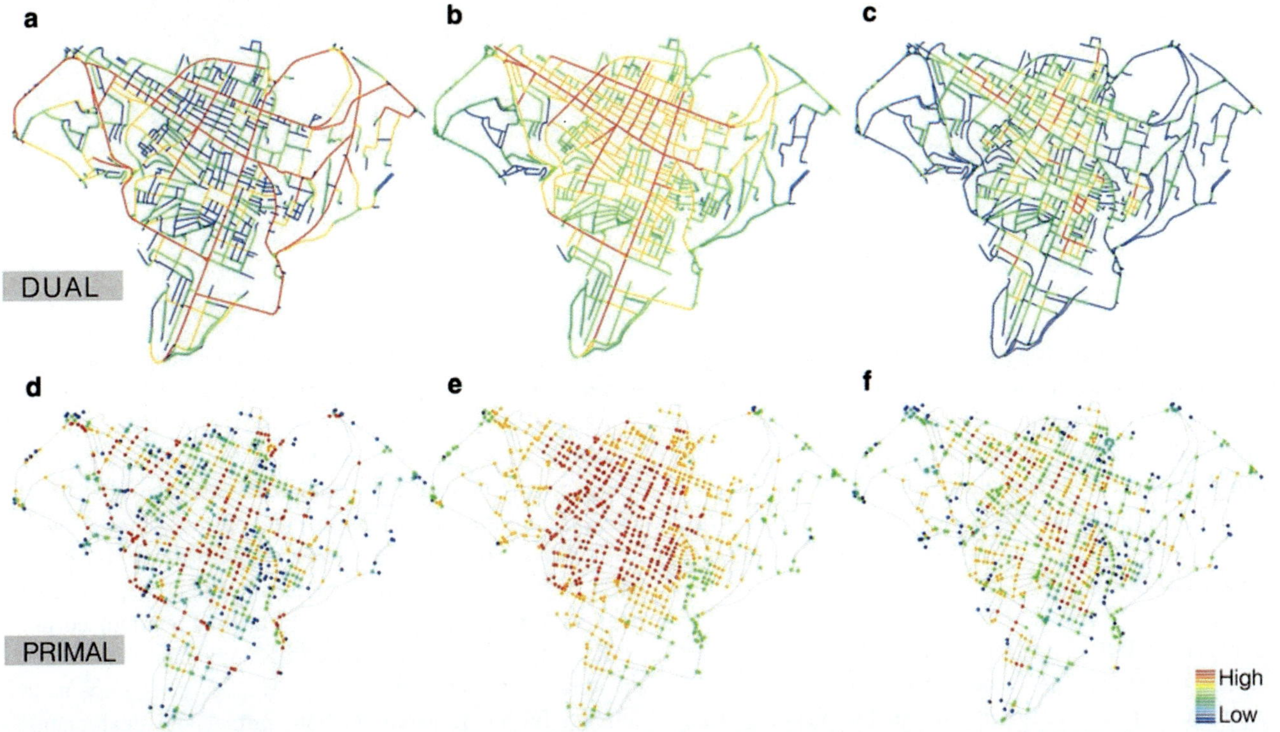

Fig. 9 L'Aquila Drive network global centrality analysis: **a** NACH and Betweenness Centrality; **b** NAIN and Closeness Centrality; **c** Connectivity and Degree Centrality

Fig. 10 Average number of connections per node (primal) and mean connectivity (dual) indices of Cali (above) and L'Aquila (below) Bike, Drive and Walk mobility networks

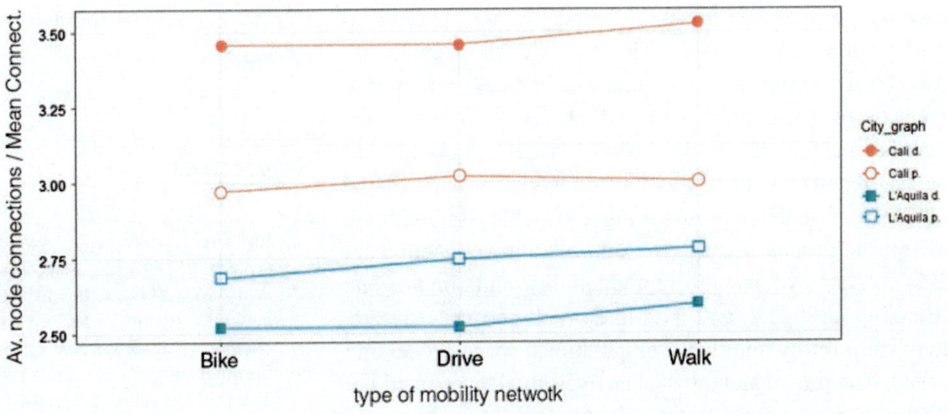

their underlying resilience concept: being hierarchical, being tree-like and having redundant connections.

In particular, Sharifi [20] suggests that scale hierarchy properties can be measured in a primal graph model by performing a K-S test to check the likely fit of *Closeness Centrality* to a power-law distribution (*p*-value). In dual graphs, Cutini [9] proposes using the *Synergy Coefficient*, defined as the R^2 of the linear regression between global (radius = *n*) and local (radius = 800 m) *Integration*, as a cross-scale configurational parameter suitable to measure spatial impedance as well as wayfinding ease.

Furthermore, by counting the number of alternative routes between two different nodes, the cyclomatic number (e.g. number of circuits in the network) and the alpha index (ratio of the actual cyclomatic number to the maximum one in a network with as many nodes) can be used to quantify how much a primal network structure resembles that of a leaf [20]. In the dual approach, a similar—despite not quite

Table 4 Resilience indices

Resilience index	Concept	Primal	Dual
Synergy coefficient	Being hierarchical		X
p-value of K-S test on Closeness Centrality	Being hierarchical	X	
Frequency index	Being tree-like		X
Cyclomatic number and α index	Being tree-like	X	
Mean connectivity	Being redundant		X
Average number of connections per node	Being redundant	X	

identical—assessment can be done using the *Frequency Index* [9], defined as the ratio of the actual maximum value of *Choice* to the maximum value *Choice* would have if the corresponding element was located on all shortest paths, which identifies tree-like network structures.

5 Conclusions

The paper presents a workflow model for street network analysis which exploits collaborative geo-data and open computational resources to better respond to planners' needs and constraints in crisis contexts. The proposal includes two different analysis approaches, namely the primal and the dual graph analysis. A set of key configurational metrics representing corresponding road network centrality and resilience concepts are identified for both approaches and reported in Tables 2 and 4, respectively. The two analysis methods are tested by retrieving four configurational indices pairs in two case studies, to identify under which conditions they are interchangeable. This is done via a visual comparison of analysis results for three mobility networks in each of the two analysed settlements and using a line chart. In both case studies, the proposed analysis workflow allowed significant gains in terms of results' reproducibility and, in the one presenting a bigger and grid-like road network system, interchangeability improved computational speed processing as well. The primal and the dual analysis models resulted not always interchangeable, potentially due to differences in grid configurations, but further tests are needed to establish thresholds and generalize results.

Despite this current limitation, the study opens doors not only to improve computational costs in preliminary planning evaluations, but also to modelling integration with other disciplines such as transport engineering, economics, etc. Additionally, it offers novel opportunities for a broad synchronic study of street network patterns, which would enable: (1) quantitative discussions about performance-based planning practices considerate of local differences; (2) the identification of deep analogies among different urban forms; (3) better communication with stakeholders trough mapping of road networks' centrality and resilience properties. Adopting such a quantitative approach would support lessons learning across different geographic regions and improve collaboration among professionals, while enhancing disaster preparedness.

Future research developments should seek to validate results to clearly establish behavioural patterns of different urban grid morphologies (and related network models) with respect to all the identified indices in both analysis approaches. To further streamline the dual graph analysis, it may be relevant to assess possible gains of making a combined use of R computational notebooks in Rstudio and of the new rdepthmap package for R [49]. Another possibility is to conduct further research on both primal and dual graphs analysis, including hybrid network models, and test models' integration and usability with target users to understand how these can be used in practice to inform decision-making. It is anticipated that professionals' needs are likely to vary on a country basis or be tied to the existing disaster management protocols.

References

1. EM-DAT: Emergency events database. https://www.cred.be/projects/EM-DAT. Last accessed March 5, 2020
2. Batty, M. (2009). Cities as complex systems: Scaling, interaction, networks, dynamics and urban morphologies. In: *Encyclopedia of complexity and systems science* (pp. 1041–1071). New York: Springer. https://doi.org/10.1007/978-0-387-30440-3_69.
3. Yi, H., & Yang, J. (2014). Research trends of post disaster reconstruction: The past and the future. https://doi.org/10.1016/j.habitatint.2013.10.005.
4. Alexander, D. (2013). An evaluation of medium-term recovery processes after the 6 April 2009 earthquake in L'Aquila, Central Italy. *Environmental Hazards, 12,* 60–73. https://doi.org/10.1080/17477891.2012.689250
5. Contreras, D., Blaschke, T., & Hodgson, M. E. (2017). Lack of spatial resilience in a recovery process: Case L'Aquila, Italy. *Technology Forecasting and Social Change, 121,* 76–88. https://doi.org/10.1016/j.techfore.2016.12.010
6. Borsekova, K., & Nijkamp, P. (2019). *Resilience and urban disasters: Surviving cities.* Cheltenham UK: Edward Elgar Publishing.
7. Pelorosso, R. (2020). Modeling and urban planning: A systematic review of performance-based approaches. https://doi.org/10.1016/j.scs.2019.101867.
8. Botequilha-Leitão, A., & Díaz-Varela, E. R. (2020). Performance Based Planning of complex urban social-ecological systems: the

quest for sustainability through the promotion of resilience. Sustain. Cities Soc. 102089 (2020). https://doi.org/10.1016/j.scs.2020.102089.

9. Cutini, V. (2013). The city, when it trembles: Earthquake destructions, post-earthquake reconstructions and grid configuration. In: *2013 International Space Syntax Symposium*. Sejong University Press.
10. Gil, J., & Steinbach, P. (2008). From flood risk to indirect flood impact: Evaluation of street network performance for effective management, response and repair. *WIT Transactions on Ecology and the Environment, 118,* 335–344. https://doi.org/10.2495/FRIAR080321
11. Mohareb, N. I. (2011). Emergency evacuation model: Accessibility as a starting point. *Proceedings of the Institution of Civil Engineers—Urban Design and Planning, 164,* 215–224. https://doi.org/10.1680/udap.2011.164.4.215.
12. Sari, F., & Sema Kubat, A.: Syntactic properties of evacuation and access routes in earthquake vulnerable settlements. In: *Proceedings of the 8th International Space Syntax Symposium*. Chile: PUC.
13. Chang, H.-S., & Liao, C.-H. (2015). Planning emergency shelter locations based on evacuation behavior. *Natural Hazards, 76,* 1551–1571. https://doi.org/10.1007/s11069-014-1557-x
14. Penchev, G. (2016). Using space syntax for estimation of potential disaster indirect economic losses. *Comparative Economic Research, 19,* 125–142. https://doi.org/10.1515/cer-2016-0041. Open access.
15. Heitor, T., Kruger, M., Muchagato, J., Ramos, T., & Tostes, A. (2000). Breaking of the mediaeval space: The emergence of a new city of enlightenment. *Urban Design International, 5,* 199–208. https://doi.org/10.1057/palgrave.udi.9000014
16. Carpenter, A. (2013). Disaster resilience and the social fabric of space. In: *Proceedings of the Ninth International Space Syntax Symposium*. Seoul.
17. Giuliani, F., De Falco, A., & Cutini, V. (2020). The role of urban configuration during disasters. A scenario-based methodology for the post-earthquake emergency management of Italian historic centres. *Safety Science, 127,* 104700. https://doi.org/10.1016/j.ssci.2020.104700.
18. Abshirini, E., & Koch, D. (2017). Resilience, space syntax and spatial interfaces: The case of river cities. *a/Z ITU Journal of Faculty of Architecture, 14,* 25–41. https://doi.org/10.5505/itujfa.2017.65265
19. Lhomme, S., Serre, D., Diab, Y., & Laganier, R. (2013). Analyzing resilience of urban networks: A preliminary step towards more flood resilient cities. *Natural Hazards and Earth Systems Sciences, 13,* 221–230. https://doi.org/10.5194/nhess-13-221-2013
20. Sharifi, A. (2019). Resilient urban forms: A review of literature on streets and street networks. *Building and Environment, 147,* 171–187. https://doi.org/10.1016/J.BUILDENV.2018.09.040
21. Platt, S., Bevington, J. S., Verrucci, E., & So, E. K. M., Pittore, M. (2014). Understanding and supporting post disaster decision making using scenario planning. In: *NCEE 2014—10th U.S. National Conference on Earthquake Engineering: Frontiers of Earthquake Engineering*, Anchorage, United States.
22. Hammond, K. R., Orasanu, J., & Kaplan, M. (1988). Judgement and decision making in dynamic tasks contracting officer's representative.
23. Kahneman, D. (2011). *Thinking, fast and slow*. Allen Lane.
24. Pezzica, C., Cutini, V., & Bleil De Souza, C. (2019). Rapid configurational analysis using OSM data: Towards the use of Space Syntax to orient post-disaster decision making. In: *Proceeding of the 12th International Space Syntax Symposium* (pp. 147.1–147.18). Beijing: Beijing JiaoTong University.
25. Hillier, B., & Hanson, J. (1984). *The social logic of space*. Cambridge University Press. https://doi.org/10.1017/CBO9780511597237.
26. Barabási, A.-L. (2002). *Linked : The new science of networks*. Perseus Pub.
27. Porta, S., Crucitti, P., & Latora, V. (2006a). The network analysis of urban streets: A primal approach. *Environment and Planning B: Planning and Design, 33,* 705–725. https://doi.org/10.1068/b32045
28. Porta, S., Crucitti, P., & Latora, V. (2006b). The network analysis of urban streets: A dual approach. *Physica A: Statical Mechanics and Its Applications, 369,* 853–866. https://doi.org/10.1016/j.physa.2005.12.063.
29. Nilsson, L., & Gil, J. (2019). The signature of organic urban growth. Presented at the (2019). https://doi.org/10.1007/978-3-030-12381-9_5.
30. Turner, A. (2007). From Axial to Road-Centre Lines: A New Representation for Space Syntax and a New Model of Route Choice for Transport Network Analysis. *Environ. Plan. B Plan. Des., 34,* 539–555. https://doi.org/10.1068/b32067.
31. Batty, M. (2017). *Space Syntax and Spatial Interaction: Comparisons, Integrations*. London: Applications.
32. Hillier, B., Iida, S.: Network and psychological effects in urban movement. In: Lecture Notes in Computer Science (including subseries Lecture Notes in Artificial Intelligence and Lecture Notes in Bioinformatics). pp. 475–490. Springer, Berlin, Heidelberg (2005). https://doi.org/https://doi.org/10.1007/11556114_30.
33. Sabidussi, G. (1966). The centrality index of a graph. *Psychometrika, 31,* 581–603. https://doi.org/10.1007/BF02289527
34. Freeman, L. C. (1977). A Set of Measures of Centrality Based on Betweenness. *Sociometry., 40,* 35. https://doi.org/10.2307/3033543
35. Crucitti, P., Latora, V., Marchiori, M., & Rapisarda, A. (2004). Error and attack tolerance of complex networks. *Physica A: Statistical Mechanics and Its Applications, 340,* 388–394. https://doi.org/10.1016/j.physa.2004.04.031
36. OSM: OpenStreetMap, https://www.openstreetmap.org/#map=6/54.910/-3.432. Last accessed May 10, 2020.
37. Krenz, K. (2017). Employing volunteered geographic information in space syntax analysis. In: *Proceedings of the 11th Space Syntax Symposium*.
38. Jilani, M., Corcoran, P., & Bertolotto, M. (2016). Machine learning for crowdsourced spatial data. In *Lecture Notes in Computer Science (including subseries Lecture Notes in Artificial Intelligence and Lecture Notes in Bioinformatics)* (pp. 294–297). Springer Verlag. https://doi.org/10.1007/978-3-319-46131-1_38.
39. Boeing, G. (2020). Urban street network analysis in a computational notebook. *Region, 6,* 39–51. https://doi.org/10.18335/region.v6i3.278.
40. Kluyver, T., Ragan-kelley, B., Pérez, F., Granger, B., Bussonnier, M., Frederic, J., et al. (2016). Jupyter Notebooks—A publishing format for reproducible computational workflows. https://doi.org/10.3233/978-1-61499-649-1-87.
41. Foundation, P.S. (2020).: Python. https://www.python.org/. Last accessed May 10, 2020.
42. Boeing, G. (2017). OSMnx: New methods for acquiring, constructing, analyzing, and visualizing complex street networks. *Computers, Environment and Urban Systems, 65,* 126–139. https://doi.org/10.1016/j.compenvurbsys.2017.05.004
43. Hagberg, A. A., Schult, D. A., & Swart, P. J. (2008). Exploring network structure, dynamics, and function using NetworkX. In: *Proceedings of the 7th Python in Science Conference (SciPy)*.
44. Matplotlib. (2020). Python plotting. https://matplotlib.org/. Last accessed March 21, 2020.

45. Varoudis, T., depthmapX by varoudis, https://varoudis.github.io/depthmapX/.
46. QGIS Development Team: Quantum GIS, https://qgis.org/it/site/. Last accessed May 10, 2020.
47. Cutini, V., Petri, M., & Santucci, A. (2004). From axial maps to mark point parameter analysis (Ma.P.P.A.)—A GIS implemented method to automate configurational analysis. In: *Lecture Notes in Computer Science (including subseries Lecture Notes in Artificial Intelligence and Lecture Notes in Bioinformatics)* (Vol. 3044, pp. 1107–1116). https://doi.org/10.1007/978-3-540-24709-8_116.
48. Jiang, B., Zhao, S., & Yin, J. (2008). Self-organized natural roads for predicting traffic flow: A sensitivity study. *Journal of Statistical Mechanics: Theory and Experiment*.
49. Koutsolampros, P., Kostourou, F., & Krenz, K.: rdepthmap: An R package for Space Syntax analysis through depthmapX. https://github.com/orange-vertex/rdepthmap. Last accessed March 7, 2020.

Camilla Pezzica is a Ph.D. candidate at the Department of Energy, Systems, Territory and Construction Engineering of the University of Pisa, and a Research Assistant at the Welsh School of Architecture, Cardiff University. She conducts interdisciplinary research in the fields of Town Planning and Architecture aimed at supporting decision-making processes of experts in different contexts, with a focus on people-centered urban modelling. Her interests are in the integration of advanced digital technologies and computational methods with performance-driven design and planning workflows to allow dealing better with the multifaceted dimensions of contemporary urban environments.

Valerio Cutini is Professor of Town Planning in the University of Pisa. Since 1996, he teaches Urban Planning at the School of Engineering of the University of Pisa. His main interests and studies are in the areas of the analysis of urban settlements, aimed at focusing on their development and the diachronic transformation of their morphology and functional consistency, investigating the way the design of the built environment affects the patterns of social and economic behavior of individuals and communities.

Clarice Bleil de Souza is a Reader at the Welsh School of Architecture, Cardiff University. Her research focuses on exploring interdisciplinary aspects of design decision-making, with research interests are in the areas of: evidence-based regenerative and sustainable building and urban design, computerized evidence-based decision-making in design, user-centric building and urban simulations and their integration in the design process, decision-making in participatory design processes and community-based design, evidence-based decision-making in post-disaster relief, decision-making in designing for harsh climates and low income communities and exploring the use of Machine Learning in design.

Alessandro Santucci has a Ph.D. degree in Sciences and Methods for the City and the European Territory from the Department of Energy, Systems, Territory and Construction Engineering of the University of Pisa, and he is a contract teacher of Territory Modelling. His main interests and studies are in the areas of Geographical Information System (GIS) coupled with territorial analysis models and the development of advanced technologies based on spatial relational databases and WebGIS platforms. He has carried out studies and research at an urban and territorial scale, in particular on spatial decision support systems (multi-criteria decision analysis and GIS).

Multiscale Configurational Approach to the Design of Public Open Spaces After Urban Disasters

Chiara Chioni, Camilla Pezzica, Valerio Cutini, Clarice Bleil de Souza, and Simone Rusci

Abstract

Housing shortage after urban disasters is often approached by governments and humanitarian organizations by building prefabricated houses that, during the transition between emergency and reconstruction, materialize the paradox of the permanency of a temporary situation. The events that followed the 2016–2017 seismic crises in Central Italy do not seem to be an exception: in absence of permanent houses, the displaced people have been accommodated in S.A.E. (housing emergency solutions) within temporary villages of heterogeneous social, spatial and environmental qualities. With the aim to explore and improve way-finding in the disaster-impacted settlements as well as accessibility and sociability of the temporary sites' public open spaces, this paper proposes a human-centred design research approach in performance-based housing recovery planning and design decision-making, combining principles and methods from Space Syntax and Public Life Studies. This research demonstrates how to coordinate different digital analysis and design tools by illustrating their application in an urban regeneration project for *Borgo1*, a temporary housing settlement in the municipality of Arquata del Tronto. Specifically, the paper shows how a multiscale and multidimensional study of the site allowed identifying the ideal location for a new public square and subsequently guided the design process towards the initial design goals' achievement.

Keywords

Temporary housing • Disaster recovery • Central Italy earthquake • Space syntax • Design decision-making

1 Introduction

1.1 Background

Housing assistance after urban disasters is often delivered to the affected population in two or more of the following stages: emergency shelter, temporary shelter, temporary housing and permanent housing [1]. This paper focuses on temporary housing (TH), which is a particularly controversial stage of post-disaster housing recovery because, despite providing dignity and privacy to the affected population while permanent homes are built [2], it often materializes the paradox of the immobility of a temporary situation becoming de facto permanent [3]. As a matter of fact, in the past, TH settlements have rarely been completely dismantled when reaching the end of their functional life at the completion of the reconstruction works, partially because of their physical constitution and extension, connected services and difficulties in restoring the original land uses as well as their cost. Table 1, which includes data from the latest eleven most destructive earthquake disasters in Italy, shows that the actual lifespan of TH sites often far exceeds the theoretical reference time of 5 years. Furthermore, among other issues, the emergence of hybrid urban realities with uncertain identities from the assemblage of temporary housing settlements and reconstructed cities in Italy has evidenced the shortcomings of implementing technocratic practices focused on quick results and lacking planning foresight [4, 5].

C. Chioni (✉) · C. Pezzica · V. Cutini · S. Rusci
D.E.S.Te.C, University of Pisa, 56122 Pisa, Italy
e-mail: c.chioni@studenti.unipi.it

C. Pezzica
e-mail: pezzicac@cardiff.ac.uk

V. Cutini
e-mail: valerio.cutini@ing.unipi.it

S. Rusci
e-mail: simone.rusci@unipi.it

C. Pezzica · C. Bleil de Souza
Welsh School of Architecture, Cardiff University, Cardiff, C10 3NB, UK
e-mail: bleildesouzac@cardiff.ac.uk

S. Eloy et al. (eds.), *Formal Methods in Architecture*, Advances in Science, Technology & Innovation,
https://doi.org/10.1007/978-3-030-57509-0_16

Table 1 Data of the most serious earthquakes that have struck Italy from 1900 to date

Event	Magnitude (Mw)	Deaths	Homeless	Permanence in TH (years)	Reconstruction time (years)
1908 Messina	7.1	120 k	–	30–110	–
1915 Marsica	7.0	33,522	15 k	100	–
1920 Garfagnana-Lunigiana	6.5	171	–	4	10
1968 Belice	6.3	296	57 k	12–20	60* (2028)
1976 Friuli	6.5	965	189 k	8–10	10
1980 Irpinia	6.9	2,735	400 k	15–37	43* (2023)
1997 Umbria-Marche	6	11	25 k	5–20	27* (2024)
2002 Molise	5.7	30	10 k	6–14	21* (2023)
2009 L'Aquila	5.9	309	70 k	Ongoing (so far 11)	14* (2023)
2012 Emilia	5.9	26	15 k	3–5	Ongoing
2016 Central Italy	6.5	299	50 k	Ongoing (so far 4)	Ongoing

The star (*) indicates a forecast of the reconstruction time based on Ruggiero [3]

According to Scamporrino [6] a disaster and a lengthy recovery generate three new cities that co-exist in parallel during the reconstruction process: (i) the historical one, destroyed and inaccessible; (ii) the present one, in transition, and (iii) the future one, in transformation. Currently, design and planning activities after urban disasters seem to follow the same tripartite structure as the following three proposals tend to be separately developed: (i) one for the reconstruction of the destroyed area of the city; (ii) one for the temporary city; and finally (iii) another for the future city as a whole. Within this framework, public spaces assume different meanings in the three cities, because the community relates to those in a different way [6]. In the destroyed city, public open spaces have an intelligible morphology, are formally determined and are known by the community. In the temporary city, such spaces are improvised, interstitial and considered precarious. In the transformed city, they are new and unfamiliar.

Nonetheless, under certain conditions another possibility exists, which involves holistically designing and planning the transitional city by triggering and supporting the transformation of both the destroyed and the temporary cities into the future permanent one. Because the main objective of disaster recovery plans is to "build back better" [7, 8], a promising strategy to foster this aim seems to be the careful (re)design of the public open space, which can work as a key reference point for local communities, hosting key social functions which define the identity of the impacted cities and territory, thus supporting the planned recovery pathway towards more sustainable urban futures. While, to date, the design of public open spaces after disasters is often concerned almost exclusively with the reconstruction of those in the destroyed city or with the proposition of new spaces for the future city-following the principle of "where it was as it was" and by proposing the realization of new areas or the relocation of previous ones-the temporary city is often overlooked. This is a missed opportunity to design for the present as well as for the future while preserving the values from the past, local community bonds as well as positive socio-economic relations. Dealing with temporariness and time as key design parameters in drafting of recovery and reconstruction plans, architects and urban planners can succeed in guiding the reconstruction of a new city whose public space supports the continuous reproduction of the key material and immaterial components of the old one [9].

1.2 Aims and Objectives

This paper seeks to explore and test novel ways in which architects and urban planners can help supporting the transition from recovery to reconstruction. To this end, it suggests coordinating various types of analyses in an urban design process whose ultimate goal is to understand and improve the quality of the open public space in the temporary settlements and encourage its ownership and enjoyment by the displaced community. Specifically, this paper will present a Design Research study [10] exploring the advantages of adopting a data-driven approach to the design of a public square, which provides opportunities for social gathering and thus improves urban and community resilience [11]. The underlying hypothesis is that the solution of such a complex urban design problem would benefit from prior and parallel research and from the use of computational methods and tools in designing for socio-spatial performance.

1.3 The Case of Central Italy

This paper applies the proposed method to the case of the latest Italian earthquake in which this multistaged housing recovery approach was implemented, i.e. the 2016–2017 Central Italy earthquake. There, the disaster affected approximately 600,000 people in 10 different districts (*province*), and 138 municipalities (*comuni*) of four Central Italy regions, thus creating a new disaster geography extending for a total of 8000 km^2. In this territory about 50,000 homeless people required housing assistance and were accommodated by resorting to different solutions, including prefabricated housing structures, hotels, rural emergency modules or containers, municipal facilities, or cash grants for private housing rental. Among these internally displaced people, 7782 requested to be housed in prefabricated temporary houses called S.A.E. (housing emergency solutions), which were assembled in 228 sites built in the proximity of the disaster-impacted towns, in locations that were deemed suitable by both local municipalities and regional authorities [12] (see Fig. 1).

To date, the physical and administrative difficulties and consequent delays of the permanent housing reconstruction process [13] have constrained local people-a predominantly elderly population-to live for an extended period of time in the temporary S.A.E. settlements with social, spatial and environmental urban qualities that can hardly match the needs of the affected communities, as these were not designed for lasting. Furthermore, a twofold mismatch is registered between: (i) temporary housing offer and demand due to dynamic needs fluctuations; (ii) use of temporary and permanent housing standards due to uncertainty in relation to the actual dwellings' life span [3]. Arguably, this involves a risk that further negative impacts will follow a short-sighted humanitarian response that is not holistic enough, i.e. not considerate that disaster recovery is a process that involves rebuilding the historical, cultural, social and symbolic identity of a place by restoring collective values also via the design of suitable public open spaces that materialize the social pact upon which a community is founded.

1.4 *Borgo1* TH Site in Arquata del Tronto

To better research and illustrate potential benefits and limitations of the proposed design approach, an applied investigation is conducted in a real case: a temporary housing site in Arquata del Tronto (district of Ascoli Piceno, in Marche Region, Italy). This was identified in [14] as a particularly relevant case, following a comparative network analysis of the post-disaster configurations of four different municipalities impacted by the 2016–2017 Central Italy earthquake. The temporary village is that of *Borgo1*, which is the largest of seven newly built temporary housing settlements around the destroyed urban historic centre of Arquata del Tronto and the closest to it (see Fig. 2).

In this specific case, based on the official planning documents, recent investments and the current situation, it is legitimate to imagine a future scenario in which the TH site will not be returned to its previous land use. Even the S.A.E. housing units are unlikely to be completely dismantled in the future, due to the permanent nature of their foundation systems (reinforced concrete slabs) and connected services.

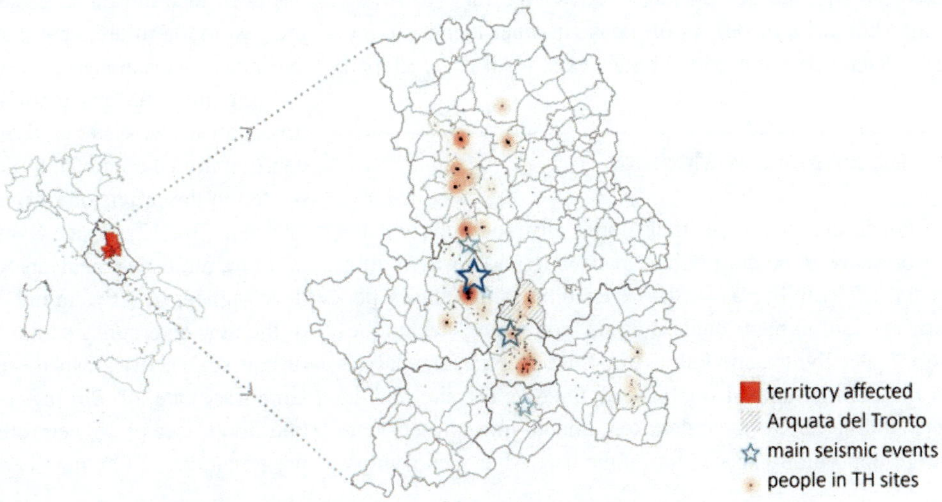

Fig. 1 Location of the disaster affected area in central Italy (with Arquata del Tronto pinpointed in a striped pattern) and heat map of the number of people hosted in the TH sites. The epicentres of the 2016–2017 earthquakes are marked with stars proportional to their magnitude

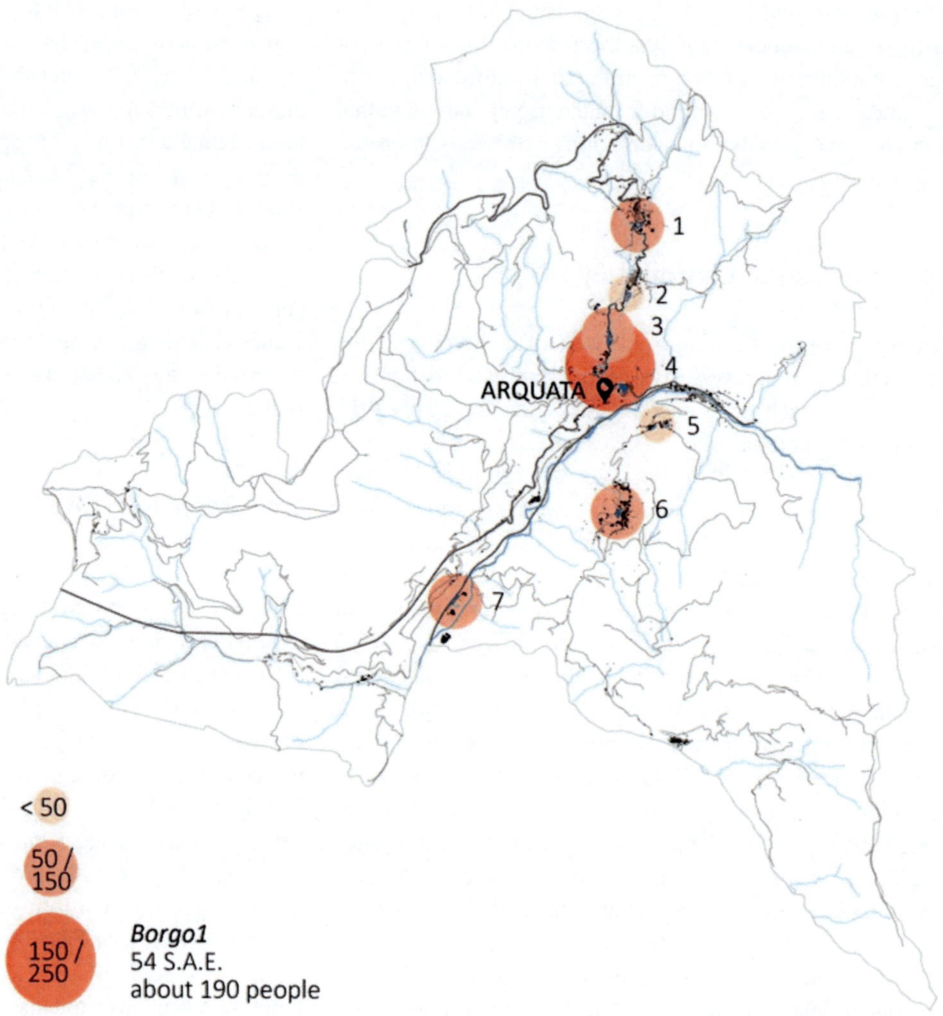

Fig. 2 Post-earthquake "temporary" demography in the municipality of Arquata del Tronto (with pinpointed historic centre). The TH sites are numbered: (1) *Pretare*, (2) *Piedilama*, (3) *Borgo2*, (4) *Borgo1*, (5) *Faete*, (6) *Spelonga*, and (7) *Pescara*

Thus, the site is likely to remain in use in the longer-term, possibly hosting other people when the local residents will return to their permanent homes and undergoing (or not) adaptions in terms of function, configuration and/or built density. Indeed, even when their intended "temporary" use ends, the site can be considered a resource for which new plans should be made. As of today, its final identity is not fully determined, but opportunities exist for it to be shaped.

2 Materials and Methods

To better deal with the transitional city and improve the performance of outdoor public space in terms of wayfinding, spatial accessibility as well as sociability, the authors perform a multidimensional analysis combining techniques from Space Syntax and Public Life Studies. First, an analysis of the configurational qualities of *Borgo1* TH site and its surrounding areas is undertaken, identifying present and future opportunity spaces for public use. After identifying a suitable place for the creation of a public open space, a design research experiment is undertaken, aimed at enhancing the socio-spatial characteristics of the transitioning residential environment of *Borgo1*. The resulting urban design proposal does not have the ambition to replace what has been lost, but rather seeks to facilitate the passage from the past to the future, acting as an interface between absence and new presence (see Fig. 3).

Assuming the articulation of the urban grid as a key driver for urban systems' dynamics [15], this paper adopts a configurational approach to guide and audit design proposals related to the provision of mid-term housing assistance after disasters [16, 17]. Space Syntax techniques are exploited as part of a multiscale analysis to: (i) understand the diachronic development of Arquata del Tronto and evaluate the impact of the new emergency settlements on its urban configuration (urban scale); (ii) grasp the relational characteristics of the temporary site of *Borgo1* to orient urban design (neighbourhood scale). Therefore, the authors analyse the pre-earthquake (T1), post-earthquake (T2), current (T3) and

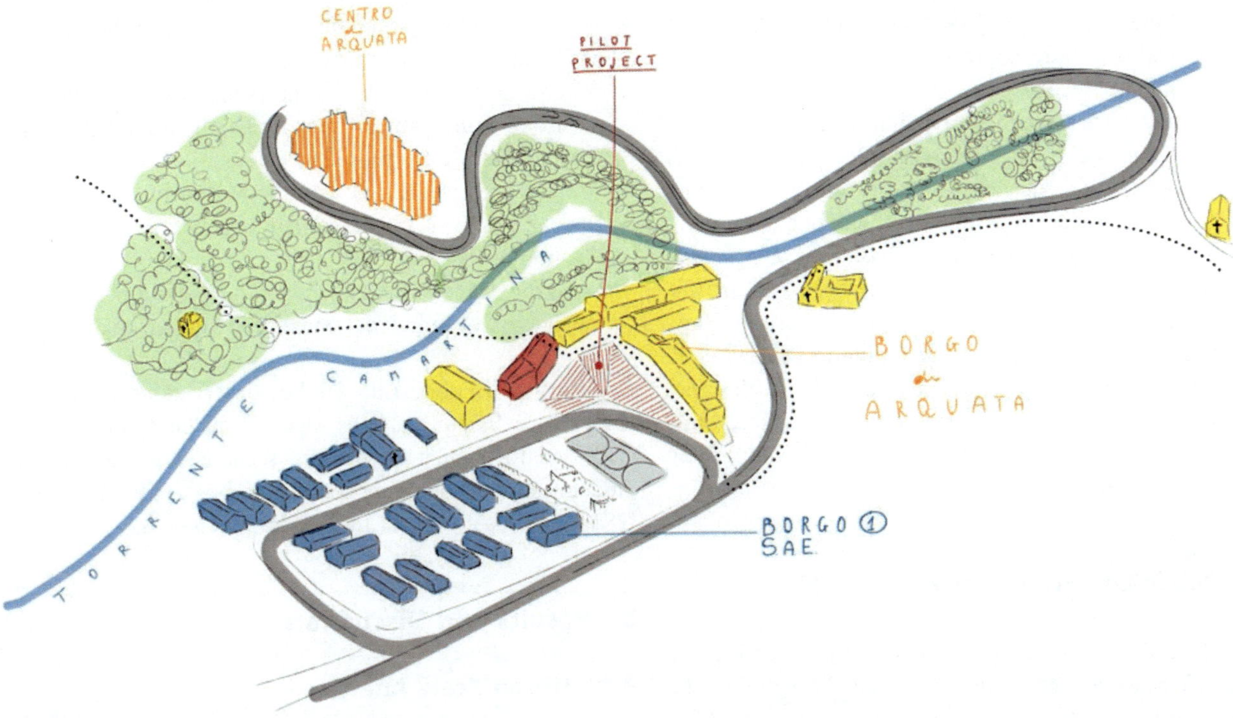

Fig. 3 Contextualization of the project

hypothetical future (T4) road network configurations of Arquata del Tronto as well as the area of *Borgo1* and its immediate context to identify a suitable space to design a new square (T5), with adequate qualities to be a successful and socially oriented public place. This research complement results and partially assesses the reliability of Space Syntax analysis models by using empirical evidence from a Public Life study [18]. The design decision-making process including how the analysis is used to iteratively set and assess design aims in a reflective manner is illustrated in Sects. 2.1 and 2.2.

2.1 Space Syntax Analysis as an Urban Design Method

In the past two decades, Space Syntax has proved to be a valuable support tool in the design of many successful public open spaces among which Trafalgar Square, Nottingham Old Market Square and Woolwich Squares [19]. Karimi [20] shows that the Space Syntax methodology can be effectively used as a formal framework in urban design, meanly because:

- It is a spatial analytical approach and thus suitable for design.
- It can create a direct link between space and people (or users).
- It can deal with different spatial scales.
- It allows investigating an urban or architectural system as a whole or just in parts.

Space Syntax analysis engages with the design process by supporting the negotiation between design problem and solution through continuous iterations of analysis, synthesis and evaluation activities [21]. When adopting this approach, a series of configurational analyses informs the design process from the outset by detecting the problems and latent potentials of a site, which refines both the brief and scope of a design project. In this sense, the analysis of the existing situation enhances the general understanding of the design context and can inspire the formation of new design ideas. When repeated to test a design hypothesis at a later stage, the analysis enhances the assessment of proposals by adding objectivity to it and thus forming a key element of the conjecture-test design cycle. In fact, when a possible design solution emerges from the negotiation with the previously identified issues, the baseline analysis can be used to formally gauge improvements in terms of socio-spatial and functional performance. This cycle can be applied at different scales (e.g. urban and micro-urban or global and local).

This paper coordinates various analysis methods: Angular Segment Analysis [22], Visibility Graph Analysis [23] and Agent-based analysis [24]. The first informs the "natural" distribution of traffic and pedestrian flows across the city, expressed by the Choice index (or NACH), and towards the urban centre, expressed by the Integration index (or NAIN). The second maps in two dimensions some relational and visual qualities of public open spaces identified by Cutini [25] and Guerreiro [26] as key spatial conditions for public squares to work as effective meeting and interaction places by using the following configurational indices: *Integration Index, Clustering Coefficient, Connectivity, Visual Control*. The Agent-based analysis simulates navigational performance and pedestrian flow intensity through a *Gate Count*. According to Pezzica et al. [18] selecting a few suitable variables can help the designer focusing on the most significant properties of a space, while increasing control over design outputs and simplifying project evaluation.

2.2 Public Life Studies as a Design Method

Public Life Studies investigate urbanity by observing the way public space is used in cities and foster the adoption of a people-centric perspective in urban research and design. This field of study became increasingly systematic with the key contributions of Gehl [27] and Whyte [28], whose works are based on first-hand observations of public life in Scandinavia and in New York, respectively. According to them, what attracts the attention and interest of those who inhabit public open spaces in cities are other people and human activities in general. Therefore, successful social spaces tend to be in the proximity of paths and roads hosting pedestrian flows because this enables people to see each other while carrying out other activities.

Gehl [27] adopts a probabilistic approach to the relationship between design and human activities and argues that, through design and within certain limits (i.e. regional, climatic and societal constraints), it is possible to influence the amount of people who use public spaces, how long individual activities will last and guess which activities people are more likely to engage with. According to him, when public spaces are of a poor quality, only strictly necessary activities tend to occur (e.g. going to school or to work, shopping or waiting for a bus). Necessary activities usually take place with approximately the same frequency—although people may choose to spend longer doing them-when public spaces are of a higher quality, but a wider range of optional (i.e. taking a walk to get a breath of fresh air, stopping for a coffee in a street café, people-watching) and social (i.e. simply seeing and hearing other people) activities also begin to occur. Furthermore, Whyte [28] highlights that successful public spaces usually possess more of the following features:

- A good location, preferably on a busy route, being both physically and visually accessible.
- A permeable threshold with the streets, without being cut off from them with railings or walls isolating it and reducing its use by people.
- Are levelled (or almost) with the pavement.
- Have places to sit, either explicit (benches, seats, etc.) and integral (steps, low walls, etc.).
- Have moveable seats, enabling choice and a better communication of individual characters and personalities.

Using field Public Life Studies observations techniques [27], Space Syntax analysis models can be validated [18] and used to simulate "what if" scenarios, while relevant guidelines for encouraging space appropriation by the public [28] can be useful to set further design choices.

3 Results and Discussion

3.1 Urban Scale Analysis

In Figs. 4 and 5, the results of the ASA configurational analysis of Arquata del Tronto's street network are illustrated, with respect to Arquata and *Borgo1*, at the four points in time mentioned in Sect. 2 (T1–T4) to disclose differences in terms of short- and long-term impact of TH plans in the urban spatial functioning. The proposal for T5 is developed by reflecting on these results. However, T5 is not analysed at this scale, because the design intentionally does not introduce a significant variation in the street network configuration.

The map highlights that the municipality of Arquata del Tronto has had a linear development along its main access roads, which is reflected by the highly polarized distributions of NAIN and NACH (T1). When the earthquake destroyed Arquata's historic centre, causing its closure for safety and reconstruction purposes (T2), the centrality values of the areas surrounding the original urban core decreased. This trend is further reinforced by the location of the TH sites during the recovery phase (T3). The analysis shows that in the T4 conservative reconstruction hypothesis, Arquata del Tronto will be significantly affected by the presence of the TH sites, because the old centre will remain a mere repository of a local level of centrality while the TH site of *Borgo1* will emerge as the new actual global centre, acquiring higher spatial accessibility, economic value and social desirability and thus competing with the reconstructed old town centre. In fact, the distribution of NAIN shows that a continued presence of the TH sites after the reconstruction is likely to redefine the spatial accessibility pattern of Arquata, creating the conditions for local economic activities to move out of the historic centre; which is likely to trigger a process of urban decentralisation [4]. Indeed, according to the analysis

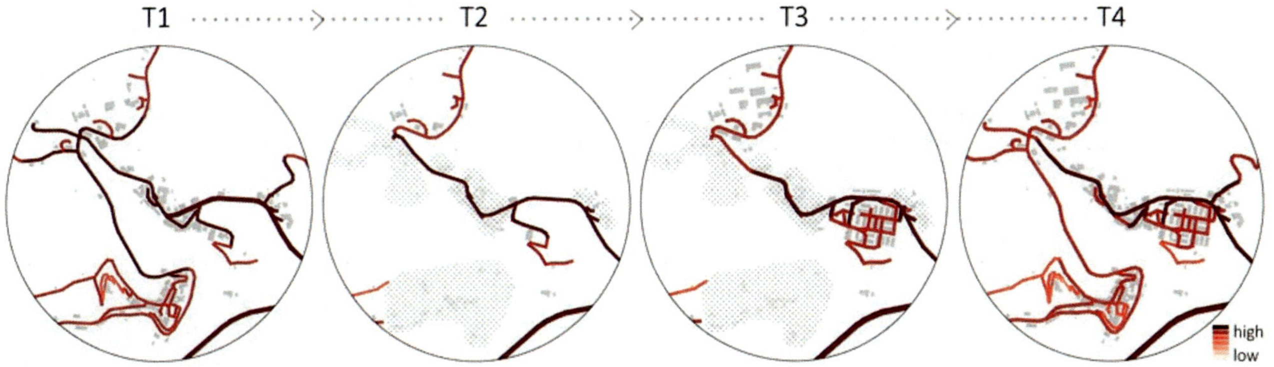

Fig. 4 Zooms on Arquata and *Borgo1* of Arquata del Tronto's ASA analysis (T1–T4): NAIN index with the integration core highlighted (the thicker lines)

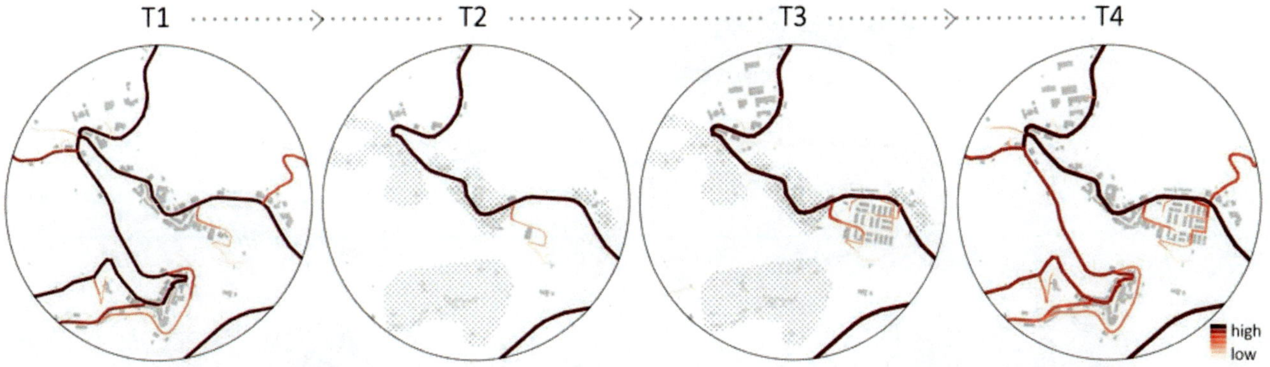

Fig. 5 Zooms on Arquata and *Borgo1* of Arquata del Tronto's ASA analysis (T1–T4): NACH index

at T4, the street running along the northern limit of *Borgo1* will result the most central and accessible.

3.2 Neighbourhood Scale Analysis

Further analyses at the neighbourhood scale have been conducted around *Borgo1*, which assumes a key social role, justifying an effort to strengthen its relational functions through urban design (see Sect. 3.4). A VGA is used to investigate properties that are closely related to manifestations of spatial perception, such as wayfinding, movement and space use. Here, the configurational metrics used are: (i) *Integration Index* (a proxy for spatial accessibility); (ii) *Connectivity* (visual capacity); (iii) *Clustering Coefficient* (identifies convex spatial pockets allowing a comfortable social interaction); (iv) *Visual Control* (identifies spaces where people tend to make navigation decisions). Indeed, the potentiality of an urban place to favour interaction depends on its level of centrality, the richness of its visual connections, its capacity to visually enclose people as if they were in an outdoor room, which make a square a "place of glances" [29].

In Fig. 6, the properties of space related to spatial navigation and visual perception have been analysed at four points in time, namely T1, T3, T4 and T5. The analysis of T2 would have been relevant to decide the location and layout of the TH site, which in this case is already known and fixed. Specifically, the figure shows the multitemporal distribution of two key VGA indices, namely the *Integration index* and the *Clustering Coefficient*.

After the construction of the TH settlement (T3), the spatial permeability of the area hosting the former playground changed dramatically: several new open spaces—from the little pockets between the TH units to the large esplanade resulting from the demolition of earthquake-damaged buildings on the south-west corner of the site—sprang up. Notably, two perpendicular axes emerged with a high value of centrality/accessibility (corresponding to a high value of the *Integration Index*), with the horizontal one crossing the temporary village from side to side. Even after the end of the reconstruction (T4), if no other changes apply, the new road axes will maintain a high local configurational importance.

The presence of the large empty space resulting from the demolition of the group of damaged houses between the

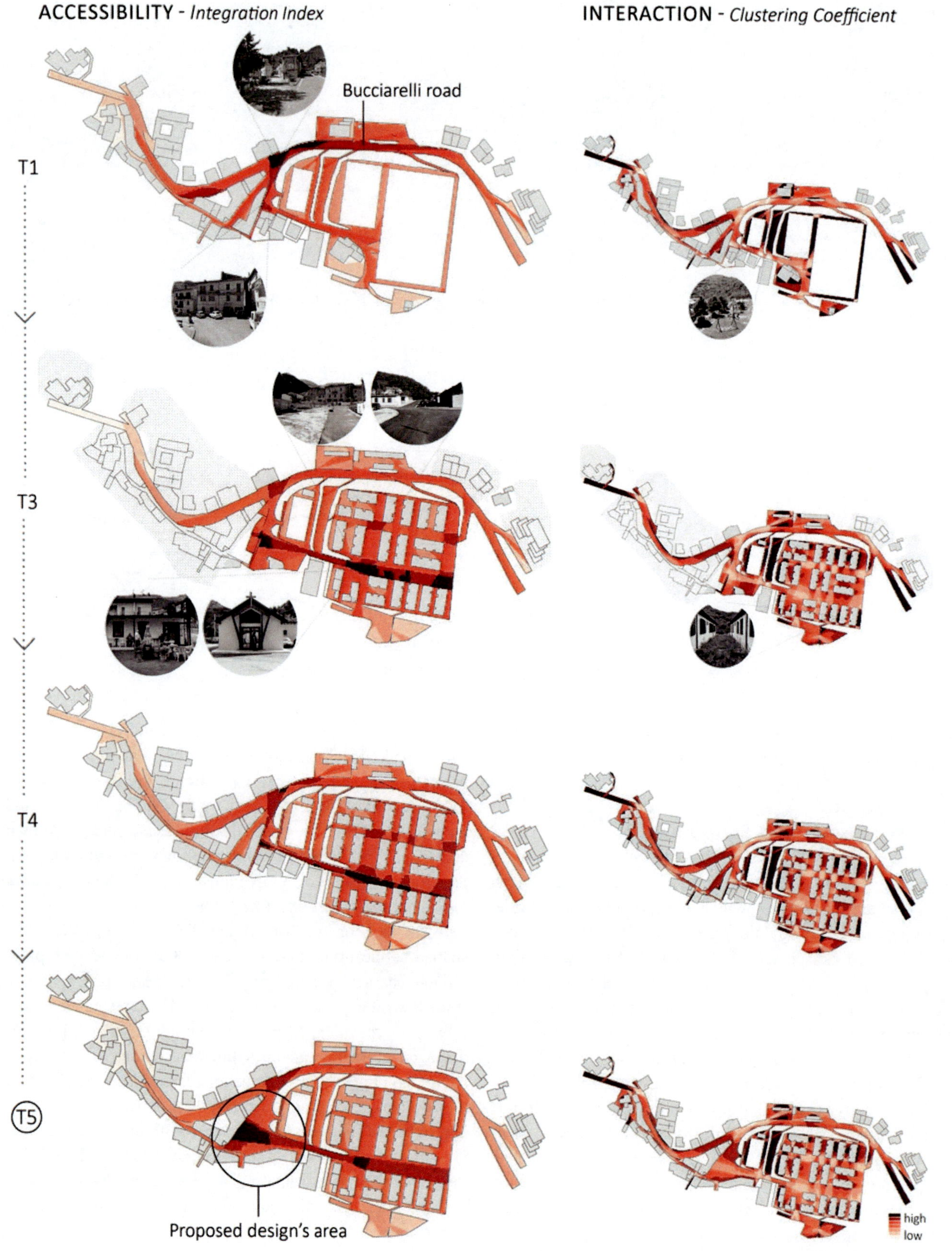

Fig. 6 *Borgo1*'s VGA analysis (T1, T3, T4 and T5): *Integration index* (left) and *Clustering coefficient* (right, miniaturized). The photos help to understand (and confirm) the VGA results

historic hamlet of Borgo and the temporary village of *Borgo1* offers a unique opportunity to regenerate the site, by better sewing up the old and the new (T5). The high values of the configurational indices emerging from the analysis indicate the potential of this urban void to work as an effective public space, which represents an urban alternative to the existing precariousness.

The analysis of the final design hypothesis at T5 shows the benefits of implementing the regenerative design proposal, whose development process is described in detail in the next two sections. In particular, the level of local accessibility and centrality increases, sign of the centralization of the stationary activities, which are the key for achieving an active occupation of the entire square's area.

3.3 Model Validation

In order to carry out a qualitative assessment of the VGA analysis model and collect complementary information useful to inform an urban design (T5), some empirical evidence was collected during a two-day field visit in September 2019, via: (i) a photographic recording with written annotations, following a direct observation of dwellers interacting with their surrounding space; (ii) a short survey, which was circulated among the residents of *Borgo1* to get their views about the performance of their environment in relation to their wishes and needs. Survey respondents were 11, which accounts for the 20% of the 54 housing units in the site. Direct observations confirmed analysis results in relation to pedestrian and vehicular flows, which concentrated in Bucciarelli road and then in the two internal axes. A few people were occupying the outdoor spaces, and the majority was found gathering at the only local bar. When asked in the survey, 59% of the respondents declared to be dissatisfied or very dissatisfied with aggregation spaces in the area and lamented the "lack of a meeting place" and that of "a multipurpose centre".

When tasked with marking on a map of the area where do they meet with other people, they pinpointed a few locations, circled in Fig. 7, which match the results of VGA analysis at T3 (see Fig. 6) and specifically the distribution of Visual Control.

The good correspondence observed between the results obtained from the model at T3 and the real social attractiveness of the space suggests the opportunity of relying on its predictive capacity for further analysis and design purposes. Project choices can then be weighted and iteratively refined according to the predicted configurational characteristics of the space so as to influence the expected behaviour of users.

3.4 The Project

The proposed design aligns with the municipality's declared priorities which include: the return of the residents to their homes; the socio-economic revitalization of the territory and the morphological and functional redevelopment of the urban system also through the redefinition of public open spaces.

The strategic framework chosen for the proposed regenerative intervention is divided into three phases with the following objectives:

- A short-term phase oriented at recovering the public and the pedestrian mobility within the municipality, between the historic hamlets and the TH areas.
- A medium-term phase ("transitional city project") involving the design of an open public space, whose boundary is defined by the construction of an anti-seismic structure meant to rehouse part of the displaced residents and permanently relocate the temporary facilities located in the TH settlement.
- A long-term phase—corresponding to a possible T6—here not explored, in which the TH site is redesigned while ensuring availability of resources to the Civil Protection Department (in view of possible future events) by transforming the TH units into something else such as (i) an holiday camp, (ii) a second-home village or (iii) a new location for the destroyed municipal hamlet of Pescara del Tronto.

The proposal for T5 specifically addresses the short-term and the "transitional city project" and is built on the outputs of the ASA and VGA analyses. The former (see 3.1, ASA analysis at urban scale, particularly at T3), allowed identifying the most accessible site to host and favour the project of a post-disaster open public space. The latter helped defining the morphology of the square and so that it maximizes its social functionality (see 3.2, VGA analysis at the neighbourhood scale, T5).

To this end, an agent analysis (see Fig. 10) was used to predict in several simulated scenarios, the likely behaviour of the local users, assuming that their behaviour in a new space would approximate that of the agents during a random walk.

The design concept is illustrated in all its multifaceted components in Fig. 8. The first step of the underlying design process was the definition of the square's boundary (see Fig. 8.1), as enclosure capacity is one of the main physical property that a square must have to work properly [30]. The southern edge is closed by introducing a new mix-uses building that, following the course of the slope, reconnects

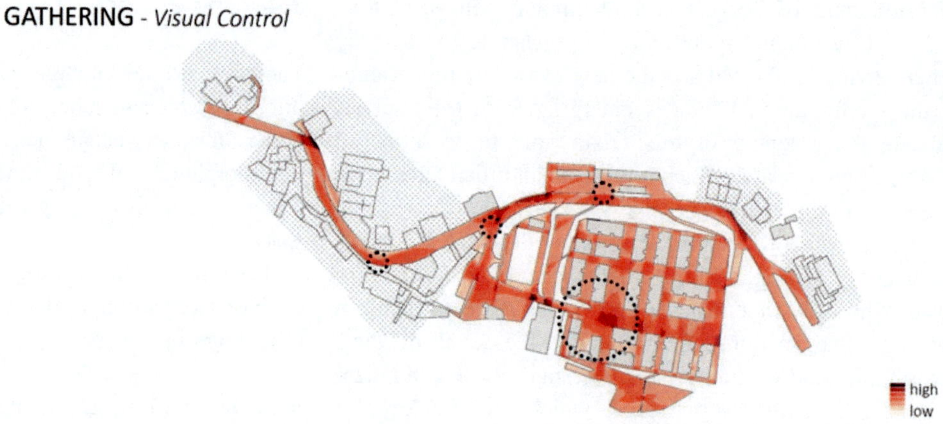

Fig. 7 *Borgo1*'s VGA analysis (T3): *Visual Control*. Indication (circles) of the most used meeting points according to the survey

the fronts of those existing buildings, which are still standing. Conversely, the front adjacent to the vertical road is left open because the street is part of the social space and cutting it out would reduce the use of the square [28]. Then, the vertical connections between the different topographic levels were re-designed in a way that the square does not present physical/visual obstacles according to the principle that the presence of people coupled with weak or no barriers creates a good public space [27]. Figure 8.2 shows that four triangular paved "slices" are inclined in order to overcome the height difference between the various topography levels. This way people in the lower spaces provide the show to the people sitting or walking above and, as a result, the orography of the site will resemble that of an amphitheatre, from which you can observe both the S.A.E. area down the street and the historic front of buildings on the top of the slope (see Fig. 8.3).

Finally, to foster the appropriation of the public space by local communities, other complementary factors relevant to public life were considered, which translated into the design of flexible sitting spaces and public lightening (see Fig. 9). Bespoke chairs were designed to fit the particular topography of the square and to be moveable [28], and a LED lighting system was embedded in the paving of the square, following the natural movement patterns retrieved from the Space Syntax analysis at T5. In fact, by exploiting the correlations existing between the Integration index—which shows a close resemblance to actual movement levels, even in night-time—and lighting parameters [31, 32], it is possible to set in advance the lighting requirements.

To verify that all the proposed interventions were effectively planned and designed to invite the occupation of the square and to promote activities that enhance social interaction, an additional comparative design assessment was made. This helped grasping the quality of the final project in relation to the before and after configurations. The conclusive test was done using a Space Syntax Agent-based analysis, shown in Fig. 10, which generated a forecast on the patterns of movement of people in the newly designed space. The result of this analysis proved that the final design proposal will lead more people (here represented by the "agents") inside the square.

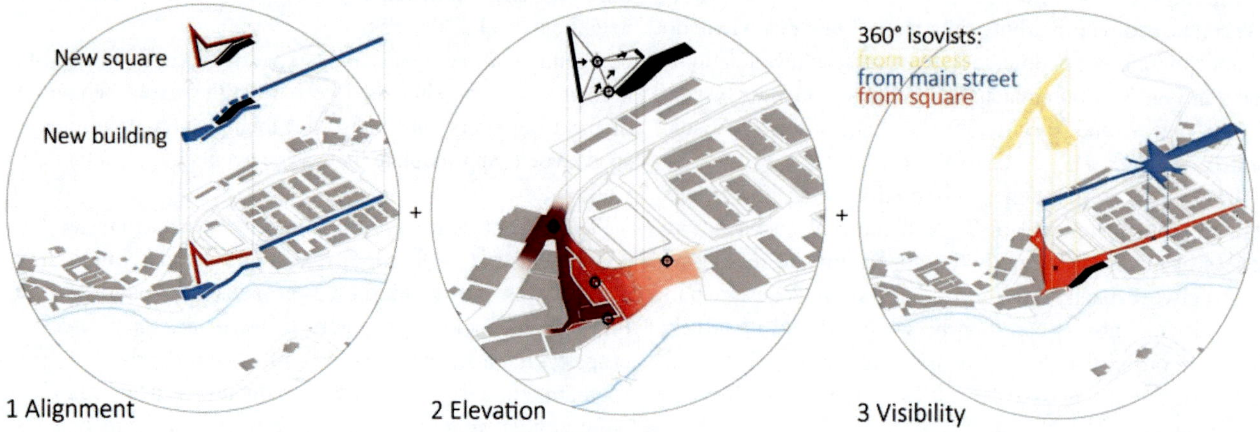

Fig. 8 Genesis of the open public space's form which is derived from: (1) boundary redefinition, (2) collimation of different elevations and (3) visual coverage

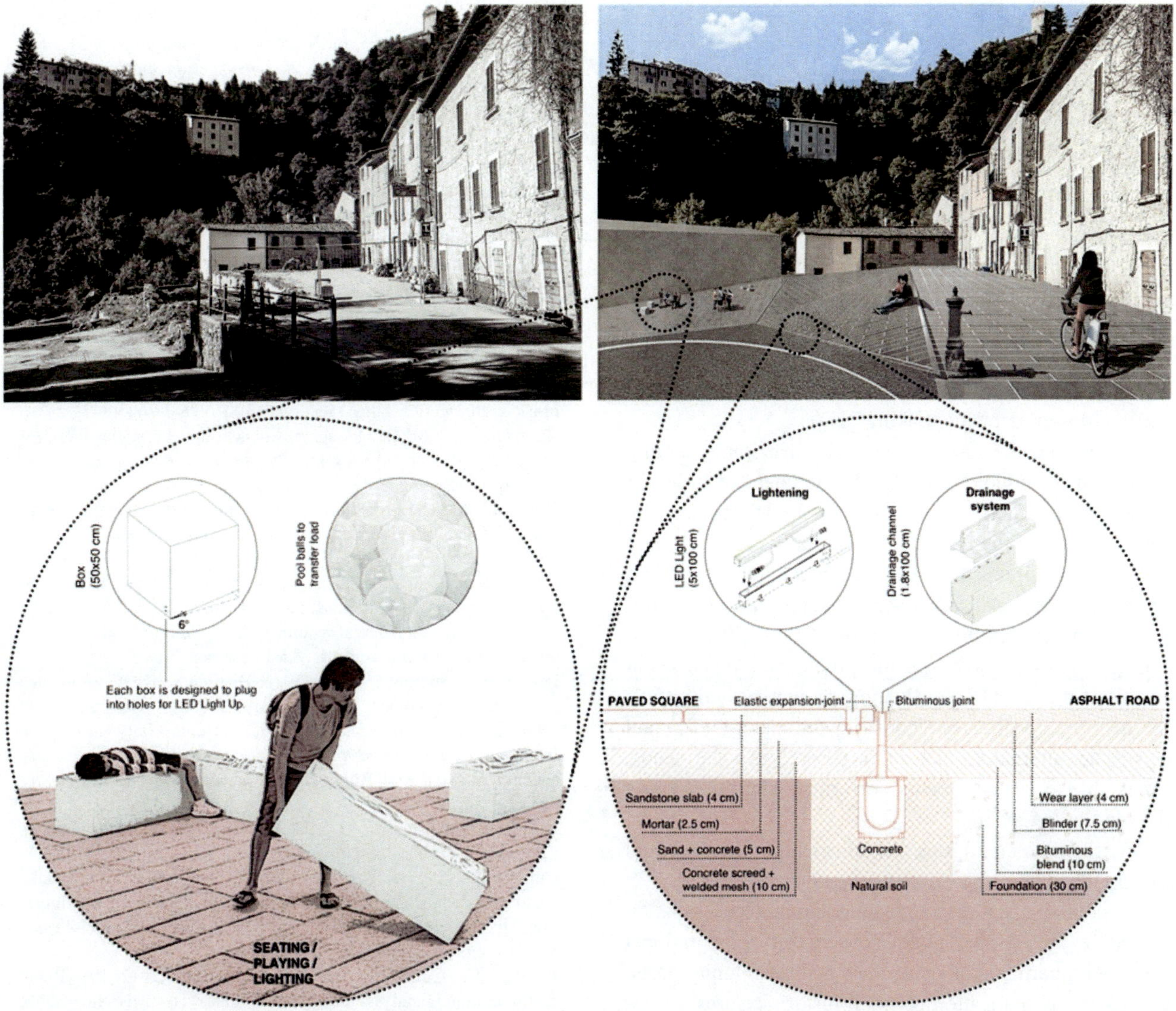

Fig. 9 View from a user perspective of the proposed design's area: as it is today (T3, left) and with the project (T5, right). In the circles, sitting spaces (left) and public lightening (right) are detailed

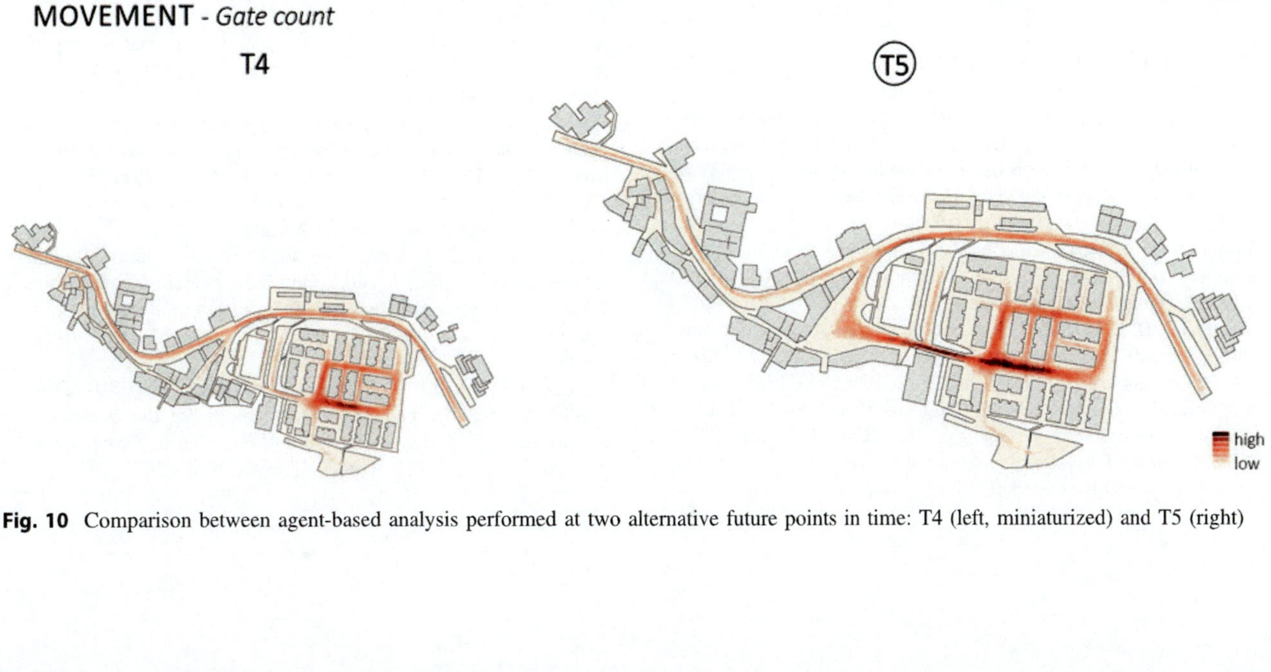

Fig. 10 Comparison between agent-based analysis performed at two alternative future points in time: T4 (left, miniaturized) and T5 (right)

4 Conclusions

The present paper proved that the solution of a complex urban design problem can benefit from prior and parallel research and from the use of computational methods and tools in designing for socio-spatial performance. This approach has a good potential to effectively support strategic decision-making and to make post-disaster planning decisions more explicit, responding to a variety of challenges posed by natural hazards at different moments in time. The design workflow was presented by reporting and explaining the parameters and indices used, and by discussing possibilities of undertaking design changes when Space Syntax is part of the design decision-making process.

By illustrating the results achieved when implementing the proposed methodology in the development of a design proposal in a real case—*Borgo1* in Arquata del Tronto—the paper showed the value of adopting a data-driven and human-centred approach to the (regenerative) design and planning of transitional urban spaces, supporting the return to a new, and possibly improved, urban "normality". In fact, the analysis allowed detecting the problems and the latent potentials of the site (T3), evaluating the unexpected pitfalls of the mainstream disaster reconstruction strategy (T4) and proposing an alternative plan (T5), tailored to local needs, and thus able to ensure a better performance in the short as well as in the longer run.

Future research could look at implementing the proposed approach for supporting strategic decision-making also in ordinary situations, addressing regeneration interventions; in particular, exploring the interplay between computational methods with participatory processes and community-based design for the revitalization of small historic centres: out of post-disaster emergency, but in the long-lasting emergency due to depopulation, abandonment and decay.

References

1. Quarantelli, E. L. (1995). Patterns of sheltering and housing in US disasters. *Disaster Prevention and Management: An International Journal, 4*(3), 43–53. https://doi.org/10.1108/09653569510088069
2. Johnson, C., & Lizarralde, G. (2012). Post-disaster housing and reconstruction. In: S. Smith (Ed.), *International encyclopaedia of housing and home* (pp. 340–346). Elsevier Science. https://doi.org/10.1016/B978-0-08-047163-1.00046-1.
3. Ruggiero, R. (2018). Temporary city. Between emergency and recovery. *International Journal of Architecture, Art and Design*, 145–152. https://doi.org/10.19229/2464-9309/4182018.
4. Alexander, D. (1989). Preserving the identity of small settlements during post-disaster reconstruction in Italy. *Disasters, 13,* 228–236. https://doi.org/10.1111/J.1467-7717.1989.Tb00712.X
5. Contreras, D., Blaschke, T., & Hodgson, M. E. (2017). Lack of spatial resilience in a recovery process: Case L'Aquila, Italy. *Technological Forecasting and Social Change, 121,* 76–88. https://doi.org/10.1016/j.techfore.2016.12.010
6. Scamporrino, M. (2013). Gli spazi della città in transizione nei processi di ricostruzione. In: V. Fabietti, C. Giannino, & M. Sepe (Eds.), *La ricostruzione dopo una catastrofe: da spazio in attesa a spazio pubblico. Urbanistica online dossier* (Vol. 5, pp. 35–38). Roma: INU Edizioni.
7. Kennedy, J., Ashmore, J., Babister, E., & Kelman, I. (2008). The meaning of "build back better": Evidence From post-tsunami Aceh and Sri Lanka. *Journal of Contingencies and Crisis Management, 16*(1), 24–36. https://doi.org/10.1111/j.1468-5973.2008.00529.x
8. UNDRR: Sendai Framework for Disaster Risk Reduction 2015–2020, https://www.unisdr.org/we/coordinate/sendai-framework. Last accessed May 31, 2019.
9. Jacobs, J. (1961). *The death and life of great American cities*. New York: Random House.
10. Chamberlain, P., Bonsiepe, G., Cross, N., Keller, I., Frens, J., Buchanan, R., et al. (2007). Design research now: Essays and selected projects. *Birkhäuser*. https://doi.org/10.1007/978-3-7643-8472-2
11. Carpenter, A. (2013). Disaster resilience and the social fabric of space. In: *Proceedings of the 9th International Space Syntax Symposium*. Sejong University, Seoul.
12. Dipartimento della Protezione Civile: I numeri del sisma in Centro Italia (dati aggiornati al 22 agosto 2018). Modulario P.C.M. 198(3) (2018).
13. Chiappanuvoli, A. (2017). Perché gli sfollati del terremoto sono ancora senza casa. Internazionale 31 July 2017, https://www.internazionale.it/notizie/alessandro-chiappanuvoli/2017/07/31/terremoto-case-italia-centrale. Last accessed July 2019.
14. Pezzica, C., Chioni, C., Cutini, V., & Bleil de Souza, C. (in press). Assessing the impact of temporary housing sites on urban socio-spatial performance: the case of the Central Italy earthquake. In: *Proceedings of the 20th International Conference on Computational Science and Applications, Lecture Notes in Computer Science*. Springer.
15. Hillier, B., & Hanson, J. (2003). *The social logic of space*. Cambridge: Cambridge University Press.
16. Cutini, V. (2013). The city, when it trembles. Earthquake destructions, post-earthquake reconstructions and grid configuration. In: *Proceedings of the 9th International Space Syntax Symposium*, Sejong University, Seoul.
17. Pezzica, C., Cutini, V., & Bleil de Souza, C. (2019). Rapid configurational analysis using OSM data: Towards the use of Space Syntax to orient post-disaster decision making. In: *Proceedings of the 12th International Space Syntax Symposium*, Beijing.
18. Pezzica, C., Lopes, J. V., & Paio, A. (2017). Informed design decision-making: From digital analysis to urban design. *Gestão & Tecnologia de Projetos, 12*(3), 9–26. https://doi.org/10.11606/gtp.v12i3.133936
19. Space Syntax Limited: Streets parks & public spaces. https://spacesyntax.com/urban-places/streets-parks-and-public-spaces/. Last accessed March 11, 2020.
20. Karimi, K. (2012). A configurational approach to analytical urban design: 'Space syntax' methodology. *URBAN DESIGN International, 17*(4), 297–318. https://doi.org/10.1057/udi.2012.19
21. Lawson, B. (1994). *How designers think: The design process demystified*. Oxford: Butterworth Architecture.
22. Turner, A. (2007). From axial to road-centre lines: A new representation for space syntax and a new model of route choice for transport network analysis. *Environment and Planning B: Planning and Design, 34*(3), 539–555. https://doi.org/10.1068/b32067
23. Turner, A., Doxa, M., O'Sullivan, D., & Penn, A. (2001). From isovists to visibility graphs: A methodology for the analysis of architectural space. *Environmental and Planning B: Planning and Design, 28*(1), 103–121. https://doi.org/10.1068/b2684

24. Penn, A., & Turner, A. (2001). Space syntax based agent simulation. Presented at: 1st International Conference on Pedestrian and Evacuation Dynamics (pp. 99–114). University of Duisburg.
25. Cutini, V. (2003). Lines and squares: towards a configurational approach to the morphology of open spaces. In: *Proceedings of the 4th International Space Syntax Symposium* (49.1–49.14). London.
26. Guerreiro, M. R., Eloy, S., Guarda, I., & Lopes, P. F. (2014). Networks and opportunistic urban design: A strategy for regeneration of public spaces in Lisbon. *New Urban Configurations*, 766–771. https://doi.org/10.3233/978-1-61499-365-0-771.
27. Gehl, J. (2006). *Life between buildings: Using public space*. Copenhagen: Danish Architectural Press.
28. Whyte, W. H. (2001). *The social life of small urban spaces*. New York: Project for Public Spaces.
29. Portoghesi, P. (1990). *La piazza come luogo degli sguardi*. Roma: Gangemi.
30. Sitte, C. (1981). *L'arte di costruire le città: l'urbanistica secondo i suoi fondamenti artistici*. Milano: Jaca Book.
31. Choi, A.-S., Kim, Y.-O., Oh, E.-S., & Kim, Y.- S. (2006). Application of the space syntax theory to quantitative street lighting design. *Building and Environment, 41*(3), 355–366. https://doi.org/10.1016/j.buildenv.2005.01.026.
32. Leccese, F., Lista, D., Salvadori, G., Beccali, M., & Bonomolo, M.: Space Syntax analysis applied to urban street lighting: relations between spatial properties and lighting levels. *Applied Sciences, 9*(3331). https://doi.org/10.3390/app9163331.

Chiara Chioni holds a M.Sc. in Architecture and Building Construction from the University of Pisa. Her dissertation, developed during a research internship at the Welsh School of Architecture, specifically addresses the topic of post-disaster temporary housing and focuses on the case study of the 2016 Central Italy earthquake. The main research themes she has explored are in the following fields: (i) housing emergency management and urban resilience, (ii) transformation of towns' image after disasters, (iii) design decision-making and spatial simulations.

Camilla Pezzica is a Ph.D. candidate at the Department of Energy, Systems, Territory and Construction Engineering of the University of Pisa, and a Research Assistant at the Welsh School of Architecture, Cardiff University. She conducts interdisciplinary research in the fields of Town Planning and Architecture aimed at supporting decision-making processes of experts in different contexts, with a focus on people-centered urban modelling. Her interests are in the integration of advanced digital technologies and computational methods with performance-driven design and planning workflows to allow dealing better with the multifaceted dimensions of contemporary urban environments.

Valerio Cutini is Professor of Town Planning in the University of Pisa. Since 1996, he teaches Urban Planning at the School of Engineering of the University of Pisa. His main interests and studies are in the areas of the analysis of urban settlements, aimed at focusing on their development and the diachronic transformation of their morphology and functional consistency, investigating the way the design of the built environment affects the patterns of social and economic behavior of individuals and communities.

Clarice Bleil de Souza is a Reader at the Welsh School of Architecture, Cardiff University. Her research focuses on exploring interdisciplinary aspects of design decision-making, with research interests are in the areas of: evidence-based regenerative and sustainable building and urban design, computerized evidence-based decision-making in design, user-centric building and urban simulations and their integration in the design process, decision-making in participatory design processes and community-based design, evidence-based decision-making in post-disaster relief, decision-making in designing for harsh climates and low income communities and exploring the use of Machine Learning in design.

Simone Rusci is a Ph.D. in Science and Methods for the European city and territory. His research activities in the School of Engineering of the University of Pisa regard urban planning with reference to its environmental and economic assessment. He is currently involved in urban regeneration and fiscal policies.

Data-Driven Spatial Analysis of Urban Renewal. Network Kernel Density Estimation of Building Renovation

Ricardo Mendes Correia, Rosália Guerreiro, and Filipe Brandão

Abstract

Local and national governments are increasingly sharing openly large amounts of georeferenced data related to city planning and administrative procedures. As such new opportunities arise for advanced data-oriented tools that are capable of providing insights into the spatiotemporal correlation of these phenomena. Kernel density estimation (KDE) appears to be an efficient tool for overcoming incomplete data, because not all urban rehabilitation needs to be reported to City Hall services. Recently, new research has proposed network kernel density estimation (NKDE) as a more accurate alternative to estimate data in urban areas. This paper aims to provide a vision of the possibilities of integrating urban renewal dispersed datasets. We propose a method to measure the intensity of renovation in a network using the spatial database of building permits from the city of Lisbon.

Keywords

Building renovation • NKDE • Urban renewal • Spatial analysis

1 Introduction

In Lisbon, building renovation transformed 45% of vacant buildings over a ten-year period, from 2009 to 2019 [1]. However, this renovation has been an asymmetric urban phenomenon, focused on the old city center and on the subsequent nineteenth-century expansions. Lisbon's renovation process differs from traditional top-down urban renewal processes, characterized by Anderson [2] as "The Federal bulldozer". It is mostly a bottom-up urban renewal, as can be demonstrated by overlapping maps of building renovation and municipal plans. That is close to Hillier's (1996) idea that cities cannot be planned but only understood.

The urban renewal process is mainly made up of building renovation initiatives led by private companies. However, it has some top-down planning characteristics because it is sponsored by public funding through low taxes. Although there is some public intervention, it does not correspond to a "Federal Bulldozer" [2], with the use of large-scale expropriation or demolition to give rise to renovation.

Thus, Lisbon's urban renewal is also a consequence of sustained public policies and low taxation to promote the renovation of old buildings. But since these policies are limited to fiscal stimulus, it becomes necessary to measure if they are achieving the desired effect. It is an urban renovation that includes a large number of small interventions (Fig. 1) in existing structures, maintaining the configuration of urban plots.

These types of small interventions consist of additions to the original volume of the building, changes in the typology of the plan, changes in the spatial distribution or the facade, maintenance or reconstruction. Those small interventions combined with the dispersion of property in small plots give rise to an extremely fragmented and interspersed renovation pattern, both in time and space. And if the idea of those who legislate was to promote housing and the repopulation of areas that started to be depopulated, the truth is that part of the renovation is housing from a legal point of view, but it is in fact accommodation, the so-called peer-to-peer accommodation. The ideal would be to assess, in real time, the effectiveness of urban policy to understand whether the planned objectives were being achieved.

We can extrapolate the hypothesis that the continuous urban renewal process can occur anywhere in the city and not in a specific parish, neighborhood or street. Traditional

R. M. Correia (✉) · R. Guerreiro
Instituto Universitário de Lisboa (ISCTE-IUL), CRIA, Lisboa, Portugal
e-mail: Ricardo.Mendes.Correia@iscte-iul.pt

F. Brandão
Instituto Universitário de Lisboa (ISCTE-IUL), ISTAR, Lisboa, Portugal

S. Eloy et al. (eds.), *Formal Methods in Architecture*, Advances in Science, Technology & Innovation,
https://doi.org/10.1007/978-3-030-57509-0_17

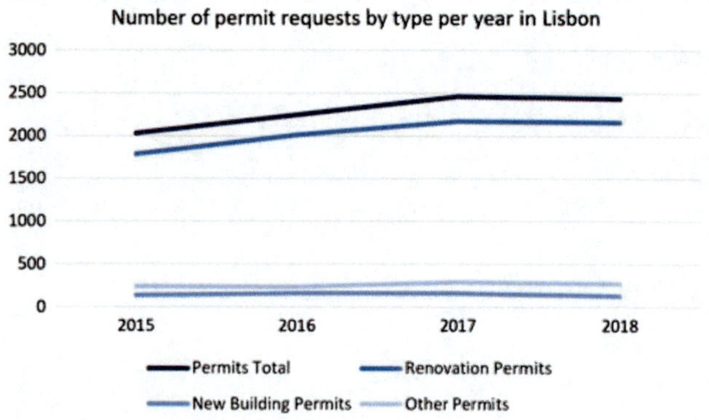

Fig. 1 Number of urban management and building permit requests by type per year 2015–2018 in Lisbon. *Source* Authors, based on Lisbon City building permits and requests database

national statistical tools or even municipal statistical instruments have serious limitations and may not provide a good picture of these highly fragmented dynamics, as they do not keep pace with the transformation or the area it involves.

The use of computational tools, such as GIS and other spatial analysis tools, can provide graphical responses by mapping the renewal statistics in space and time. GIS can provide a good visualization of the phenomena. Thus, computational tools can be used to generate maps that correlate the georeferenced databases of building permits with the street network. Tools that represent urban renewal events in the city network and that can relate the renovation to urban morphology.

Statistical methods such as kernel density estimation (KDE) [3, 4] or Getis-Ord Gi* [5] can provide a good balance between analytic and synthetic representations without the loss of the underlying spatial information and allow a better reading of the phenomena. KDE is a statistical method for spatial smoothing and interpolation that is easy to understand and implement and produces a continuous field of densities, highlighting peaks of areas with a greater concentration of events from areas with lower densities [6–8].

KDE requires two parameters, one is the bandwidth or window, which is the distance of influence, and the other being the weighting function *K*, often a normal function. The bandwidth choice has a great effect on the results, while the weighting function choice is less critical [7, 9, 10]. The point *x* kernel density is estimated to be the sum of hills within the bandwidth [8].

$$kde(x) = \frac{1}{nh^2}\sum_{i=1}^{n} K\left\{\frac{1}{h}(x - X_i)\right\} \tag{1}$$

The implementation of open data policies by City Halls in major cities in Europe and North America facilitated the estimation of a large volume of events in a city. This way the KDE (Fig. 2) is a method already used to determine the distribution of properties on a large volume of data [11, 12], and in previous work [13] we have used a double KDE on databases of building permit applications.

However, KDE can have limitations when applied to urban space, because the buildings are not evenly distributed but are concentrated along a one-dimensional network of streets. Building entrances are located on the streets, which means that buildings in cities can be represented by points along a network of streets. This way we can calculate distances according to the hypothesis formulated by Miller [14] that the planar space is too vast for the analysis of events which take place in only the part of that space that is the one-dimensional network [7]. In fact, many times it is the national laws that impose that building construction can only be authorized in lots that confine with the network of streets or roads.

The study of theoretical methods using spatial analysis to calculate distance in networks has been the focus of several research papers [12, 15–19, 19–21] in addition to the aforementioned Miller [14]. Specifically addressing KDE on a network (NKDE), there were some important works mainly in the second part of the 2000s [6, 7, 16, 22–30]. Okabe, Satoh and Sugihara [28] used the basis for this study and developed through the GIS software; they coauthored SANET 4.1—Standalone 1.0 Beta version [31].

This paper presents preliminary research on applying the kernel density function in a network to determine the density of building renovation on the city of Lisbon. We measured that density using a KDE on a network [28], discontinuous at nodes, which for briefness sake can be simplified to network kernel density estimation (NKDE). We used GIS tools [31] to estimate the density of points on a network [28], comparing NKDE with KDE results and discussing NKDE limitations as well as suggesting alternatives for estimation accuracy improvement.

2 Methodology

In order to measure the renovation density of buildings within the city limits of Lisbon, we chose NKDE which allows us to compare the density of building renovation through a network of streets in space and time.

2.1 Dataset and Building Renovation Permit Types

The georeferenced databases used in this research were based on open data from the Lisbon City Hall Web site https://geodados.cm-lisboa.pt/, which are composed of vector information of points, in shapefile format. These are

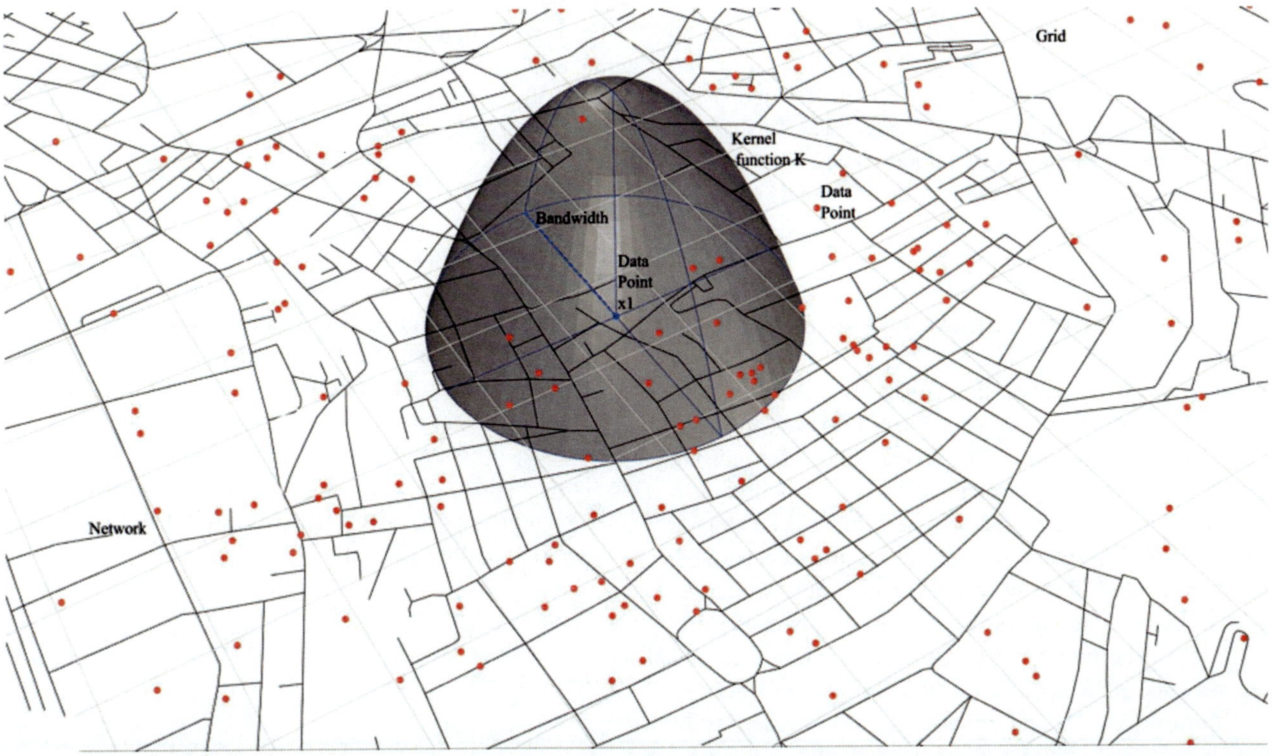

Fig. 2 Spatial representation of KDE with data from this research [8, 9]. *Source* Authors, based on Lisbon City building permits and requests database

georeferenced databases of buildings with information on the building permit dates and type, type of administrative proceeding and an exclusive building ID, among other types of administrative-specific data.

With this method, only points that represent the geometric centers of buildings, the centroids, were considered. In the point database, the most important information corresponded to the type of building permit, as this allowed us to exclude the new building permits, leaving only renovation permits. The database of renovation permit points contains 12 categories in the field building permit type that were divided into three groups for this research: *construction, renovation* and *others*. Only the *renovation* group was considered in this research. It contains four subcategories: *alteration*, *expansion* or *addition*, *reconstruction* and *conservation*. By considering only those four subcategories, we have filtered the buildings with renovation procedures. The other building permit type groups, *construction* and *others*, contained, respectively, new building permits and administrative procedures related to ongoing permits changes or other less relevant procedures that could not be included in the previous groups.

It is also important to mention that other kinds of interior refurbishment do not need a permit to be issued and that it is also the case for smaller procedures like changes in compartment dimensions. These types of renovation can only be tracked by other types of permits, not necessarily related to the construction of buildings, such as scaffolding permits. Therefore, these types of renovation work are not in the building permit database and were not included in this analysis.

Subsequently, the City Hall database provides us with only a partial view of the complete building renovation activity across the city even though it represents the most significant part. Facing the data availability, we assumed that this dataset is an unbiased representation of the global renovation work.

Like previously stated, the City Hall georeferenced databases have a building ID, an exclusive identifier for each building. That unique GIS ID relates the shapefile geometry with the building permits database. In this research, four years of records from 2015 to 2018 were used, with a total of 9171 requests and an average of 2292.75 requests per year that corresponded to an exclusive record in the requests and permits database requests. The permits database not only includes permits, but also includes other procedures that do not end with issuing permits, such as prior communication requests.

There were duplicate GIS ID entries in the renewal permit database, corresponding either to simultaneous permit requests for the same GIS ID or to permit requests for two separate units in the same building by the same proprietor, or even to different privately held units of the building, all sharing the same building identifier, the GIS ID.

Our datasets contain information for all the renovation permits within the Lisbon City Hall area. The final set of issued permits for the four-year period has 6276 renovation entries comprised of 4944 unique GIS IDs and 1332 duplicated GIS IDs.

2.2 Kernel Density Estimation Over a Network

NKDE is a kernel density estimation method that estimates the density of points on an environment [28]. According to Okabe, Satoh and Sugihara [28], several kinds of events are possible to be analyzed through a density of points on a network, like car crashes on streets, seabirds located along a coastline or WiFi signal strength along the streets. This estimate considers that the events occur only within an assumed network and not in the neighborhoods of the network. Unlike the ordinary KDE measured on an area and other network KDE approaches like Borruso and Schoier [22], Borruso [23, 24] and Porta et al. [12], the density of points is estimated only inside the network and in this case only in the street sections and not in on the surrounding region [28].

NKDE is a method for analyzing and estimating the density of points on a network. If in our previous research [13] the renovation KDE was measured in an area, in this research, the renovation density is measured in the street network of Lisbon. A finite network embedded in a plane made of line segments corresponding to street sections between crossings [28].

In the finite network $N = (V, L)$ in which there are a set of nodes, V and a set of links L. N is assumed to be planar with links in L not intersecting each other except at nodes in V and also non-directed with two-way links.

$\tilde{L}$ is the union of all links in L including the nodes in V.

y is an arbitrary point on $\tilde{L}$, L_y is a subset of $\tilde{L}$.

The shortest-path distance between y and any point on L_y is equal or less than h - segments shown in bold on Fig. 3 [28].

NKDE is a function that can be defined for an arbitrary y and the bandwidth h, $K_y(x)$ satisfying.

$$\begin{aligned} &K_y(x) \geq 0 \text{ for } x \in L_y \\ &K_y(x) = 0 \text{ for } x \in \tilde{L} \backslash L_y \end{aligned} \tag{2}$$

$$\int_{x \in L_y} K_y(x) \mathrm{d}x = 1 \tag{3}$$

where $\tilde{L} \backslash L_y$ in Eq. (2) is the complement of L_y considering $\tilde{L}$ and the integral over $x \in L_y$ (3) along all the line segments of L_y [28, 30].

For n points independently distributed on $\tilde{L}$, we have the kernel estimator $K(x)$ for $f(x)$ at x and the estimation of $f(x)$ (4) [28, 30].

$$K(x) = \frac{1}{n} \sum_{i=1}^{n} K_{yi}(x) \tag{4}$$

Equation (4) used the equal split discontinuous kernel function at each node because this kernel type required less computational time for a 15 m bandwidth when applied to all of Lisbon's 22,098 street segments than the equal split kernel function continuous at the node [28, 30] (Fig. 4).

In the equal split discontinuous kernel function, $v1, v2, \ldots vs$ are nodes on the path p when traveling from y to x along the shortest path p, n_i is the degree of the node v_i, for $i = 1, 2, \ldots, s$, and let $d(y, x)$ be the shortest path distance from y to x being the function [28, 30]:

$$\begin{aligned} K_y(x) &= \frac{K(d(y,x))}{(n_1 - 1)(n_2 - 1)(n_s - 1)} \text{ for } 0 \leq d(y,x) \leq h \\ 0, d(y,x) &\geq h \end{aligned} \tag{5}$$

The equal split discontinuous kernel density function was applied to a network of 22,098 segments consisting of 75,472 nodes, comprising all 13,358 street intersections of all the 4399 streets of Lisbon. The segments we used come from an open data line layer of the Lisbon City Hall Website, vector information, in shapefile format. The bandwidth used was 15 m, the grid had the value of 1.5 m, and all streets were given the same weight.

3 Results

The method of data-driven analysis, adopted in this research, provided us with the use of a discontinuous network kernel density function with equal division, using a GIS software environment. We have produced NKDE estimates for the renovation data for all the Lisbon City Hall area. The renovation dataset was used in its entirety but was also divided into annual periods producing four annual datasets. We used a bandwidth h of 15 m and a grid of 1.5 m due to computing times. Then, the results of the entire renovation permit dataset and of each annual dataset are presented and compared.

Figure 5 shows the NKDE for the renovation permit request events for the 2015–2018 period. It is shown that there is a great building renovation intensity concentrated in the group of streets composed by the historic center and its late nineteenth-century expansions but also by the Lapa parish. There are four distinct focus areas: The first is Baixa area including Chiado; the other is on Bairro Alto; the third is on the Avenida da Liberdade and surroundings. All of

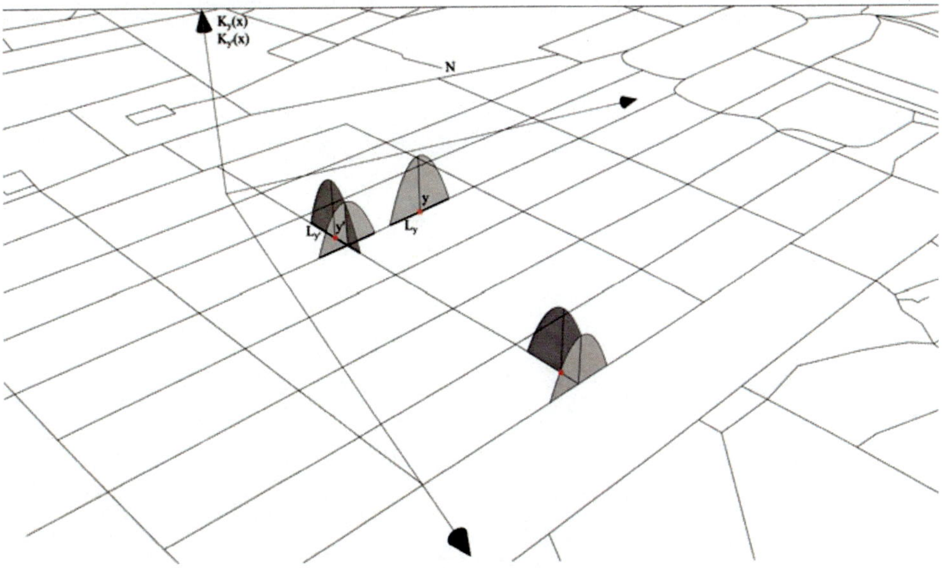

Fig. 3 NKDE - kernel functions with centers at y and y' on $L^{\sim}$ of N and the street network used in this research [28]. *Source* Authors, based on Lisbon City building permits and requests database

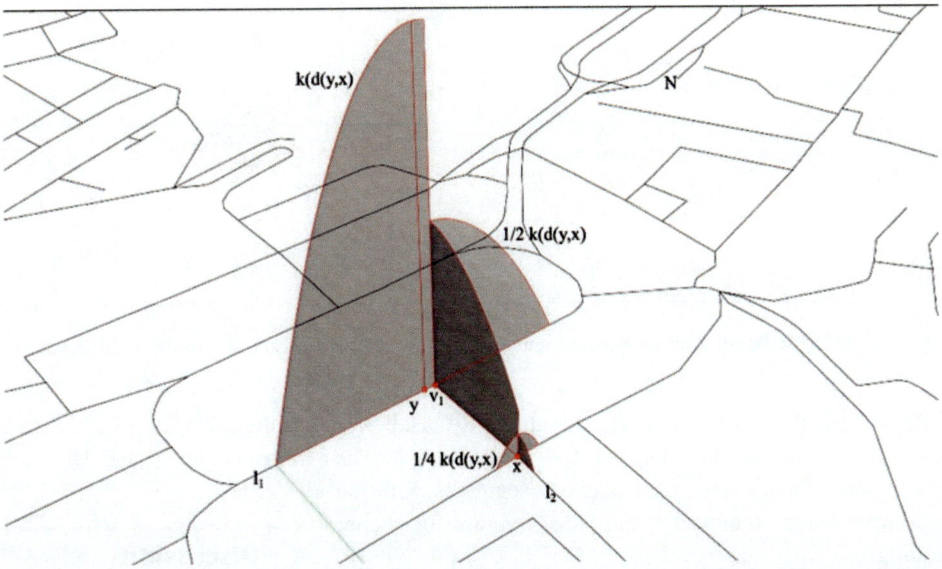

Fig. 4 NKDE - an equal split discontinuous kernel density function centered at y, where $K(d(y,x))$ is a base kernel density function with the street network used in this research [28]. Source: Authors, based on Lisbon City building permits and requests database

these three areas are not separated with more than one kilometer between them. The fourth area is the upper-class Lapa neighborhood that is more than one kilometer away from the less central Bairro Alto areas. The Baixa–Bairro Alto–Avenida da Liberdade focus spreads particularly toward Avenidas Novas area, an urban expansion of the mid-twentieth century, creating wide avenues. There are two other areas with small spots of relatively high values in Belém and Bairro Azul, areas with museums and department stores, respectively. The Marvila and Santa Clara parishes are areas with very low building renovation density, with values close to areas that have almost no buildings at all like Monsanto Park or the Airport.

From the comparison of NKDE with KDE, Figs. 5 and 6, respectively, it should be noted that NKDE is able to clearly identify the streets or the set of these. On the other hand, the KDE has a greater tendency to identify patches with larger areas giving greater emphasis to the great Baixa–Bairro Alto spot and smaller to Avenida da Liberdade and Lapa, although it identifies these as well as the less intense spots of Belém, Bairro Azul and Avenidas Novas.

The renovation permit requests highest densities are largely related to the most expensive and most demanded neighborhoods. Nevertheless, the higher concentration of renovation permits can be partially explained by the morphology of the city with some streets of the network having more building centroids than others. These city central areas, especially Baixa e Bairro Alto, have streets with smaller lots, while in the newer areas, which came after the second part of the twentieth century, the lots are more widely spaced.

Figure 7 presents the yearly NKDE estimations for permit requests. Each year contains all the permits

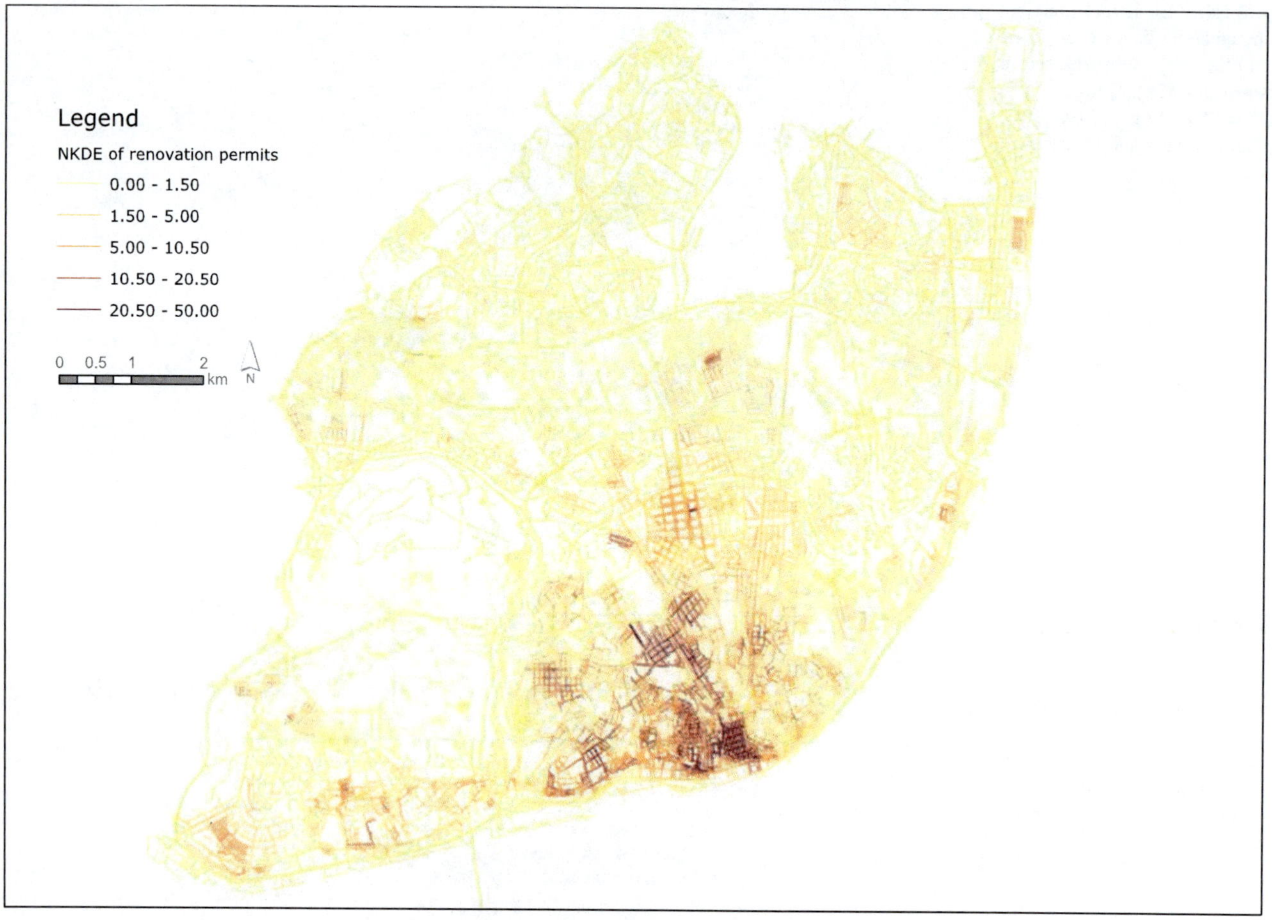

Fig. 5 NKDE of building renovation permit requests in the 2015–2018 period, bandwidth = 15 m and grid = 1.5 m. *Source* Authors

requests which were submitted to the City Hall for building renovation according to the methodology. The image only contains the street network segments with the buildings being removed from the figures for better reading.

The dataset in Figs. 7 and 8 is the same used in Fig. 5, and all the yearly sets use the same NKDE estimate of building density used in the 2015–2018 period. In the yearly sets, the number of events is smaller, and an average of 1569 per year against 6275 permits for the four years total. As it is shown in the figures, the last year, 2018, has more remote areas with high densities with areas like Belém, Bairro Azul and Avenidas Novas with highest values when compared with 2015 when almost all hotspots were concentrated only within Baixa and Avenida da Liberdade area with fewer hotspots in the Bairro Alto area. From 2015 to 2018, the number of renovation permit requests increased by 23% and the increase in requests may have led to the emergence of new high-density spots. However, it can be concluded that there was an increase in the density of permits in other areas and therefore a dispersion of urban renovation by new independent focuses. This suggests that the spread is not due to the contiguity of the areas but may be caused by other factors (Figs. 9 and 10).

4 Discussion

Cities are constantly changing, and we have an increasing amount of big digital data to understand this transformation. Taking advantage of this availability, we can investigate new ways of determining the existence of areas in the city where the renovation of buildings is more intense than others, and these new ways can serve to enrich the patterns of social and urban analysis.

In an urban area, building renovation can only take place in buildings located on a network of city streets, and this paper presents a method of cross-referencing City Hall's alphanumerical and spatial databases, frequently kept apart, producing a measure of intensity of renovation that can be visualized on a map or other digital representation of the urban area.

Although we can predict part of the renovation phenomena by simply mapping the building permits, great part

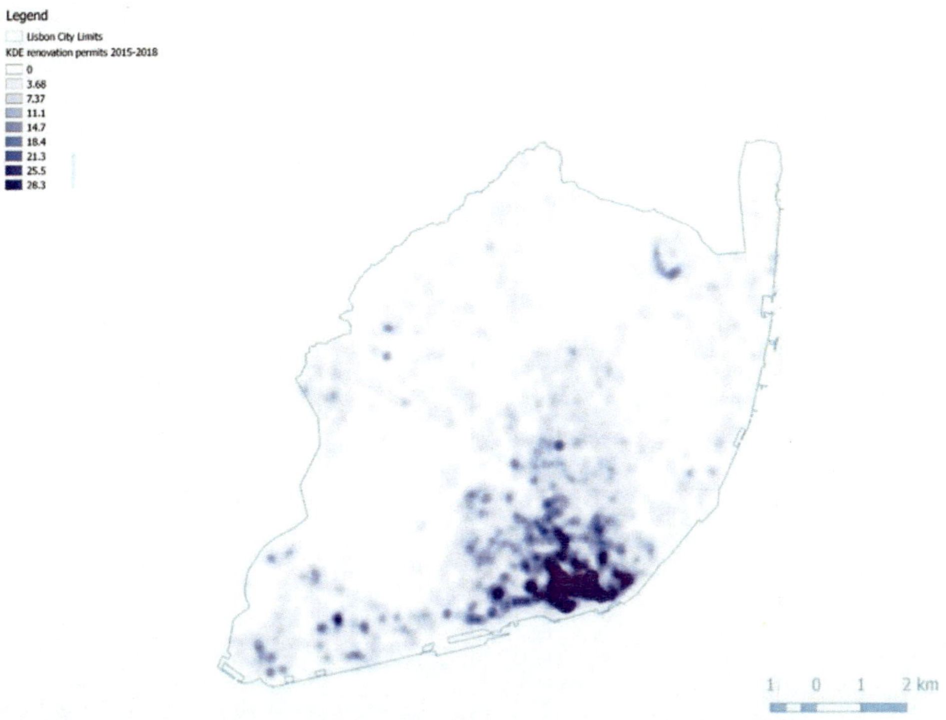

Fig. 6 KDE of building renovation permit requests (Epanechnikov Kernel) in the 2015–2018 period, bandwidth = 150 m and grid = 15 m. *Source* Authors

of the phenomena can only be understood by using statistical methods that show density like NKDE. The density estimation on a network provides a measurement of building renovation linear hotspots on the street network that gives a more adequate idea of the phenomena than the mapping of points over the segments of that network.

By giving us a measure of density that overcomes incomplete data areas, NKDE provides us a way of having data on all the streets, including the ones without a single building permit request.

The results indicate that the renovation of buildings in Lisbon is focused on the streets of three main areas not far from each other: Baixa including Chiado, Bairro Alto and Avenida da Liberdade, where the intensity of NKDE is greater and, also, the fourth area farther away - Lapa, as shown in Fig. 5. Our results obtained with NKDE largely coincide with the empirical perception and market prices of real estate agents and other city dwellers.

Note that, although there is an increasing number of permit requests, from year to year, during the period studied, as shown in Fig. 1, this increase does not correspond only to the growth of the four main zones with new hotspots emerging, distant from the initial main axis zone Baixa/Bairro Alto/Avenida da Liberdade, Fig. 5.

When comparing NKDE with KDE, Figs. 5 and 6, NKDE clearly identifies streets while KDE identifies larger areas, in general, the same intensity is shown; however, NKDE more accurately identifies a street and, as a consequence, the buildings on that street. Although it is not the purpose of this article, to prove the robustness of KDE, we can say that, in both cases, the highest densities are in the same areas, demonstrating the robustness of the two methods. NKDE is more specific in indicating the streets, but basically shows the highest densities in the same areas as KDE. This, in part, demonstrates the importance of using different methods to analyze the same complex urban phenomena.

Another important idea is related to the issues mentioned in the introduction that NKDE has fewer restrictions than KDE, being related to various limitations of KDE, detected in our previous research, namely the fact that distances in KDE are measured in two-dimensional space.

As we have already mentioned, when estimating building renovation in segments, NKDE does not create hotspots as extensive as KDE, limiting the occurrence of some problems when there are neighboring areas of high density and low density. Thus, NKDE appears to adapt better to block-level estimates.

In the proposed method, only the distance between the points was considered as a measure of density. We did not analyze the gross construction area of the building to which the renovation permit belonged. In addition, the underlying density of city buildings was not considered. In future work, in which the construction area is considered, weights may be

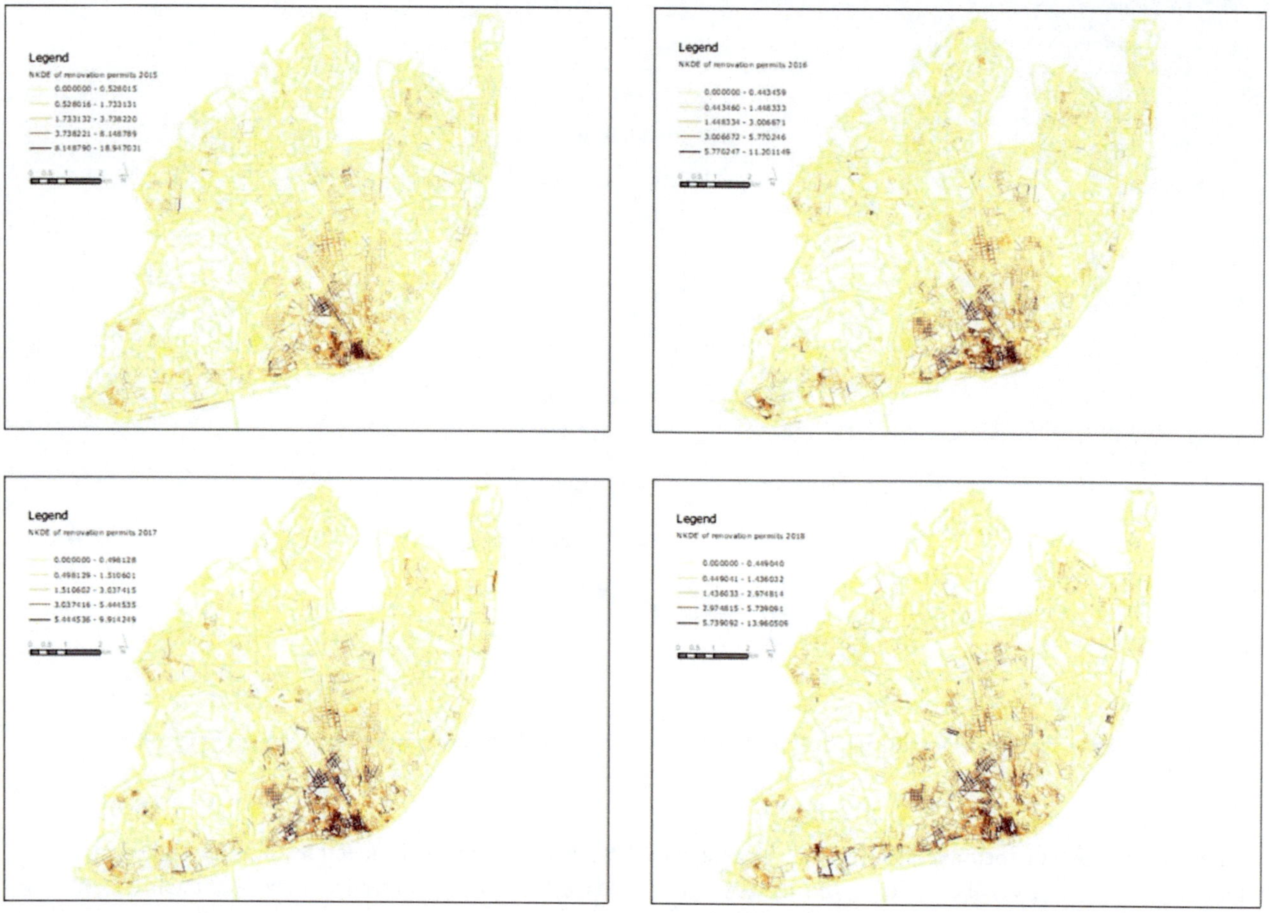

Fig. 7 Yearly NKDE of building renovation permit in the 2015–2018 period, bandwidth = 15 m and grid = 1.5 m. *Source* Authors

assigned based on that area. However, in the municipality's open data database, there was no information about the construction area of the renovated buildings, nor whether the renovation was in the entire building or just in a subdivision of the property. These areas can be estimated for a building, multiplying the area of the building's polygon by the number of floors, but this can produce gross overestimations, especially in buildings with higher number of floors.

It is important to emphasize that the set of points of the renovation permits corresponds to a sample that represents four subcategories in which two of them, *reconstruction* and *alteration*, have the vast majority of permits. Both subcategories have more than 90%, and the *alteration* subcategory appears to have greater variations of the number of permits than *reconstruction*.

It should also be noted that this City Hall database is more favorable to show the numbers of new construction activities in the city than renovation. Due to the laws that regulate permits, all new buildings must be licensed and are all in the database, while in renovation not all construction work have to be licensed and are not all in the database. Renovation works such as refurbishments that do not alter the structure or infrastructure of the building and interior layouts do not need to be licensed. Thus, this sample, can be in a small part, considered a biased sample, at least at the level of some of the renovation subcategories, for example in the *alteration* subcategory.

To minimize this partially biased sample of the renovation and improve the accuracy of the results, indirect methods of collecting more points could be used to obtain renovation worksites that do not require building permits but that require another type of permission.

These indirect methods may well include activities associated with that permit-free renovation works that could need another kind of licensing, such as the location of scaffolding permits for refurbishments and minor alterations exempted from building permit requirements or the case of activities licensed under noise regulations. However, the georeferenced databases on the Web site of the Lisbon City Hall, on which this research was based, did not have this type of information.

In summary, this article presents a method of cross-referencing the city's alphanumeric and spatial databases, which are often kept separate, producing a measure of

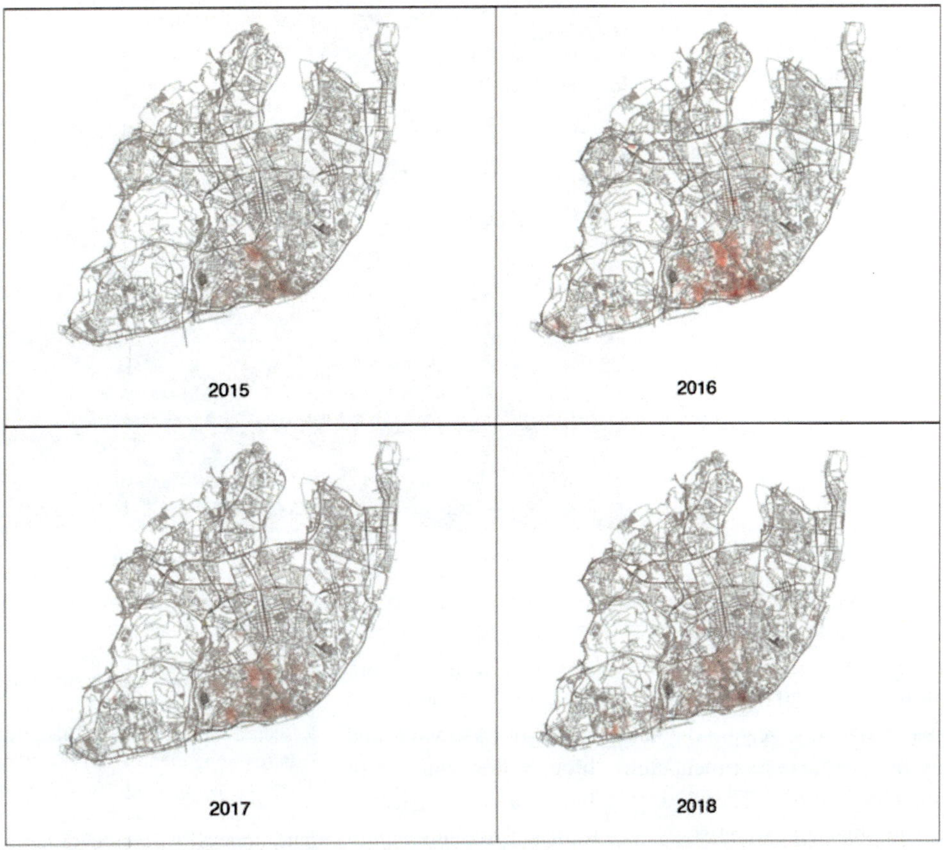

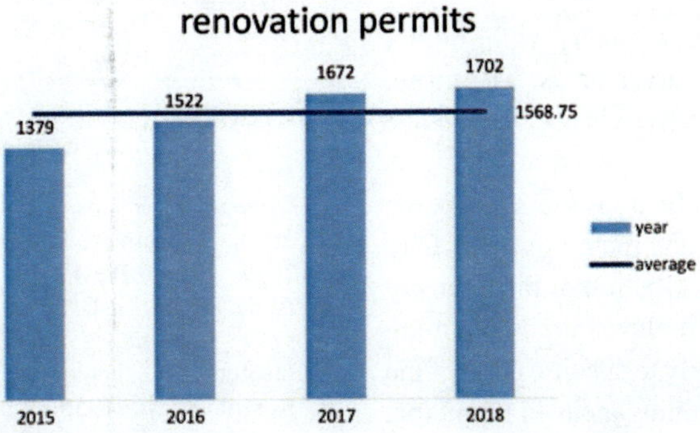

year	renovation permits
2015	1379
2016	1522
2017	1672
2018	1702
total	6275
average	1568.75

Fig. 8 NKDE hotspots per year in the 2015–2018 period, bandwidth = 15 m and grid = 1.5 m and number of building renovation permits per year in the same period. *Source:* Authors

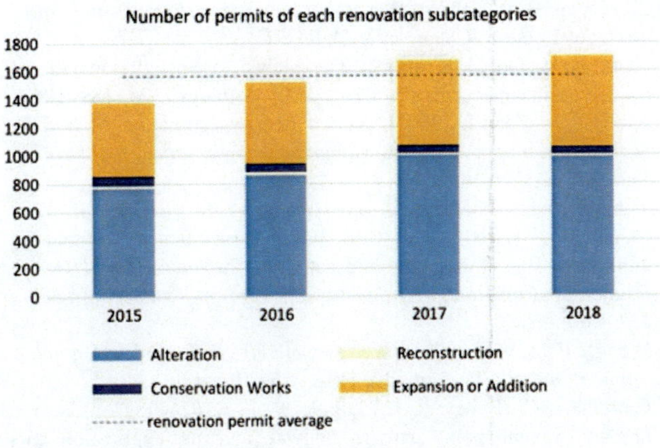

Fig. 9 Number of permits of each renovation subcategories: alteration, expansion or addition, reconstruction and conservation

renovation intensity that can be easily visualized. In addition, the same methodology can be applied to different types of information, such as the distribution of specific types of services, offices or stores.

This article presents a method of cross-referencing City Hall alphanumeric and spatial databases using NKDE. This network kernel estimates densities in a network of line segments without the loss of spatial information, deriving the density of points in the network from the density along the line, estimating the underlying density of a given phenomenon represented by these points. The points are finite random samples, and they are points that give us representations along the lines, vector information.

Like the KDE we used in the previous research, NKDE does not require an in-depth knowledge of statistics to be

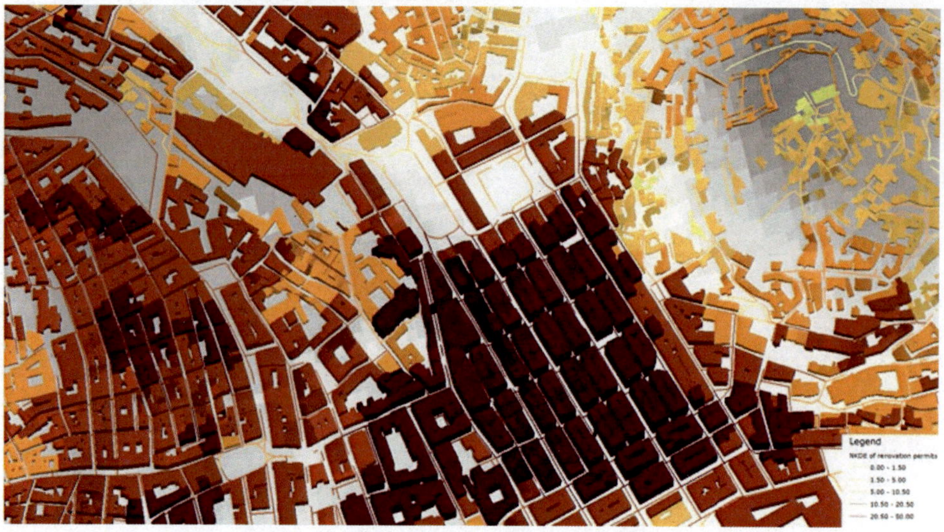

Fig. 10 3D view of the Baixa NKDE hotspot over a digital terrain model, where the height of the building corresponds to the value of the NKDE and its color too, these values being assigned by spatial join. Corresponding to the period 2015–2018—bandwidth = 15 m and grid = 1.5 m. *Source* Authors

used. Despite the existence of vast literature to support the width of the bandwidth and the grid, the choice of values was limited by the sample and the large size of the network with more than 20,000 segments and more than 7500 nodes. It was possible to reach acceptable values of bandwidth and grid by simple experimentation, through the analysis of computation times. The choice of these values considered that the objective of the study was to raise awareness about urban trends on an urban scale.

NKDE is able to clearly identify streets with higher densities, and KDE only identifies larger areas. Therefore, one of the advantages of NKDE over KDE is accuracy, since its rigor reaches the street section.

The conclusions of this study can be useful to support the application of public policies and urban planning at the City Hall level. By providing a spatial distribution of the intensity of a phenomenon across a network, it allows urban planning and policies to be adapted more quickly to specific parts of the city network. Thus, urban renewal zoning areas in municipal plans can have an exact list of streets with their geographic location. Likewise, we can also more accurately outline public financing through low taxes in the areas of renewal.

This precision of the NKDE that reaches the street section has other advantages for public policies, as it allows, in real time, to assess whether the top-down process produces more indirect effects over time and whether those effects are the desired ones. The availability of large amounts of data to download on the Web sites of City Halls provides new ways to support decision making by other stakeholders, in addition to politicians, such as private investors, real estate agents, architects and urban designers. On the one hand, technicians may have reasons to change projects and plans; on the other hand, the process of analyzing prices and market trends can be altered by the existence of more detailed information.

Acknowledgements We would like to thank Professor Atsuyuki Okabe for sharing the SANET software.

Funding Acknowledgements Ricardo Mendes Correia is funded by FCT-Fundação para a Ciencia e a Tecnologia, IP through the PhD Research Scholarship SFRH/BD/146858/2019. Filipe Brandão research is partly supported by a merit scholarship granted by ISTA-Escola de Tecnologias e Arquitetura of Instituto Universitário de Lisboa (ISCTE-IUL) for the development of his PhD project and by national funds through FCT-Fundação para a Ciencia e a Tecnologia, IP with reference UIDB/04466/2020 e UIDP/04466/2020.

References

1. Oliveira, J., Santos, T., & Soares, N. P. (2019). A reabilitação urbana e dinâmica comercial na Faixa Oriental de Lisboa (2009–2019). Estudo Prévio.
2. Anderson, M. (1964). *The Federal bulldozer. A critical analysis of urban renewal, 1949–1962*. Cambridge, MA: Massachusetts Institute of Technology.
3. Rosenblatt, M. (1956). Remarks on some nonparametric estimates of a density function. *The Annals of Mathematical Statistics, 27,* 832–837.
4. Silverman, B. W. (1986). *Density estimation for statistics and data analysis*. London: Chapman & Hall/CRC.
5. Getis, A., & Ord, J. K. (1992). The analysis of spatial association by use of distance statistics. *Geographical Analysis, 24,* 189–206.
6. Xie, Z., & Yan, J. (2008). Kernel density estimation of traffic accidents in a network space. *Computers, Environment and Urban Systems, 32,* 396–406.
7. Timothée, P., Nicolas, L.-B., Emanuele, S., Sergio, P., & Stéphane, J. (2010). A network based kernel density estimator applied to Barcelona economic activities. In *International Conference on Computational Science and Its Applications* (pp. 32–45). Springer.
8. Wang, F. (2015). *Quantitative methods and socio-economic applications in GIS* (2nd ed.). USA: CRC Press.
9. Gatrell, A. C., Bailey, T. C., Diggle, P. J., & Rowlingson, B. S. (1996). Spatial point pattern analysis and its application in geographical epidemiology. *Transactions on the Institution of British Geographers, 21,* 256–274.

10. Row, J. R., & Blouin-Demers, G. (2006). Kernels are not accurate estimators of home-range size for Herpetofauna. *Copeia, 2006,* 797–802.
11. Thurstain-Goodwin, M., & Unwin, D. (2000). Defining and delineating the central areas of towns for statistical monitoring using continuous surface representations. *Transactions in GIS, 4,* 305–317.
12. Porta, S., Strano, E., Iacoviello, V., Messora, R., Latora, V., Cardillo, A., et al. (2009). Street centrality and densities of retail and services in Bologna, Italy. *Environment and Planning B: Planning and Design, 36,* 450–465.
13. Brandão, F. J. S., Correia, R. M., & Paio, A. (2018). Measuring urban renewal: A dual kernel density estimation to assess the intensity of building renovation—Case study in Lisbon. *Urban Science, 2.*
14. Miller, H. (1999). Measuring space-time accessibility benefits within transportation networks: Basic theory and computational procedures. *Geographical Analysis, 31,* 187–212.
15. Okabe, A., Yomono, H., & Kitamura, M. (1995). Statistical analysis of the distribution of points on a network. *Geographical Analysis, 27,* 152–175.
16. Okabe, A., & Kitamura, M. (1996). A computational method for market area analysis on a network. *Geographical Analysis, 28,* 330–349.
17. Okabe, A., & Yamada, I. (2001). The K-function method on a network and its computational implementation. *Geographical Analysis, 33,* 271–290.
18. Spooner, P. G., Lunt, I. D., & Briggs, S. V. (2004). Spatial analysis of anthropogenic disturbance regimes and roadside shrubs in a fragmented agricultural landscape. *Applied Vegetation Science, 7,* 61–70.
19. Yamada, I., & Thill, J.-C. (2004). Comparison of planar and network K-functions in traffic accident analysis. *Journal of Transport Geography, 12,* 149–158.
20. Lu, Y., & Chen, X. (2007). On the false alarm of planar K-function when analyzing urban crime distributed along streets. *Social Science Research, 36,* 611–632.
21. Xu, Z., & Sui, D. Z. (2007). Small-world characteristics on transportation networks: A perspective from network autocorrelation. *Journal of Geographical Systems, 9,* 189–205.
22. Borruso, G., & Schoier, G. (2004). Density analysis on large geographical databases. search for an index of centrality of services at urban scale. In A. Laganá, M. L. Gavrilova, V. Kumar, Y. Mun, C. J. K. Tan & O. Gervasi (Eds.), *Computational science and its applications—ICCSA 2004* (pp. 1009–1015). Berlin, Heidelberg: Springer Berlin Heidelberg.
23. Borruso, G. (2005). Network density estimation: analysis of point patterns over a network. In O. Gervasi, M. L. Gavrilova, V. Kumar, A. Laganà, H. P. Lee, Y. Mun, et al. (Eds.), *Computational science and its applications—ICCSA 2005* (pp. 126–132). Berlin Heidelberg, Berlin, Heidelberg: Springer.
24. Borruso, G. (2008). Network density estimation: A GIS approach for analysing point patterns in a network space. *Transactions in GIS, 12,* 377–402.
25. Xie, Z., & Yan, J. (2013). Detecting traffic accident clusters with network kernel density estimation and local spatial statistics: An integrated approach. *Journal of Transport Geography, 31,* 64–71.
26. Borruso, G., & Porceddu, A. (2009). A tale of two cities: Density analysis of CBD on two midsize urban areas in northeastern Italy. In *Geocomputation and urban planning* (pp. 37–56). Springer.
27. Okabe, A., & Satoh, T.(2009). Spatial analysis on a network. In *The SAGE handbook on spatial analysis* (pp. 443–464). Sage.
28. Okabe, A., Satoh, T., & Sugihara, K. (2009). A kernel density estimation method for networks, its computational method and a GIS-based tool. *International Journal of Geographical Information Science, 23,* 7–32.
29. Sugihara, K., Satoh, T., & Okabe, A. (2010). Simple and unbiased kernel function for network analysis. In *2010 10th International Symposium on Communications and Information Technologies* (pp. 827–832).
30. Okabe, A., & Sugihara, K. (2012). *Spatial analysis along networks: Statistical and computational methods* (2nd ed.). Hoboken, NJ: Wiley.
31. Okabe, A., Okunuki, K.-i., Satoh, T., Shiode, S., Okabe, K., Sugihara, K., et al. (2015). *SANET. A spatial analysis along networks* (Ver.4.1). Tokyo, Japan: Centre for Spatial Information Science, University of Tokyo.

Ricardo Mendes Correia is a Research Assistant with Ph.D. Scholarship at CRIA, Instituto Universitário de Lisboa (ISCTE-IUL) in Lisbon.

Rosália Guerreiro is an Assistant Professor at the Department of Architecture and Urbanism at Instituto Universitário de Liabo (ISCTE-IUL), researcher at CRIA.

Filipe Brandão is a Ph.D. candidate at Instituto Universitário de Lisboa (ISCTE-IUL) and a Research Assistant at ISTAR in Lisbon.

Urban Forms of Informality: Decoloniality as a Perspective for Morphological Studies

Silvia Spolaor and Vítor Oliveira

Abstract

Recent decolonial studies explore the possibilities for planning and urban investigation, promoting context-based research that acknowledge the distinctive and spontaneous practices of territorial occupation and questioning the universality of 'modernity' models. Considering the informal settlements as objects of study—constrained by poor conditions of living and, at the same time, characterized by collaboration and creativity—the paper addresses the lack of methodological tools for decolonial research and insufficiencies in morphological tools to tackle informality. It is discussed how urban morphology can operate in a decolonial framework drawing from previous research on classic and reconceptualized morphological principles. The possibilities for innovative paths and tools in urban planning practice using morphological knowledge from different schools of thought, such as space syntax, process-typological, and historico-geographical approaches (and their combination), are explored in the spatial and non-spatial features found in the autoconstructed scenario of different cities around the world. Traditional knowledge, cultural patterns, and everyday practices are brought into discussion as important resources to enrich morphological analysis and inform planning practices.

Keywords

Urban planning • Urban morphology • Informality • Decoloniality

S. Spolaor (✉) · V. Oliveira
Centro de Investigação do Território, Transportes e Ambiente, Faculdade de Engenharia, Universidade do Porto, Rua Dr. Roberto Frias S/N, 4000-465 Porto, Portugal
e-mail: up201711437@fe.up.pt

1 Introduction

Recent decolonial studies explore the possibilities for planning and urban investigation to promote context-based research that acknowledges the distinctive and spontaneous practices of territorial occupation while questioning the universality of 'modernity' models [1–4]. This perspective suggests a specific vision emerging from the Global South in which investigation is based on plural ways of thinking [5] and everyday practices of the urban spaces [6–9]. Challenges faced by planning practice to organize, control, and modify urban spaces for people are often synthetized in top-down planning initiatives and model solutions based on EuroAmerican contexts. Such models are not capable to grasp subtle differences within settlements in order to plan, and follow up the future transformation of those cities that face the greatest demands in massive processes of informal urban growth [10, 11].

Drawing on decolonial and urban morphological theories, this paper raises concerns related to urban form, shedding light on the understudied physical aspects of those cities. Urban morphology [12, 13] provides tools to understand the many different aspects intrinsic to the urban forms. Elements of urban form—such as streets and squares, street-blocks, plots, buildings—that are structural components of human settlements are analyzed in typological studies and historical readings of the urban landscape. An analytical exploration of the role performed by different actors reveals the relationship between agents and form. Urban morphology is then proposed as a supportive discipline for studying, planning, and designing cities.

Considering that most cities in the world suffering major urban transformations are located in the Global South and that their growth is related to informality, the focus is to propose methods and tools to study and plan those settlements. However, studies indicate that central morphological concepts should be enriched for the analysis of informal settlements [14]. This paper proposes a decolonial perspective to do so. Taking into account the lack of tools for spatial

S. Eloy et al. (eds.), *Formal Methods in Architecture*, Advances in Science, Technology & Innovation,
https://doi.org/10.1007/978-3-030-57509-0_18

thinking and for understanding the sophisticated forms and adaptation of informal settlements [15, 16] and from reported challenges in describing and representing urban forms in those contexts, it is discussed the possibilities of innovative methods for morphological investigation, highlighting essential elements to be observed and analyzed.

The challenging definition and tracing of urban elements—due to the lack of proper cadastral maps [17] or a system of property boundaries—hamper morphological analysis and the graphic representation of informality. Decolonial thought adds the importance of considering multiple perspectives for theory building and the lack of methodological contributions for a decolonial approach for planning emerging from the Global South and not only for the South [1–3, 18]. Morphological theory already argues for the importance of recognizing cultural patterns of dwellings [19], however, investigations in the field centered in Southern cities are in lesser number than in Northern cities. In this way, important features for producing morphological analysis in the context of informality are presented based on five previous research. They are synthetized through non-spatial analytical patterns—such as cultural behavior, ways of living, affordability, availability [14]—and physical aspects of the settlements, as in streets, plots, and buildings, as well as the availability of data and the timespan of urban transformation.

The studies reinforce the relevance of understanding local economic, historic, cultural, and social characteristics of the settlements to create accurate representation of the reality that suits the everyday practices of their dwellers. The adoption of the indicated analytical patterns—that are intrinsic to the formation and development of urban settlements—is considered necessary for morphological analysis to be developed in the context of informality. Drawing on the assumption that a new agenda for urban studies and planning should include a critical perspective that cares for the value of local experiences and ways of living, exploring the possibilities of plural scenarios, the paper proposes enabling reconceptualization and adaptations of long-standing theories and principles toward building a decolonial way of thinking. To avoid planning concepts based on 'ideas out of place and places out of ideas' [20] and inspired by the lack of planning theories and research methodologies grounded on empirical research from the South [9, 10], the challenge is to find answers on the built spaces of Southern cities, developing new knowledge and new ways of learning [21].

2 Decoloniality: Advances in the Planning Field

Scholarship agrees that despite suggestions that colonial times have formally finished, societies are still built on and operate within legal, institutional, governmental, and decision-making systems that are a legacy of colonial times [3]. The recognition of rights in land and claims for political autonomy challenges the existing planning domain demanding a reconceptualization of the role it could perform in readdressing injustices from colonial times [3]. Yet, planning in practice is still colonialized [22]. Considering the current debate on post-coloniality and further discussions on decolonial thoughts, unfolding ideas for theories in the planning field are discussed.

Examples of planning paradigms that have caused more harm than solutions have been reported from different countries. This paper is more interested in its consequences for the physical space—besides social and economic issues—that removing houses or entire communities have impacted, injuring cultural patterns of living and turning even more vulnerable the poor population due to massive evictions and/or to the change of building characters. In Abuja, Nigeria, Watson [18] describes major evictions that deleted the character of whole areas of the city. In the state of Gujarat, India, Datta [23] reports the impact of implementation of Smart Cities' projects. McFarlane [24] points out the substitution of traditional architecture typology, the *chawls*, by towers of office blocks in Mumbai. In Brazil, Sánchez et al. [25] criticize planning and architectural solutions for the implementation of the FIFA World Cup facilities that removed many residents from favelas and changed the landscape of the city, just to cite a few examples.

In a scenario in which these global tendencies of homogenization and the loss of identities dominate architectural, urban and territorial practice and planning projects, it is important to understand how international standards and the concept of modernity and development is absorbed and implemented by contexts outside their references [26]. Different authors have questioned the aesthetics of urbanism in this globalization era, professing that some cities are so similar to one another that it is possible to describe the internationalization of its aspect [4] or that cities could lack particular identities to the point they become a 'Generic City' [27].

Therefore, the discussion of new paradigms to plan the future of cities, control development, and regulate growth at this moment seems essential. Moving beyond colonial strains is still necessary for planners to adopt a deconstructive approach, capable to challenge planning's theoretical, epistemological, and normative assumptions [3]. Current literature suggests a possible path in decolonial thought, based on the idea of multiplicity, developing a theoretical knowledge based on plural contributions [5]. In the field of urban studies and planning, Robinson [28] argues that being open to ideas from elsewhere, dislocating the traditional EuroAmerican centered academic debate, raises challenging questions about the specificity or limited scope of some concepts; this is also discussed in urban morphology by Whitehand [29].

Following Connell [30], that suggests there is a lack of blueprint to change and challenge current northern knowledge, the proposed aim is to investigate theories and principles to tackle the particularities of informal settlements, contributing with a methodological perspective. The history of colonization do not delete memories and stories from traditional population [5] and therefore, the evidences should be used to produce context-based academic research that is respectful and engaged to local knowledge. Learning between sites across the world is important [18], and it adds essential contributions to improve planning effects, especially in conditions of informality.

3 Urban Morphology: Addressing Informal Settlements

Suggesting decolonial lenses for urban morphology means that its fundamental principles can, and should, be applied to different contexts. The morphological research line on comparative studies, involving the application of different morphological approaches in the same case study [31–33] and of the same morphological concept or method in different case studies, in different parts of the world [34–36], is crucial for this purpose. The dissemination of urban morphology and the range of regional networks of the International Seminar on Urban Form (ISUF)—the international organization of urban form for researches and practitioners present in Cyprus, Italy, Poland, Brazil and Portugal, Serbia, Spain and Mexico, and Turkey—shows how much the discipline is spread and well-accepted across the world and how scholars coming from different backgrounds study its fundamental concepts to perform analysis and to design proposals for their own cities [37]. ISUF's publication, the Urban Morphology Journal, receives contributions from many non-European countries such as Brazil, China and Turkey.

However, there is a shorter list of contributions in the field dealing with informality [14], and, so far, no mentions for 'decoloniality' is found among main publication sources. Considering the size of Southern cities—in which most of the global informal growth is concentrated—and the negative impact of current planning efforts, it is clear that they are the most vulnerable to and yet, with the strongest need of, massive physical transformation to improve the quality of their urban spaces. The hypothesis is that a more informed research, based on the way of living of local inhabitants and on their everyday practices is essential to produce an optimistic future to improve their physical spaces. Relevant studies exploring urban forms in conditions of informality are addressed further. First, the fundamental principles of the discipline are described and the variety of possibilities in the field explored by research located in different geographical settings, other than Europe, is presented. The examples show the adaptations and reformulations of those principles to produce a context-based investigation and some important contributions are extracted for a decolonial approach for urban morphology.

3.1 Fundamental Principles of Urban Morphology

Oliveira [12] defines urban morphology as the study of urban forms and of the agents and processes that shape them. He delineates the main physical elements that structure the urban tissues of the city: streets and squares, street-blocks, plots and buildings. Traditions in the field analyze the processual transformation of urban form through the support of different perspectives, drawing from a variety of concepts and parameters to understand spatial relations, to characterize urban aspects and to represent the urban space. The variety of its analytical tools and the possibilities of adaptation of its main contributions according to each specific context makes urban morphology an essential discipline for understanding spatial organizations, relations, and patterns.

The use of three of the main approaches in urban morphology is going to be explored as important methodological resources. The process-typological approach mostly focuses on understanding tissues and building types over time; recent work explores the reading of urban forms as a design method [38, 39]. The historico-geographical approach [40–42] is based on the comprehension of the morphogenetic characteristics of urban settlements, the main geographic, spatial and built features of a certain space. Space syntax [43, 44] develops studies based on the relationships between space and movement, assuming spatial configuration as a key concept for analysis.

The different aspects of urban form elements might be analyzed according to hierarchical levels of resolution, with varying specificity [13]. In any way, morphological studies that investigate the origins, the process of transformation, and the spatial organization of a determined human settlement depend on the availability of resources to conduct accurate analysis. Information gathering is an essential step for a successful morphological research [13]. Bearing in mind the difficulties in gathering data of spontaneous and informal settlements, the next section discusses important contributions that address the lack of precise spatial data and morphological definitions to perform morphological analysis.

The role of agents and agencies can also be explored. In the case of informal settlements, the state participation might be—sometimes deliberately [45]—shallow to attend the demand of the poor and the contribution of self-build efforts come into place. The lack of planning, regulation, and control of the urban landscape leaves room for environmental, sanitary, accessibility concerns but, at the same time, paves the way for creativity, cooperation, and, often, insurgency [16, 46–48]. Here lays a learning opportunity for methodologies to search for more effective analytical tools in urban morphology and planning, coming from multiple sources, traditional and local knowledges expressed through the urban forms.

According to Kropf [13], any analytical method is culturally rooted and therefore, not universal. The author sheds light on the role of urban morphology as a supporting discipline—amid anthropology, archeology, and cultural geography—to evidence cultural structure, values and practice, comparative studies and to undertake cultural critique. For that purpose, the concepts developed in different morphological schools are useful to explore the cultural aspects embedded in dwelling traditions. For the scope of this paper, some of them are evidenced. From process-typological approach, the concepts of type (and typological process), polarities and routes describe the processual transformation of base and special buildings in relation to the social structure of a certain area [49]. From historico-geographical approach, morphological regionalization groups the physical features of a certain area as a unit, contributing to a comprehensive understanding of the structure of the city and its particular historical development [50]. And, finally, from space syntax, the spatial configuration is observed to analyze movement and space use and the influence of the urban layout on social life [12].

3.2 Urban Morphology, Informality, and Decoloniality: Conceptual Contributions

Urban morphology as a discipline emerged in Europe [40, 51] and so far, most morphological studies are based on geographical contexts with long-established urban forms and precise legal property delimitations [14]. Some authors highlight that there is still a gap on a sophisticated knowledge of how informal urbanism works, empirically or theoretically. They evidence the lack of methodological tools to analyze them [14, 16, 52] and suggest that some conceptual reformulations and additions might be necessary for that purpose. Decolonial thoughts and contributions from morphological research are presented to address this issue, tackling the need to understand the materiality of cities beyond the copy/unique [24]; beyond singular models of modernity and development [26]; beyond European aesthetics and standards of beauty [5]; and beyond a normative idea of national identity [4].

A claim to include cultural diversity and popular local knowledge on building, living, and everyday practices to plan and analyze spaces is already part of the scope of planning [6–9] and morphology literature [39, 48, 53]. Kropf [13] suggests even that a potential application of urban morphology could include also the recording of ethno-diversity as expressed through built form. Considering specifically the context of informal settlements, further concerns on working with morphological concepts within their complexity are addressed by McCartney and Krishnamurthy [14].

Proposing to expand the methods of analysis, the authors delineate five-scaled analytical patterns. The first one, non-spatial factors (a), include social, economic, and political aspects; as affordability of urban land and building materials in addition to political and ethnical constraints that can affect conditions for people to own land, to build and to maintain their settlements in certain urban spaces. Situational factors (b) determine location of affordable land over time. Site factors (c) relate to topography and the innovative morphological solutions for inhabiting risky areas. Ambiguities of ownership (d) addresses the circulation spaces and the importance of negotiations between public and private areas. Because access paths might have blurred boundaries, establishing plot lines or land uses in informal settlements is challenging. Building typologies permanence (e) is also a category that needs specific analysis, as the indication of the building material might suggest its longer permanence or substitution (brick house vs. tarpaulin, for instance).

Further on the non-spatial factors, the role of grassroots organizations, neighborhood collaborations, traditional knowledge, and cultural identity in building and planning spaces is discussed by different authors [1–3, 54, 55]. They explore how knowledge building on cultural and religious values and social relationships might change the way urban forms and its transformation are studied and interpreted. A decolonial research on the physical space should search for plural understandings of what the built space means, and on how different concepts and tools from urban morphology can be appropriated by planning. The transformation of cities, and mostly the needed transformation of informal settlements, still lacks directions on how to deal with their dense, complex, contested, and often neglected urban tissues. Further findings are addressed on next section by studies carried on by empirical research.

4 Decolonial Tools and Methods: Paths for Future Morphological Research

Winkler [2] suggests that decolonial research still lacks innovative philosophical lenses and methods to collect, analyze, synthetize, and interpret research findings. Through the findings of recent research, the potentiality of understanding traditional patterns and working within cultural standards of informal settlements is suggested as a path for future studies of informality. The objective is to demonstrate how to use and/or adapt concepts to understand local patterns of build environments located in different geographical settings, outside Europe. Five relevant studies were selected in which one or more morphological traditions are used to analyze the urban aspects of informal settlements. Structural elements are assessed through different methodologies and units of analysis, with the support of a diversity of data and tools by these investigations, in which fieldwork or extensive drawing and redrawing processes were an important resource to collect information.

4.1 Five Empirical Studies on Urban Forms of Informality

The studies selected present methodological strategies to perform analysis in informal settlements developed within the last decade. Bearing in mind the conceptual contributions by McCartney and Krishnamurthy [14], the following studies add up empirical research, with inputs from fieldwork and/or from extensive mapping and drawing efforts. In a general way, they developed methodological tools in order to quantify and to qualify urban forms and physical features of the slums, to understand the relationship between urban form and activities and to describe the dynamics of transformation in these contexts with the aim to inform policy making.

The thesis elaborated by Loureiro [56] focuses on the streets' networks. She uses space syntax to systematize spatial patterns, recognizing tendencies, patterns, and distinctions of slums in 120 cities. The author demystifies the idea of disorganization attributed to informal settlements, describing the simple localized actions from which rules emerge organizing the whole urban fraction. Using accessibility and connectivity concepts, the work evidences the potentiality of slums' urban grids in relation to the whole urban system. Karimi and Parham [57] also use space syntax approach to analyze the informal settlements in the city of Jeddah. Through the reading of the existing spatial configuration, and its distinct local structure, the study recognizes a number of axis with integration potential to improve social cohesion. The authors prove that with small changes on the physical space it is possible to minimize segregation of slums. The two cases show how space syntax might inform decision making on the connectivity and accessibility of slums to the rest of the city and improve the settlements' accessibility conditions with minor interventions.

Iovene [52] explores process-typological and historico-geographical approach to analyze the informal settlement of San Pedro de Ate, Lima, Peru. On the comparative analysis between the informal settlement and the historic fabric of San Bartolomeo in Venice, Italy, the author finds similarities in the processes of development, despite the speed of transformation that is much faster in the first. Through a pattern analysis methodology—in different scales of streets, plots and buildings—the author evidences the lack of clear morphological definition to describe the processes of urban change in informal contexts. The process-typological approach is also used by Maretto et al. [53] in the study developed for Villa 31, located in Buenos Aires, Argentina. Patterns of the settlement development are explored to provide a context-based analysis capable to guide masterplan and design proposals, based on the existing urban forms and patterns of living of local inhabitants. The methodology includes tracking housing patterns back to the typical Argentinian housing unit and finding its reproductions and distortions in the contemporary fabric. Those studies are relevant for decolonial studies because they seek to understand how cultural behavior is manifested in the physical space, evidencing the important parameters to be analyzed in these contexts: the origins and historical evolution, the timeframe of transformation and the specificities of patterns of buildings, plots and streets.

The study of Maputo, Mozambique, by Viana [46] explores not only the physical space but also the relationship between urban forms and activities. With a combination of approaches, the adaptable methodology captures top-down and bottom-up planning initiatives, grasping more than space itself, but activities and actions that shape spatial life. This approach to the urban space offers the opportunity of tracking what happens in between urban elements and showing how everyday practices of inhabitants might affect and change the urban environment. Moreover, it opens the possibility for bottom-up and top-down planning efforts to cocreate and collaborate, enabling also a dynamic and hybrid planning possibility.

4.2 Methodological Contributions for Future Research

The studies have presented how different morphological approaches were used to operationalize analysis of informal urban systems. They demonstrate the structural elements that are essential to be analyzed in relation to their specific contexts and their historical, geographical, and cultural settings. They also evidence that the configuration of streets performs a high connectivity profile in slums and highlight the importance of the public space as a space of livability and work. Other researches that are not addressing specifically urban forms, but rather planning efforts through a decolonial approach, are also relevant for the scope of this paper.

In these investigations, non-spatial factors include—besides affordability and political constraints—cultural and symbolic values that might come into place. Urban elements can have a variety of symbology and meanings regarding different cultural believes and religion, affecting even the way property ownership is understood. The case of Xhosa people's settlements in two informal settlements in Cape Town [2] and the citizenship situation regarding the Nubians inhabiting Kibera, in the city of Nairobi [58], are some examples. Those questions are not only reported from the Global South; in Canada and in the USA, traditional planning struggled with the indigenous' understanding of jurisdiction, property, and consultation [3].

Concerning spatial units, the plot is probably the most contested element for informal settlements. Issues of land tenure, boundaries' tracing, and definitions between public and private spaces are debated as well as the lack of proper cadastral maps [17] to operate analysis on that level of resolution. Moreover, it is necessary to discuss the lack of epistemological categories in planning that account for the multiple ways of living and using spaces in the popular areas of the cities, such as the multifunctional spaces present in the slums [11]. The building unit and its uses are subjected to specific analysis to understand the origins of their typology and how much the cultural setting and relationships might influence their shape and use [19, 48, 53].

Agents of change are relevant here because most of the housing stock in slums are constituted by self-build initiatives. Residents are the main agents of urbanization, building their houses step by step in a process of autoconstruction[1], in a way that buildings are always being transformed. If, in one way, informal settlements suffer a process of 'distinctive temporality, [since] homes and neighborhoods grow little-by-little, in long-term processes of incompletion and continuous improvement led by their own residents' [54], on the other hand, the timeframe of the settlement's development and growth—compared to historic cities—happens in a much faster pace [52]. A more precise analysis might be necessary to understand typological variations over time because time periods in slums vary according to levels of investments in housing and also to tenure security [14].

Table 1 systematizes the contributions combining the findings of both the theoretical and the empirical investigations addressed. The morphological approaches and their most explored concepts are listed together with their difficulties to be applied at, and necessary adaptations for, informal contexts and further assets for planning practice.

5 Conclusions

The paper has evidenced some possible directions for morphological investigation. The proposal is based on the consideration of context-based morphological knowledge when investigating informality and on understanding how socio-cultural practices are expressed in terms of space. Moreover, it has demonstrated the possibility of adding different methodological actions—without denying or excluding other ways of analyzing and understanding territories—to include knowledge on bottom-up, autoconstructed urbanisms for informing planning practice. When proposing so, the paper aims at understanding whether there is a possibility to engage with decoloniality when studying the physical aspects of space. It indicates paths, tools, and methods through which it is possible to understand the reasoning behind the form—the physical space—and the history, social relations, and cultural settings—the non-spatial aspects—of urban tissues.

The methodological contributions from the studies presented offer an opportunity to read in between the lines of urban fabrics, interpreting the visual space together with the context, to understand what is represented and what it is not, in planning terms and conditions. Morphological studies show that medieval historic cities and slums have similarities but planning history assumes different approaches for both. Despite the urban qualities, as in connectivity and accessibility, offered by the urban tissues of informal settlements, its recognition as a part of the city is many times denied, and therefore their forms are deliberately erased from the formal gaze of the city [59]. Many slum upgrading projects do not respect or listen to the voice of the inhabitants, changing completely the environment of the place when subjected to urban regeneration [2, 25].

More than understanding space itself, the general objective is to unveil the process of transformation, and what is retained from it to apprehend the contemporary space and to predict and plan the future scenario. Decolonial thought entails grounding urban theory, observing history from different starting points. Seeking for an effective theory, building on a decolonial urban morphology, further empirical research is still necessary to understand to what extent patterns of living in 'most of the world' [60] can be generalized or need to be specified according to each case. Exploring morphological concepts that have not yet been associated with the phenomenon of informality—such as the fringe belts—could be an interesting path for understanding its spatiality within the formal city.

There is still a need to establish strategies for knowledge production that put into dialogue academic investigations and traditional and local knowledges. Some positive examples are present, for instance, in Bolivia and

[1]Caldeira [54] refers to Latin American name used for the process.

Table 1 Morphological approaches, their concepts, necessary adaptations, and main assets for studying and planning informal settlements

Approach	Concept/tool	Difficulties/adaptations	Assets for studying/planning informality
Space syntax	Spatial configuration analysis (spatial accessibility)	Lack of data on street network	To understand the grid organization and influences on social interactions; To evidence the potentiality of the informal grid in relation to the whole system; To highlight possibilities of minor change to improve the system; To test possible scenarios; To improve social cohesion;
Process-typological	Type, typological process	Substitution or permanence of original buildings due to construction quality; Lack of clear morphological definition to address building and uses (multifunctionality); Monitoring of form additions conditioned to tenure security;	To describe process of fabric formation and transformation; To understand socio-spatial implications of multifunctionality and neighborhood relations;
	Polarities, routes	Identification of the value of specialized buildings; Identification of dominant cultural values for the establishment of the grid;	To identify cultural patterns and dwelling options for guidelines and masterplans;
Historico-geographical	Morphological regionalization	Lack of proper cadastral maps; Definition of plot boundaries and property ownership standards; Observation of timeframe of transformation;	To define boundaries for urban transformation and masterplan guiding; To establish evolution patterns throughout time to predict growth.

Equator in which the indigenous *buen vivir* has been included in the criteria—at least theoretically—to be part of their constitutions [4]. Nevertheless, there is still plenty of space for cooperation between bottom-up and top-down planning.

Moreover, there is still an economic point of view. Economic solutions based on the way people live might come up from these analysis. Karimi and Parham [57] show how minor infrastructure interventions might improve the whole system. Also, contributions could be essential to improve autoconstruction standards. In order to build a more optimistic future in plans for informal settlements, the first step is to rethink the choice of case studies and the way analysis has been proposed, understanding the city beyond physical plans and formalization standards. Secondly, there is a possibility to propose new ideas, principles or to reconceptualize existing theories, in the light of grounded analysis.

References

1. Patel, S. (2016). A decolonial lens on cities and urbanisms: Reflections on the system of petty production in India. Singapore.
2. Winkler, T. (2018). Black texts on white paper: Learning to see resistant texts as an approach towards decolonising planning. *Planning Theory, 17,* 588–604. https://doi.org/10.1177/1473095217739335
3. Ugarte, M. (2014). Ethics, discourse, or rights? A discussion about a decolonizing project in planning. *Journal of Planning Literature, 29,* 403–414. https://doi.org/10.1177/0885412214549421
4. Farrés Delgado, Y., & Matarán Ruiz, A. (2014). Hacia una teoría urbana transmoderna y decolonial: una introducción *13,* 339–361. https://doi.org/10.4067/s0718-65682014000100019.
5. Mignolo, W. D. (2011). *The darker side of Western Modernity: Global futures, decolonial options* (1st ed.). Durham & London: Duke University Press.
6. Cirolia, L. R., & Scheba, S. (2019). Towards a multi-scalar reading of informality in Delft, South Africa: Weaving the 'everyday' with wider structural tracings. *Urban Studies, 56,* 594–611. https://doi.org/10.1177/0042098017753326
7. Rokem, J., & Boano, C. (2018). *Urban geopolitics: Rethinking planning in contested cities* (1st ed.). New York: Routledge.
8. Inzulza-Contardo, J. (2012). Contemporary urban culture in Latin America: Everyday life in Santiago, Chile. In T. Edensor & M. Jayne (Eds.), *Urban Theory beyond the west: A world of cities* (pp. 207–216). London: Routledge.
9. Streule, M. (2020). Doing mobile ethnography: Grounded, situated and comparative. *Urban Studies, 57,* 421–438. https://doi.org/10.1177/0042098018817418
10. Watson V (2016) Shifting approaches to planning theory: Global North and South. *Urban Planning, 1,* 32–41. https://doi.org/10.17645/up.v1i4.727.
11. Rolnik, R. (2015). *Guerra dos Lugares: A colonização da terra e da moradia na era das finanças* (1st ed.). São Paulo: Boitempo.
12. Oliveira, V. (2016). *Urban morphology: An introduction to the study of the physical form of cities* (1st ed.). Cham: Springer.
13. Kropf, K. (2017). *The handbook of urban morphology* (1st ed.). West Sussex: Wiley.
14. McCartney, S., & Krishnamurthy, S. (2018). Neglected? Strengthening the morphological study of informal settlements. *SAGE Open, 8.* https://doi.org/10.1177/2158244018760375.

15. Kamalipour, H. (2017). Forms of informality and adaptations in informal settlements. *International Journal of Architectural Research ArchNet-IJAR, 10*, 60. https://doi.org/10.26687/archnet-ijar.v10i3.1094.
16. Dovey, K., & King, R. (2011). Forms of informality: Morphology and visibility of informal settlements. *Built Environment, 37*, 11–29. https://doi.org/10.2148/benv.37.1.11
17. Perez, L. P., Bischof Dos Santos, R., De Almeida, G. M. J. A., & Carvalho, G. C. (2017). Spatial data in the Global South: A case study of alternative land management tools for cities with limited resources. In *Global Humanitarian Technology Conference (GHTC)* (pp. 1–10). San Jose: IEEE.
18. Watson, V. (2014). Learning planning from the South: Ideas from the new urban frontiers. In S. Parnell & S. Oldfield (Eds.), *The Routledge handbook on cities of the Global South* (1st ed., pp. 98–108). New York: Routledge.
19. Caniggia, G., & Maffei, G. L. (2017). *Interpreting basic buildings: architectural composition and building typology* (2nd ed.). Firenze: Altralinea Edizioni.
20. Maricato, E. (2000). As ideias fora do lugar e o lugar fora das ideias. In E. P. Giachini (Ed.), *A cidade do pensamento único* (3rd ed., pp. 121–192). Petrópolis: Vozes.
21. Connell, R. (2014). Using southern theory: Decolonizing social thought in theory, research and application. *Planning Theory, 13*, 210–223. https://doi.org/10.1177/1473095213499216
22. Vainer, C. (2017). Disseminating "best practice"? The coloniality of urban knowledge and city models. In S. Parnell & S. Oldfield (Eds.), *The Routledge handbook on cities of the Global South* (1st ed., pp. 48–56). New York: Routledge.
23. Datta, A. (2015). New urban utopias of postcolonial India: 'Entrepreneurial urbanization' in Dholera smart city, Gujarat. *Dialogues in Human Geography, 5*, 3–22. https://doi.org/10.1177/2043820614565748
24. McFarlane, C. (2008). Urban shadows: Materiality, the "Southern city" and urban theory. *Geography Compass, 2*, 340–358. https://doi.org/10.1111/j.1749-8198.2007.00073.x
25. Sánchez, F., Oliveira, F. L., & Monteiro, P. G. (2016). Vila Autódromo em disputa: Sujeitos, instrumentos e estratégias para a reinvenção do espaço. *Revista Brasileira de Estudos Urbanos e Regionais, 18*, 408–427.
26. Robinson, J. (2006). *Ordinary cities: Between modernity and development* (1st ed.). New York: Routledge.
27. Koolhaas, R. (1998). Generic city. In J. Sigler (Ed.), *S, M, L, XL* (2nd ed., pp. 1248–1264). New York: The Monacelli Press Inc.
28. Robinson, J. (2016). Comparative urbanism: New geographies and cultures of theorizing the urban. *International Journal of Urban and Regional Research, 40*, 187–199. https://doi.org/10.1111/1468-2427.12273
29. Whitehand, J. W. R. (2009). Urban morphological classics? *Urban Morphology, 13*.
30. Connell, R. (2018). Decolonizing cociology. *Contemporary Sociology, 47*, 399–407. https://doi.org/10.1177/0094306118779811.
31. Kropf, K. (2009). Aspects of urban form. *Urban Morphology, 13*, 105–120.
32. Oliveira, V., Monteiro, C., & Partanen, J. (2015). A comparative study of urban form. *Urban Morphology, 19*, 73–92.
33. Monteiro, C. (2019). Morfologia Urbana na Prática de Planeamento: para uma metodologia integrada de análise e proposta de cidade. Universidade do Porto.
34. Conzen, M. P. (2009). How cities internalize their former urban fringes: A cross-cultural comparison. *Urban Morphology, 13*, 29–54.
35. Whitehand, J. W. R. (2009). The structure of urban landscapes: Strengthening research and practice. *Urban Morphology, 13*, 5–27.
36. Hillier, B. (2002). A theory of the city as objects: Or, how spatial laws mediate the social construction of urban space. *Urban Design International, 7*, 153–179. https://doi.org/10.1057/palgrave.udi.9000082
37. ISUF. (2020). ISUF—International Seminar on Urban Form. www.urbanform.org. Accessed March 2, 2020.
38. Cataldi, G. (2018). Towards a general theory of urban morphology: The type-morphological theory. In V. Oliveira (Ed.), *Teaching urban morphology* (1st ed.). Cham: Springer International Publishing.
39. Strappa, G. (2018). Reading the built environment as a design method. In V. Oliveira (Ed.), *Teaching urban morphology* (1st ed.). Cham: Springer International Publishing.
40. Whitehand, J. W. R. (1981). *The urban landscape: Historical development and management*. Papers by M. R. G. Conzen. London: Academic Press.
41. Conzen, M. R. G. (1960). Alnwick, Northumberland: A study in town-plan analysis. *Transactions and Papers (Institute of British Geographers), 27*, iii+ix-xi+1+3+122. https://doi.org/10.1017/CBO9781107415324.004.
42. Oliveira, V. (2019). *J.W.R. Whitehand and the historico-geographical approach to urban morphology* (1st ed.). Cham: Springer.
43. Hillier, B., & Hanson, J. (1984). *The social logic of space* (1st ed.). Cambridge: Cambridge University Press.
44. Hillier, B. (1996). *Space is the machine* (1st ed.). Cambridge: Cambridge University Press.
45. Roy, A. (2005). Urban informality: Toward an epistemology of planning. *Journal of the American Planning Association, 71*, 147–158.
46. Viana, D. L. (2019). *Maputo: (auto)organização e forma-dinâmica urbana* (1st ed.). Porto: U.Porto Editorial.
47. Roy, A. (2009). Why India cannot plan its cities: Informality, insurgence and the idiom of urbanization. *Planning Theory, 8*, 76–87. https://doi.org/10.1177/1473095208099299
48. Maretto, M., & Scardigno, N. (2016). Muratorian urban morphology: The walled city of Ahmedabad. *Urban Morphology, 20*, 18–33.
49. Maretto, M. (2018). Teaching urban morphology in a sustainable perspective. In V. Oliveira (Ed.), *Teaching urban morphology* (1st ed., pp. 243–264). Cham: Springer International Publishing.
50. Whitehand, J. W. R. (2001). British urban morphology: The Conzenian tradition. *Urban Morphology, 5*, 103–109.
51. Moudon, A. V. (1997). Urban morphology as an emerging interdisciplinary field. *Urban Morphology, 1*, 3–10.
52. Iovene, M. (2018). *Understanding the morphology of informal settlements* (Unpublished Ph.D. thesis). University of Strathclyde, Glasgow.
53. Maretto, M., Amato, A. R., Boggio, N., et al. (2014). Morphological analysis of the informal city. The 'Villa 31' in Buenos Aires, Argentina. In V. Oliveira, P. Pinho, L. Mendes Batista, & T. Patatas (Eds.), *International Seminar on Urban Form*, Porto.
54. Caldeira, T. P. R. (2017). Peripheral urbanization: Autoconstruction, transversal logics, and politics in cities of the global south. *Environment and Planning D: Society and Space, 35*, 3–20. https://doi.org/10.1177/0263775816658479
55. Whitehand, J. W. R., Conzen, M. P., & Gu, K. (2016). Plan analysis of historical cities: A Sino-European comparison. *Urban Morphology, 20*, 139–158.
56. Loureiro, V. R. T. (2017). "Quando a gente não tá no mapa": A configuração como estratégia para a leitura socioespacial da favela. Universidade de Brasília.
57. Karimi, K., & Parham, E. (2012). An evidence informed approach to developing an adaptable regeneration programme for declining informal settlements. In *8th International Space Syntax*

Symposium (pp. 1–27), Pontificia Universidad Católica de Chile, Santiago.
58. Elfversson, E., & Höglund, K. (2018). Home of last resort: Urban land conflict and the Nubians in Kibera, Kenya. *Urban Studies, 55,* 1749–1765. https://doi.org/10.1177/0042098017698416
59. Kamalipour, H. (2016). Urban morphologies in informal settlements. *Contour Journal Agency/Agents of Urbanity, 1,* 1–10.
60. Chatterjee, P. (2004). *The politics of the governed: Reflections on popular politics in most of the world* (1st ed.). New York: Columbia University Press.

Silvia Spolaor is a Ph.D. student and researcher in Territorial Planning at Research Centre for Territory, Transports, and Environment (FEUP). She graduated in Architecture and Urban Planning (Universidade Federal do Espírito Santo) and holds a Master of Science degree in Urban Planning and Policy Design from DASTU-Politecnico di Milano. The author's research involves understanding how a decolonial view could contribute for morphological studies and for more informed decision making in planning activity, in conditions of informality.

Vítor Oliveira is Secretary General of the International Seminar on Urban Form (ISUF) and the President of the Portuguese-language Network of Urban Morphology (PNUM). He is Principal Researcher at the Research Centre for Territory, Transports, and Environment (CITTA/FEUP) and Professor Auxiliar of Urban Morphology and Urban Planning at ULP. He is Associate Editor of 'Urban Morphology' and Founding Editor of the 'Revista de Morfologia Urbana.' His latest books are 'Urban morphology, an introduction to the study of the physical form of cities' (2016), 'Teaching urban morphology' (2018) and 'J.W.R. Whitehand and the historico-geographical approach to urban morphology' (2019).

Non-intrusive Semi-automated Modelling of Ordinary Buildings' Intermediate Floors: Collective Housing in Porto, Portugal

David Pereira-Martínez, Gisela Lameira, Plácido Lizancos, Isabel Raposo, and Virgílio Borges Pereira

Abstract

Contemporary ordinary collective housing buildings constitute quantitatively the greater built heritage of many compact cities, and in the near future they will need to be rehabilitated *en masse*. Their repetitive patterns promote the application of GIS/BIM modelling technology. Thus, this research develops a shape grammar to semi-automatically estimate the layout of the intermediate floors of residential buildings from their external shape. The grammar is based on the geometric capacity of floors to accommodate rooms in contact with the façade and to be divided into several dwellings. Grammar Part I (rooms) uses the topological concept of 'cells' to estimate the spaces behind the façades. Grammar Part II (dwellings and stairs) divides the floor into dwellings and locates the common staircase. This shape grammar is programmed in Arcpy (ArcGIS' Python library) with variations in accordance with its tools, restrictions, and processing performance. Finally, this shape grammar's script is applied to one hundred collective buildings in Porto, studied in a previous Ph.D. thesis. The results of this sample illustrate that the shape grammar can estimate reality in many cases, revealing an underlying ordinary model with varying standards. Moreover, it offers a relevant analysis of the buildings' capacities and morpho-typological characteristics, even for irregular geometries. Its general applicability, non-intrusive methods, and *en masse* processing can be useful in quantitative analyses and preliminary modelling for urban regeneration studies and policies.

Keywords

Ordinary buildings • Layout estimation shape grammars • GIS processing in Arcpy • Python

1 Introduction

This study focusses on the research of geometrical patterns in the floor plans of ordinary collective housing in order to preview their characteristics and model them more easily in the context of their necessary massive rehabilitation, after their energetic, functional, and cultural obsolescence. For this, a shape grammar has been developed to estimate their main spaces from their external shape, which has been programmed to process them semi-automatically, obtaining interoperable information models. Finally, it has been applied to buildings in Porto, designed between 1930 and 1970, with relevant results.

Our research is focussed on the ordinary built environment, an important issue due to its quantitative relevance, which has normally been eclipsed by other aspects of the design. Several authors have highlighted the necessity of global solutions aimed at a quantitative response, working from the ordinary production [15]. These works have analysed the schemes of collective buildings, producing important contributions to the understanding of their origin and characteristics [14]. Nevertheless, as these studies are focussed on the characterisation of different common types, the theoretical problem about an 'ordinary' category still remains unclear: Does a common category exist that includes several types? What are the characteristics of such a category? If the 'ordinary' exists and is empirically determinable, rehabilitation solutions adapted to many cases can be applied. Thus, our study is a continuation of that research

D. Pereira-Martínez (✉) · G. Lameira · V. B. Pereira
Center for Studies in Architecture and Urbanism (CEAU), University of Porto (UPorto), Porto, Portugal
e-mail: david-pereira.martinez@arq.up.pt

D. Pereira-Martínez · P. Lizancos
Group of Territorial Studies (GET), University of A Coruña (UDC), A Coruña, Spain

I. Raposo
Research Centre for Architecture, Urbanism and Design (CIAUD), University of Lisbon (ULisboa), Lisbon, Portugal

S. Eloy et al. (eds.), *Formal Methods in Architecture*, Advances in Science, Technology & Innovation,
https://doi.org/10.1007/978-3-030-57509-0_19

into the repetitive geometric patterns of ordinary construction—free floor plan, tree-based layout (corridor and rooms), repetition (similar size of rooms), habitability standards, etc. This series of basic characteristics is used to introduce the existence of a deep spatial structure which guides the design of housing layouts in more or less ordinary buildings (Fig. 1).

This research into the deep structure of buildings follows the contributions of structuralism in architecture, such as studies into habitability and housing programmes. These studies were designed for and anticipated the current application of technologies to study the deep structures of architectural form. Concepts like the 'metaproject' [16], and the related 'metaprogramme', are used to understand the deep and widespread reasons below the particular forms and elements of the 'project'. These conditions belong to 'cultural models' which are shared and internalised in society through traditions and which survive the reasons for which they were created [13]. Naturally, these ideas and requirements are expressed by 'patterns' in the process of design [2], and these patterns are especially relevant and repetitive in the case of collective housing. They are so repetitive that the perimeter of an ordinary floor plan is highly related to these patterns and limited types of composition, as we found empirically. Inversely, the perimeter greatly restricts its layout, and our study explores to what extent the perimeter and the exterior conditions determine ordinary housing layouts.

On the other hand, contemporary ordinary collective housing buildings quantitatively constitute the greater built heritage of many compact cities, but as they are becoming obsolete—in their energetic, functional, and cultural aspects—they will need to be rehabilitated *en masse*. Due to the standard and repetitive patterns of these buildings, technologies are particularly suitable to manage the information in these cases of renovation processes [7]. Specifically, the use of shape grammars is proposed by these authors to deal with graphical and analytical information in rehabilitation processes.

Shape grammars are methods which process objects and repetitive patterns based on their geometry [7]. This methodology was principally developed to generate new forms and compositions in the work of designers, e.g. its application in residential construction systems [5] or in urban creations and plans [3]. In contrast, we want to highlight the possibilities of shape grammars to analyse,

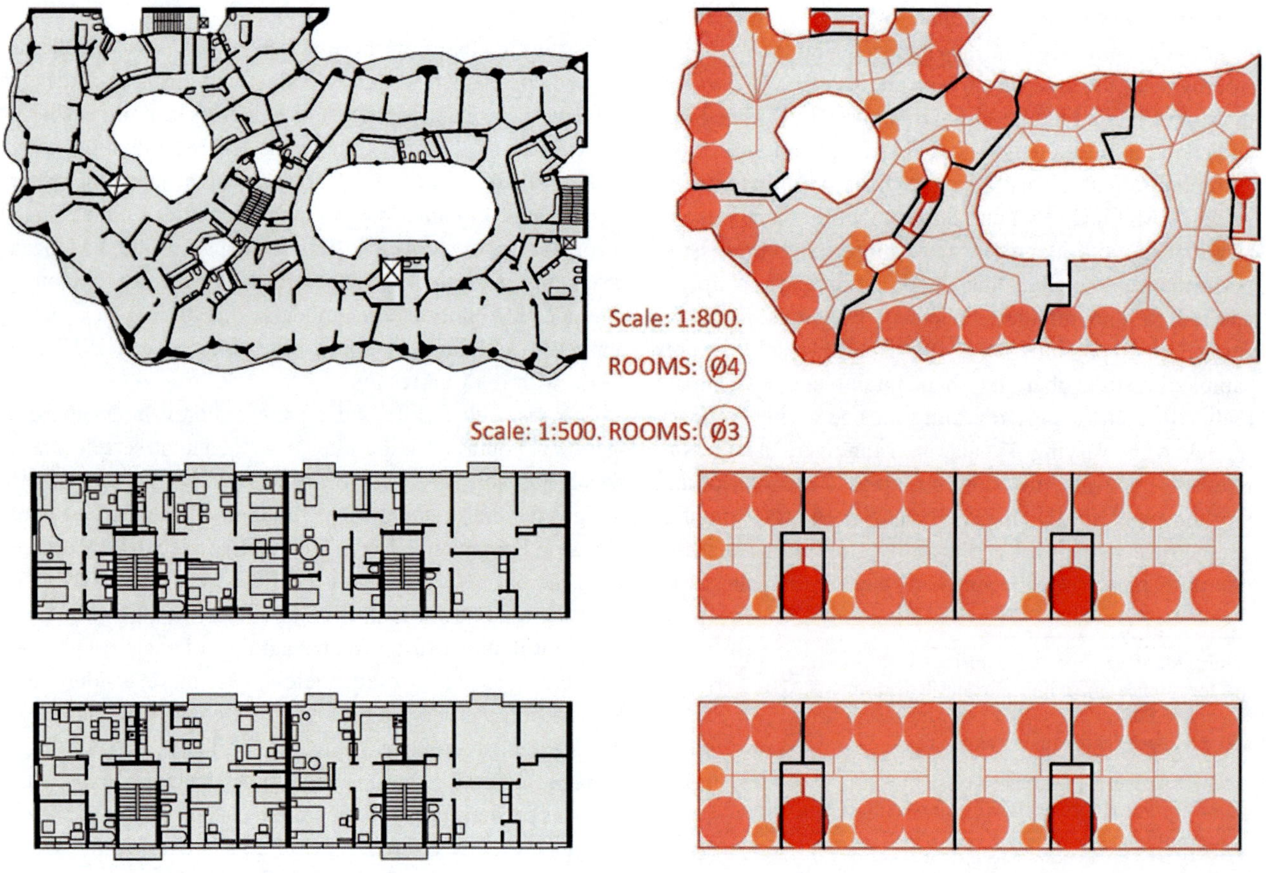

Fig. 1 Spatial structure of Casa Milà (Gaudí 1912) and Weissenhofsiedlung collective housing (Mies van der Rohe 1927)

deconstruct, and find patterns in existing buildings [6]. This research is thus located one step before the use of parametric systems to transform the current constructions [7, 775–800], and our approximated results can serve as an input to feed those algorithms or others in order to test preliminary strategies in rehabilitation.

As an analytical tool, shape grammars can take into account the procedures of many morpho-typological approaches [11, vol. 2, 14, 15] and surpass them. The classic quantitative or graphical methods based on distances and proportions have many problems with irregular and non-canonical cases—not classified within a type. In addition, these methods are not sufficient to fully describe and then emulate complex contemporary phenomena affected by hybridisation, randomness, and generic design. Instead, shape grammars can better deal with them because these concepts are at the beginning of their development. In addition, shape grammars have the great advantage of having the potential to natively use ontologies [4], which impose limitations to prevent incoherent or impossible situations and enable a better control of the process. The ontological concepts of the classes of boundaries—façade, partition, party wall—and the cells—rooms and closed spaces—produce a meaningful modelling process because they embed in the system the conditions of adjacency and then introduce the capacities of topology [8].

In the literature, several shape grammars work on architectural layouts, even specialising in ordinary collective housing like in our case [1], albeit focussing on a specific geometry or only using a certain geometry in their applications. Many shape grammars are based on an orthogonal paradigm, and they cannot be applied in other ways; others use hexagons or triangles, and they do not work with orthogonal objects. In contrast, our approach is based on topology [8], which enables much more flexibility and more possibilities for its application because it deals with the formal relationships of entities and does not depend on their shape. Thus, our shape grammar analyses floor plans through the topology and size of their spaces (Fig. 1, right), rather than based on geometry (Fig. 1, left); this enables its compatibility with both rigid grids and irregular curved shapes, e.g. Fig. 1.

We highlight the versatility of this topological (and non-geometrical) design of the shape grammar and its general applicability. This makes it able to process many different cases, including the chaotic shapes of the irregular and non-canonical urban fabrics (too complex to apply classic methods). Thus, the shape grammar makes it possible to obtain quantitative information for the rehabilitation of residential buildings: approximated number of dwellings, rooms, façade per room, windows, etc. This data is relevant in the preliminary studies of buildings, e.g. energetic simulation [8] or calculation of costs and materials—not only in the case of their demolition [1], but also for their rehabilitation.

We explain hereafter the development of the shape grammar, its programming, and its application, for a better understanding of this approach.

2 Development of the Shape Grammar to Estimate Layouts

In order to obtain information about current ordinary buildings in a massive and non-intrusive way (which is relevant in urban rehabilitation programmes), we created a shape grammar to estimate the main elements of housing layouts from external information (perimeter and environmental conditions). This type of remote and non-intrusive approximated analyses from exterior data [1] is useful in many contexts when the number of buildings or other circumstances does not make it possible the visit and the analysis on site. The necessity of these methods is highlighted by episodes such as the coronavirus crisis when decisions about buildings had to be carried out remotely or stopped.

This research is based on the hypothesis that ordinary housing layouts are especially determined by their perimeter and its access to air and light (the adjacency conditions that allow the buildings' rooms to have a façade, or not). For this, other hypotheses are needed, and they constitute the definition of the elements of the grammar. Main rooms—cells like bedrooms, kitchens, living and sitting rooms, offices—must constitute separate spaces of similar sizes, and they have to maintain independent contact with the façade. Rooms must be internally in contact with a corridor in a tree-based compact structure. The assembly of these spaces forms dwellings (flats and apartments), and each dwelling must be connected to the common staircase, using its corridor, and in a compact spatial scheme.

The approach of this grammar is limited to certain conditions: it applies to intermediate independent floors of ordinary buildings, designed using a free plan, a tree space structure, internal corridors, independent and similar rooms, and following habitability principles. Outside of the remit of this hypothesis are, for instance, attics with windows on the roof, duplex solutions consisting of several floors, and open plan layouts without a corridor. Cases with balconies and service entrances can be processed, not taking into account these spaces.

The shape grammar uses the topological concept of cells to represent the studied spaces [8]. Mathematical 'cells' refer to areas with different types of boundaries. In contrast to polygons or solids (objects with homogenous boundaries), cells better represent built spaces because they distinguish ontologies such as 'façade', 'party wall', and 'partition' which will be needed in the grammar. This affects the rooms and also the dwellings (flats and apartments), which are cell complexes [8]: an assembly of cells with its façade and partitions as boundaries.

The shape grammar is based on the dependence of housing layouts on their external perimeter and the conditions of the façade. It focusses essentially on (I) the capacity of a building's floors to accommodate rooms in contact with the façade and (II) to be divided into several dwellings in contact with the common staircase. These differences justify the division of the grammar: Part I, referring to the perimeter and habitability, and Part II about its area and the efficient connections of spaces. This is explained in its workflow (Fig. 2) and the detailed description found hereafter.

2.1 Shape Grammar Part I: Rooms at the Façade

Part I of the grammar tries to locate as many rooms of a certain size as possible on the façade. Size is represented by a circle which refers to both width and a minimum depth from the façade—which is a basic rule found in empirical surveys and a method implemented in some habitability norms. This is then a mathematical problem of feature allocation and the partition of a plane adapted to housing architecture.

Essentially, Part I follows the logic of buffers of polar systems, but not necessarily from a centre. It uses the façade line as a curvilinear main axis, with variable direction, and curvilinear coordinates to calculate new positions based on a straight-line distance to the façade and the distance along the curvilinear equidistant lines.

For a better explanation of this curved coordinated system and the shape grammar, we used as an example the perimeter of a 'rabo de bacalhau' specimen [7], a variation of a block type with a protuberance at the back. This perimeter (Fig. 3) is challenging for geometric algorithms because the front façade acts as a linear block, its 'tail' as a convex semi-tower, and the concave parts as tiny courtyards. Moreover, the specimen used as example has some irregularities and entrances to which the shape grammar must adapt.

In Figs. 2 and 3, the main rules of Part I are explained:

- 0. Façades need to be separated from the rest of the boundary.
- 1a. The façade (*A*) is used to draw equidistant lines (*B*, *C*) at certain distances (e.g. 1.5 m, 3 m).
- 1b. Following the lines, circles of a certain size (e.g. Ø3 m) are placed.
- 1c. Tangent to the circles, partition walls are drawn perpendicular to *A*, *B*, *C*.
- 2. Vectors are drawn from current partition to the next one along façade (*A*) and lines (*B*, *C*).
- 3a. If one of the vectors is smaller than a certain distance (e.g. 1 m), there is not enough space for either a window (façade *A*) or a door (line *C*); thus, the partition has to be displaced.
- 3b. The process is repeated along all the façades, and finally, the circles' centres are drawn.

The result is a collection of cells and centroids representing the rooms.

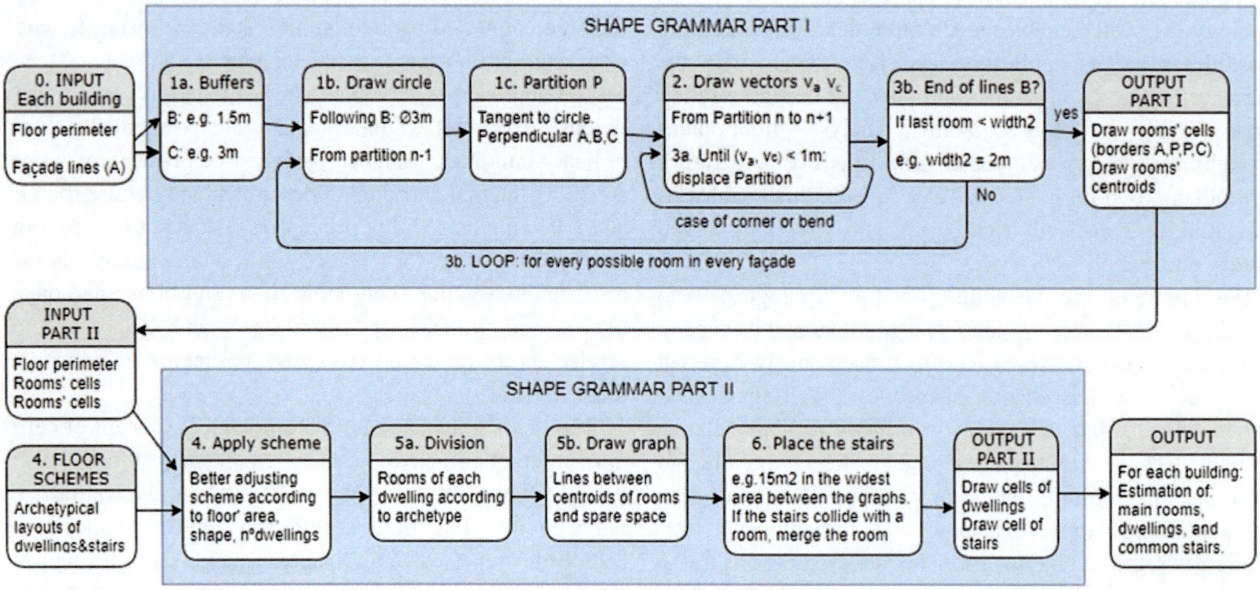

Fig. 2 Simplified workflow of the shape grammar which processes one building

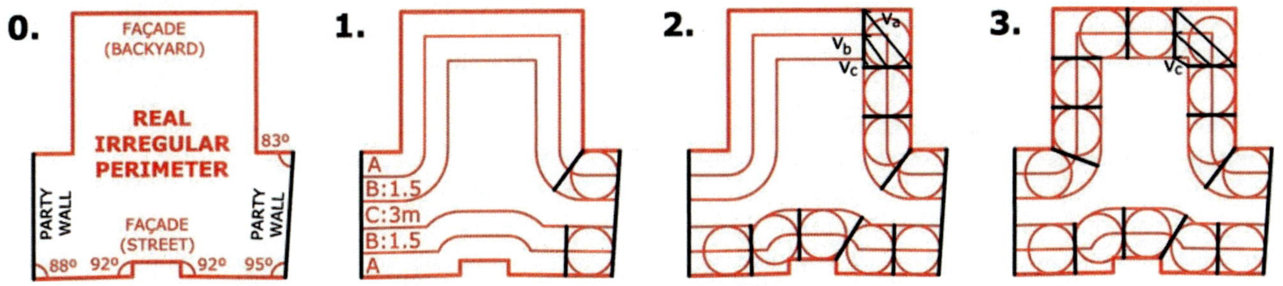

Fig. 3 Shape grammar part I: Ø3 m circles in contact with both the façade and corridor are used to estimate possible rooms

2.2 Shape Grammar Part II: Dwellings and Stairs

Part II of the grammar estimates the dwellings and common stairs using heuristics based on the shape of the floor and the number of dwellings. It also solves a mathematical problem of the partition of a plane, trying to apply the best possible, compact, homogenous layout.

In Figs. 2 and 4, the main rules of Part II are explained:

- 4. Apart from the results of Part I (cells and centroids), the input is a series of heuristics: archetypical layouts with a number of dwellings per floor: less than one (a house, no common staircase), one (lateral staircase), two (front/back, right/left scheme), etc.
- 5a. According to the number of dwellings (calculated by floor area or a given datum, e.g. 2), the most appropriate scheme is applied to divide the rooms' centroids.
- 5b. The area of the staircase (e.g. 15 m^2) is located in the biggest free central area.
- 6. If there is not enough space for the stairs (collision with a room) or they should be on the façade if required by law or due to a cultural model or (habitability) improvement, the staircase merged with the nearest room with façade.

The result is a layout with staircase, rooms, and spare space (the latter two grouped in their dwelling).

2.3 Discussion of the Proposed Shape Grammar

In Fig. 4, we include the real layout of the example with its main spaces. Firstly, we need to clarify that this is a probabilistic method for preliminary estimations focussed on ordinary buildings. Thus, in buildings such as this example (which follows the hypothesis and has ordinary patterns), the shape grammar generally produces results that are quite close to reality in terms of quantity and position of spaces. For instance, in this case there is an error of one room, the centroids are located less than 1 m away from the real ones, and the accuracy is maintained in the case of irregular geometries and curves. Note that the grammar does not necessarily draw symmetrical layouts, because it starts from one end of the façade. This produces a more accurate probabilistic result in the number of rooms than forcing the symmetry (e.g. starting in the middle of the façade) because it produces less non-whole rooms. In addition, the grammar is prepared for irregular and anti-symmetrical schemes of dwellings (e.g. 2 + 3 rooms on the street façade and 3 + 2 rooms on the courtyard), which are quite frequent schemes, as explained in the empirical application (Sect. 4).

Thus, this grammar can be a useful method to estimate capacities or obtain preliminary information models in a non-intrusive way in the context of urban regeneration plans. This is relevant in both the first studies and the discussion of renovation policies, before having the layouts accurately designed. This would enable preliminary decisions for certain rehabilitation solutions and then optimise the tedious (and expensive) task of redrawing every building, which is evidently necessary for their renovation.

On the other hand, this algorithm is more limited in special cases, where they do not follow the mentioned ordinary patterns. Observing Fig. 1, the grammar estimates the spaces in Mies' block quite accurately—despite the different sizes of the dwellings and the open plan layouts in some of them. In

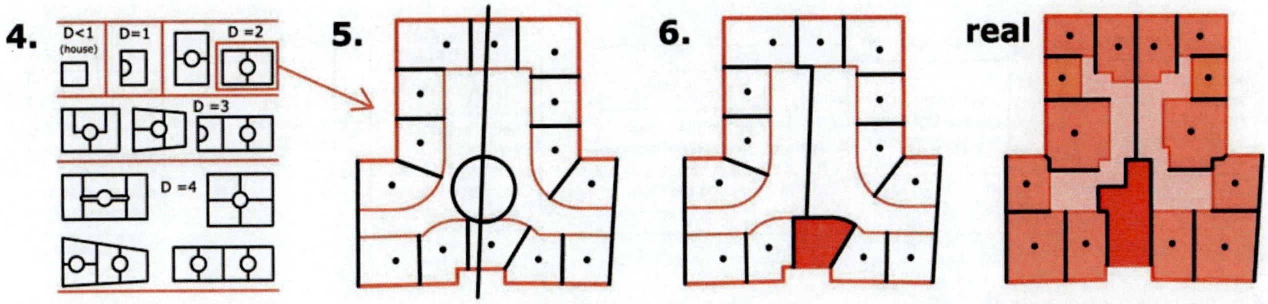

Fig. 4 Shape grammar part II: an archetypical scheme is used to divide the rooms between the dwellings and locate the staircase

Gaudí's building, the grammar works perfectly on the complex curved façade; nevertheless, the results are very different for the interior because the layout is not compact. Moreover, the grammar highlights the superior standards of Gaudi's example—Ø4 m rooms. Thus, the shape grammar could be used to discriminate between special specimens (or cartographic errors)—which should be treated manually—and the other ordinary cases—which can be processed massively. A double debate between both exceptions versus patterns and processing of individual versus recursive cases is then crucial. This discussion is very fruitful after having the information derived from programming the grammar and testing it on a large number of real cases.

3 Shape Grammar's Adaptation to Be Programmed in Arcpy

Being aware of the possibilities of computers, the algorithm was conceived from the beginning to work with known accessible data—perimeter and external conditions—in order to have the capacity to feed the algorithm massively and take real advantage of the shape grammar with technology. In order to apply this grammar, a script is programmed in Arcpy (ArcGIS' library in Python) with certain adaptations. The processing in geographical information system (GIS) was chosen because it enables the processing of large amounts of geometric and analytical data in an accurate and reliable workflow. In addition, the results can be converted in building information system (GIS/BIM) via the importation through both Grasshopper and Dynamo.

3.1 Differences Between GIS Processing and Shape Grammars

As operations in GIS affect the whole dataset, the shape grammars' inductive approach to each element is inverted. The script is designed to process the whole sample of buildings, and it contains a loop to isolate every building, process it using the two parts of the grammar, and reassemble the results (Fig. 5).

3.2 Changes of the Shape Grammar to Be Implemented in Python and ArcGIS

The grammar is programmed in Arcpy (ArcGIS' library in Python) in accordance with its tools, restrictions, and processing performance. ArcGIS does not work natively with topological cells and ontologies, so they are emulated by intersecting polygons and lines before operations. This means more instructions and complex code, which is not explained in this paper and can be seen directly in the scripts, freely available in GitHub.

We explain in Figs. 5 and 6 the main changes in the grammar (in relation to the original manual conception), which also produces a slight difference in its results:

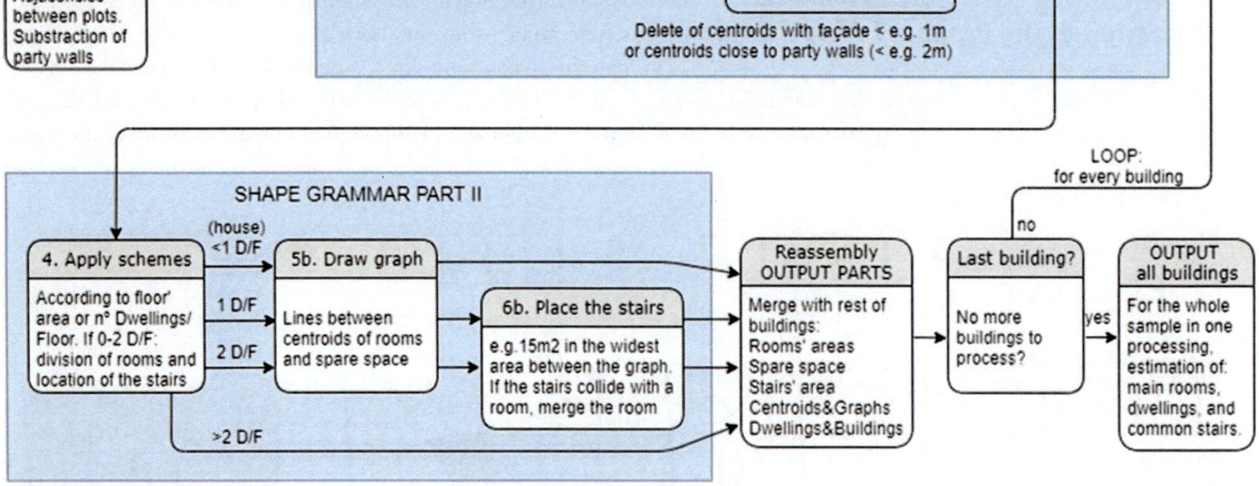

Fig. 5 Simplified workflow of the programmed shape grammar for an assembly of buildings

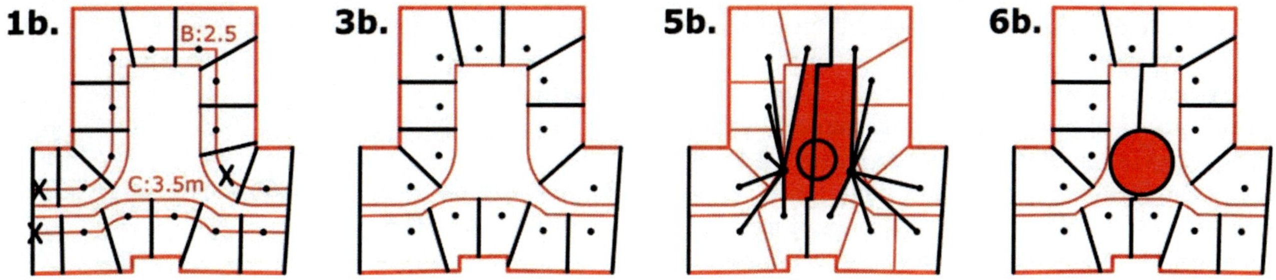

Fig. 6 Geometric changes of the programmed grammar adapted to Arcpy tools and processing

- 1b. In Part I, the sequential drawing of circles for rooms is substituted with a tool that sets all the points on an equidistant line, which is deeper to ensure enough space in convex corners.
- 3b. The partitions are created from the points using Thiessen tessellation (Voronoi diagrams); the rooms with small façades are simply deleted, and the tessellation is re-done.
- 5b. Part II of the grammar is currently only programmed for up to two dwellings per floor; a function draws the graphs, which are used to find the biggest free space to place the staircase.
- 6b. The fusion between the staircase and a room on the façade is not yet programmed.

These changes in Arcpy mean a slight loss of geometric accuracy, but the improvement in efficiency and the possibility of processing large amount of cases compensate for it. This implementation was developed in the Arcpy library of ArcMap 10.5, compatible with Python 2.7. The programme is not optimised yet, and it takes 3–7 min to process every building (tests done in Win10 64 bits, normal session, PC with Intel Core i7, 8 GB RAM, SSD 250 GB, NVIDIA GeForce 1050 Ti), so we advise not exceeding one thousand buildings. The scripts' code and supporting material are freely available on GitHub's project repository (github.com/DavidPereira-Martinez/ordinary-buildings-characteristics) with licence CC BY-NC 2.0. We encourage researchers to use these materials, providing them with our help if necessary.

This is a linear classic programme, and the main possible errors are managed, so it always produces one consistent output, and result is the same with equal parameters. It can be used for a preliminary automated analysis and to test different parameters. The results can be revised afterwards and edited by researchers in ArcGIS before exporting them to BIM. This workflow can produce a better control of ordinary buildings in urban regeneration in preliminary phases, optimising the tedious and fruitless work of refining, redrawing, and updating cartography and databases.

4 Application of the Shape Grammar to Porto's Buildings

Finally, this shape grammar's script is applied to a sample of collective buildings in Porto, whose layouts were drawn in a previous Ph.D. thesis [11], producing an inventory of multi-family housing buildings designed between 1930 and 1970 in Porto. The selected buildings constitute a relatively homogeneous corpus, albeit with different characteristics in terms of scale, façade, languages, and typological solutions. These buildings are located in different types of housing blocks, some of them in the core area of the city and others found in expansion areas of the compact city that were built before the 1970s [10].

The multi-family housing buildings gathered in the referred inventory were built by private promoter and targeted a new urban bourgeoisie. This mass construction paradigm was different at different levels from that observed in Lisbon and other European capitals, in terms of genesis and references, types, models, and, above all, chronology. In Porto, the first examples of new multi-family housing appeared only at the beginning of the twentieth century with the expansion of the city centre, which broadly defines its identity. These typological models have varied configurations and remarkable plasticity and adaptability. This characteristic is normally the result of the shape of the urban plots, the particularities of the programme, and the urban regulations to date [9].

Porto's housing building tradition also has a relevant specificity. In 1956, Porto Municipal Services introduced legislation that prohibited the use of small inner yards expressly and limited the building's depth to 15 m. This mandatory regulation led to the fast disappearance of building designs with an irregular configuration of the perimeter [9]. A building type was being imposed, with a regular shape and aeration conduits, establishing a housing model that was particularly popular during the 1970s and is still in use in contemporary real estate housing construction [12].

This sample of buildings is dominated by specimens of one and two dwellings per floor. In the latter case, the dwellings generally present a similar residential programme with the same number of rooms—a hypothesis of Part II of the grammar—but this does not mean a symmetrical scheme in the majority of cases (only in 18 out of 46 cases of two dwellings per floor). The presence of corners of the blocks caused most of the buildings to have a non-symmetrical scheme due to the urban location. A certain number of them (4 out of 46 cases) have a deliberate anti-symmetrical scheme (e.g. 2 + 3 rooms on the street façade and 3 + 2 rooms on the courtyard).

A graphical and analytical repository was created to apply and compare the results of this shape grammar, as well as to share the data set of the thesis. All the information is freely and fully available on the Zenodo open-access platform (https://doi.org/10.5281/zenodo.3723591), and we encourage other researchers to benefit from this work.

4.1 Characteristics of the Geometric and Analytical Database of the Layouts

The graphical information from the Ph.D. thesis [11] with Porto's buildings' layouts was re-elaborated, and characteristics were included as metadata in order to increase its interoperability. For this, some criteria were introduced, like the understanding of 'building' as any construction or part with its own communications (staircase), even if it has shared party walls and was designed with another building. The repository includes 109 buildings, having discarded 7 specimens due to their lack of ordinary patterns: either they were not intermediate housing floors, they had the main communication through an exterior corridor, or they had an open plan with no divided rooms.

The database contains different types of georeferenced information (Fig. 7): the CAD drawings of the thesis (Fig. 7(1)), a geometrical and analytical GIS/BIM database (Fig. 7(2)), and other interoperable information derived from the latter. The data is stored using three hierarchical levels: buildings, dwellings, and rooms. The functioning of every level is independent, but the superior levels are assemblies of entities on the level below. Spaces are drawn in a lightweight modelling system [8], without the thickness of the boundaries (masses in BIM). With this information, we can automatically produce more detailed plans (the scripts for the database are not explained in this paper, but they are available on Zenodo), for example: the drawing of possible windows and doors (Fig. 7(3)) using the spaces' adjacencies, or the drawing of walls with different thicknesses, doors, and windows according to rooms' intersections (Fig. 7(4)). Alternatively, the database information could be simplified, representing spaces with graphs and points (Fig. 7(5)), or showing only the main rooms (Fig. 7(6)), which is the part of the repository used for comparison in the next section.

4.2 Methodology of Application, Results, and Discussion

Figure 7(6) (an amplification of Fig. 8) shows a simplification of the database with the main rooms (light red) and secondary rooms (bathrooms, orange) on the façade, spare space (corridor and other spaces, pink), staircase (red), and dwellings (bold black line). This is used for comparison with the automated results of the algorithm, but it also reveals the deep spatial structure of these cases. Balconies and service entrances were not taken into account.

The programmed shape grammar was applied to the known floor perimeters of the repository in several tests. Façades were calculated using the adjacencies of building plots, and some small interior courtyards were not treated as façades if they were not used by main or secondary spaces.

Several tests have been carried out on the sample, leading to some variations (Table 1) in the application of the grammar. Firstly, we carried out empirical research into the average room's width using Part I of the grammar. In this sample, we obtained the highest coincidence in the number of rooms (88%) using 3 m as the average of rooms' width. Nevertheless, on small floors (<99 m^2), a 2.8-m width performed better, and 3.6 m on large floors (>250 m^2). The bigger the floor, the more numerous and wider the rooms of the dwellings—especially in some buildings in the central part of Porto, an aspect which is related to social groups. Curiously, there was no correlation in this sample between this width and the date of construction (no improvement in access to air and light through time). Thus, we tested an application of Part I separately with a constant room width = 3 m (Variation 1) and with the previously mentioned three widths, depending on floor's area (Variation 2), which is shown in Figs. 9 and 10, and also used for Part II.

In Part II, we tested the estimation of the number of dwellings by floor area, obtaining in this sample:

- <75 m^2: probably <1 dwelling per floor (one-family house, not a collective building)
- 75–180 m^2: probably 1 dwelling per floor
- 180–640 m^2: probably 2 dwellings per floor
- >640 m^2: probably >2 dwellings per floor.

This sample is especially dominated by cases of 1 and 2 dwellings per floor, with a great variability in surface area (some floors with 1 dwelling are bigger than other buildings' floors with 2 dwellings). Thus, we carried out a separate analysis using the number of dwellings according to the

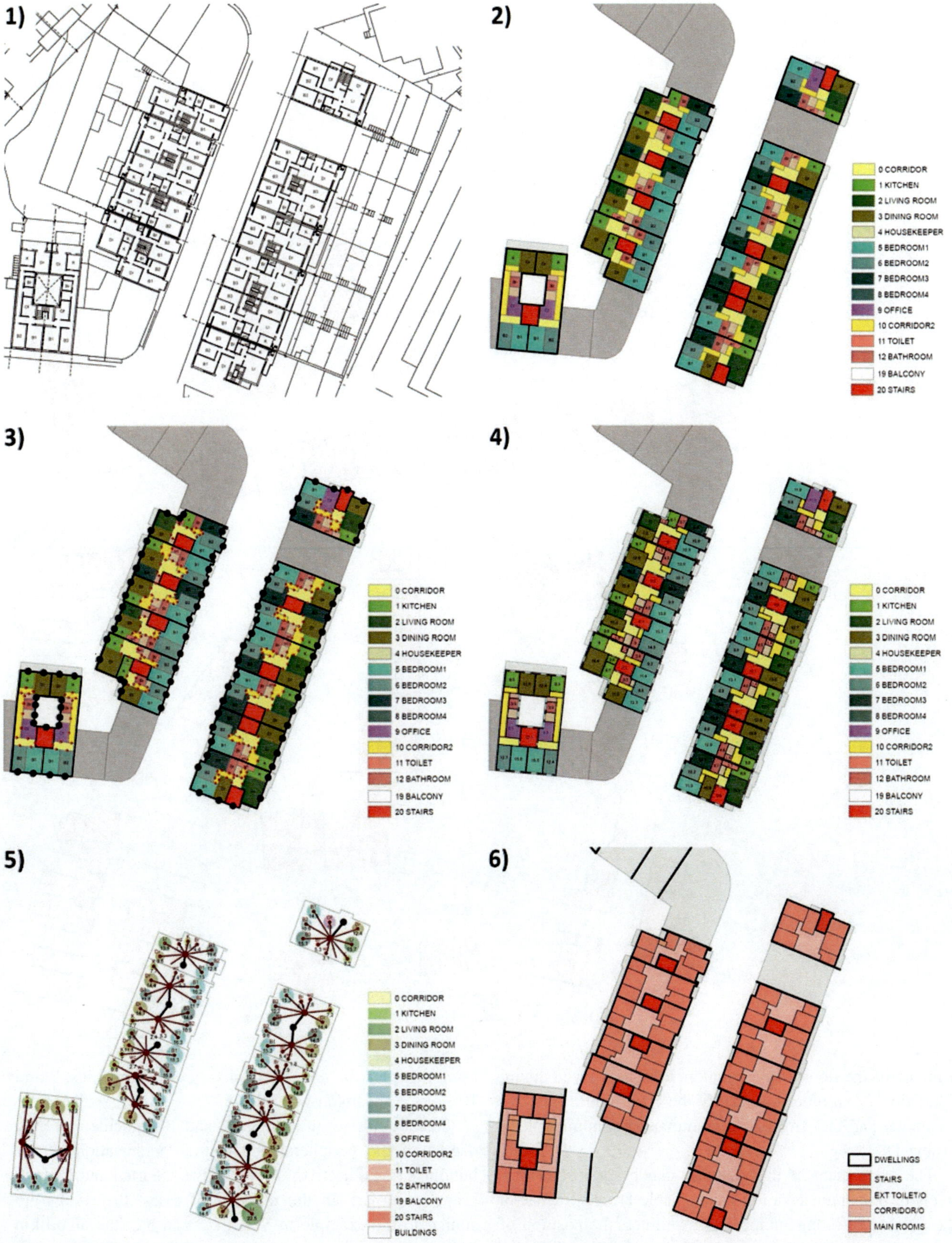

Fig. 7 Detail of repository information: (1) CAD from the thesis, (2) geometric and analytical GIS/BIM database, (3) windows and doors generated with adjacencies, (4) emulation of detailed layout, (5) simplification by rooms' centroids and graphs, (6) staircase and main rooms for comparison. The last four drawings are derived from the GIS database

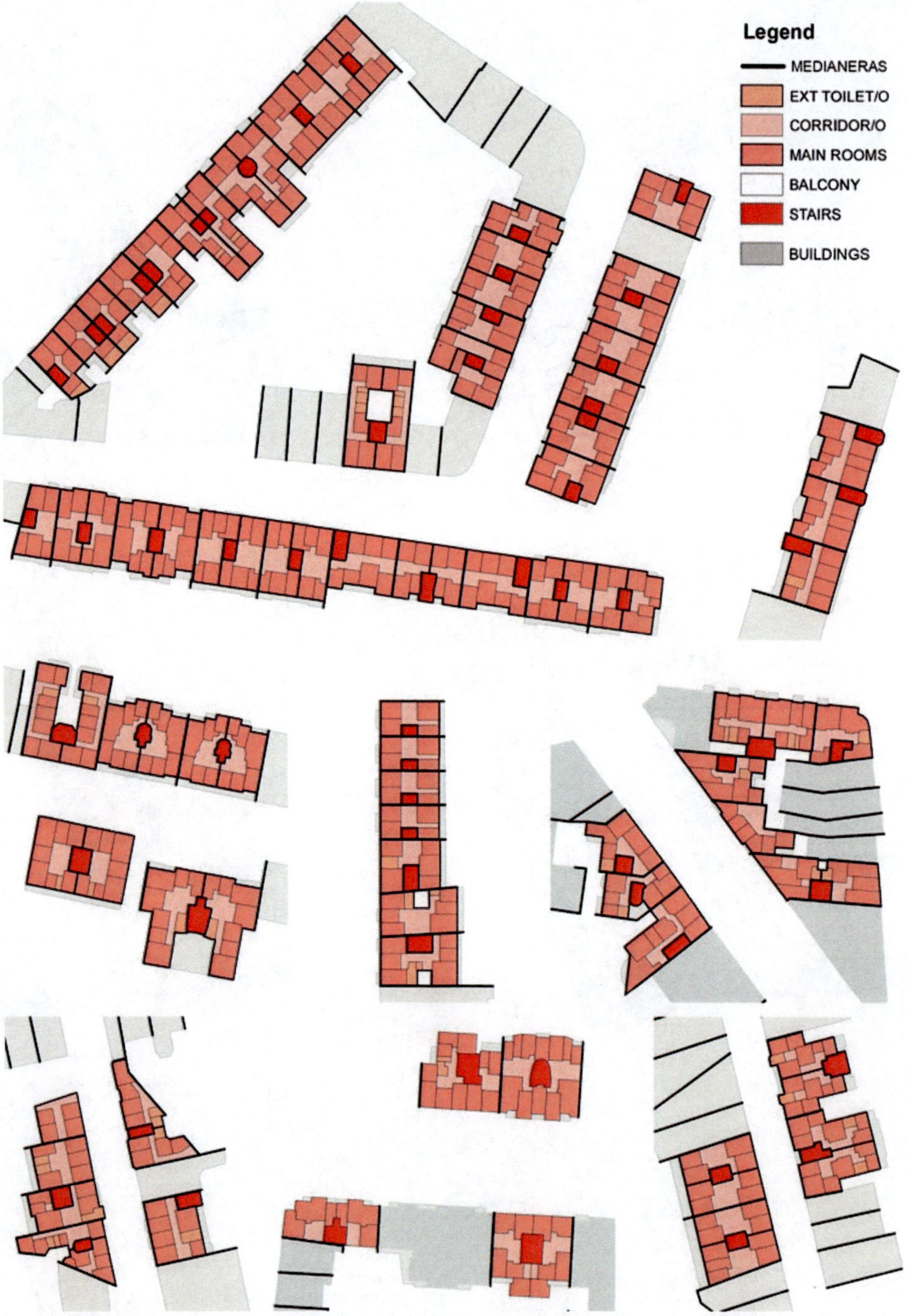

Fig. 8 Simplified layouts of the Porto's case studies: staircases, dwellings, main exterior rooms, and spare space

estimate of the floor area (Variation 1) or as a given datum (Variation 2), in order to quantify the error induced by this parameter and the error of the grammar-programming itself (Table 1).

The assessment of the grammar can be done quantitatively using parameters of measure (Table 1): coincidence in the number of rooms, number of dwellings, the position of the staircase, and the accuracy of the position of rooms (average of distances of segments between proposed and real centroids, like in Fig. 9 centre). On the other hand, assessment can be done qualitatively observing Figs. 9 and 10 in its comparison with Fig. 8.

Globally, the results in Figs. 8 and 10 coincide to a high degree in the peripheral and central organisation of the buildings (the objectives of, respectively, Parts I and II of the shape grammar). In the majority of cases, the results are similar to the example in Sects. 2.2 and 3.2: the algorithm estimates the majority of the rooms correctly (with a difference of ±1 room) and, with more difficulty, the dwellings and the staircase. In some cases (from 10% in Part I to 30%

Table 1 Assessment of the results

Part of the shape grammar	Parameter of measure	Variation 1	Variation 2
Part I: Rooms	Coincidence nº rooms	88.0%	**90.9%**
	Accuracy position rooms	1.1 m	**1.0 m**
Part II: Dwelling and stairs	Coincidence nº dwellings	**87.1%**	(100%)
	Coincidence position stairs	**70.6%**	78.0%

In Part I, Variation 1 corresponds to a constant room width = 3 m, and Variation 2, to a variable width = 2.8, 3.0, 3.6 m. Part II uses a variable width; in Variation 1, the number of dwellings is estimated according to the floor area, and in Variation 2 it is given. The option retained, shown in figures, is in bold

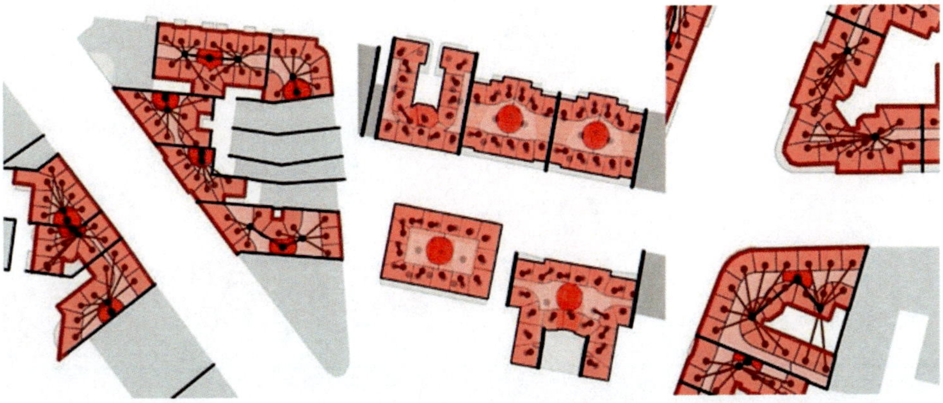

Fig. 9 Detail of results. Coincidence with reality (left), comparison of rooms' centroids (centre), grammar's errors (right)

in Part II), the programmed grammar produces errors and incorrect predictions (Fig. 9, right). A part of these errors needs to be taken into account in further developments; nevertheless, another is intrinsic to the lack of homogeneity of the sample, and by extension of reality (irregularity, exceptions and differences impossible to preview). The main differences to real layouts are in the resulting geometry and the differences in rooms' depth, both aspects beyond the purposes of the research. The results constitute useful and interoperable information for preliminary studies for renovation process. The graphical and analytical data can be exported to an LoD1 model (lightweight mass model in BIM) and serve as an input for topological BIM workflow (e.g. Dynamo-Topologic-OpenStudio-EnergyPlus) [8].

We highlight the reliability of the sample, which has been originally chosen for other purposes. As mentioned, the sample of buildings in Porto was selected to study the variety of solutions in those decades, was carefully analysed in the Ph.D. research, and was validated in its presentation and publication. The information is fully available in the original format of the thesis [11] and in the repository on the Zenodo platform. The exclusion of 6% of specimens for the aforementioned reasons is irrelevant to the global conclusions, which would remain the same, even with slight different values in the results. The sample seems representative of an average reality (further studies could confirm or contradict this circumstance), but, even so, the sample is illustrative of many realities, and then theoretically relevant. Thus, its results imply empirical evidence for the general discussion.

This larger application of the shape grammar enables us to study its implications. The high results of this estimation (90% Part I and 70–80% Part II), as long as this sample is representative of other ordinary contexts, support the two main hypotheses of the research. Firstly, the shape grammar can estimate and emulate ordinary reality to a high degree. Secondly and more importantly, the real ordinary built environment mainly follows tight spatial patterns. These patterns do not necessarily produce low architectural quality if the adaptation of the layout is previewed (e.g. Weissenhofsiedlung collective housing by Mies in Fig. 1), but this underlying framework highlights an excessive homogeneity in the built environment, and probably a lack of creativity. Our hypothesis that the internal layout is very constrained by the exterior perimeter and conditions is supported by this sample. The layout is then mainly determined by the building's footprint and its façades.

In this sample, the majority of the buildings fulfil the initial hypotheses and conditions, using the same ordinary spatial patterns, even if they belong to different morphological types and they have different standards (classes). This supports the initial hypothesis of the existence of a main ordinary (cultural–geometrical) model for buildings and confirms this theoretical category. This ordinary model would be above types and standards (as it appears in this

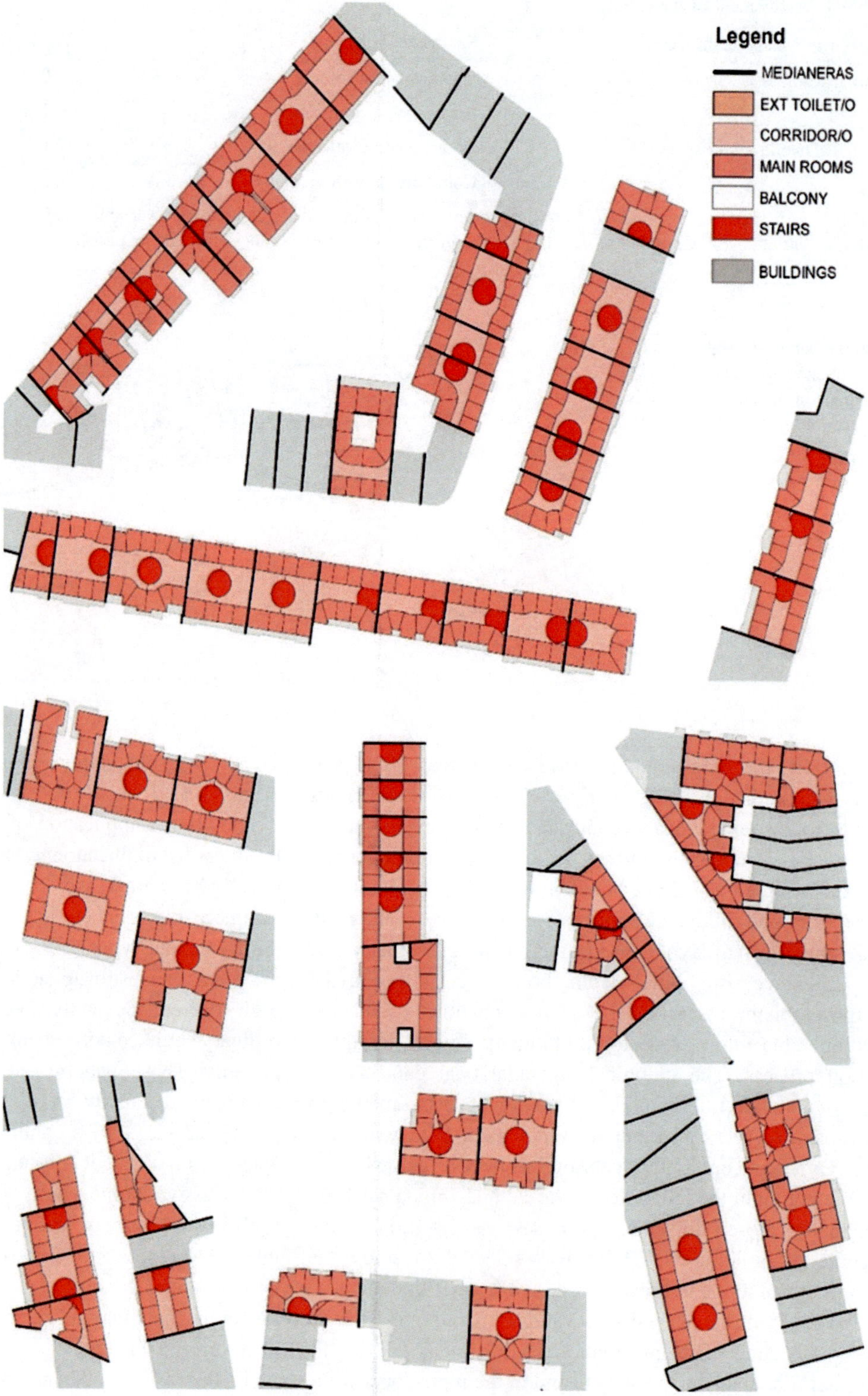

Fig. 10 Results of the programmed shape grammar from the case studies' perimeters: main rooms, spare space, and staircase

sample), and within its framework we can precisely discuss their parameters. Certain buildings with equivalent parameters have different spatial conformations, and others, belonging to the same type, have different standards. Differences in parameters, such as the average room size from 2.8 to 3.6 m in the sample (or in Mies' and Gaudì's examples), are particularly evident when they are analysed through the model framework. These differences need to be analysed in their relationship to social groups and in their possibilities of improvement.

5 Conclusions

As a first conclusion, as long as this sample is representative of reality, the application of this programmed shape grammar would contribute to clarify the aforesaid theoretical problem around the 'ordinary' category of buildings. This study supports the existence of patterns [2], a shared cultural model [13], some standards of habitability, and an underlying metaproject [16] for ordinary housing buildings. This sample also confirms our hypothesis of the large impact of the perimeter on the layout of collective housing, as well as the impact of other parts of the grammar: maximum number of rooms on the façade and a compact structure of spatial connections. Finally, the results of the sample open a wide discussion about this ordinary model, and its use as framework to analyse buildings' parameters and their improvement in different types and classes. The fact that a building has ordinary characteristics does not mean it has lower architectural quality (e.g. Weissenhofsiedlung collective housing by Mies in Fig. 1); nevertheless, this theoretical framework enables a better comparison of certain basic parameters and levels of architectural quality: standards, size, access to air and light, etc. This study illustrates the applicability and utility of this approach.

The result of the shape grammar's script coincides with the configurations of the vast majority of these cases, emulating reality almost at the limit of their patterns. Most importantly, it offers a quite useful analysis of the capacities of buildings and their most basic morpho-typological characteristics, even in the case of irregular geometries. The two parts are developed using a two-dimensional approach, although they are prepared for three-dimensional modelling (following LoD1 CityGML standards) and can be exported to BIM using both Grasshopper and Dynamo (this is in fact a current line of research). Thus, a preliminary but complete analysis can be carried out in early phases of rehabilitation plans, in order to have the information for their renovation, as proposed by other authors [7]. The use of a shape grammar has shown its possibilities and versatility as an analysis tool, its limitations, the potential of its results for urban regeneration, and its possible development.

Our current line of research is focussed on a workflow to export results into 3D BIM systems and then execute a cost and energy analysis [8]. Ongoing research into energetic performance by the FEUP (University of Porto) will assess the difference in results between the estimated and real layouts. We expect differences of up to ±30% derived from the geometrical diversions. Thus, this estimative shape grammar would be relevant in preliminary renovation studies because it can be applied en masse with indicative results. We are also working on estimating other floors of the building with the information about the main floor. Thus, we could have preliminary holistic modelling of every building with its characteristics and possibilities of renovation, which can be useful for the policies of the rehabilitation programmes in the municipality of Porto. Finally, this non-intrusive and remote methodology has contributed (with other sources, such as cadastral information) to the estimation of interior layouts of problematic buildings in Spain when a site visit was not an option. This is part of an ongoing research project in the ETSAC (University of A Coruña) that aims to assess buildings' conditions during the coronavirus lockdown to help the users to adapt their spaces.

As possible evolutions, we can point out the use of other figures to estimate the rooms on the façade, such as ellipses and rectangles; nonetheless, this will notably increase the complexity and difficulties in programming. Some functions should be implemented to control the size and shape of the spare space in relation to the rest of the dwelling. Machine learning (AI) can work from the centroids to estimate the depth of the rooms and distinguish their functions (kitchen, bedrooms, living room, etc.).

The data set and the code are freely available, and we encourage other researchers to do other analyses with the data and use the scripts in other situations. The comparison between the results would be remarkably interesting for the current discussion of the built environment.

Acknowledgements The research is based at the University of Porto, Faculty of Architecture (FAUP), Center for Studies in Architecture and Urbanism (CEAU) which is co-financed by the European Regional Development Fund (ERDF) through COMPETE 2020—Operational Programme Competitiveness and Internationalisation (POCI) and national funds from the Foundation for Science and Technology (FCT, Lisboa, Portugal). Gisela Lameira's research has also been funded by FCT, through a Ph.D. scholarship (SFRH/BD/72996/2010). David Pereira-Martínez has been funded by the Iacobus programme (application 0051) of the Galicia–North Portugal Euroregion through a 3-month stay in CEAU, as part of the project *Modelling of collective buildings for energetic renovation and urban regeneration in Galicia–Northern Portugal.*

Resources Scripts in GitHub: github.com/DavidPereira-Martinez/ordinary-buildings-characteristics. Graphical and analytical sample data set on Zenodo: https://doi.org/10.5281/zenodo.3723591.

References

1. Aksamija, A., Yue, K., Kim, H., Grobler, F., & Krishnamurti, R. (2010). Integration of knowledge-based and generative systems for building characterization and prediction. *Ai Edam, 24*(1), 3–16. https://doi.org/10.1017/S0890060409990138; https://journals.cambridge.org/abstract_S0890060409990138.
2. Alexander, C. (1964). *Notes on the synthesis of form*. Cambridge, MA: Harvard University Press.

3. Beirão, J. (2012). *CItyMaker: Designing grammars for urban design (Dissertation/Thesis)*. Delft University of Technology.
4. de Klerk, R., & Beirão, J. (2016). Ontologies and shape grammars: A relational overview towards semantic design systems. In *eCAADe 34 spatial reasoning and ontologies* (Vol. 2, pp. 305–315). Oulu, Finland: eCAADe and University of Oulu.
5. Duarte, J. P., & Correia, R. (2006). Implementing a description grammar: Generating housing programs online. *Construction Innovation, 6*(4), 203–216. https://doi.org/10.1108/14714170610713890; https://www.emerald.com/insight/content/doi/10.1108/14714170610713890/full/html.
6. Eilouti, B. (2019). Shape grammars as a reverse engineering method for the morphogenesis of architectural façade design. *Frontiers of Architectural Research, 8*(2), 191–200. https://doi.org/10.1016/j.foar.2019.03.006.
7. Eloy, S., & Duarte, J. P. (2015). A transformation-grammar-based methodology for the adaptation of existing housetypes: The case of the 'rabo-De-Bacalhau'. *Environment and Planning B: Planning and Design, 42*(5), 775–800. https://doi.org/10.1068/b120018p; https://journals.sagepub.com/doi/full/10.1068/b120018p.
8. Jabi, W. (2016). Linking design and simulation using non-manifold topology. *Architectural Science Review, 59*(4), 323–334. https://doi.org/10.1080/00038628.2015.1117959; https://www.tandfonline.com/doi/abs/10.1080/00038628.2015.1117959.
9. Lameira, G. (2012). Hygienism, salubrity and regulations. The Oporto's collective dwelling in the 1st half of the 20th century. In R. Amoeda, et al. (Eds.), *Heritage 2012—3rd International Conference on Heritage and Sustainable Development* (pp. 1255–1265). Portugal: Green Lines Institute.
10. Lameira, G. (2017). Contemporary Oporto fragments: Oppositions within the morphological relationship between collective housing and the city. In *Rethinking, Reinterpreting and Restructuring Composite Cities* (pp. 138–155). New Castle, UK: Cambridge Scholars Publishing.
11. Lameira, G. (2017). *O Prédio de Rendimento Portuense. Topologias, Tipologias e Modelos de Habitação Plurifamiliar na 1ª Metade do Século XX (Ph.D. thesis)*. University of Porto, Faculty of Architecture. https://hdl.handle.net/10216/109984.
12. Lameira, G., & Rocha, L. (2018). Barrios Residenciales En Las Áreas De Expansión De Oporto: La Nueva Modernidad (1950–1970). *Ciudad Y Formas Urbanas. Perspectivas Transversales, 10* (Teorías, Historia Urbana y Metodologías Urbanísticas), 67–78. https://doi.org/10.26754/uz.9788417358891; https://hdl.handle.net/10216/115689.
13. Lévi-Strauss, C. (1962). *La Pensée Sauvage*. Paris: Plon.
14. Moley, C. (1991). *L'Immeuble En Formation*. Liège: Mardaga.
15. Panerai, P. R., Castex, J., & Depaule, J.-C. (1977). *Formes Urbaines: De L'Îlot À La Barre*. Paris: Dunod.
16. Portas, N. (1969). *A Cidade Como Arquitectura. Apontamentos de Método e Crítica*. Lisboa: Livros Horizonte.

David Pereira-Martínez is Pre-doctoral Researcher in the Center for Studies in Architecture and Urbanism (CEAU-UPorto) and Group of Territorial Studies (GET-UDC). His ongoing Ph.D., which is being carried out under international joint supervision, focusses on the ordinary patterns of collective buildings and their physical/social requalification. He has participated in research projects and international seminars in Portugal, Spain, and France and has written several articles. He completed a degree in architecture (ETSA-UDC) and has professional experience in rehabilitation, building retrofitting, and sustainability (with prizes in competitions).

Gisela Lameira is Architect and Researcher at Atlas da Casa (housing architectural design and forms of dwelling) of the Center for Studies in Architecture and Urbanism, Faculty of Architecture of the University of Porto (CEAU, FAUP). She completed a Ph.D. in architecture from FAUP. Her research includes the development of studies in architecture and urban theory and history, specifically the genesis and transformation of multi-family housing in Portugal. She was a research fellow of the research project Mapping Public Housing: a critical review of the State-subsidised residential architecture in Portugal (1910–1974), hosted by FAUP/CEAU [P2020-PTDC/CPC-HAT/1688/2014]. Currently, she is carrying out research for the exploratory projects Ageing in Place/Architecture4Ageing, and ICAVI. Independent Living model, at FAUP, where she also teaches History of Contemporary Architecture.

Plácido Lizancos is Senior Professor, Researcher, and Director of the School of Architecture of A Coruña (ETSA-UDC). He is Architect (University of A Coruña) and Doctor with Honours for his thesis about the configuration of the contemporary house Galicia, which is a system of processes rather than an object production strategy. He approaches thorough analytical tools to architecture committed within social issues. Volunteering with Architecture Sans Frontiers provided him expertise on critical situations, which feeds teaching and researching activities in Europe, Caribe and Africa. He is Professor teaching architectural analysis at the School of Architecture of A Coruña, since 1993.

Isabel Raposo is Architect-Urbanist and Associate Professor at the Faculdade de Arquitectura of the Universidade de Lisboa (FAUL), teaching there since 2002. Architect by ESBAL in 1976, in 1986 she completed a master's and in 1999 a Ph.D. in Urbanisme et Aménagement, in the Université de Paris XII, and, in 2002, she did postdoctoral studies in the Centro de Estudos sobre África (ISEG-ULisboa, then UTL). In 2017, she is awarded the title of Honorary Member of the Order of Architects. Since 2007, she has been Coordinator of Gestual, a Socio-Territorial and Urban Studies and Local Action Study Group, at CIAUD/FAUL.

Virgílio Borges Pereira is Tenured Associate Professor at the Department of Sociology of FLUP and Researcher at the Institute of Sociology of the University of Porto. Since 2003, he collaborates with the Faculty of Architecture of the University of Porto at the Center of Studies in Architecture and Urbanism. He has a degree in sociology, a master's course in sociology: local power, development and social change, a Ph.D. and Habilitation in sociology. His research focusses on the production of social and cultural inequalities in different spatial contexts of Northern Portugal through Pierre Bourdieu's work. He is Member of different national and international editorial boards.

Understanding a Bastion Castle in Pidhirtsi Using the Space Patterns System for Spatial Analysis

Olha Tikhonova

Abstract

In recent studies, scholars started to pay attention to space grammar and building organization more than to the building function, as the form and space of the building are more permanent than function. This paper expands the scope of an already used tool based on spatial analysis theory in residential buildings to more complex structures, such as bastion castles. This study aimed to investigate the use of the Space Patterns application to provide a comprehensive analysis of historic buildings regardless of their primary function. The method demonstrates the potential for the general applicability of this system to improve the understanding of building structure. It is a necessary step to a future adaptation of buildings to new functions, which became inevitable and urgent for bastion castles. The results showed that bastion castle in Pidhirtsi had well-lit spaces, 50–80% of the spaces have direct daylight access on each floor. The pedestrian circulation view captured the enfilade planning, characteristic feature for Renaissance villas.

Keywords

Bastion castle • Spatial analysis • Pidhirtsi

1 Introduction

Our history is long enough that it has left us a significant built heritage to manage. Most of these heritage structures need adaptation to enable them to serve new functions. In particular, this applies to buildings of historical military significance, such as castles, whose primary functions are no longer relevant or required. In the case of castles, usually, architects use the historical descriptions and field observations for analysis, before rehabilitation. However, this approach can create some prejudices against analyzed architecture, creating limitations in our imaginations regarding the possibilities these buildings present.

To be able to do interventions and adaptations to meet these new functions, we need to understand a building's morphology, regardless of its style and even its type. To do that, we must establish some distance from usual building concerns, such as primary building function, social role, material, and patronage, instead of examining the geometry and spatial relations of these structures.

This study shows how to approach this pure analysis of historic buildings using the theory of space, introduced by Hillier et al. [1], and Faulkner's method of "planning diagrams" [2]. Although the idea of spatial analysis was introduced in the second half of the twentieth century, its application to historical buildings is recent, a first for a modern bastion castle evaluation.

Hiller and Hanson using their method demonstrated that random groups of spaces have specific rules of order. In most cases, these rules are associated with cultural principles. They show how important is a door placement in the plan for occupants and visitors' circulation in the space. The authors presented four plans with the same geometrical configuration and spatial organization in general but with different doors location. Their justified gamma maps that show spatial patterns were different when the adjacency graphs were identical [3, pp. 149–151].

Besides, researchers applied their methodology for analysis and comparison of two large historical complexes: Ambo kraal and Ashanti palace. Using these examples, they showed that more complex spaces are harder to evaluate while relying just on visual comparison. In such cases, the abstract model can help to create "a sociological picture of a particular type of building" [3, p. 163]. They found that even if the number of spaces in buildings is almost equal, the

O. Tikhonova (✉)
Faculty of Architecture, University of Lisbon,
1349-055 Lisbon, Portugal
e-mail: olechka.tikhonova@gmail.com

S. Eloy et al. (eds.), *Formal Methods in Architecture*, Advances in Science, Technology & Innovation,
https://doi.org/10.1007/978-3-030-57509-0_20

structures are different. In the Ambo kraal case, the building has an elaborate non-distributed structure. By contrast, the Ashanti palace has a far simpler distribution structure [3, p. 167]. The most distributed space was the deepest one in the case of Ambo kraal, whereas in Ashanti palace, it was the shallowest one. The differences are straightforward regarding the analysis of individual spaces too. Comparing to the Ambo kraal, the Ashanti palace does not have a substantial distinction of spaces according to the depth index.

As a next step, authors associated the spatial relationship with their inhabitants, dividing them into two categories: spaces for men and women. In the Ashanti case, women's quarters were in the shallowest spaces. By contrast, men's quarters were in a more superficial position comparing to other occupants, in the case of Ambo kraal.

Patrick Faulkner [2] was the first scholar who decided to analyze a medieval castle using a schematic simplification of its plan before instigating any developmental interventions. He did his work for the Ministry of Public Building and Renovation [4, p. 351]. However, he did not use an entirely schematic model. His diagrams show a general layout of buildings and underline the functional relationships from the viewpoint of the user, more than from the actual adjacency of rooms. This method has certain disadvantages in the comparison between different examples. It is easier to make a comparison when the layouts have unified rules of representation.

One of the most straightforward diagrams, which decent illustrate Faulkner's work, is Goodrich Castle in Herefordshire, England [2, p. 222]. The diagram consists of boxes representing the shape of rooms connected by lines. The lines show the possible circulation of people between the spaces.

Faulkner was able to make a theoretical reconstruction of room functions using the position of spaces in the diagram. He defined the possible location of the essential parts of the castle, such as the main halls, the owner's quarters, and the rooms for principal guests. This example of Faulkner's methodology clearly shows how we can cognize the rooms' functions and the social life in the building looking into the connection of spaces and their size.

Previously described methods can be instrumental in combination with more recent works, especially with the power of automatization of analysis using computer processing of information. A few years ago, Georg Suter at TU Wien developed a Space Patterns system [5] to analyze the spatial structure of residential buildings. This software is used for the case study analysis in this paper, namely analysis of the Pidhirtsi bastion castle.

We can interpret different spaces in the building using the graphs that represent the spatial configuration of the space and indicate the integration value of each room. For example, the value of depth in such graphs represents power [6, p. 16]. Also, we can compare the results of graph evaluation with available documentation, field observations, and old photographs of a case study to make a comprehensive analysis of the monument. Moreover, such an in-depth analysis can help to validate or reinterpret available assumptions of castle history or enrich the knowledge about it. Although the method cannot be used as a guide in the assignment of possible new functions yet, the suitability of this method for improving the understanding of buildings is undoubted.

2 Methodology

2.1 Demonstration of the Method

To demonstrate the application of the proposed approach for analyzing the historical buildings, the author of this paper chose a bastion castle in Pidhirtsi (Ukraine) as a case study. The building itself is a complicated case; it is a complex that sees a residential area embedded in its military section. The approach proposed by the author provides an explanation of the design logic using a computer-based method for spatial analysis, namely the Space Patterns Web app, based on graph theory, created by Georg Suter [5]. The results of such an examination not only help the understanding of the building but also can help validate historical pieces of evidence, underlining building transformations when comparing findings from different periods.

The study consists of two main parts. The first one is data collection, which includes historical documentation, scaled drawings processing, and modeling. After that, the analysis of spatial structure is done using the Space Patterns Web application, specifically space access analysis and access to the daylight. Local and global metrics are automatically calculated using the system [5]. This system creates heat map style visualizations ("views") of space access ("Pedestrian Circulation" view) and daylight access graphs ("Natural lighting" view). Also, it produces other views, such as natural ventilation, zoning, and general view. However, the author of this paper considered "Natural lighting" and "Pedestrian Circulation," the most crucial for historic building analysis.

The "Pedestrian Circulation" view is a heat map style visualization automatically generated from an input floor plan. In this view, a space access network is modeled with spaces and doors as nodes. Edges represent access relations between doors and spaces. A unit door is an entrance to the building from outside or to the floor from the main staircase. In visualizations, it is represented by a distance value "0" and a light color (light blue in color version). The darkening of the color represents the distancing of the space further away from the unit door.

The "Natural lighting" view [5, p. 4] represents the evaluation of access to daylight, where the light zones on the layout (yellow filling in color representation) represents the direct daylight access, middle ton (light green filling)—indirect light via other spaces and dark tone (red filling) means that space does not have access to the light.

3 Compositional Analysis Based on the Current State and Historical Background

Pidhirtsi Castle represents an integrated structure that combines the civil and military functions in a robust architectural conception. The residential part of the castle falls into two main divisions: the private lodgings (main body of the palace, which consists of three floors) and additional rooms in the curtains (casemates). According to the description from 1717, the rooms on the ground floor were used for residential and auxiliary purposes [7, p. 13]. The gateway is connected to two bastions from both sides through two rooms.

The majority of rooms on each floor are planned as an enfilade. They are formally aligned with each other through doors that lay on the same axis. Also, great care was taken to achieve the architectural symmetry of the complex.

Considering all three levels, we can observe that the significant rooms do not have a wide variety of geometry and express a uniformly high standard—especially the rooms in the main corpus. The main corpus is made as a palace, raised on the bastion fortification platform. The more significant part of the accommodation is arranged above the ground floor. It can mean that accommodation within the palace was dedicated to those of comparatively high rank as family members and principal guests. These observations are supported by the description of the ground floor from 1754. The description says that rooms on the ground floor primarily were used for guests and partly were formed by two-room apartments [7, p. 13]. The presence of many rooms for guests emphasizes the information from one of the most important treaties in the Republic of Poland [8, p. 9]. The treaty underlines that Polish nobility was very hospitable; that is why their houses needed many guest rooms. The next two levels consist of bedrooms, cabinet, table room, boardroom, and a two-level chapel [7, p. 13].

Wladyslaw Krzyczynski made the most detailed description of the room functions on the ground floor of the main corpus. According to his report, the main door on the ground floor led to the main hall. On the right side of the hall was the entrance to the buffet. There was an old second hall straight through the hallway from the main hall. The second hall was used as a dining room in the period of description. An owner's parlor was on its left side, and it led to the owner's bedroom. On the left of the first hall, there was a door, which led to the guest suite. The guest suite was adjusted to the owner's suite [9, pp. 48–49]. The author presumed that this guest suite destined for the principal guest considering the location of the space and its size. The main living room was situated on the right from the dining room [9, p. 49]. This living room led to the guests' rooms [9, p. 50].

The functions of the rooms on the second floor have not been clearly defined yet. However, it is possible to make their theoretical reconstruction using the description of Wladyslaw Krzyczynski [9] and inventory documents from 1717 and 1729 presented by Jan Ostrowski and Jerzy Petrus [7].

Primary, according to the documents described by Jan Ostrowski and Jerzy Petrus dating from 1717 [7, p. 28], the second floor was just above central and lateral pavilions. Neither space above antechamber of yellow room nor antechamber of table room had livable rooms. According to the same authors, the plan of the second floor was the same as nowadays except one difference according to the inventory of 1729. The difference was in the south rooms aligned to the chapel. They had bay windows, which were later eliminated [7, p. 28]. In 1717, according to authors [7, p. 29], a residential suite was above the mirror room and mosaic cabinet. It consisted of the antechamber above the mirror room and bedroom above the mosaic cabinet.

The description of the second floor, by Wladyslaw Krzyczynski [9, pp. 87–89], confirms the same amount of rooms on the first and second floor in 1729 and 1894. There was a vast ballroom on the second floor. An alchemical laboratory of Seweryn Rzewuski Targowicznin was next to it and above the crimson room. A cabinet was adjusted to the alchemical laboratory and located above the Chinese cabinet. The room next to the cabinet above the gold room was well decorated. It had a beautiful table laid out with tiles from Delft [9, p. 89]. More likely, it was a meeting room considering its decoration and location. The theater room was above the yellow room, and a room with some instruments from the chapel was above antechamber of the yellow room [9, p. 89].

4 Spatial Analysis

4.1 "Pedestrian Circulation" Analysis

The outcomes of the automatically performed spatial analysis (see Fig. 1) of pedestrian circulation in Pidhirtsi Castle based on all three levels of the current state are as follows:

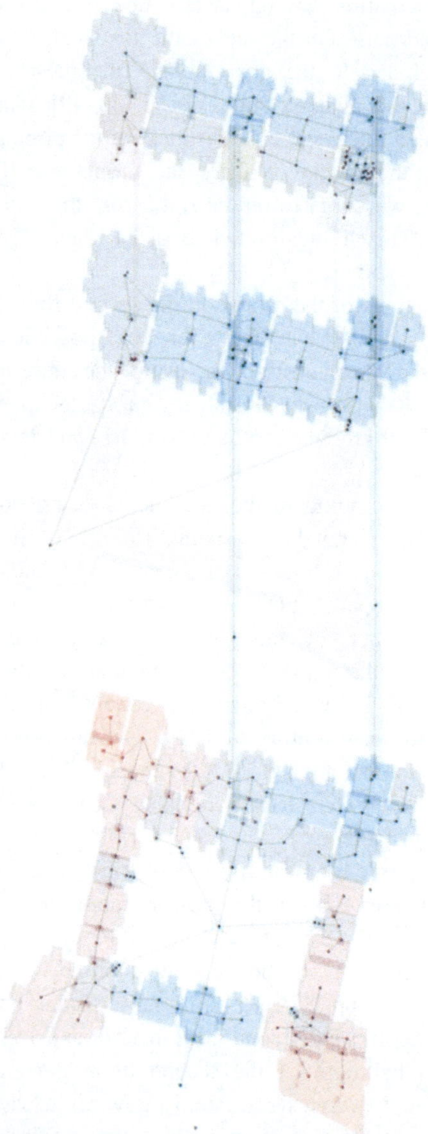

Fig. 1 Pedestrian circulation view for the current state of Pidhirtsi Castle concerning the nearest staircase

- the space access network has many cycles (main part: 10, courtyard: 4, fortification: 1). It is caused by enfilade planning, which was a common feature in the design of this period.
- cycles are extensive, including four or more spaces.
- the main part has a centrally located staircase.

The functions of the rooms of Pidhirtsi Castle were changed several times. In some cases, rooms were partitioned, or, in most cases, doors were walled up or added. It led to changes in pedestrian circulation during the period of castle existence. Additional modeling of a ground plan of Pidhirtsi Castle, which reflects the state of the castle in 1956 [7], helps to trace the most significant changes in pedestrian circulation on the ground floor (see Fig. 2). Furthermore, the author of this paper decided to overlap the theoretical reconstruction of ground-floor functions with a plan of pedestrian circulation, based on the previous description of the ground floor (see Fig. 2).

We can notice that there is just one entry point to the whole structure through the main guarded gate in both pedestrian circulation views of the ground floors (see Figs. 1 and 2). The most connected space in both cases is an inner courtyard. It provides direct access from the outside into the private chambers. Nowadays, the inner courtyard is part of four cycles and seven access points to the rooms, two of which give direct access to the main corpus, and three of them give indirect access through casemates. There were eight external cycles and eight access points through the inner courtyard to the building in 1956. Three of them provided direct access to the main corpus, and five of them provided indirect access through casemates.

In general, pedestrian views from both periods are relatively similar; however, they have differences in the close analysis (see Table 1). Nowadays, we have four significant cycles of pedestrian circulation on the ground floor of the main corpus. Those huge cycles connect four smaller cycles linking the staircase, main hall, and dining room. In 1956, there were 21 cycles. Eleven of them were in the central area: Five primary cycles were between the main rooms, four passed through the staircase, and two cycles joined the living room and the guest room. A major difference in pedestrian access is visible in the southeast bastion, where the pedestrian circulation diagram has changed extensively, from two cycles in 1956 to a relatively simple, tree layout today.

The pedestrian plan of the ground floor from 1956 shows a guest suite on the left, with two points of access from the vestibule (one direct and one through the owner's bedroom) and one access from an area adjoining the courtyard, whose function is unknown. The owner's suite with a private parlor also had two access points: through the same area of unknown purpose, or through the lobby, stairwell, and dining room. This part of the building could function as a separate structure as it has a private exit. The guest suite had access to the owner's parlor through a narrow corridor that connected it with the owner's bedroom. Both suites had access to the dining room through the same hallway and parlor. Alternatively, guests could reach the dining room through the lobby and the hall that contained the staircase. The dining room, the second of two halls, was a well-connected space. It had direct access to the living room, through which the other guest rooms could be reached. The kitchen opened directly onto the courtyard, which was logical, as it needed to receive provisions. Today, there is a window in the kitchen instead of a door. This change is reflected by comparing the diagrams of pedestrian circulation routes in Figs. 1 and 2.

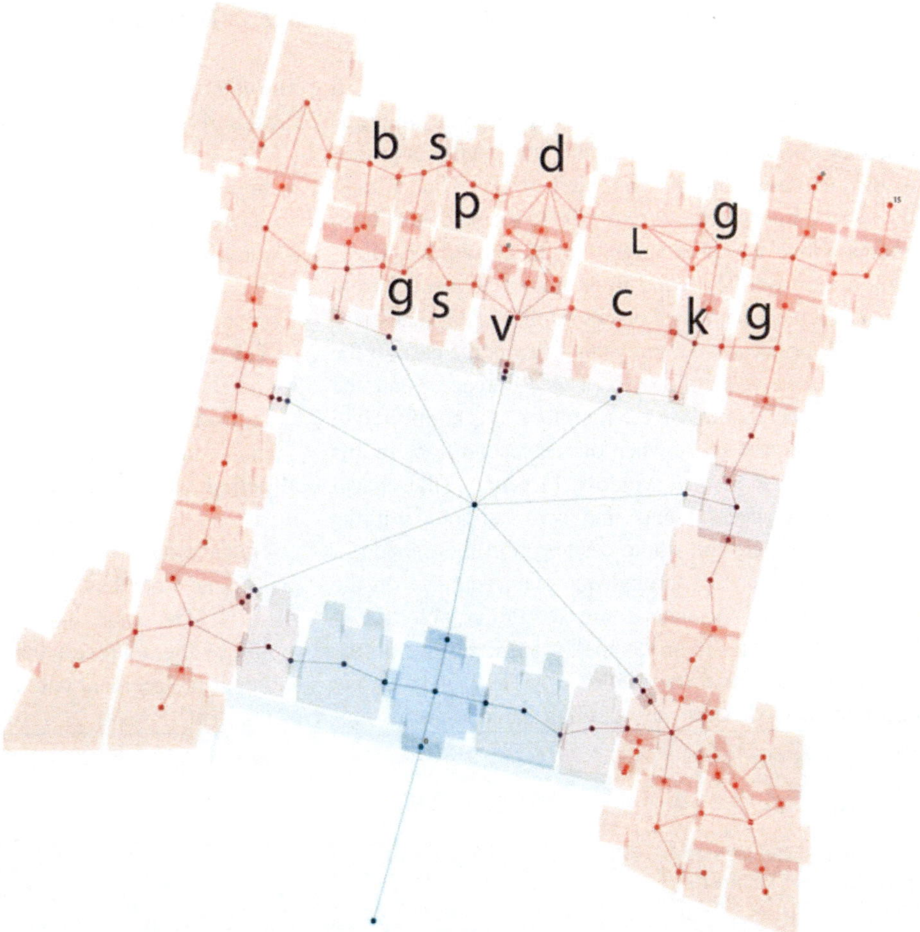

Fig. 2 Pedestrian circulation view with functions (1956): bs—bedroom suite; p—parlor; d—dining room; L—living room; g—guest room; gs—guest suite; v—vestibule; c—closet; k—kitchen

Table 1 Metrics comparison of current pedestrian circulation with circulation in the plan from 1956

Title of metrics for comparison	Current pedestrian circulation	Pedestrian circulation in 1956
Number of cycles	8 cycles: • 4 cycles in the main part • 4 cycles in the courtyard	21 cycle: • 11 cycles in the main part • 8 cycles in the courtyard • 2 cycles in the bastion
Most connected space	Courtyard (degree: 7)	Courtyard (degree: 8)
The depth of the deepest space	15 (7 spaces)	15 (7 spaces)

The first and second floors have the same configuration of spaces and very similar pedestrian circulation characteristics. Moreover, the values of maximum distance to the deepest space are the same—9 edges. The most significant difference is in the external cycle and the southeast space on the second floor.

Nowadays, the southeast space on the second floor is the most connected. It is located above the antechamber of the yellow room. The number of rooms on the second and the third floors was the same in 1729 and 1894. From this, and from the thickness of the partition walls, it can be inferred that this space was not previously divided and was not the most connected space. Therefore, the most connected spaces were the antechambers and the mirror room on the first floor and part of the residential suite on the second floor. The staircase on the first floor has the same amount of connection edges as the mentioned rooms.

Wladyslaw Krzyczynski claims that there were two entry points to the second floor. The first way was through the main central staircase and the second through lateral spiral stairs from both antechamber rooms of the first floor [9, p. 87]. Therefore, the author can suppose that the staircase next to a mosaic cabinet leading to the second floor was added after July 1894.

On the first floor, there is one external cycle, which provides a variety of primary access points to the rooms and, at the same time, affords less privacy to the occupants. It connects two antechamber rooms through the pass along the open terrace above the casemates. We can reach this terrace using the staircase from the courtyard of the ground floor. On the second floor, there are two exits to the balconies from the rooms above the antechambers of the first floor. However, the rooms on the second floor are not connected from outside, as they use separated balconies. In the internal space of the first floor, we have four cycles when in the case of the second floor, we have just two. The difference is in the connection of the chapel. On the first floor, it is connected to a door with staircase space when on the second floor in this place we have just a small window. In general, the chapel had three main access points and two additional on the second floor, serving the whole castle. From the outside, it was possible to access it through the main entrance on the ground floor, taking the stairs connected to the lobby (main hall), which leads on the first floor. From the inside (on the first floor), it connects to two bedrooms. The gallery on the second floor has two entrances through the side rooms too.

4.2 "Natural Lighting" Analysis

The automatically performed analysis of natural lighting (see Figs. 3, 4 and 5) shows:

- high daylight access (50–80% of spaces with direct access) on all levels (floors), including ground level.
- multilateral spaces (mainly in the south part of the first and second floor) facilitated by mass with recesses, which open exceptional views on the garden and lowland (see Table 2).
- the staircase on the second floor is the only space with indirect daylight access, which was a standard feature for inner staircases in Palladio's villa design from the same period, when the main stairs were outside [10, pp. 48–50].

Nowadays, as we can notice from the chart (see Fig. 6), the ground floor has better lighting than in 1956, which is 74–57%. The most significant changes were made in the rooms of the southeast bastion. The demolishing of walls and their replacement with columns led to the better lighting of this part.

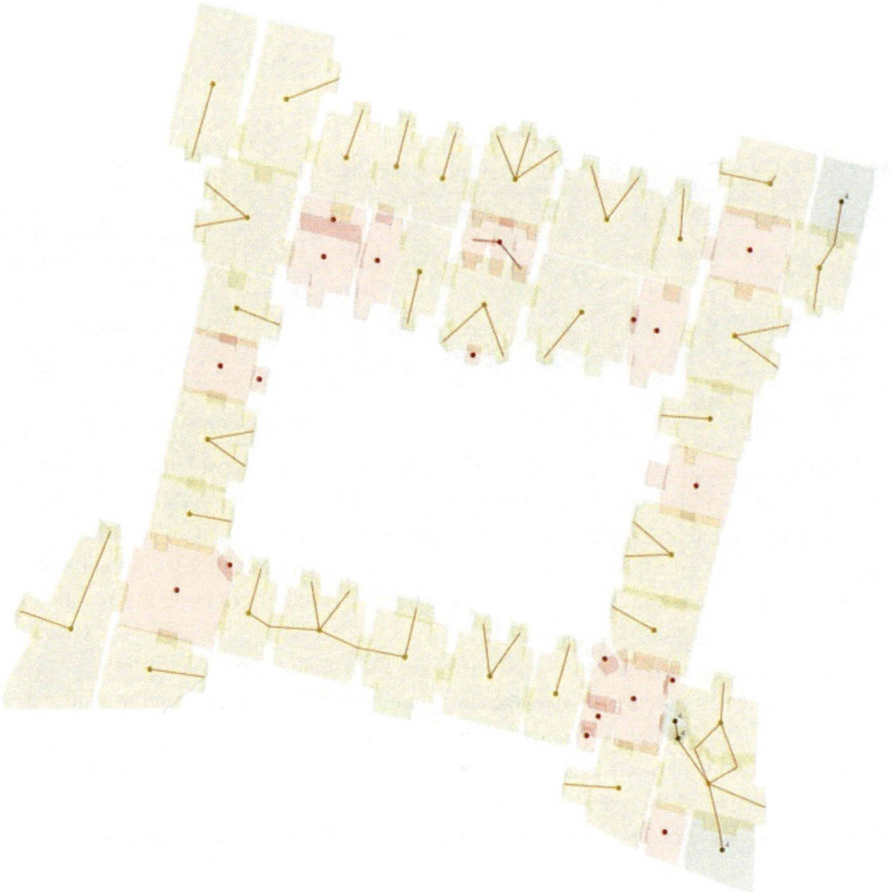

Fig. 3 "Natural lighting" view of the ground floor of Pidhirtsi Castle in 1956

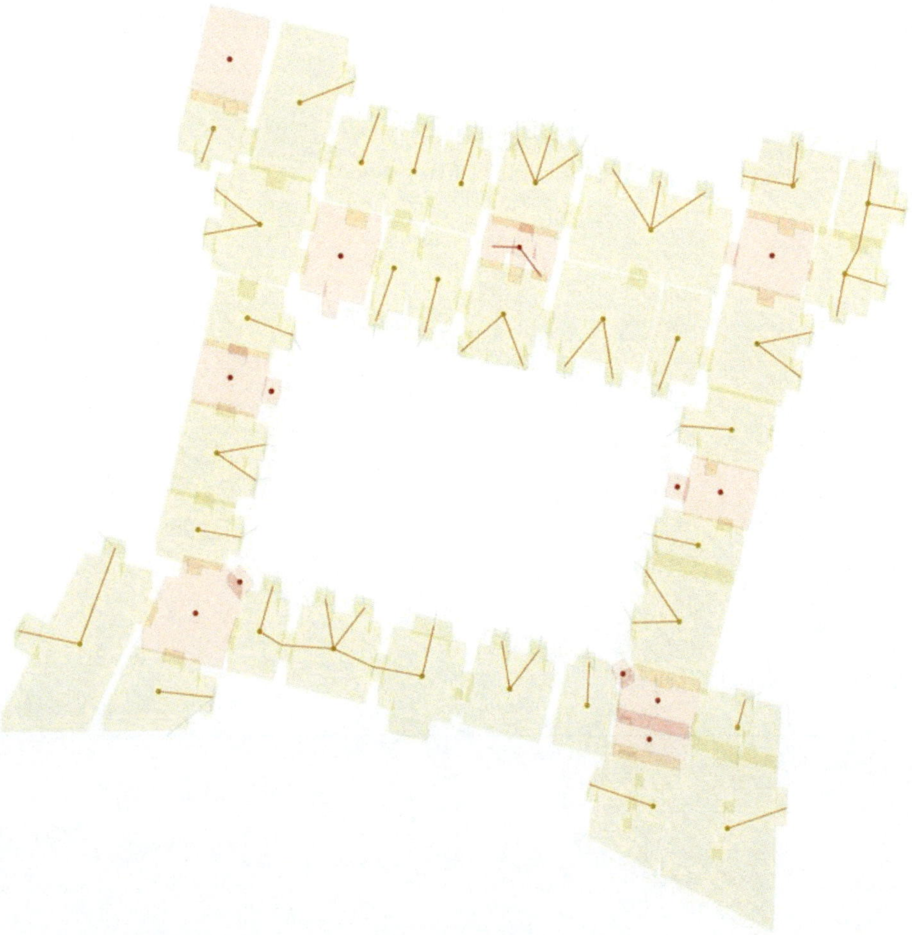

Fig. 4 "Natural lighting" view of the ground floor of Pidhirtsi Castle in 2018

The visualization shows that windows in the part of the guest suite and a closet on the ground floor do not pass over any light. Although there are niches for windows in those spaces, the openings are walled up. They are hidden behind the stone stairs that lead from the courtyard to the terrace above the casemates. This fact proves the researcher Kossak's theory that outer staircases were added later [11, p. 147].

The second floor nowadays has worse lighting access than before (54–80%). The author can claim this using the data on the first floor, which has not changed since its construction.

5 Results

The author of this paper used "Pedestrian Circulation" and "Natural Lighting" analysis to approach the evaluation of the changes to the design of the Pidhirtsi Castle; several interventions caused those changes during its existence.

As a result, it was found that although the building in Pidhirtsi had a defensive purpose, compared to other modern residential houses, it has well-lit spaces and a strong connection with the landscape. Including the ground level, 50–80% of the spaces have direct daylight access on each floor. Moreover, the pedestrian circulation view captured the enfilade planning, characteristic of Renaissance villas. A staircase on the second floor is the only space with indirect daylight access, a common feature for inner staircases in Palladio's villa designs of the same period, with their main stairs outside. These findings show that although the Pidhirtsi Castle had a defensive component, its residential quarters shared many similarities with well-designed villas; they were open to the landscape. Moreover, the studied case is comparable to well-daylit houses built in modern times.

6 Concluding Remarks and Future Work

There are different strategies to analyze the building. In the case of castles, people usually use the historical description of monuments or geometrical analysis of fortifications. The proposed comprehensive approach to Pidhirtsi Castle has enabled us to describe the complex building with some

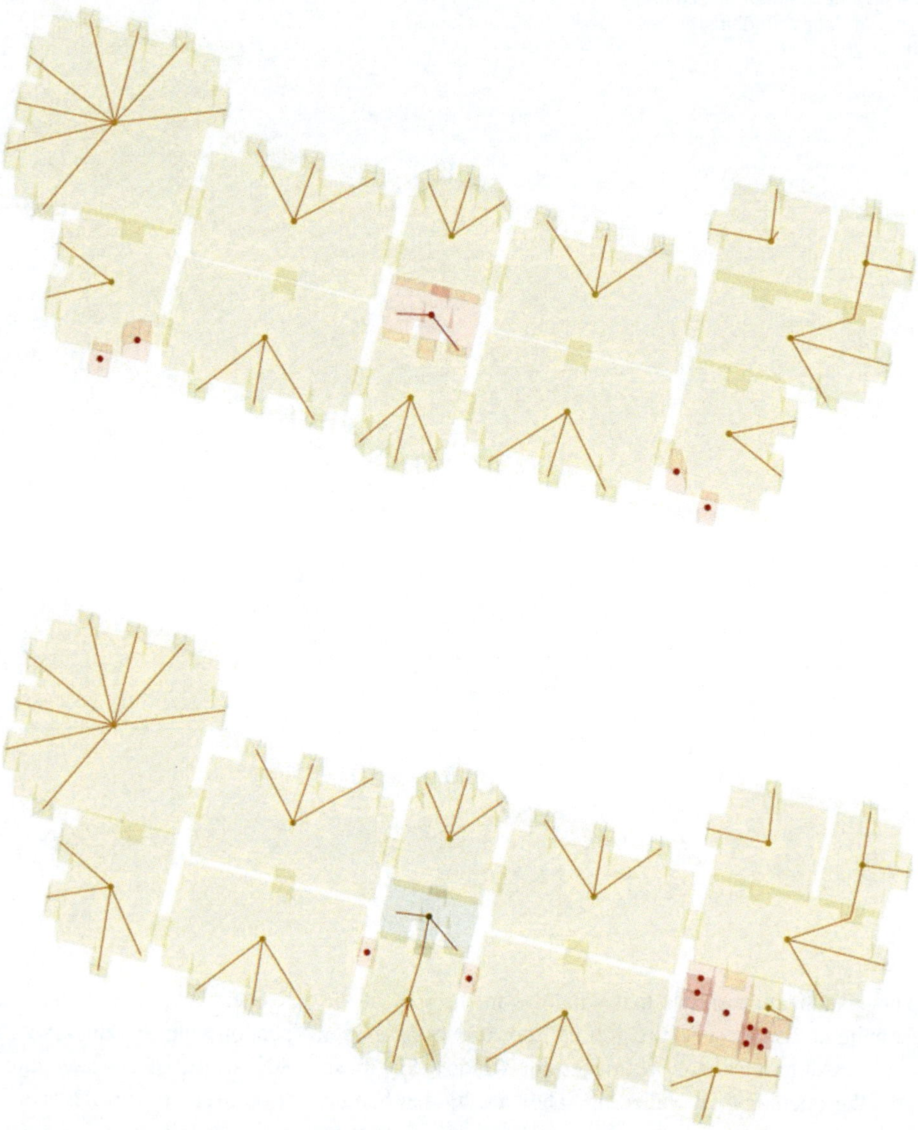

Fig. 5 "Natural lighting" view of the first and second floor of Pidhirtsi Castle in 2018

Table 2 Relationship between orientation and function on the first and second floor

Orientation	Number of spaces	Functions
Quadrilateral	2	d, L
Trilateral	4	v, bs
Bilateral	8	bs, c
Unilateral	10	g, bs, c

precision. The knowledge of complex structure from compositional analysis correlated with an understanding of building power using a pedestrian view, and the results from the lighting view provide a better picture of the building characteristics.

The resulting charts of purely mathematical graph analysis, when compared with the results of more modern buildings, can provide some advice regarding possible new functions for the historic building. Unfortunately, this approach is still limited as the current database includes residential houses and several bastion castles; as yet, not many other building types have been studied using spatial analysis. The quality of comparison is expected to improve as the number of reference projects in the database increases. However, some parts of it, including seven bastion castles, are already available on the Space Patterns Web site.

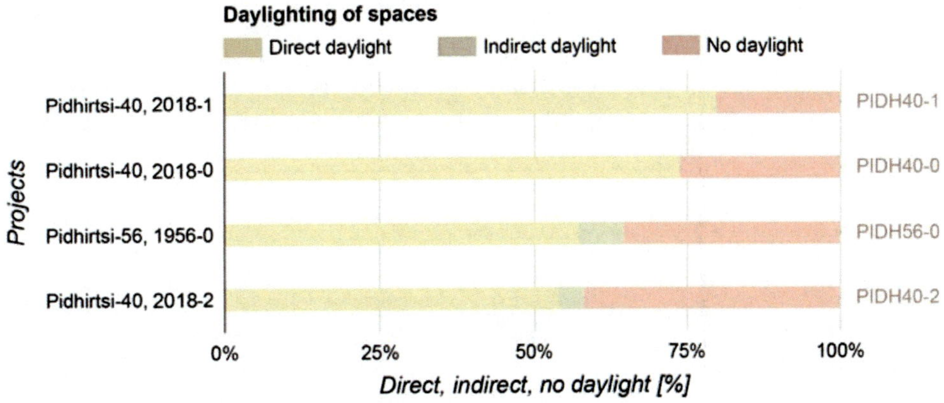

Fig. 6 Daylighting chart of spaces of Pidhirtsi Castle floor plans

Acknowledgements The Austrian Agency (OeAD-GmbH) supported this research. A big thanks to Prof. Oresta Remeshylo-Rybchynska (Lviv Polytechnic University) for scanned, scaled drawings of Pidhirtsi Castle. A special thanks to a developer of Space Patterns application, Prof. Georg Suter (TU Wien), whose comments greatly improved this research paper.

References

1. Hillier, B., Hanson, J., & Graham, H. (1987). Ideas are in things: An application of the space syntax method to discovering house genotypes. *Environment and Planning B: Planning and Design, 14* (4), 363–385.
2. Faulkner, P. (1963). Castle planning in the fourteenth century. *Archaeological Journal, 120*(1), 215–235.
3. Hillier, B., & Hanson, J. (1984). *The social logic of space*. Repr. Cambridge. Cambridge: Cambridge University Press.
4. Fairclough, G. (1992). Meaningful constructions—Spatial and functional analysis of medieval buildings. *Antiquity, 66*(251), 348–366.
5. Suter, G. (2015). Definition of views to generate, visualize, and evaluate multi-view space models of schematic building designs. In *22nd Workshop of the European Group for Intelligent Computing in Engineering* (pp. 1–9), Eindhoven, Netherlands.
6. Markus, T. A. (1993). *Buildings & power: Freedom and control in the origin of modern building types*. New York: Routledge.
7. Ostrowski, J. K., & Petrus, J. T. (2001). *Podhorce: Dzieje wnętrz pałacowych i galerii obrazów*. Kraków: Zamek Królewski na Wawelu.
8. Anonymous. (1659). *Krótka nauka budownicza dworów, pałaców, zamków podług nieba i zwyczaju polskiego* (1st ed.). Krakow: Wdowy y Dziedzicow Andzeia Piotrkowczyka.
9. Krzyczynski, W. (1894). *Zamek w Podhorcach*. Złoczów: Bookshop of Wilhelm Zukerkandl.
10. Palladio, A. (1997). *The four books on architecture*. Cambridge: MIT Press.
11. Коссак, М. О. (2006). Особливості фортифікації підгорецького замку. Вісник Національного Університету "Львіська політехніка" 2006(571), 144–147.

Olha Tikhonova holds a Ph.D. from the Faculdade de Arquitetura da Universidade de Lisboa. Her Ph.D. work concentrates on the typological and morphological studies of fortified palaces of the Renaissance period. Currently, she is interested in novel approaches to the analysis of architectural monuments and historical urban spaces. During her research fellowship in Austria, Olha carried out spatial analyses of some surviving bastion castles in the historical region of Galicia, using Space Patterns software developed by Prof. Georg Suter. She is also working on a project using machine learning techniques for the typological analysis of historical cities.

Public Space Use: An Approach to Livability in Urmia Bazaar, Iran

Momen FoadMarashi and Francisco Serdoura

Abstract

The public liveliness based on how friendly an area is to walking reaches an important role for the Iranian culture, particularly when the social interaction takes place in spaces like the 'Bazaar,' or in the 'Mosque.' For the Iranian urban society, the 'Bazaar' is a place where the public space works as a 'hub' where public life occurs as a result of social and cultural interaction. A new urban proposal developed by the municipality (2010) changed the Bazaar surroundings, building a large public space. The new square interacts with the ability to move around the central area of Urmia, and it can affect the liveliness in the area, penalizing the pedestrian mobility. The study developed, on one hand, compares the soft mobility based on proportion of urban pedestrian movement determined by the grid configuration itself 'natural movement' as Hillier et al. (Enviro Plann B Plann Des 20(1):29–66, 1993) [17] defined, which can be related with the syntactic measure 'integration' (Hillier and Hanson in The social logic of space. Cambridge University Press, Cambridge, 1984) [13]. On the other hand, the paper also studies how the streets are to be passed through on all shortest routes from all spaces to all other spaces (Hillier et al. in Archit Comportement/Archit Behav 3(3):233–250, 237, 1987) [16], in the central area of Urmia, and within a predetermined distance, to analyze how the new public space increases the public liveliness on the surroundings and within the covered space (Bazaar). The study allowed to conclude how the new urban structure can affect the pedestrian mobility through public space, and the changes made on the urban configuration can affect the public space liveliness.

M. FoadMarashi (✉) · F. Serdoura
Lisbon School of Architecture, CIAUD, Universidade de Lisboa, Lisbon, Portugal
e-mail: mfoadmarashi@fa.ulisboa.pt

F. Serdoura
e-mail: fs@fa.ulisboa.pt

Keywords

Bazaar • Public space • Urban form • Vitality • Urmia • Space syntax

1 Introduction

Public spaces, their use, human activities, vitality and livability in the built environment are not unfamiliar terms in urban studies. Moreover, various points of views about the built environment and its components such as form, public and private, social, cultural and political issues, the relation and interaction between these terms are some of the interesting debates among the urban theorists. Subsequently, dynamic of the public spaces as a result of different use concerning the reflection of various forms and contexts in built environment and interaction between different socio-cultural and economic components is one of the motivating subjects in urban studies.

This paper based on spatial study of the 'Bazaar,' as the most important and centric also covered social meeting site in Moorish/Islamic/Arab/Iranian culture in past, tries to illustrate how this space and its surroundings used in a period of time. For this purpose, Urmia Bazaar was selected and analyzed based on the space syntax theory and methodology.

Urmia is the capital of the West Azerbaijan, located in the northwest of Iran, and is one of the longstanding cities in the region which now has more than 750 thousand populations. The city located in the flat area close to Urmia Lake with the origin of 2000 years B.C. (see Fig. 1). The fabric of the city similar to most of the old cities is monocentric and compact but 20 centuries modernization movements affected the inner city and later growth and developments in city configuration.

S. Eloy et al. (eds.), *Formal Methods in Architecture*, Advances in Science, Technology & Innovation,
https://doi.org/10.1007/978-3-030-57509-0_21

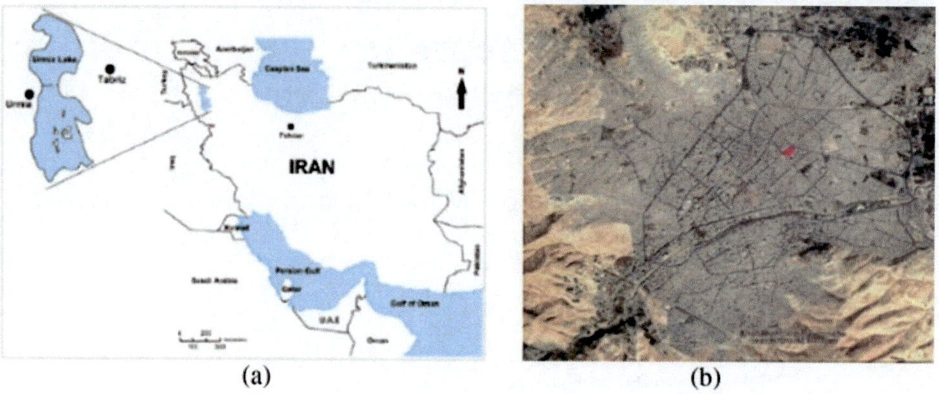

Fig. 1 Location of the Urmia (**a**) and its Bazaar (**b**)

2 Theoretical Framework

The use of public space in former urban fabrics is one of the inspiring issues for both academics and professionals and also policy and/or decision makers. Although many scholars such as Sennett [36], Arefi and Meyers [1], Kingwell and Turmel [22] and Parkinson [30], among others, debate the public space and its uses more based on politics and commerce and/or as an arena for social interaction for economic productions, other academics such as Tibbalds [40], Madanipour [27], Carmona et al. [4] and Gehl [7, 8] more focused on the term public space as a scene for social, cultural creativity and community interaction between humans and built environment. Also, there are few discussions about understanding the form of public space and its transformation through the time, considering what Kostof [23] emphasized about the production of the public spaces (design, planning and built) and their construction (use and appropriating). Since the meanings, use and all transformation of the public space created through its journey through the time. Also as Hillier and Hanson [13], Penn et al. [31], Hillier and Iida [14] and Madanipour et al. [26] discussed through reading of different public spaces, urban movements, human behaviors and their relations and interactions between public space transformation and people activities, it can be clearly understood that there is a strong relation between public spaces and human activities and how people are using them.

As another point of view, regarding the use of the pubic space, it can be addressed to livability, sociability and highly in accessible, comfort and safe spaces which attract more people to be present there; as Jacobs [19], Whyte [45], Lang [25] and Gehl [8] confirmed. Following this, the term public space vitality can be distinguished as successful spaces which as Montgomery [29, p. 97] said '[…] refers to the number of people in and around the street (pedestrian flows) across different times of the day and night, the uptake of facilities, the number of cultural events and celebrations over the year, the presence of an active street life, and generally the extent to which a place feels alive or lively.'

By definition, public spaces are places for social interaction or modeling of the social network [28]. Those places became an experience for people [10] which can help the development of the society as well as social and cultural identity of the community [1], it is clear that even though social activities resulted by different use of the space, but it is obvious that any changes in form planning and design of the built environment usually affect the public activities in urban spaces. Whyte [45] emphasized any small changes in built environment affect the people presence in the city. Also, Gehl [8] on his experience in redesigning of pedestrian street for Copenhagen reached that physical characteristic of the space can change the number of outdoor activities.

Trend of the people's outdoor activities in medieval cities considering the urban makeover shows that streets and squares were shaped to support residences' activities and passageways [8, 24]. Even the modernist looks to the urban fabric and society of Jacobs [19] which verified variety of activities in the streets and looks to the block size and how it affects the use and presence of the people in the public space by short blocks which will show more accessibility which meanwhile provides more activities and social interactions. Correspondingly, new urbanism design guidelines more highpoints the pedestrian streets and mixed-use neighborhoods to integrate more vitality in the cities [21].

Furthermore, spatial configuration of built environment shapes potential movement spaces that might collect or diverse people [5, 32]. Diversity of use can address the urban vitality, and public spaces can show strong interface between themselves as socio-cultural productions and built environment [6, 8, 19, 41, 43].

While these values are intuitive and easy to access or be employed or practiced, it seems that there is a lack of quantitative evaluation considering the term vitality, its factors and use of the space and often the existing approaches are qualitative and related to the limited scales [39, 47]. Following this, many of the morphological analysis of the cities studied historical neighborhoods as small-scale studies [20, 44, 47]. At the same time, different studies try to understand and make links between urban morphology, urban and public space transformation and urban design to modeling the cities in quantitative manner by calling new theories, methods and techniques [35, 46, 47]. As Batty [3], Hillier and Hanson [13] and Hillier et al. [17] confirmed new data environment by using spatial analysis and modeling methods and techniques as space syntax based on the '*natural movement*' theory or georeferenced modeling with GIS tools help to collect and gather more information considering the form of the cities, their functions their transformation and interactions between them and people and their outdoor activities.

Following this, it seems that space syntax success is due to at least two factors: '[…] a concise definition of space and a high degree of falsifiability and testability […]' [42, p. 278]. Space syntax theory and methodology examined the relationship of the human behavior and events and spatial configuration [17, 31, 37, 11]. Embedded in graph theory and the idea of urban morphology, space syntax theory describes and measures quantitatively the configurational properties of the urban space [13]. Space syntax is, also, a set of techniques for representing and analyzing the street networks of cities in such a way to bring to light underlying forms and structures which influence patterns of activity in space, most notably movement and land use [12]. This study can be counted as an attempt to this trend of quantitative analysis in urban design.

Subsequently, some syntactic measures such as permeability can be read how dynamic can a space be. For this purpose, permeability of the public space or generally the system as Hillier [15] emphasized can be classified into two main aspects: 'Potential to movement and potential through movement […]' [5, p. 4]. While, the first one as potential to movement, namely as integration, which shows how accessible the space is for the other spaces in the system that more related to the flow of visitors; the potential through movement, namely as choice, defines the betweenness that addresses how often a specific space can be chosen as a bypass or shortcut in an origin-destination travel and seems it is related to the residents' flow [5, 15].

In the same stream, since Kostof [23] stressed that the city spaces are social-cultural products which constructed in the coherence of the traditions and meanings of the social formulations, the integration of the space use, streets, joints and concatenation of various public spaces '[…] locked together in interdependent systems expressive of social conventions […]' [23, p. 64]. It can be debated that the vitality of the space and the use can be more affected, defined and explained by these aspects.

Bazaar as an important historical complex and social and economic meeting public space in most of the Middle Eastern countries as well in Iran has a great role in the urban fabric mostly in past, considering the cultural identity of the Iranian, regional and local understanding of the social, cultural and economic interactions between people and built environment. Indeed, although streets act as a connectors and transit passes as well, Bazaar with its subdivisions and enclosed neighborhood, and land uses such as mosques and public spaces was/is the heart of all of the social, commercial and cultural activities, and so its role in connecting different urban spaces as market passes and space for people's movement [34] was so bold; While, modern urbanism in twentieth century, which produced unconnected fabrics and neighborhoods comparing to past, affects this role and also the spatial vitality and use. This research tries to investigate the use of the space of the Bazaar and its surroundings as the most centric and covered public space of the city, in Urmia in two different periods with descriptive analytical study based on a comparison between the old urban layout with new plans which presents new public spaces (streets and squares) in Urmia inner city.

3 Case Study and Methodology

3.1 Case Study Area

Urmia Bazaar is one of the former parts of the inner city of Urmia and the oldest public space of the city connected to the surrounding neighborhoods and services not only for locals but in all city. Bazaar neighborhood is a big area in the city center which contains residential, markets and covered public spaces (which is famous as 'Bazaar' too) and usually considers cultural issues, variety of activities and land uses. Bazaar as biggest and oldest covered public space of the city is traditionally more coherent to the enclosed direct spaces and neighborhoods and not only acts as passes but it is the social, cultural and economic center of the city [9, 33, 34, 38]. This study more focuses on the Bazaar as covered public space and as the largest traditional mall in the city (see Fig. 2a). The latest layout of the Bazaar and passes is presented in Fig. 2b.

But the study area faced various changes related to the modernization and new planning during the last 90 years. The most recent intervention was new changes in the layout of the public spaces close to Bazaar.

The new existing configuration (which is almost the same for more than 8 years) is just including a new plaza and vast vacant area that is used as a large parking plot (see Fig. 3a).

These changes in the last ten years forced by demolishing some of the streets and buildings that built in 70–90 years ago and remained until 2010 when the new proposal by municipality tried to make a new adjustment for the neighborhood (see Fig. 3a, b) by transforming a street to a new square. The history of this area with that kind of streets and blocks belongs to the 1930s until mid-twentieth century in which some early modernization changed the fabric.

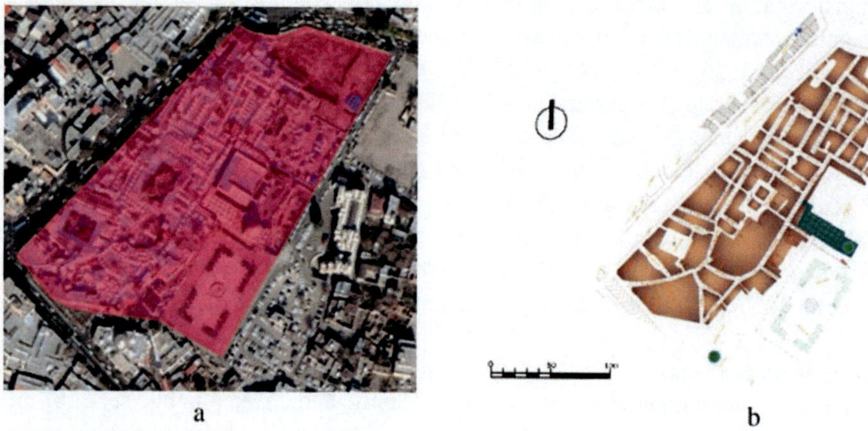

Fig. 2 Location of the Urmia Bazaar (**a**) and the last layout of the Bazaar (**b**)

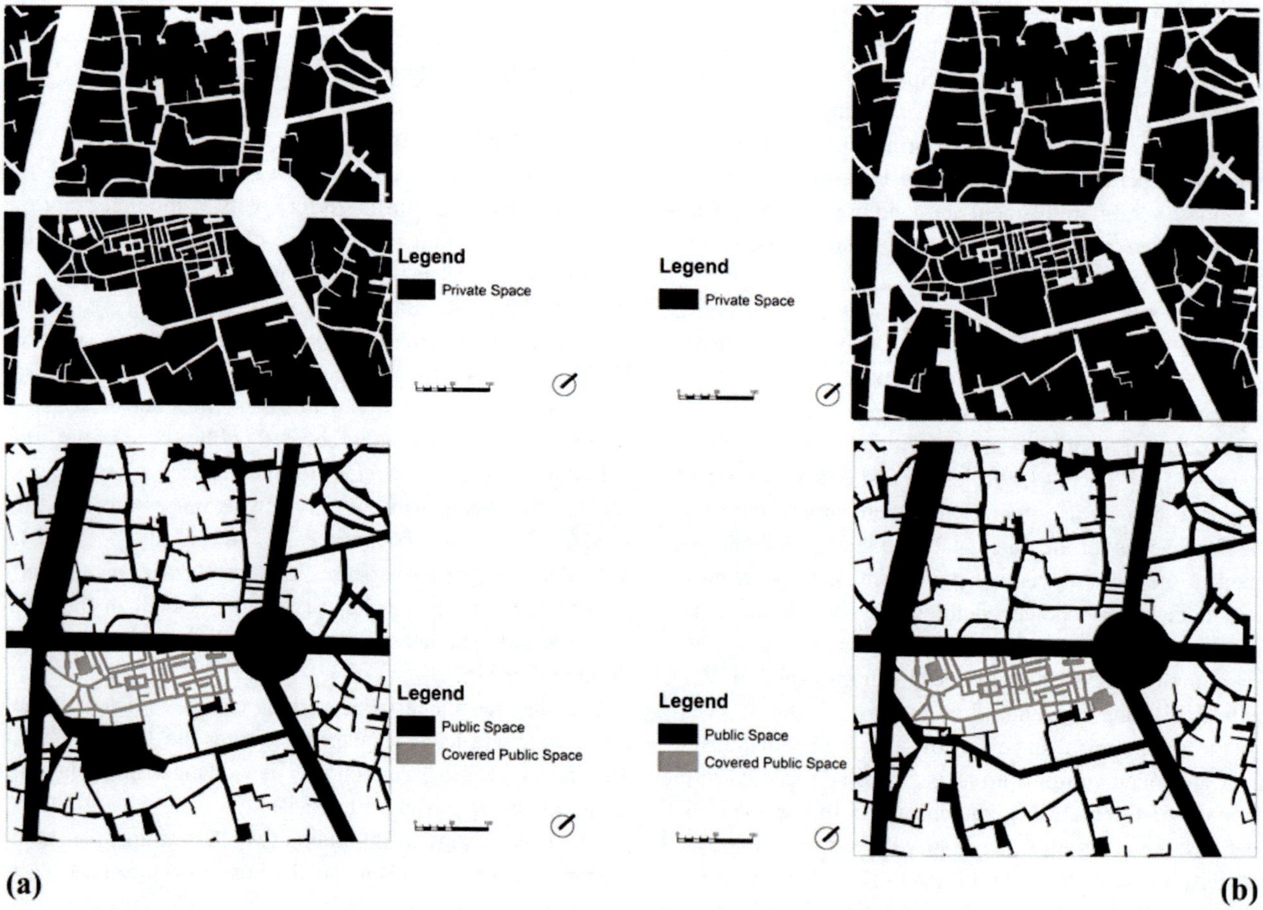

Fig. 3 Comparison of the changes in enclosed spaces to Urmia Bazaar in 2018 (**a**) and 2003 (**b**)

3.2 Methodology

The present study is a small-scale quantitative approach to understand the relationship between urban transformation (form) and vitality. For the spatial configuration analysis, a linear representation of the city based on two different periods maps, first, in 2003 (prepared by the municipality) and second, in 2017 (Tarh-o Amayesh latest detail plan) in georeferenced platform (ArcGIS and CAD), was built. These plans and representations were updated with the help of orthophoto maps (Google Earth) and supported by some field surveys in 2018. The final representations were processed in Depthmap X.

Considering the study scale and size of Bazaar blocks, in order to remove the impact of edge effects, the axial lines have been drowning for a buffer area of 2000 m surrounding the Bazaar block, for each period. The angular segment analysis (ASA) maps processed for variety of local metric radius: 400, 800, 1200 and 2400 m. Also, the meaning of 'integration' as a potential of explaining the pedestrian movement and/or as the normalized measure of distance from any space of origin which calculates how close the origin space is to all other spaces [13], in one hand; and the syntactic measure 'choice' as potential of local flow of inhabitants and as a measure of how likely a street segment is to be passed through all routes from all spaces to all other spaces considering the pedestrian walking distances [16], on the other hand; were employed.

Hillier et al. [18] discussed the normalized values of integration, and choice can better represent the pedestrian flows. Normalized angular integration (NAIN) aims to normalize angular total depth by comparing the system to the urban average [18], and normalized angular choice (NACH) aims to solve the paradox that segregated designs add more total (and average) choice to the system than integrated ones. It divides total choice by total depth for each segment in the system. This adjusts choice values according to the depth of each segment in the system, since the more segregated is, the more its choice value will be reduced by being divided by a higher total depth number. This would seem to have the effect of measuring choice in a cost-benefit way [13]. Indeed, normalized values of integration and choice as better representation of the urban form were processed, and so NACH and NAIN for each period and scale were calculated and examined.

Concerning built form analysis, syntactic measures of the Bazaar, for both period analysis (2003 and 2018) and also, the correlation between NACH and NAIN and the relation of the NACH and segment length in different metric radius were compared.

4 Analysis and Results

Urmia is basically monocentric city with very compact urban fabric in the inner city. Bazaar as one of the old fabrics in the city characterizes itself as one of the centric economic locations and connected space to surroundings. Axial lines as a representation of the city shows that the Bazaar public space is one of the most used through the time. Axial maps also addressed the compactness of the fabric mainly in the center and inner city with small blocks and dead ends (see Fig. 4).

Figure 4a and c shows how pedestrian mobility happened around the old Urmia Bazaar, before the municipality promoted for the surroundings a new urban design layout.

It is possible to verify that people tend to make the shortest journeys in their daily commutes, which means taking advantage of the covered public spaces of the Bazaar to move in the center of the city, both in the northeast/southwest and in the northwest/southeast directions. The same reading can be achieved (see Fig. 4b, d) for 2018, after the new intervention in the urban fabric by designing a new big square in southeast of the Bazaar.

In order to represent the pedestrian movement of the urban system, a segment map with angular segment analysis was produced (see Fig. 5).

Comparison between calculated global NACH that measures the probability of the people use of street segments for their travels and NAIN which measures how people try to pass through streets or part of them where other people also pass, in 2003 and 2018, shows significant increase for both measures (see Figs. 5 and 6).

Although the global value for NAIN and NACH in 2003 and 2018 is similar, the correlation between them announces an increase from $R^2 = 0.5784$ to $R^2 = 0.5982$ which defines more use of the space and stronger relation between the spaces people pass more and people's movement through public space in both periods.

Figure 6 shows correlation between the global NAIN and global NACH, for different radius, that can explain the increasing of the use of the public space.

For the 400, 800, 1200 and 2400 m radius, the correlation between NAIN and NACH has shown more coherence in the current time which in some cases shows around 10% increase. It expressed more people uses and more people movements in the public space. In 2018, the 2400 m radius, (see Fig. 6d, e) presents a $R^2 = 0.5884$ as the maximum correlation between NACH and NAIN comparing with the same analysis made for 2003, is the 1200 m radius for walking distance that seems to be the route distance more use for the pedestrian in the city core. Also, it can explain the new square built on the eastern side of Bazaar and radius of the people's movement in the space increased by the time. As obtained from Table 1, global NACH values generally show an increasing trend.

NAIN as a representation of the movement, people mobility and more presence of the people in the public space, from 2003 to 2018 shows an increase in all series (see Table 2) that confirms more walkability in the Bazaar and

Fig. 4 Axial map representation of Urmia announcing integration HH for 2000-m buffer area surround Bazaar in 2003 (**a**, **c**) and 2018 (**b**, **d**)

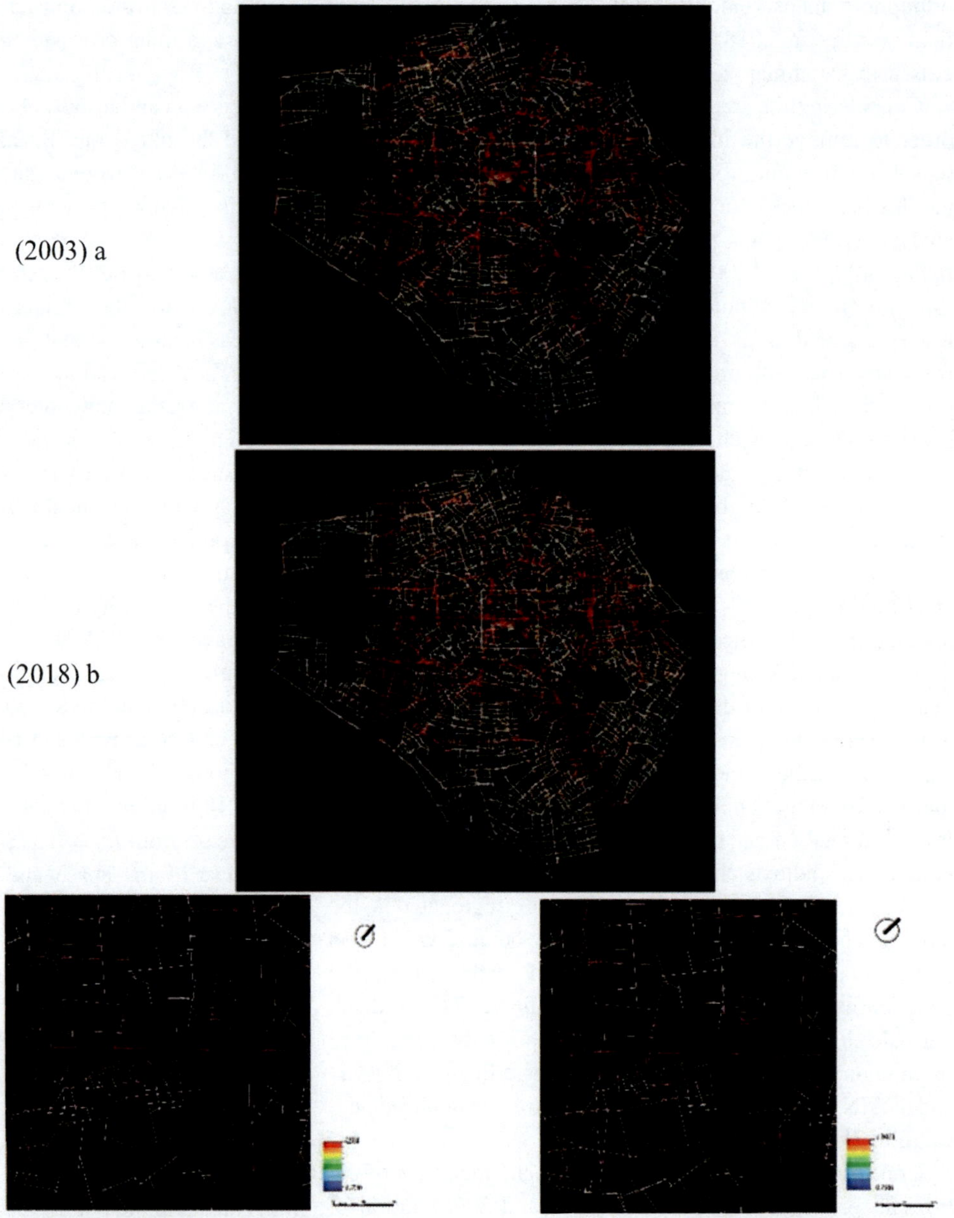

people's presence in the public space. Comparison between the NAIN values in all series shows a significant increase. It explains an improvement in people mobility as pedestrians in the public space in different radius.

Clearly, the study of the global NACH and NAIN in various radius in the fabric announced that people are more presenting in Bazaar and the main routes are more in used in different scales. Indeed, there is an improvement growth on people movement in their daily commutes throughout the system or within a predetermined distance (radius) from any part of the Urmia city core. Considering the mobility type in Bazaar, it represents more walkability, use or presence and activities in the new layout of the neighborhood. On the other hand, the study of the NACH and NAIN in different scales in 2003 and 2018 presents that R 1200 m can be distinguished as the most radius for the local people mobility in Bazaar and surrounding neighborhood in 2018.

5 Discussion

The axial map representation of the Urmia Bazaar announced a compact and more connected spaces in inner city which was cut by modernization planning. Following the literature presented before as Hillier [15], Hillier et al. [17], Hillier et al. [11], Penn et al. [31] and Donegan et al.

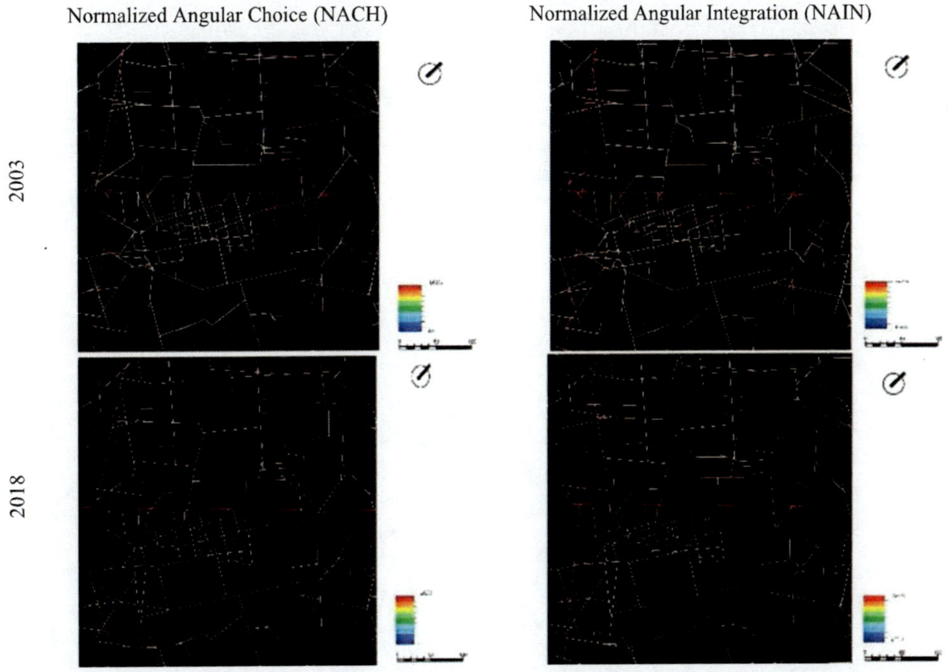

Fig. 5 Comparison of the normalized global angular choice (NACH) and normalized global angular integration (NAIN) for angular segment analysis (ASA) in Urmia Bazaar in 2003 and 2018

[5], synthetic measure values achieved by space syntax method for different times, Bazaar was always one of the more used and attracted spaces of the Urmia central area.

'Choice' (NACH) measures the probability that people use street segments for their travels on all shortest routes from all spaces in their daily commutes throughout the system or within a predetermined distance (radius) from any part of the Urmia city core; it confirms more mobility through Bazaar covered public space by people in their walking routes.

On the other hand, 'NAIN' (integration) shows how people, when choosing their most direct routes in their daily commutes, try to pass through streets or part of them where other people also pass, to push socialization activities. It can also be assumed that these walking routes through the Bazaar are made using the street segments where the main business activities are located.

Following Gehl' [8], Gehl and Svarre [6], Vaughan et al. [43] and Donegan et al. [5] idea which addressed the changes in the built environment configuration affects the people movement through the space, and this study, considering the integration value which can help to explain the movement in the public space as Hillier [15], Hillier et al. [18] expressed, announced that although new urban layout in the city center changed the coherence and connection of the Bazaar to its neighbor spaces in recent years, the new spaces seem to prepare more accessibility to the public space and improve people mobility through the public space. Also, it confirms that more spaces are selected by users in their walking routes which can define more social and economic activities in public space.

While Urmia Bazaar as an interconnected covered public space was one of the important accessible spaces in former of social meeting and people movements and one of the pedestrian walking paths in the city center, it expects that the current urban layout as an imposing to the last modernization changes of the fabric can affect the accessibility of this central area. Following the meaning of 'construction of the form of the city' as expressed by Kostof [23] and how Bazaar formed and interacts as an important socio-economic public space [9, 33, 34, 38] also counting the results of people mobility through and in the spaces in this area in different radius, it expects that the Bazaar can have its role in attracting more people to use and present in the inner city.

6 Conclusion

This study distinguished that the highly used spaces of the Bazaar are more used locally than globally; indeed, the vitality of such space could be achieved more with local residents. Also, it confirms the reflections of the more coherence and correlation between local users and chance of the spaces to be chosen and it increased again by new layout, but even in this case, the maximum correlation shows more local interaction and confirms the more use of the space.

Spatial characteristic of the Bazaar as a traditional public mall of the city was the same even after new urban layout imposed by municipality as designing a new big square in the surrounding neighborhood in 2010. This study shows an increasing growth on people movement in their daily travels throughout the spaces or within a predetermined distance

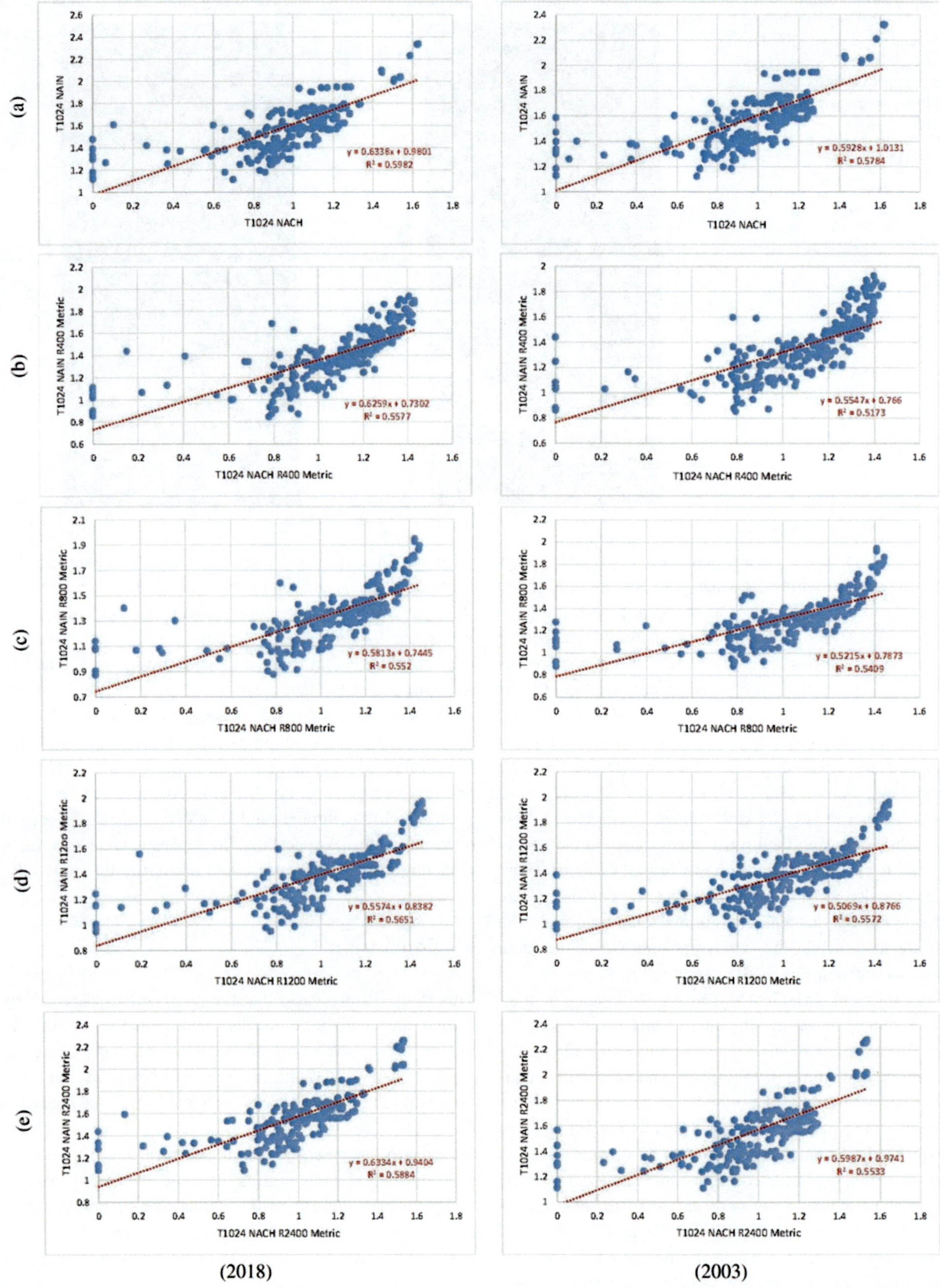

Fig. 6 Correlation between normalized angular integration HH (NAIN) and normalized angular choice (NACH) in various *R n*, 400, 800, 1200 and 2400 m in 2003 and 2018

Table 1 Normalized angular choice (NACH) values for angular segment analysis (ASA) in various *R n*, 400, 800, 1200 and 2400 m radius

		NACH	NACH400	NACH800	NACH1200	NACH2400
2018	Average	1.0204	1.0658	1.0519	1.0370	1.0269
	Max	1.6231	1.4375	1.4434	1.4630	1.5355
	Min	0.0000	0.0000	0.0000	0.0000	0.0000
2003	Average	1.0365	1.0725	1.0614	1.0472	1.0388
	Max	1.6290	1.4317	1.4453	1.4635	1.5361
	Min	0.0000	0.0000	0.0000	0.0000	0.0000

Table 2 Normalized angular integration (NAIN) values for angular segment analyses (ASA) in various *R n*, 400, 800, 1200 and 2400 m radius

		NAIN	NAIN400	NAIN800	NAIN1200	NAIN2400
2018	Average	1.6180	1.3572	1.3358	1.4022	1.5889
	Max	2.3238	1.9269	1.9453	1.9661	2.2783
	Min	1.1264	0.8537	0.8818	0.9617	1.1098
2003	Average	1.6371	1.4015	1.3615	1.4219	1.5984
	Max	2.3372	1.9352	1.9480	1.9736	2.2587
	Min	1.1148	0.8469	0.8747	0.9478	1.0872

(radius) from any part of the Urmia city core, especially in Bazaar paths as the centric social, cultural and economic routs of the city center.

Subsequently, more use signifies more participation in the space, following the literature presented, and addresses increasing livability and vitality of the Bazaar as a social and economic urban space. On the other word, Urmia Bazaar as a cultural meeting space in the former fabric with small blocks (which serves different land uses) supports more people mobility through its paths and main routes for more people presence in the public space not only for local neighborhoods, but also for the people from other city districts. And so, the new urban design in the city center expected to increase more people presence in Bazaar and its neighborhood spaces and improve dynamic and vitality of the public space.

Acknowledgements To the Portuguese Foundation of Sciences and Technology (FCT) for the Ph.D. scholarship of the first author.To Mrs. Aida Sedighinanloo and Mr. Edison Khenanisho for their accompany on geo-database and base map preparation.

References

1. Arefi, M., & Meyers, W. (2003). What is public about public space. The case of Visakhapatnam, India. *Cities, 20*(5), 331–339.
2. Baran, P. K., Smith, W. R., & Toker, U. (2006). Conflict between space and crime: Exploring the relationship between spatial configuration and crime location. Paper presented at the 37th Annual Conference of the Environmental Design Research Association (EDRA), Atlanta.
3. Batty, M. (2013). Big data, smart cities and city planning. *Dialogues in Human Geography, 3*(3), 274–279.
4. Carmona, M., Heath, T., Oc, T., & Tiesdell, S. (2003). *Public places BE, urban spaces: The dimensions of urban design*. Oxford: Architectural Press.
5. Donegan, L., Silveira, J. A. R., & Silva, G. J. A. (2019). Under and over: Location, uses and discontinuities in a centrally located neighbourhood in João Pessoa city reflecting current urban planning effects. *urbe. Revista Brasileira de Gestão Urbana, 11*, e20180123. Availabel from https://doi.org/10.1590/2175-3369.011.e20180123.
6. Gehl, J., & Svarre, B. (2013). *How to study public life* (2 ed.). Island Press/Center for Resource Economics.
7. Gehl, J. (2002). *Public space and public life city of Adelaide*. Adelaide, Australia: City Council of Adelaide. Retrieved from https://www.cityofadelaide.com.au/assets/documents/ACC_edited_Adelaide_PSPL2011_01Introduction.pdf.
8. Gehl, J. (2011). *Life between buildings: Using public space* (6th ed.). Washington, DC: Island Press.
9. Habibi, S. M. (2000). *Az Shar ta Shahr. A historical analysis of the concept of the city and its physical appearance*. Tehran: University of Tehran Press.
10. Hajer, M., & Arnold, R. (2001). *In search of new public domain*. Rotterdam: NAI Publishers.
11. Hillier, B. (1996). *Space is the machine*. Cambridge: Cambridge University Press
12. Hillier, B., & Sahbaz, O. (2009). An evidence based approach to crime and urban design. Bartlett School of Graduate Studies, University College London.
13. Hillier, B., & Hanson, J. (1984). *The social logic of space*. Cambridge: Cambridge University Press.
14. Hillier, B., & Iida, S. (2005). Network and psychological effects in urban movement. In A. G. Cohn & D. M. Mark (Eds.), *Proceedings of Spatial Information Theory: International Conference, COSIT 2005* (pp. 475–490), Ellicottville, September 14–18, 2005. Berlin: Springer.
15. Hillier, B. (2009). Spatial sustainability in cities. Organic patterns and sustainable forms. In *7th International Space Syntax Symposium*.
16. Hillier, B., Burdett, R., Peponis, J., & Penn, A. (1987). Creating life: Or, does architecture determine anything? *Architecture et Comportement/Architecture and Behaviour, 3*(3), 233–250, 237.

17. Hillier, B., Penn, A., Hanson, J., Grajewski, T., & Xu, J. (1993). Natural movement: Or, configuration and attraction in urban pedestrian movement. *Environment and Planning B: Planning and Design, 20*(1), 29–66.
18. Hillier, B., Yang, T., & Turner, A. (2012). Normalising least angle choice in depthmap and how it opens up new perspectives on the global and local analysis of city space. *Journal of Space Syntax, 3*, 155–193.
19. Jacobs, J. (1961). *The death and life of great American cities*. New York, USA: Random House.
20. Jin, W. (1993). The historical development of Chinese urban morphology. *Planning Perspective, 8*(1), 20–52.
21. Katz, P., Scully, V. J., & Bressi, T. W. (1994). *The new urbanism: Toward an architecture of community*. New York, NY, USA: McGraw-Hill.
22. Kingwell, M., & Turmel, P. (Eds.). (2009). *"Rites of Ways". The politics and poetics of public space*. Waterloo, Ontario, Canada: Wilfrid Laurier University Press.
23. Kostof, S. (1991). *The city shaped: Urban patterns and meanings throughout history*. London: Thames and Hudson Ltd.
24. Kostof, S. (1999). *City assembled: The elements of urban form through history*. Bulfinch Press (1750).
25. Lang, J. (2005). *Urban design: A typology of procedures and products illustrated with over 50 case studies*. Oxford, UK: Architectural Press.
26. Madanipour, A., Kenierbein, S., & Degros, A. (2014). *Public space and the challenges of urban transformation in Europe*. New York: Routledge.
27. Madanipour, A. (2003). *Public and private spaces of the city*. London, England: Routledge.
28. McMillan, D. W., & Chavis, D. M. (1986). "Sense of community". A definition and theory. *Journal of Community Psychology, 14*, 6–23.
29. Montgomery, J. (1998). Making a city: Urbanity, vitality and urban design. *Journal of Urban Design, 3*(1), 93–116. Available from https://doi.org/10.1080/13574809808724418.
30. Parkinson, J. R. (2012). *"Democracy and Public Space". The physical sites of democratic performance*. New York, USA: Oxford University Press.
31. Penn, A., Hillier, B., Banister, D., & Xu, J. (1998). Configurational modeling of urban movement networks. *Environment and Planning B: Planning and Design, 25*, 59–84.
32. Peponis, J. (1989). Space, culture and urban design in late urbanism and after. *Ekistics, 56*(334–335), 93–108.
33. Pourjafar, M. R., & Pourjafar, A. (2013). Urban space typology. *City-Life-Beauty Journal*, Special Issue: Public Space Tehran Beautification Organization, 212–220.
34. Pourjafar, M. R., & Pourjafar, A. (2013). Iranian—Islamic semipublic covered urban spaces in comparison to French passages. *Armanshahr Architecture & Urban Planning, 13*, 169–182.
35. Samuels, I. (2013). ISUF task force on research and practice in urban morphology. An interim report. *Urban Morphology, 17*(1), 40–43.
36. Sennett, R. (1992). *The fall of public man*. New York: W. W. Norton & Company.
37. Shu, S. (1999). Housing layout and crime vulnerability. In F. R. B. Holanda, L. Amorim, & F. Dufaux (Eds.), *2nd International Space Syntax Symposium* (pp. 1:25.1–25.12), Brasilia.
38. Soltanzadeh, H. (2011). *A brief history of the city and urbanization in Iran (ancient Era to 1976)*. Tehran, Iran: Chahar Tagh Publication.
39. Sung, H., & Lee, S. (2015). Residential built environment and walking activity. Empirical evidence of Jane Jacob's urban vitality. *Transportation Research Part D Transport and Environment, 41*, 318–329.
40. Tibbalds, F. (2001). *Making people-friendly towns. Improving the public environment in towns and cities*. London, England: Spon Press.
41. Van Nes, A., & López, M. J. (2007). Micro scale spatial relationships in urban studies. The relationship between private and public space and its impact on street life. In *6th International Space Syntax Symposium* (pp. 023). Istanbul: ITU.
42. Van Nes, A. (2014). Indicating street vitality in excavated towns. Spatial configurative analyses applied to Pompeii, Spatial analysis and social spaces. De Gruyter. Available from http://resolver.tudelft.nl/uuid:072645d0-b6ff-4ffb-b755-c0f4145d5bfd.
43. Vaughan, L. S., Dhanani, A., & Griffiths, S. (2013). Beyond the suburban high street cliché: A study of adaptation to change in London's street network: 1880–2013. *Journal of Space Syntax, 4* (2), 221–241.
44. Whitehand, J., & Gu, K. (2006). Research on Chinese urban form. Retrospect and prospect. *Progress in Human Geography, 30*(3), 337–355.
45. Whyte, W. H. (1980). *The social life of small urban spaces*. New York, NY: Project for Public Spaces (PPS).
46. Ye, Y., & Van Nes, A. (2014). Quantitative tools in urban morphology. Combining space syntax, spacematrix and mixed-use index in a GIS framework. *Urban Morphology, 18*, 97–118.
47. Ye, Y., Li, D., & Liu, X. (2018). How block density and typology affect urban vitality. An exploratory analysis in Shenzhen, China. *Urban Geography, 39*(4), 631–652. Available from https://doi.org/10.1080/02723638.2017.1381536.

Momen FoadMarashi For 15 years has been working in practical fields of urbanism in master and detailed plans, regeneration and truism planning. He was an academic member of the Elm-o-Fan College of Higher Education for 10 years; and 12 years of experiences as special tour guide on the field of architecture in Iran. He won his MSc from Isfahan Art University as top Student on Urban and Regional Planning. Now, he is Ph.D. candidate in Urbanism on the field of Urban Design in Lisbon School of Architecture, University of Lisbon, on the theme of public spaces, their use and vitality.

Francisco Serdoura has been working on public space livability and urban design, with special interest on sustainable mobility; public space safety; space and societal analysis; and tourism. Coordinates the Group of Urban and Environmental Dynamics, in the Research Centre for Architecture, Urbanism and Design. He won his Ph.D. in Urban Planning with emphasis on patterns of public life from the Instituto Superior Técnico. He holds a M. Sc. in Urban and Regional Planning from Technical University of Lisbon. He is Assistant Professor, Department of Urbanism, Lisbon School of Architecture, University of Lisbon, where teaches in undergraduate and postgraduate courses.

Design Tools and Technics

Spectral Clustering and Integration: The Inner Dynamics of Computational Geometry and Spatial Morphology

Tasos Varoudis and Alan Penn

Abstract

Deviating from common evaluation strategies of spatial networks that are realised through numerical comparison of single floating-point numbers such as global and local space syntax measures (centralities, connectivity, etc.) we aim to present a new computational methodology for creating detailed topo-geometric encodings of spaces that encapsulate some of the fundamental ideas about spatial morphology by Hillier (Space is the Machine: A Configurational Theory of Architecture, London, UK, Space Syntax, 2007 [1]). In most cases, space syntax measures try to capture a particular quality of the space for comparison but they lose much of the detail of the spatial topo-geometry and morphology by mainly aggregating graph path traversals and not retaining any other information. This research explores the use of weighted graph spectra, in a composite form, for the purpose of characterising the spatial structure as a whole. The new methodology focuses on the three primary space syntax graph modelling concepts, 'angular', 'metric' and 'topological', from the point of view of the resulting spatial geometries and develops new computational innovations in order to map spatial penetration of local neighbourhood spectra in different scales, dimensions and built environment densities in a continues way. The result is a new composite vector of high dimensionality that can be easily measured against others for detailed comparison. The proposed methodology is then demonstrated with the complete road-network dataset of Great Britain. The main dataset together with subsets is then used in a series of unsupervised machine learning analyses, including clustering and a form of Euclidian 'spectral integration'.

T. Varoudis (✉) · A. Penn
Bartlett School of Architecture, University College London (UCL), London, UK
e-mail: t.varoudis@ucl.ac.uk

A. Penn
e-mail: a.penn@ucl.ac.uk

Keywords

Graph analysis • Graph spectra • Unsupervised clustering • Space syntax • Spectral integration

1 Introduction

In the space syntax field of research the evaluation of spatial networks is done using numerical comparison of streets or space with scalar measures such as graph centralities or connectivity. Through various graph types , including, axial and segment graphs [2, 3] there have been a number of effective representations of built environment that try to capture the particular quality of the space. These techniques though fail to retain much of the detail of the spatial topo-geometry and morphology by mainly aggregating graph path traversal measures. This is particularly discouraging if the aim is to employ machine learning techniques in your analysis of spatial dynamics. In space syntax practice, the overall understanding and classification of space based in a combination of secondary data, with spatial analysis measures representing only a very small number of data features in the analytical model.

This research explores the use of a new form of graph spectra [4, 5] in order to characterise the spatial structure as a whole. The new methodology presented here creates a unique composite spectra following on the three primary space syntax graph modelling concepts, 'angular', 'metric' and 'topological'. With this new computational model we aim to map spatial penetration of local neighbourhood spectra in different scales and dimensions focusing purely on geometric features. The result is a vector of high dimensionality that can be easily measured against others for detailed comparison. This feature space is then used in a series of unsupervised learning analyses building the first GB wide spatial model of this kind.

S. Eloy et al. (eds.), *Formal Methods in Architecture*, Advances in Science, Technology & Innovation,
https://doi.org/10.1007/978-3-030-57509-0_22

The project uniquely achieves full scale analyses of Great Britain where large metropolitan and regional areas' detailed characteristics of the special morphology give intriguing results. Ultimately the focus of this stream of work is to develop novel approaches for the next generation of space syntax driven machine learning systems.

2 From Space Syntax to Multi-dimensional Datascapes

This project represents a set of steps in the direction of creating computational and research ground for space syntax theory and methodology towards machine intelligence driver models. While on first read the methodological and theoretical approach seems completely different, the aim is to reinvent the foundational building block of geometry and spatial configuration while linking to the ideas about built environment morphology pioneered in space syntax by Bill Hillier.

Space syntax [1] is a set of techniques for analysing spatial layouts and human activity patterns in buildings and urban areas. Some of the basic ideas proposed in the literature is that space is not a background to human activity, but is intrinsic to it and that space is first and foremost configurational [6]. That is to say, what happens in any individual space is fundamentally influenced by the relationships between that space and the network of spaces to which it is connected. Elements of space are considered geometrically derived like an axial line, segment or an isovist polygon area. In this project we look at GB wide road network models thus the concept of space syntax's street segment lines is used [7]. Relationships between street elements result from their configuration. These relationships can be objectively analysed using various graph-based measures [8]—open source depthmapX [7], were the graph represents the configurational relations between those spatial elements (example in Fig. 1).

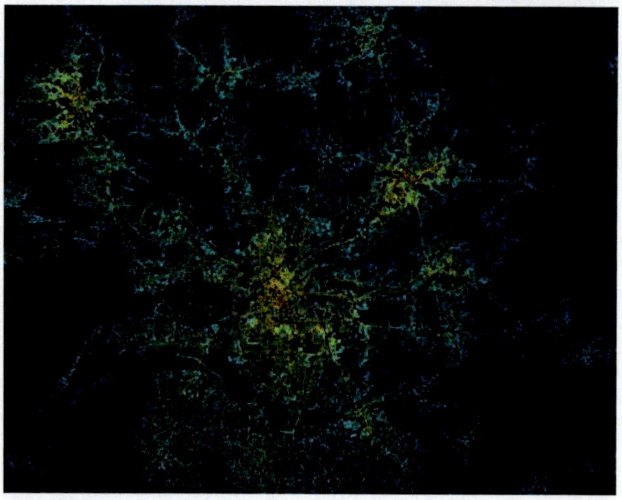

Fig. 1 Greater Manchester road network—space syntax analysis of integration or closeness centrality in graph sciences. Red is high, blue colour is low values

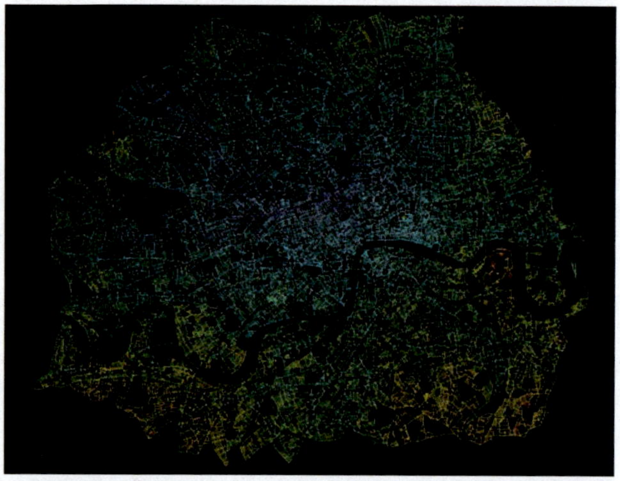

Fig. 2 Angular distance from The Bartlett School of Architecture to every other location in London. Simplified map, 50 k segments. Blue represents low distance values, red high

Moreover, space syntax heavily relays on three basic ways of measuring 'distance' in the graph based relational models. In measuring spatial relations the concept of depth is widely used in the literature, meaning the distance between any pair of spatial elements [9]. The three definitions of distance used are: topological distance, the number of turns/steps from one space to another, angular distance, the angular change from one space to another; and the metric distance, the Euclidean distance in metres from one space to another. Different spatial patterns will be generated by assessing the three types of distance from a single street segment.

Figures 2, 3 and 4 depict the different spatial structures when the street network is assessed based on: angular distance (Fig. 2), metric distance (Fig. 3) and topological distance (Fig. 4) from The Bartlett School of Architecture to every other location in London. Blue represents low distance values, red high. We see how certain morphological complexities force areas to be more segregated in different distance definitions (London's street network is simplified for this visualisation; total of 50,000 streets).

In the following sections we will decompose our model and present how we can incorporate all three ideas of 'depth' creating a highly detailed multi-dimensional dataset that can provide insights of spatial dynamics beyond what traditional space syntax models could achieve.

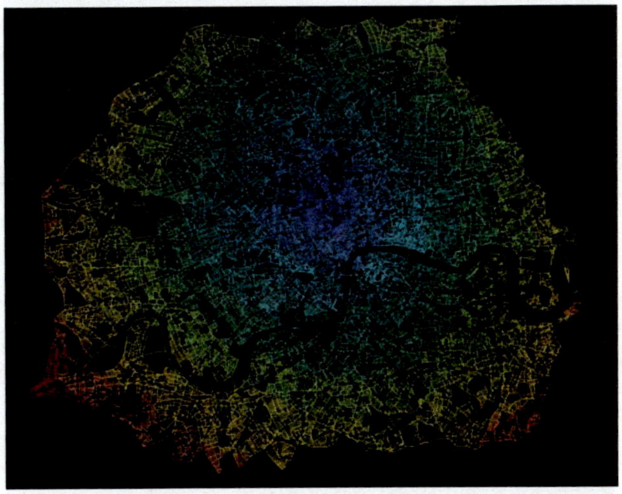

Fig. 3 Metric distance from The Bartlett School of Architecture to every other location in London. Simplified map, 50 k segments. Blue represents low distance values, red high

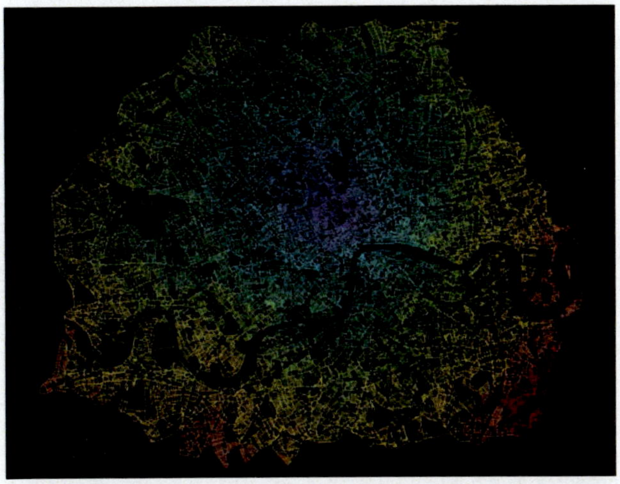

Fig. 4 Topological distance from The Bartlett School of Architecture to every other location in London. Simplified map, 50 k segments. Blue represents low distance values, red high

2.1 Local to Global—Neighbourhood Spatial Penetration

The project started with the dataset design and the aim to create comparative knowledge between the local and global spatial morphology of Great Britain (GB) in a continues feature space, from urban to rural as one thing. Our GB dataset consisted of 4 million road segments. Basic cleaning and pre-processing were applied like de-duplication of multiple lanes and non-destructive simplifications. Each straight road or road segment is represented by a single linear element. The completed road network is then read as a graph, with each node corresponding one of the segment lines and connections to each road line it intersects, similar to the space syntax literature [1, 7].

From the dataset we extracted one local subgraph per street segment of a set size n (n = 200 in this publication), which ensures that any computational analysis is always made between graphs of equal size [4, 5]. Starting from street segment *V_origin*, defined as the centre of the subgraph, we use the angular graph distance measure depicted in Fig. 2 [7, 9] to calculate the angular distances to all other segments in the graph, where D(origin, destination) is the angular distance from segment *V_origin* to *V_destination*. These are then sorted in descending order such that $D(o, d1) < D(o, d2) < \ldots < D(o, \ldots)$, and a new subgraph is composed of the closest n (i.e. 200 here) segments.

This creates local neighbourhoods of street segments that represent the closest 'angular shortest path' graph, a method not considered in past studies and that traditionally has found strong correspondence in explaining aggregate movement and morphological understanding of built environment by many space syntax studies [9]. Additionally, it creates a scale-agnostic local penetration understanding of the built environment were natural break or significant dissimilarities can be easily identified.

Throughout the project we investigated different values of neighbourhood sizes, n = 50, 200, 500 and others, but here we present the findings from local sub areas of size n = 200 (i.e. the 200 angularly closest streets around a central segment).

2.2 Graph Spectra

The graph spectrum of a closest 'angular shortest path' neighbourhood extracted with the method described above consists of the ordered set of eigenvalues of its connectivity matrix. The standard method for deriving the spectral representation begins with an unweighted graph given by a set of nodes V and edges/links E, as presented by Luo et al. [10], Robles-Kelly and Hancock [4], and Hanna [11], and can be obtained from an adjacency matrix representing all connections between nodes based on the following algorithm:

$$A(i,j) = \begin{bmatrix} 1 \text{ if } (i,j) \in \text{E} \\ \text{or} \\ 0 \text{ otherwise} \end{bmatrix} \quad (1)$$

Eigenvalues λ and eigenvectors φ for A are derived: $A = \Phi\Lambda\Phi^{T}$ where the matrix $\Phi = (\varphi 1|\varphi 2|\ldots|\varphi|\text{V}|)$ contains the eigenvectors as columns and the matrix $\Lambda = \text{diag}(\lambda 1, \lambda 2, \ldots, \lambda|\text{V}|)$ contains the eigenvalues as diagonal elements. The spectrum is defined as the set of ordered eigenvalues: $S = (\lambda^{1}, \lambda^{2}, \ldots, \lambda^{\upsilon})$. At this point, eigenvalues are sorted to form a feature vector by actual value such that $\lambda 1 > \lambda 2 > \ldots > \lambda |V|$ [5].

Graph spectra extracted from topological adjacency matrices have been widely demonstrated in the literature [4, 10–13] to reliably capture differences between the overall topo-geometric structure of compared graphs and they are almost unique to a particular spatial graph, especially of the size presented here ($n = 200$).

The first clear distinction of the work presented here and all past studies is that the local sub-graph and subsequent methodology is based on a closest 'angular shortest path' neighbourhood and not a closest topological neighbourhood. Linking back to core space syntax research, we feel that by using this method of neighbourhood extraction to explore micro similarities or differences we created a dataset that more closely encapsulates the embodied understanding of someone's immediate environment that can be captured in higher dimensionality data.

2.3 Composite 'Angular-Scalar' Graph Spectra

While past studies focus on cities in isolation or small urban areas within a single city and only use the angular deviation between segments to describe the geometry of a topological graph [5, 11] our aim is to create a GB-wide cooperative analysis of urban and rural areas together in a continuous space. We see the evaluation of topo-geometric transition between different urban densities and rural towns as the most significant step in understand spatial integration and segregation as a function of natural 'breaks' between areas. This seamless continuity also avoids some parallels to the space syntax 'problem' of edge-effect, when urban networks are artificially cropped from the rest of the dataset [1, 14, 15].

This unified spatial dataset means that graph spectra extracted from a small part of a town in Wales, using past methods, could have similar topological or angular spectra to a graph extracted from a rural area in Cornwall, with the only difference being their relative scale.

To overcome this problem, we propose the use of a new composite 'scale-angular' geometric spectra that can extract high dimensionality data based only on the pure geometric characteristics of the GB's road network. Seeking to encode the angular deviation between street segments and their relative scale create a complete set of geometric data not consigned in the past. Additionally, the fact that the scale is encoded in the spectral model enhances the 'local' to 'global' comparative model significantly.

Angular and scale weighting. The base process of creating the spectra of a graph described earlier captures the topology only, as the geometry is not represented in the adjacency (connectivity) matrix of a graph.

To create the composite spectral vector we start by extracting the angle weighted spectra with a similar approach to Hanna [11]. The 'continuity' of segments in a straight line is weighted highest (1.0) while values are decreasing with greater angles between the adjacent segments. In all data fields of the binary matrix (1) that represented a connection with a value of 1 and all unconnected nodes with 0, an updated value for connected nodes can be extracted from feeding the adjacent unit vectors of the segments in:

$$A(i,j) = \begin{bmatrix} a\cos(i_{\text{vect}} \cdot i_{\text{vect}})/\pi \, \text{if} \, (i,j) \in E \\ \text{or} \\ 0 \text{ otherwise} \end{bmatrix}$$

which results in a maximum $A(i, j) = 1$ when i and j are part of a single, straight line, and decreasing to zero as the turn angle becomes more acute [11].

We do the same process again for the 'scale' weighted adjacency matrix but this time we substitute the binary value of 1 in (1) with a normalised value of the average length of the two connected segments $(L1 + L2)/2.0$, which also represents the standard way of calculating the travelled distance between segments in depthmapX and in essence throughout space syntax research.

Composite spectra. Finally, we concatenate the two ordered spectra, first the spectra extracted from the angular weighted adjacency matrix followed by the one extracted from the scale weighted matrix. The single composite scale-angular geometric spectra of a local neighbourhood of $n = 200$ segments is a vector of 400 values (shape = (1 400)). This vector represents a single data point in a 400-dimensional space projection.

Starting from the GB data set of 4 million unique road segments (data elements) we created a 'table-looking' dataset with 4 million rows and 400 column, 1.6Billion floating point numbers ($\sim$9GBytes).

3 Spectral Integration: Topo-geometric Similarities and Characteristics

Generating the angle-scale composite spectra has given us an extremely detailed view of the pure geometric configuration of Great Britain's road network. This is particularly significant because angle-scale composite spectra change gently with gradual changes in the geometric configuration of a spatial graph [15] making each vector an excellent proxy to quantify morphological differences across different scales and densities. Additionally, spatial graph spectra are nearly unique to a particular graph [13] especially in this composite form and size.

Projecting every spectral vector (with shape (1, 400) for $n = 200$ closest segments) in a 400-dimensional space creates a data cloud of topo-geometries were the distance between points can characterise the differences in angle-scale morphology between the extracted neighbourhoods [15].

Our initial intuition drove us to generate a Euclidian distance integration model of all data point with all other. Understanding the clusters of close proximity spectra we could identify a special form of 'spectral integration' that can be evaluated as 'uniqueness' analysis of spatial morphologies.

In this model we focused on pure differences in geometry between each neighbourhood and the average difference between all other neighbourhoods and for simplicity we will only look at London's urban network comprised of roughly 300,000 segments. In Fig. 5 purple represents local spectral topo-geometries that are closely situated in this feature space, morphologically very similar to each other. White represent unique morphological structures or 'outliers' in London.

This model gives a unique perspective in understanding spatial morphologies through graph spectra as the creation of the sub-graphed neighbourhood spectra follows the 'angular shortest path' that is widely regarded to create a link between the immediate spatial morphology of a place and the pedestrian's understanding of it [1]. Using these composite angle-scale spectra that depict the internal micro-dynamics of local geometric representations from a human perception enhances traditional network theory techniques like space syntax that, unlike graph spectra, generate certain graphs metrics by turning them into aggregate measures of a single scalar value.

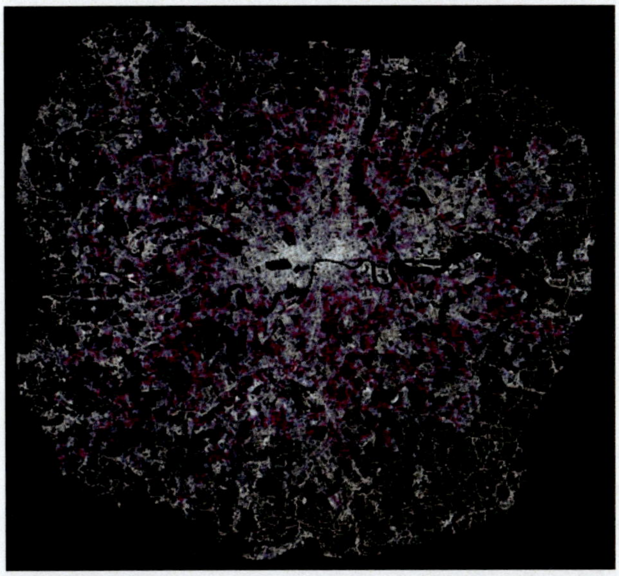

Fig. 5 Composite scale-angular geometric spectra of 300 k road segments in London—'spectral integration—Purple represents local geometries that are closely situated in this feature space, morphological very similar to each other. White represent unique morphological structures

Looking at these 'purple' areas of London in Fig. 5 are what you could call 'typical London looking neighbourhoods', crescents, Victorian houses, in some cases parts of a movie set about London. We also need to note that two different 'white' neighbourhoods are not similar to each other by definition, they can be outliers from different direction in this 400-dimensional space.

4 Unsupervised Clustering of Composite Graph Spectra

Having the angle-scale composite spectra projected into a 400-dimenional feature space that changes gently with gradual changes of the geometry of the neighbourhoods we can easily use an unsupervised clustering model like K-Means [16, 17] to extract similarities. By using an unsupervised clustering algorithm we try to make inferences regarding the morphology of the spectral dataset using only the input vectors (shape (1,400) per local neighbourhood) without referring to any other known, or labelled, data.

K-Means algorithm fed with a point cloud of encoded morphological data can identify a number of clusters through defining the number k, which refers to the number of centroids or clusters you need in the dataset. In our case these will be clusters of similar morphological or spatial characteristics in terms of angularity of the connected segments and scale across the complete GB landscape. In Fig. 5 you can see an example of how every spectral vector, representing the 200 angularly closest segments from the origin, is allocated to one of the clusters through reducing the in-cluster sum of squares of the spectra. In Fig. 6 we can see that an unsupervised extraction of a morphological cluster can lousily reassemble the foreground network of London as traditionally described by Hillier [1, 18]. Depicting only one cluster we can also observe how different the local spatial penetration patterns can be even though are extracted from similar topo-geometric graph spectra.

Moving to the full GB dataset, Figs. 7, 8 and 9 depict the K-Means results of 5 angular-scale spectral clusters that all seem to have a very distinct character. Spatial morphologies extracted from a sub-graph with a focus on angular geometry and scale naturally link to timeless ideas from space syntax but also uniquely introduce new propositions for understanding spatial communities and natural breaks in the built environment fabric.

Each of the clusters represent a distinct set of morphological or geometric characteristics. Unsupervised models like these can be used for understanding the natural

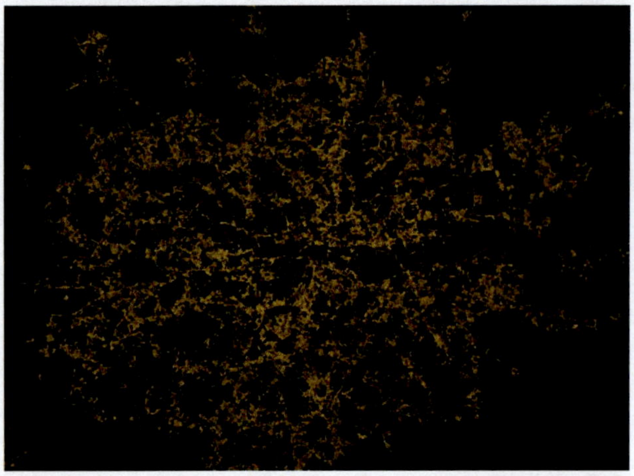

Fig. 6 Initial experiments in London—Composite scale-angular geometric spectra of roughly 300 k road segments projected into a 400-dimensional point cloud and fed into a K-means unsupervised clustering model with $K = 6$. Depicting one cluster only

morphological breaks between wards or apparent communities. Comparing individual sub-graphs and the associated composite spectra with can also explore network penetration patterns as depicted in Fig. 6. This computational modelling also allows new regional spatial analytics and spatial isolation metrics to be developed.

5 Discussions and Future

The work presented ultimately deals with the accelerated paradigm shift in the built environment towards machine intelligence. Traditionally evaluation of spatial networks is done using numerical comparison of spaces with floating point measures such as global and local space syntax measures. The scalar measures fail to retain much of the detail of the spatial topo-geometry and morphology. This is particularly discouraging if the aim is to create data models for subsequent analyses through machine learning techniques. Therefore, we revolved around creating new analyses models and datasets in order to facilitate further research in spatial machine learning.

Gaining insights from our background, the methodology's centre of attention are the three primary space syntax graph modelling concepts, 'angular', 'metric' and 'topological' that create a link between geometry, morphology and human centric perception of the built environment. While looking into existing concepts we identified that in order to create a continues model that uniquely flows between scales and built environment densities we had to develop a new composite graph spectra. Angular change and scale give a complete topo-geometric characterisation of the morphology of the road network. The addition of the "scale" component and the creation of the neighbourhood subgraph following space syntax's 'angular shortest path' can establish a novel GB wide model with comparative power between urban and rural areas.

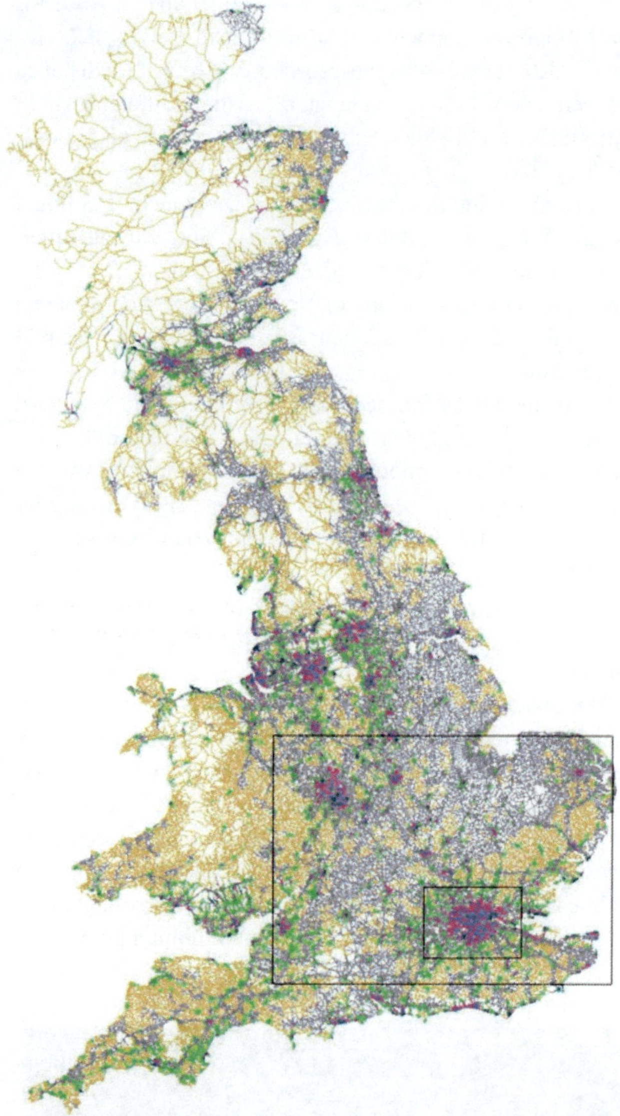

Fig. 7 Great Britain: Composite scale-angular geometric spectra projected into a 400-dimensional point cloud and fed into a K-means unsupervised clustering model with $K = 5$

Having found a way: (a) to encode spatial morphology without losing much of the detail of the spatial topo-geometry including scale and angular geometry, (b) to create several GB wide models of topo-geometric similarities or dissimilarities of neighbourhoods with the presented form of spectral integration, and finally, (c) characterise the topo-geometric characteristics by including the differentiation of scale, we aim to use the new dataset in a number of existing new directions.

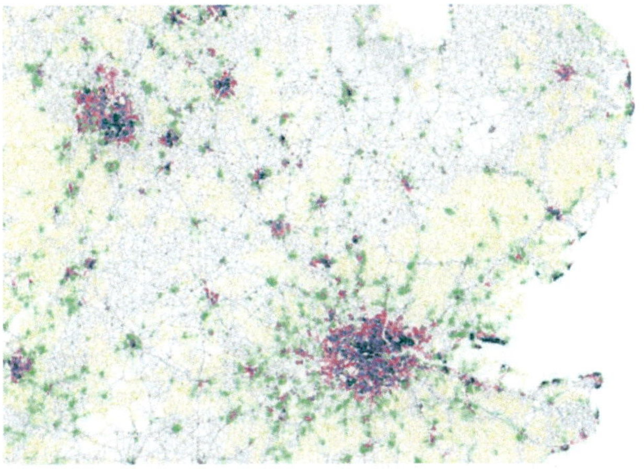

Fig. 8 Composite scale-angular geometric spectra projected into a 400-dimensional point cloud and fed into a K-means unsupervised clustering model with $K = 5$

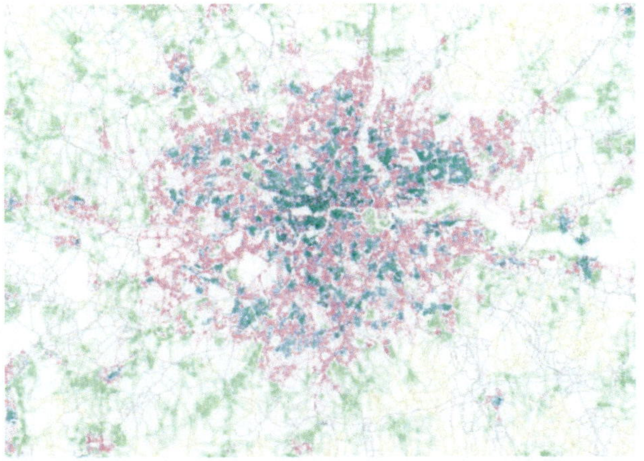

Fig. 9 Composite scale-angular geometric spectra projected into a 400-dimensional point cloud and fed into a K-means unsupervised clustering model with $K = 5$

From an analysis point of view, we currently work on models understanding how the nature of some inter- and intra-regional questions are represented in the new composite spectra. Moreover, detailed mapping of clustered morphological similarities could address challenges of interdependency or spatial isolation of cities, councils or neighbourhoods. Mapping dependencies in a continues form can be a proxy to a numerical fitness model based on the morphology of the road infrastructure.

Finally, from a proposal or design augmentation perspective we focus on how performance-based space syntax analytics can be hybridised with the help of models based on novel machine learning techniques and composite spectra. This is increasingly needed in order to greatly impact research and accelerate practice, local authorities and policy makers.

Acknowledgements The research wouldn't be possible without the support by UK's Engineering and Physical Sciences Research Council (EPSRC) fund EP/M023583/1 and mentoring by Sean Hanna and Bill Hillier at Bartlett's Space Syntax Laboratory.

References

1. Hillier, B. (2007). *Space is the machine: A configurational theory of architecture* (pp. 120–121). Space Syntax: London, UK.
2. Hillier, B., & Hanson, J. (1984). *The social logic of space*. Cambridge University Press.
3. Turner, A. (2005). An algorithmic Definition of the axial map. *Environment and Planning B: Planning and Design, 32*(3), 425–444.
4. Robles-Kelly, A., & Hancock, E. R. (2003). Edit distance from graph spectra. In: *Proceedings of the ninth IEEE International Conference on Computer Vision (ICCV 2003)*.
5. Hanna, S. (2009). Spectral comparison of large urban graphs. In D. Koch, L. Marcus, & J. Steen (Eds.), *SSS7: Proceedings of the 7th International Space Syntax Symposium*. Stockholm, Sweden: Royal Institute of Technology (KTH).
6. Hillier, B., & Vaughan, L. (2007). The city as one thing. *Progress in Planning, 67*(3), 205–230.
7. Turner, A. (2007). From axial to road-centre lines: A new representation for space syntax and a new model of route choice for transport network analysis. *Environment and Planning B: Planning and Design, 34*(3), 539–555.
8. Varoudis, T. (2012). depthmapX multi-platform spatial network analysis software. OpenSource, https://varoudis.github.io/depthmapX.
9. Hillier, B., & Iida, S. (2005). Network effects and psychological effects: A theory of urban movement. In A. van Nes (Ed.), *2005, Proceedings of the 5th International Symposium on Space Syntax* (pp 553–564), TU Delft, Delft, Netherlands.
10. Luo, B., Wilson, R. C., & Hancock, E. R. (2003). Spectral embedding of graphs. *Pattern Recognition, 36,* 2213–2233.
11. Hanna, S. (2012). Comparative analysis of neighbourhoods using local graph spectra. In *Eighth International Space Syntax Symposium*. Santiago, Chile.
12. Van Dam, E. R., & Haemers, W. H. (2002). Spectral Characterizations of Some Distance-Regular Graphs. *Journal of Algebraic Combinatorics, 15,* 189–202.
13. Gil, J. (2015, July). Examining "edge effects": Sensitivity of spatial network centrality analysis to boundary conditions. In *Proceedings of the 10th International Space Syntax Symposium* (p. 147).
14. Okabe, A., & Sugihara, K. (2012). *Spatial analysis along networks: Statistical and computational methods* (p. 41). Wiley.
15. Zhu, P., & Wilson, R. C. (2005). A study of graph spectra for comparing graphs. In *British Machine Vision Conference 2005*.
16. Lloyd, S. P. (1957). Least squares quantization in PCM. Technical Report RR-5497, Bell Lab, September 1957.
17. MacQueen, J. B. (1967). Some methods for classification and analysis of multivariate observations. In L. M. Le Cam & J. Neyman (Eds.), *Proceedings of the Fifth Berkeley Symposium on Mathematical Statistics and Probability* (Vol. 1, pp. 281–297). California: University of California Press.
18. Hillier, B. (2001). A theory of the city as object: Or, how spatial laws mediate the social construction of urban space.

Tasos Varoudis is a qualified architect and computing engineer. He is a Senior Fellow at UCL and since 2011 he drives the spatial and architectural computation research for the Space Syntax Laboratory at the Bartlett School of Architecture (UCL) where he is developing new methodological and computational innovations combining spatial data-driven models with machine learning and agent-based models. He leads the development of 'depthmapX' spatial network analysis software that has attracted more than 100,000 downloads since its open source release in 2012. He co-founded and leads the Machine Learning Urbanism Research Cluster (RC14) for the MArch Urban Design at UCL and teaches Spatial Dynamics and Computation at the M.Sc. in Space Syntax: Architecture and Cities. His personal research interests are driven by the concept of 'spatial dynamics', the interaction between architectural space and machine intelligence.

Alan Penn is a former Dean of the Bartlett faculty of the Built Environment (2009–19), he is a founding director of Space Syntax Ltd., a UCL knowledge transfer spin out with a portfolio of over 100 applied projects per year, including whole city mas-terplans, neighbourhood development plans and individual buildings. He is a board member of UCL Consultants Ltd. He is a member of the Space Syntax Laboratory within The Bartlett School of Architecture. He was the Chair of the Architecture, Built Environment and Planning sub-panel 16 and a member of Main-panel C for the Research Excellence Framework 2014. He is Principal Investigator on the £5m five year EPSRC funded Digital Economy Hub: Urban Dynamics Lab. He is a founding trustee of the Shakespeare North Trust a charity which is constructing a new Shake-spearian theatre and educational centre in Prescot outside Liverpool. He has a part time role as Chief Scientific Advisor at the Ministry of Housing, Communities and Local Government.

ScavaTools: Strategies and Tools for the Mass Dissemination of SCAVA Techniques

Catarina Ruivo, David Leite Viana, Franklim Morais, Isabel Carvalho, and Jorge Vaz

Abstract

This paper presents ScavaTools—a set of digital tools integrated into COTS software for architecture and urbanism (A&U). Its scope is on spatial configuration, accessibility, and visibility analysis (SCAVA), correlating different methodologies and underlying theories such as 'isovists', 'space syntax', 'visibility graph analysis' and 'agent-based analysis' to A&U practitioners (with ease-of-use and facilitated access). ScavaTools overall aim is to overcome the user interface shortness of the current stand-alone software for this scientific area, in use within academic communities, but much less widespread in A&U professional activity. ScavaTools specific goals are (i) to translate the usual CAD project to a SCAVA project, (ii) to translate among themselves of the traditional five SCAVA models—(1) 3D and (2) 2D metric models, (3) segment and (4) axial line networks and (5) condensed spaces graphs, (iii) to execute the necessary calculations to get SCAVA quantities (i.e. visibility, distance, depth, control, skewness, accessibility, entropy, etc.) and, finally, (iv) to present to A&U practitioners a set of reports in a language easily readable by them. Those reports are functionally orientated, meaning solving A&U problems the A&U professionals asked ScavaTools to do. The paper describes the conceptual and discrete specifications of each of the five models and the eleven tools (which integrate ScavaTools software) developed to perform the four tasks described earlier (i–iv). Although the paper is a generic presentation of the conceptual structure of ScavaTools and not an operational handbook, it also presents some brief ideas of its operation.

Keywords

3D space syntax • New methods of spatial analysis • Formal methods in architecture

C. Ruivo
Centro de Investigação em Arquitetura, Urbanismo e Design (CIAUD), Faculdade de Arquitetura da Universidade de Lisboa, Pólo Universitário, Rua Sá Nogueira, Alto da Ajuda, 1349-063 Lisbon, Portugal
e-mail: catarinaruivopereira@gmail.com

D. L. Viana
Instituto Universitário de Lisboa (ISCTE-IUL), ISTAR, Avenida das Forças Armadas, 1649-026 Lisbon, Portugal
e-mail: david.viana@iscte-iul.pt

F. Morais (✉) · J. Vaz
Laboratório de Investigação em Arquitetura e Design (LIAD), Escola Superior Artística do Porto, Largo de São Domingos, n.º 80, 4050-545 Porto, Portugal
e-mail: franklim-esap@cyberoikos.com

J. Vaz
e-mail: iorge@hotmail.com

I. Carvalho
Centro de Investigação em Artes e Comunicação (CIAC), Universidade Aberta, Palácio Ceia, Rua da Escola Politécnica, n.º 141, 1269-001 Lisbon, Portugal
e-mail: isabel.cristina.carvalho04@gmail.com

1 Introduction

Space configuration, accessibility, and visibility analysis (SCAVA) constitutes a set of methodologies with an underlying set of theoretical assumptions, with profound and ancient roots in classical architectural theory (the interconnection between form and function of spaces—their configuration and their human uses), but developed in recent and formal languages since the 1970s, like 'isovists', 'space syntax', 'visibility graph analysis' and 'agent-based analysis'. However in dynamic progress, SCAVA is already a proven and academically disseminated instrument in architectural and urban studies (A&U). Nevertheless, those set of methodologies are still very far to be a current tool for the common A&U practitioner. Many profound considerations could be held to justify such a state of the world. Similarly,

S. Eloy et al. (eds.), *Formal Methods in Architecture*, Advances in Science, Technology & Innovation,
https://doi.org/10.1007/978-3-030-57509-0_23

many broad technologies, such as Industry Foundation Classes (IFC), have still confidential dissemination among architects. The team of this project (ScavaTools) has been involved in the creation of DepthSpace3D (DS3D), a SCAVA digital tool for analysis of 3D A&U spaces, freely available for academic use. Although this new tool overcomes an existing failure in the domain of SCAVA (the analysis of 3D spaces), it has done almost nothing in favour of the massive widening of SCAVA users. ScavaTools project was driven by the idea of not only to make it possible (through the availability of tools), but also to compel many architects to use them.

2 Strategies

2.1 Analysis Versus Design

Two very different solutions have been considered, taking into account the old quarrel between 'analysis' and 'design'. With analysis, we start from a given situation (in SCAVA—a given geometric space, built or in a project), and then we calculate the hidden theoretical quantities that characterize the situation (in SCAVA—distance, visibility, depth, among others) that allow us to determine its desirability for the accomplishment of our desired functions (accessibility, security, visibility, etc.). With the design, we start from desires and currently existing environment and resources; then, we construct the actions (the project) to produce a new situation that fulfil those functional desires. It is not unusual to use formal analysis tools in the process of global design: initially, we produce a project, using our heuristic skills; secondly, we analyse it with the automatic tools; finally, we use the results to improve the initial project—and then, we recursively use these procedures, till we get a satisfactory project. Not only is it not unusual, but it is very commonplace to do so. Even in technologies with much older theoretical developments, such as structural engineering, this procedure is widely used. Roughly speaking, the main geometry of the edification's structure is defined by the engineer, and only the geometry of the details (e.g. dimensions of the sections of the structural elements) is refined after formal analysis. 'Synthetic' design tools that provide an automated geometric solution for the structure are in use in aeronautics, but rarely in construction. This happens because formal methodologies for design are usually much more intellectually demanding than their analysis counterparts. For example, the design methodologies have the tendency to produce a large space of solutions that have to be reduced with foreign restrictions and optimizing parameters. Automated design tools seem better to disseminate the theory because the mathematical intricacies of the calculations are transparent to the user. Nevertheless, this fact introduces the problem of human control of the architect over the result of automated design. Given this situation, and as our team is involved in another project (OntoArchi: Architectural Design Automation Based on Ontologies), we are producing tools for the two approaches to SCAVA methodologies—ScavaTools for analysis and OntoArchi for automated design. This paper relates only to the first of them, which is in a much more advanced state of development. All the current SCAVA software belongs to the 'analysis' paradigm. This means that the analysed space is either a given real situation or a project designed by the architect at a certain stage of development.

2.2 CAD-SCAVA Integration

The 'psychological' means to induce the A&U practitioners to use SCAVA relate to accessibility and ease-of-use (ease-of-use will be developed later in this paper). Regarding accessibility, the strategy is as follows: current SCAVA software is stand-alone software application, with some import/export from other tools—the main strategy of ScavaTools endeavour is the availability of SCAVA inside widely disseminated IT tools that architects are in current and assiduous contact (the commercial off-the-shelf CAD tools). In this same sense, other teams have already made very good efforts, with greater integration in the GUI's of the COTS CAD/GIS software than pure import/export (e.g. [3]). The ultimate goal is to bring access to SCAVA just a click away. ScavaTools complement the three main current fields of add-ons to the CAD software: (1) plumbing, (2) structural and (3) climatic. But while they deal with several constructions engineering specialties, ScavaTools deal with the core business of architecture in construction: space.

2.3 Overcoming the Ease-of-Use Issues of Today's SCAVA Software

The SCAVA stand-alone software has two main issues regarding their use by common A&U practices:

The first—the project to be analysed has to be modelled inside the stand-alone software, usually importing most of the project modelled in a CAD tool and then improving the model, using the internal editing features of the stand-alone software, which are not as powerful as those of CAD. That means that the project has to be modelled twice, although with some common work. It must be stressed that the models are different—what is important for A&U project is not the same thing that is important for SCAVA analysis. The A&U project has a much greater level of detail. Usually, more than 99.9% of the A&U model is not relevant to SCAVA. And some sets of information needed for SCAVA are not in the A&U project.

The second—the results of the SCAVA calculations are hermetic to the common A&U practitioner. The concepts and quantities of SCAVA (depth, integration, choice, control, closeness, among other sorts of measurements) that integrate the internal knowledge network of SCAVA (which we will call the connotational semantics of SCAVA) are not understandable to those not familiar with SCAVA, and therefore are useless.

The solutions of ScavaTools for these two issues are—the largest possible integration of A&U and SCAVA models of the project and the introduction of a denotational semantics of SCAVA—a sort of a translation between SCAVA connotational semantics and A&U semantics (the knowledge of actual A&U practitioners). The result outputs of SCAVA analysis have to be readable by architects.

Each of these goals is very difficult to implement, and several partial and gradual evolution of the solutions had to be considered.

Automation brings always the problem of loss of control by the human partners in the process. Even in current stand-alone SCAVA software, the embroilments of the SCAVA numerical calculations are hidden from the user, but usually, their users are academics familiar with SCAVA. Broad automation levels and unskilled users increase the danger of misuse of SCAVA. So, we stress that we are by no means advocating that users know nothing about SCAVA. Nevertheless, they can be spared profound technical SCAVA knowledge that is not relevant to their goals: better A&U projects.

3 Connotational Semantics—SCAVA Concepts

3.1 Topological SCAVA Models of Space

The current five topological representations of SCAVA, organized in descendent order of quantity of information and upward degree of abstraction, are:

- 3D metric space;
- 2D metric space;
- segment line networks;
- axial line networks;
- graphs of condensed spaces.

Each of them has its uses, pros and cons, and all of them are addressed by ScavaTools. The five representations are related to each other. The most abstract/general representations are semantic constructions made from the most concrete ones.

3D Metric Spaces. The initial model, created from the CAD project, is the 3D metric space—the general concepts assumed in ScavaTools global model of the world are as follows (the options involved in this set of semantic decisions are developments of the semantic model of DepthSpace3D, discussed elsewhere [4]) (Table 1).

Besides the geometric definition, SCAVA spaces can have **attributes** or properties, allowing the differentiation and aggregation of segments of those spaces. The denotation

Table 1 ScavaTools concepts for the 3D metric space model

Spaces	Denotation
• AS—Active space	
– ASV—Volume	The space where the agents can stay, view and move
– ASP—Paths	Movement trajectories of agents in the active volume
• PS—Passive space	
– PSV—Global volume	The entire global volume The volume can have two distinct functions in the model –
PSVV—Viewed volume	– It can be seen This must be seen potentially, as the visibility of an object that was virtually located in each position of the space
PSVO—Volume opacity	– It can be an obstacle to visibility. A continuous linear loss of visibility throughout the global volume (i.e. fog and optical effects)
– PSS—Objects	Objects in the global volume The objects can have three distinct functions in the model The objects as obstacle to staying and movement, as obstacles to visibility and as viewed space, although conceptually different can be seen as having the same physical or material referent
PSSV—Viewed surfaces	– Their surfaces can be seen, from either side
PSSO—Surface opacity	– It can be an obstacle to visibility. Surfaces of objects have a certain degree of opacity from each side
PSSI—Surface permeability	– It can be an obstacle to staying and mobility. Objects prevent the use by the active agents of the space they occupy

of attributes is left open to the users, although some are used by several of the ScavaTools. For example:

- convex spaces;
- functional attributes.

The third concept is the active **agent**, whose denotation is people, vehicles, video cameras, etc. An agent has certain properties, such as:

- regarding staying and movement:
  - dimensions;
  - slope range (static and dynamic);
  - step range (static and dynamic).
- regarding visibility:
  - window of vision.
- regarding movement:
  - behaviour parameters for agent-based analysis.

The operational model for deployment of the software implied the discretization of each of the SCAVA spaces (Table 2).

ScavaTool_BETA is the machine that takes the PS definition and creates the ASV (Table 3).

Table 3 Short description of ScavaTool_BETA machine

3D metric space created	From	Manual inputs	Operation
ASV	PS Agent attributes	Types of desired agents	Analyses space ability to accommodate the agents, regarding their dimensions, clogging in the space and range of static and dynamic slope and step admissible for each type of agent

2D Metric Spaces. 2D metric space is similar to 3D metric space, but presents some differences:

- The 3D global volume is now assumed to be only the 'ground' flat surface.
- The 'ground' has two disjoint parts:
  - the object occupation of the ground, defined by the cut lines of the objects with the ground (roughly the project plan lines);
  - the free and viewable ground. A part of the free ground is active and can be occupied by agents (Table 4).

Table 2 ScavaTools concepts for 3D metric space discretized model

Spaces	Discretization for the operational model
• AS	
– ASV	Set of points. Each point (*X*, *Y*, *Z*) represents a possible amount of space where people can stay. Although tendentiously all the points will represent the same amount of space, they can have a different volume. And, as there can be some differences in the density of agents per volume, there is also a weight attribute. For each point → ASV = {*x*, *y*, *z*, volume, weight, [set of attributes]}
– ASP	Ordered set of ASV points For each path → ASP = {ASV_1, ASV_2, ASV_3, …}
• PS	
– PSVV	Set of points. Although data representation does not impose any other restriction, the user interface guides to equal amount of space for each point. Global space is cut in axial (*X*, *Y* and *Z*) slices of equal thickness in each axis. The thickness can be different from axis to axis For each point → PSVV = {*x*, *y*, *z*, volume, [set of attributes]}
– PSVO	Global value → PSVO = {opacity}
– PSSI, PSSO	Set of triangles {defined by three points), modelling the surfaces of the frontier of the object. For PSSO, two opacity attributes are added. For each triangle → PSSI = {P_i, P_j, P_k}; PSSO = PSSI + {opacity front, opacity back} For each of P_i, *j*, *k* = {*x*, *y*, *z*}
– PSSV	Set of points, centroids of the little homomorphic triangles in which the triangles that shape the surfaces are divided. Each point is actually two points, one for each face of the surface For each point → PSSV = {*x*, *y*, *z*, area, [set of attributes front], [set of attributes back]}

Table 4 ScavaTools concepts for the 2D metric space model

Spaces	Denotation
• AS—Active space	
– ASV—Ground	The part of the ground where the agents can stay, view and move
– ASP—Paths	Movement trajectories of agents in the ground
• PS—Passive space	
– PSV—Ground	The entire ground The ground can have two distinct functions in the model:
• PSVV—Viewed ground	– It can be seen
• PSVO—Space opacity	– It can be an obstacle to visibility. A continuous linear loss of visibility throughout the global space (i.e. fog and optical effects)
– PSS—Objects	Objects in the global volume
• PSSV—Viewed surfaces	– Does not exist, the only viewable object is the 'ground'
• PSSO—Surface opacity	– It can be an obstacle to visibility. Surfaces of objects have a certain degree of opacity
• PSSI—Surface permeability	– It can be an obstacle to staying and mobility. Objects prevent the use of the space they occupy by the active agents

The discretization of 2D metric space is similar to 3D metric space discretization:

- There is a general reduction of 3D to 2D spaces (global volume to a flat surface, 3D points to 2D points).
- The level of detail is equal for all the surfaces, not particular to each point.

We no longer have three different metrics for the three independent conceptual spaces (volume, active and objects). We define a single metric (a division of the flat global surface of the ground in a tessellation of equal rectangles—pixels) that they have to fit into. This is the major difference from 3D. The equivalent to the triangle discretization of the objects would be straight lines. Instead, point discretization is used in 2D (Table 5).

This model for ScavaTools is almost equal to DepthMap's model [1]. The main difference is the distinction between active space (the ground where the agents are) and the viewed space (the free ground). This distinction is present in ScavaTools and not in DepthMap.

ScavaTool_GAMMA is the machine that takes the 3D metric space model definition and creates the 2D metric space model (Table 6).

Segment and Axial Line Networks. The segment and axial line network models are condensations of 3D or 2D metric space models. They condensate parts of active spaces in single entities. This simplification can be done if the condensed information does not lead to a lack of analysis power for a given purpose. For example, if we want to analyse the connectivity of a certain urban place to the city, we have the same situation for each point in a street section. We do not need to study all the points in the section (Table 7).

Table 5 ScavaTools concepts for the 2D metric space discretized model

Spaces	Discretization for the operational model
• AS	
– ASV	Set of points. Subset of PSVV For each point → ASV = PSV + {weight}
– ASP	Ordered set of ASV points For each path → ASP = $\{ASV_1, ASV_2, ASV_3, \ldots\}$
• PS	
– PSV	Set of points The points are the centres of the rectangles that tessellate the ground For each point → PSV = {*x*, *y*, [set of attributes]} PSV has two disjoint subsets: PSSI/PSSO and PSVV
– PSVV	Set of points. Subset of PSV
– PSVO	Global value → PSVO = {opacity}
– PSSI, PSSO	Set of points. Subset of PSV The cut of the objects in the ground defines lines. If any of those lines crosses the pixel, that pixel is of type PSSI/PSSO. If not, it is of type PSVV For PSSO, the opacity attribute is added For each point → PSSI = PSV; PSSO = PSSI + {opacity}
– PSSV	Does not exist. Visibility of the objects is almost the same as that of the points of the 'ground' just adjacent to them

Table 6 Short description of ScavaTool_GAMMA machine

2D metric space created	From	Manual inputs	Operation
All	3D metric space model	Desired elevation	Reduces the 3D geometry to a plan (2D geometry in a horizontal flat surface cut) at the desired elevation

Table 7 ScavaTools concepts for the segment and axial line network models

Spaces	Denotation
• AS—Active space	
– ASV—Metric graph	Streets and crossroads. These are metaphors for any reality that can be modelled by this type of metric graph
– ASP—Paths	Movement trajectories of agents in the graph

The discretization of this model is (Table 8) as follows.

An axial line is a set of segments with the same axial line attribute. In classical formulation, all the segments are in a straight line.

ScavaTool_DELTA is the machine that takes the 3D metric space model definition and creates the segment and axial line network model (Table 9).

Graphs of Condensed Spaces. The graph of condensed space model is also a condensation of the 3D or 2D metric space models. But they are suitable for condensing parts of the passive spaces in single entities. The simplifying purpose is the same, as it allows for much fewer analysis resources, with little loss in analysis accuracy.

One of the most used cases is the definition of convex spaces, or functional spaces (Table 10).

The discretization of this model is (Table 11) as follows.

ScavaTool_EPSILON is the machine that takes the 3D metric space model definition and creates the condensed space graph model (Table 12).

Table 8 ScavaTools concepts for the segment and axial line network discretized models

Spaces	Discretization for the operational model
• AS	
– ASV	Set of nodes. For each node → ASVP = {*x*, *y*, *z*, weight, [set of attributes]} Set of lines. For each line → ASVL = {ASPi, ASPj, weight, [set of attributes]}
– ASP	Ordered set of ASV points For each path → ASP = {$ASVP_1$, $ASVP_2$, $ASVP_3$, …}

Table 9 Short description of ScavaTool_DELTA machine

Segment and axial line created	From	Manual inputs	Operation
Segment network —all	3D metric space model—accessibility graph (Sect. 3.2.1)		Condensates the ASV points to a limited number of lines and nodes. Uses a technique of machine learning—classification through clustering [5]
Axial lines network —all	Segment network model		Condensates the segments in axial lines, defined by belonging to the same straight line

Table 10 ScavaTools concepts for the graphs of condensed space model

Spaces	Denotation
• (A + P)S—Active + Passive Space	
– (A + P)SV—Graph	Convex spaces and connections between those spaces. These are metaphors for any reality that can be modelled by this type of non-metric graph
– ASP—Paths	Movement trajectories of agents in the graph

Table 11 ScavaTools concepts for the graphs of condensed space discretized model

Spaces	Discretization for the operational model
• (A + P)S	
–	(A + P)SV
Set of nodes. (A + P)	SVN = {[set of attributes]} Set of lines. For each line → (A + P) SVL = {(A + P)SVN_i, (A + P)SVN_j, [set of attributes]}
– ASP	– Ordered set of (A + P)SVN – For each path → ASP = {(A + P)SVN_1, (A + P)SVN_2, (A + P)SVN_3, …}

SCAVA Calculations. ScavaTools have calculation tools that provide values for the SCAVA quantities. Many quantities may be calculated, mostly through means of algebraic, statistical or graph theories, beginning with some basic quantities. These quantities are meant to describe the space, with an advanced semantics. The three main semantic fields are:

Table 12 Short description of ScavaTool_EPSILON machine

Condensed spaces graph created	From	Manual inputs	Operation
All	3D metric space model—accessibility and visibility graphs (Sect. 3.2.1)		Condensates the PSVV points to a limited number of lines and nodes. Uses a technique of machine learning—classification through clustering [5]

- physical accessibility, inside the active space (AS) taking into account the PSxI (obstacle to visibility spaces);
- visual accessibility or visibility, between the AS and the PSxV (viewed spaces) taking into account the PSxO (obstacle to accessibility spaces);
- configuration, inside each type of space, the most important being the PSVV (volume space).

SCAVA Basic Quantities. ScavaTools basic quantities are functions which domain is a set of points or the Cartesian product of two sets of points belonging to several discretizations of SCAVA spaces. The counter-domain is usually a number. As, with discretization, the domains are numerable sets, those functions can be represented with a graph language, but not always a 'flat' one. Some special graphs are used, for example when two different types of nodes are needed. For example, the visibility function is no longer bijective but is always directed from the active space to the viewed spaces.

VisibilityFunction (ASV, PSxV) = continuous quantity varying from 0 (not visible) to 1 (full visibility).

Some primitive functions in 3D or 2D metric space model are (Table 13) as follows.

The machine that performs the basic SCAVA calculations is the ScavaTool_ZETA. Two tools are available, for 3D and for 2D metric space models (Table 14).

Table 14 Short description of ScavaTool_ZETA machine

xD metric space created	From	Manual inputs	Operation
Basic quantities	xD metric space model		As described earlier

The graphs created by these tools, expressing the basic quantities of SCAVA, are the base for the machines that perform the creation of the 'segments' and 'axial line network' and the 'condensed space graph' models (Sects. 3.1.3 and 3.1.4, respectively).

SCAVA-Derived Quantities. Derived quantities are calculated using basic quantities, through means of algebraic, statistical, graph theories, data mining, artificial intelligence and so on (Table 15).

The machine that performs the SCAVA calculations for derived quantities is the ScavaTool_ETA. Five tools, one for each of the SCAVA models, are available (Table 16).

From 3D Metric Spaces to Condensed Models. The 'segment and axial line networks' and the 'graphs of condensed space' SCAVA models are already in the graph form and have the basic quantities in their initial definition. In 3D and 2D metric space models, the graphs are constructed from geometry. In condensed models, the graph is geometry.

The five topological types of SCAVA representations are related to each other. The most abstract/general concepts must be constructed from the most concrete ones, so that the explanatory/predictive degree of the model is increased and does not reflect the loss in the quantity of information.

The involved techniques may vary:

- 2D metric spaces are traditionally constructed with projections of slices of the 3D global space. It is a trivial task. And the discretization in small rectangles makes the operation of reduction of the level of detail an easy task, compared to the same reduction from the CAD project to the 3D metric space model.
- Segment lines are aggregations of parts of the active space that have little variations in some attributes, so all

Table 13 some basic quantities in 3D or 2D metric space models

Functions	Domain ('Cartesian product')	Counter-domain	Connotation
Visibility distance	ASV * PSxV	Set of real numbers	Distance from any point of the AS to any point of the PSxV
Accessibility	ASV * ASV	Boolean	The possibility of movement between two points of the ASV
Visibility	ASV * PSxV	Set of real numbers from 0 to 1	The degree of visibility of each point in the ASV to any point in the PSxV through the PSSO

Table 15 Some SCAVA-derived quantities

Functions (domain: 'Cartesian product' AS * VS)	Domain	Counter-domain	Connotation
Accessibility depth	Cartesian product ASV * ASV	Set of integers	Minimum number of straight lines in the ASV that need to be traversed from any point in the ASV to any other point in the ASV
Visibility depth	Cartesian product ASV * PSxV	Set of integers	Minimum number of straight lines in ASV that need to be traversed from any point in the ASV to see any point in the PSxV
Mean depth of functional uses	Functional use attributes	Set of real numbers	Medium of the depths of every PSxV, aggregated by their attribute—Functional use

Table 16 Short description of ScavaTool_ETA machine

Model created	From	Manual inputs	Operation
Derived quantities	SCAVA model and basic quantities		As described earlier

the space of each part can be treated as one indistinct space. This is achieved by employing a technique of machine learning—classification through clustering [5].

- Axial lines are aggregations of segment lines following some common attributes.
- Condensed spaces are the same thing as segment lines for condensation purposes, but, in this case, the condensation of the space is made so that the condensed space is represented by a node in the graph, instead of a segment between nodes in the segment line case. Classification through clustering is also the used technique to reduce a large number of points of the space to a small set of points.

Creation of Paths. The paths in the active space (ASP), *ipsum est* the movement trajectories of agents in the active volume, can also be created by automated processes. The machines that perform the SCAVA calculations for path definitions are the ScavaTool_THETA.

Two different machines are available (Table 17 and Table 18).

Creation of Attribute Assignments to Spaces. The attributes for each of the spaces that define the model can also be created by automated processes.

The machine that performs the SCAVA calculations for attribute assignment is the ScavaTool_IOTA (Table 19).

This machine is similar to ScavaTool_DELTA and ScavaTool_EPSILON machines, but its results are assigning aggregation attributes to spaces inside the same SCAVA model. DELTA and EPSILON machines do not stay at that level and aggregate the parts of space in single entities, creating new SCAVA condensed models.

Table 17 Short description of ScavaTool_THETA_SCAVA machine

3D metric space created	From	Manual inputs	Operation
ASP—Paths in the active space	AS—Active space Basic and derived quantities		Calculates optimized paths using only SCAVA techniques. Attractors and repulsors can be defined manually or by SCAVA considerations (e.g. high visibility)

Table 18 Short description of ScavaTool_THETA_AGENT_BASED machine

3D metric space created	From	Manual inputs	Operation
ASP—Paths in the active space	AS—Active space, Basic and derived quantities, Agent behaviour attributes	Attractors and repulsors	Calculates optimized paths using techniques of agent-based analysis. Attractors and repulsors can be defined manually or by SCAVA considerations (e.g. high visibility). Agents may have several types of behaviour [2]

Table 19 Short description of ScavaTool_THETA_IOTA machine

3D metric space created	From	Manual inputs	Operation
All spaces—Attributes	3D metric space model—accessibility and visibility graphs (Sect. 3.2.1)		Condensates the PSVV points to a limited number of lines and nodes. Uses a technique of machine learning—classification through clustering [5]

Denotational Semantics. A&U practitioners are not aware of the internal semantics of SCAVA. If the intention is disseminating SCAVA throughout the entire classes of practitioners, then a friendly approach is desirable. The current semantics of the architect includes many concepts that can be fully taken into account:

- a full formal description of the A&U space, in CAD language–the survey of a real building environment or the project of a not yet built one;
- a set of informal or semi-formal theories the architect uses to analyse functional features and desirability of the project.

The solution for this 'back turned semantics' problem lies in the creation of translation machines.

From A&U Project to Full Model SCAVA Project. The first translation machine must transform the architectural project in a SCAVA space model. It is not difficult to make the geometrical translation of the architectural project objects in CAD to the SCAVA surfaces, although it carries some harsh intricacies:

- Not all the CAD software applications have a complete and univocal description of the space—for example, old 2D CADs represent drawings that represent objects. In those CADs, the translation from points and lines to construction objects relies on human codification and there is no easy way to automate that translation. This problem orientated the ScavaTools team to produce the software only for 3D CADs based on objects.
- The architectural project has a much higher level of detail that the one needed for SCAVA. For operational reasons (time and memory resources), more than 99% of the information in the architectural project must be semantically disregarded as not relevant.
- The current generation of CAD is dedicated to the representation of physical objects of the building—that means construction objects. Nevertheless, the concepts that matter for the core business of architectural thinking are rather more abstract—spaces and functions. Those concepts have no simple representation in CAD. They must be constructed from the inside of SCAVA.
- The same problem arises with the definition of the SCAVA AS (active space). Nothing in the CAD project says explicitly what are the physical actual paths (e.g. the eye level of the active agents).
- In most cases, there is no explicit representation of the degree of transparency/opacity of some objects.

ScavaTool_ALPHA is the machine that takes the CAD project and creates part of the 3D metric space model (Table 20).

From SCAVA analysis to A&U practitioners: Reports. The opposite movement is necessary to complete the process. Now, ScavaTools have to gather all the information relevant to the goals of the A&U practitioner and deliver it. Although all the usual SCAVA scientific data is available, predefined templates with easy-to-understand A&U meaningful reports are available. Several functional fields are represented, for example:

- security (CCTV placement, agent placement, children surveillance);
- economy (value assignment to space, product placement, publicity);
- traffic (accessibility, clutter, jams prediction);
- aesthetics (monumentality, space diversity, remarkable places);
- urbanism (skyline protection, centralities, equipment distribution);
- social studies (social segregation, control, functional uses of space).

Table 20 Short description of ScavaTool_ALPHA machine

3D metric space created	From	Manual inputs	Operation
PSVV		Global volume Level of detail	Simple voxelization of the global volume regarding desired level of detail
PSVO		Volume global opacity	Equals
PSSI, PSSO	CAD objects in the CAD project	Global volume Level of detail	Geometry data reduction, through retessellation of object surfaces with a smaller level of detail Surface opacities are deduced from the properties of the objects in the CAD project (not always available)
PSSV		Level of detail	Simple subdivision of PSSI/PSSO triangles
All spaces: attributes	Annotations in CAD project	Attributes	Assignment of manual input and, eventually, available CAD project annotations

4 Operational Model

4.1 ScavaTools

The global steps of the process are:

- The user (a A&U practitioner) creates the CAD project.
- The user declares to ScavaTools the desired objectives, substantiated in a set of reports.
- ScavaTools consult the report template database and decide the calculations to make and the tools to use. It gathers the set of needed data to input from the user.
- ScavaTools prepare a GUI form with the necessary human data for the user to complete.
- ScavaTools get the CAD project and perform the calculations.
- The intermediate results are available to the user, so that he can change any of this data.
- Recalculations with changed data are available.
- ScavaTools deliver the reports to the user.
- The user, regarding those reports, can change his CAD project and restart the process.

4.2 ScavaTools Development Accessibility

ScavaTools do not intend to exhaust all the possibilities that SCAVA methodologies can provide to the A&U activity. SCAVA nowadays has hundreds of applications, and each year dozens of new developments are proposed. ScavaTools open to new developments by third-party developers through two ways:

- All the data of each project is available in SQL form. It can be used by external algorithms.
- The underlying theories of many of the calculating machines of ScavaTools have patterns registered in database structures that can be populated with new ones.

Patterns are available for (Table 21).

ScavaTools operational diagram.

See Table 22.

ScavaTools rely on very highly numerical calculations that are not prone to formalization with ontological languages. Although its calculus is much more related to the math of quantities than to the math of qualities and logic, many of its theoretical concepts fit perfectly in a theoretical ontology of architecture and urban planning. Set on this premise, architects will be able to offer new improved theoretical analysis on their projects and can produce building projects with programming aids. To do so, ScavaTools operational diagram is as follows (Fig. 1).

Table 21 ScavaTools patterns' availability

Pattern	Used by ScavaTools
CAD objects	
Agents	BETA THETA_AGENT_BASED
Formulas for derived quantities	ETA_xx
Aggregation techniques	DELTA EPSILON IOTA
Report templates	KAPPA
Attributes	ALPHA DELTA EPSILON IOTA

Table 22 Summary of ScavaTools machines

ScavaTools	Does...	Table
ALPHA	Translates CAD project in SCAVA's 3D metric space model	20
BETA	Creates the Active space from the passive space and agent attributes	3
GAMMA	Creates 2D metric space model from the 3D metric space model	6
DELTA	Creates 'segment and axial line network' model from the 3D metric space model	9
EPSILON	Creates 'condensed space graph' model from the 3D metric space model	12
ZETA_xx	Creates basic quantities from space model	14
ETA_xx	Creates derived quantities from basic quantities	16
THETA_SCAVA	Creates path definitions through SCAVA quantity analysis	17
THETA_AGENT_BASED	Creates path definitions through AGENT-BASED analysis	18
IOTA	Assigns attributes	19
KAPPA	Produces reports	21

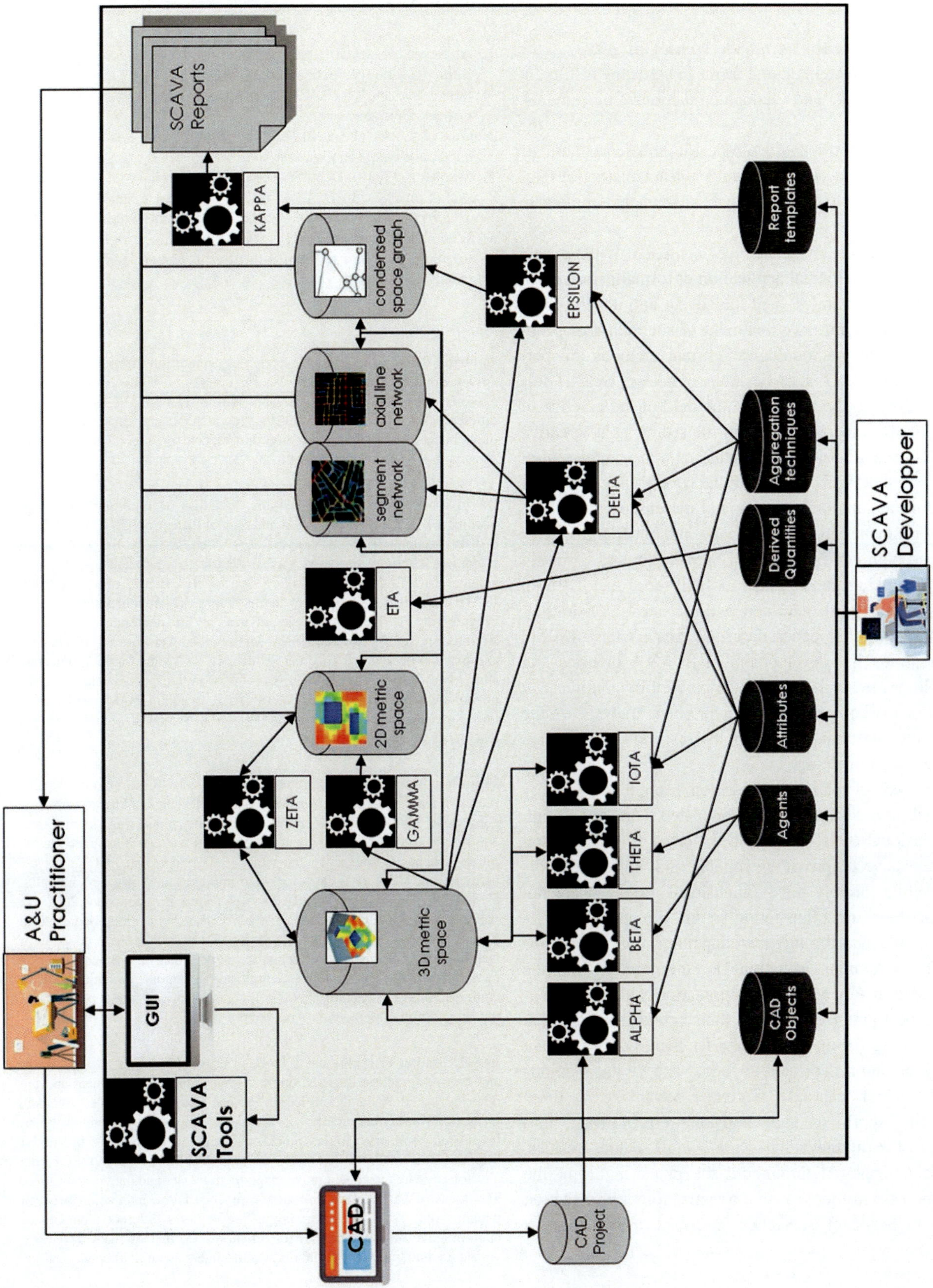

Fig. 1 ScavaTools operational diagram

5 Discussion

The design process (in its broader sense) in A&U is the optimization of the architectural forms towards the fulfilment of natural, social and human functions/requirements, including the representational/significant/ideological/linguistic/aesthetical ones. For that, ScavaTools embrace formal methodologies (those that stem from the mathematical sciences) applied to A&U.

Formal methods must not be confused with digital methods (as the millennial application of formal geometry to architecture demonstrates), and the advent and development of ICTs led to the massive treatment of such class of problems with digital methodologies. Formal methods are pervading all the A&U activities. Nevertheless, ScavaTools direct its efforts to the research in the field of the practice of A&U in its own specificity within the sector, as a 'creative industry'. ScavaTools' understanding of architectural specificity is that A&U deals with the overall interactions of the building actions of humanity and the environment. The built environment has a large scope and profound consequences on both environment and human life, and it should be designed taking into consideration all those factors, both the physical and the social and psychic ones. Considering this framework, this paper describes ScavaTools' overall outline assuring the development of SCAVA theories and methodologies to set digital tools to be used by architects. It was said ScavaTools can bring a brighter analysis of the architectonic or urban situations that are not available in traditional ways of doing.

The analysis of spatial configuration through accessibility and visibility has been conducted mostly with two-dimensional tools, relying on the capacity architectural plans often have of providing reliable and sufficient information. While forcibly three-dimensional problems are not the norm, these are often found in the literature and frequently approached through the adaptation of existing software. In a similar situation is the ability of the third-generation CAD to bring automatically to the practitioner the best practices saved in their knowledge base that the architect may be unaware. ScavaTools introduce relevant work of innovation dedicated to formal and digital methods for A&U, which can take a greater advantage of these methods in a short-, medium- and long-term horizon, contributing to the digitalization of the overall sector. Besides, the contents explored throughout the paper highlight the need to increase further semantic organization of information used in higher-level semantics inscribed in theories as SCAVA.

References

1. Al_Sayed, K., et al. (2014). *Space syntax methodology* (4th ed.) Bartlett School of Architecture, UCL.
2. Batty, M., et al. (2012). *Agent-based models of geographical systems*. Springer.
3. Gil, J., et al. (2017). https://github.com/SpaceGroupUCL/qgisSpaceSyntaxToolkit, last ace.
4. Morais, F. (2018). DepthSpace3D: A new digital tool for 3D space syntax analysis. In D. L. Viana, et al. (Eds.) *Formal methods in architecture and urbanism*. Cambridge Scholars Publishing.
5. Witten, I. H., et al. (2011). Data mining. In *Practical machine learning tools and techniques* (3rd ed.). Elsevier—Morgan Kaufmann Publishers.

Catarina Ruivo has a Master's degree in Architecture from the Faculty of Architecture, University of Porto, and is currently a Ph.D. candidate in the Lisbon's School of Architecture, University of Lisbon. She has worked in different research projects and develops research focusing on spatial analysis of architecture and the urban environment and on the relation between society and the built-environment.

David Leite Viana holds a post-doc. in Urban Morphology/Civil Engineering (FEUP), a Ph.D. in Urban and Spatial Planning (IUU-UVa), a DEA in City Planning (ETSA-UVa) and Dipl. Arch. (ESAP). He develops his professional activity in the Municipal Planning Department at Porto City Council and at the Digital Living Spaces Research Group at ISTAR, ISCTE-IUL. He is visiting fellow in the Theories of Architecture Fellowship Program at TU Delft, Ph.D. supervisor in the doctorate program on Architecture for the Contemporary Metropolitan Territories (ISCTE-IUL) and external reviewer at Chalmers University of Technology (Architecture and Urban Design Program). He is professor in the specialization program on Collaborative Territories (IPPS, ISCTE-IUL) and in the Master Program on Geographic Information Systems Applied to Spatial Planning, Urbanism and Landscape at UPV.

Franklim Morais is a Portuguese civil and electronic engineer, Ph.D. professor and researcher. He was the leader of development teams of private commercial software for scientific applications: pioneering, in the early 1980s, the use of finite element methods for structural engineering on micro-computing; and, since the early 1990s, developing COTS private hardware and software systems for smart buildings, urban spaces and cities. He is an invited professor in ESAP's Architecture Program since 2001. He participates in LIAD, an academic research team, in a group dedicated to Formal Methods in Architecture. A major achievement is the production of DepthSpace3D, a digital tool for space syntax analysis. Since 1991, he participated in the organization (scientific and operational committees) of 11 conferences on his scientific fields of interest, including the 5 symposia of the series Formal Methods in Architecture (since 2011).

Isabel Cristina Carvalho has a Ph.D. in Digital Media Art, a Dipl. Arch. and two postgraduate degrees: one in Urban Centre Management and the other in Spatial Planning and Urban Project. She is a researcher at CIAC, UAb. In 2019, her research project Gender in Map—women empowerment in planning and urban development was selected for funding in the 1st Stimulus Contest for the Scientific and Individual Employment of FCT. She develops research in collaborative mapping processes and data visualization. The focus of her research and artistic practice is the interaction between people, urban space and technology. She explores different processes of appropriation and apprehension of urban spaces, part of them developed with the community in a collaborative and intergenerational way.

Jorge Vieira Vaz is a registered architect at Portuguese Architectural Board since 1991, creative designer and technologist. He has a Diploma in Architectural Design from the University of Porto and an Advanced Studies Diploma in Architecture from University of Valladolid. He is professor in Construction and Technologies at the Integrated Master Program in Architecture, ESAP, since 1993. Building Thermal Behavior qualified expert of the Portuguese Energy Agency. Virtual Design to Construction Project Manager (AAVV) and BIM (Building Information Modelling) methodologist. Develops research in taxonomies, ontologies and classification systems, in the scientific field of architecture and construction technology, at LIAD, ESAP.

Self-learning Agents for Spatial Synthesis

Pedro Veloso and Ramesh Krishnamurti

Abstract

Over the last decades, a vast repertoire of computational methods has been employed for the synthesis of spatial configuration. Many of these techniques, such as the long-standing black-box optimization or the recent generative adversarial networks, enable a quick exploration of the design solutions based on destructive operations, but encapsulate the generative process, promoting disruptive turn-taking between computer and designer. In contrast, techniques based on agents naturally provide partial design information and enable fine-grained interaction. However, existing agent-based models originate from non-architectural problems, so it is not straightforward to adapt them for spatial design. To address this gap, we present a method to create custom spatial agents that can satisfy architectural requirements. While the method can be adapted to a diversity of representations and goals, we focus on a proof of concept where agents control spatial partitions (represented as polyominoes with no holes) and interact in an environment represented as a grid. The agents learn how to satisfy its individual (shape, area, etc.) and collective goals (adjacency) using multi-agent deep reinforcement learning. In this paper, we focus on the formulation of the environment, agents, and goals and present simulations of trained agents to illustrate possible variations.

Keywords

Space planning • Interactive generative systems • Multi-agent deep reinforcement learning

P. Veloso (✉) · R. Krishnamurti
Carnegie Mellon University, Pittsburgh, PA 15213, USA
e-mail: pedroveloso13@gmail.com

R. Krishnamurti
e-mail: ramesh@andrew.cmu.edu

1 Introduction

Computational generative systems are constructs that, given a proper input of information and a program, can automatically synthesize design solutions. Early computational generative systems were developed to automate the generation of architectural layout, which at the time was called spatial synthesis or space planning [1, 2]. In this context, generative systems were largely influenced by two fields concerned with problem-solving and decision-making—operations research and symbolic artificial intelligence—and their respective methods: optimization and search [1, 3–5].

Following the incorporation of black-box methods, which enabled the solution of general design problems with custom objectives and constraints, optimization became the dominant category for generative systems. For example, genetic algorithms became a widespread technique in the recent debate on generative systems [6–12] and designers today have access to many black-box optimization algorithms encapsulated in plug-ins for CAD software. In this context, the combination of parametric modeling and optimization to explore design alternatives became a mainstream generative workflow [13].

However, the availability of recent computational techniques, such as agent-based modeling and deep learning, provides an opportunity to research and investigate novel approaches to generative systems in order to support different aspects of spatial exploration.

1.1 Interactive Generative Workflows

In design theory, the critical engagement and circular reasoning of designers supported by direct interaction with the design media are considered central elements of design ideation [14–19]. Particularly in the early stages of design, the exploration of design alternatives supports not only the discovery of provisory solutions but also the reformulation

S. Eloy et al. (eds.), *Formal Methods in Architecture*, Advances in Science, Technology & Innovation,
https://doi.org/10.1007/978-3-030-57509-0_24

of the design problem itself [17, 20, 21]. Interaction with the partial states of design representation during the synthesis is important for acquiring knowledge about the problem, imposing design biases, and to support intuitive behaviors such as the leap of insight [22].

However, a large part of the generative techniques relies on strong assumptions about the problem and, consequently, focuses on the process of synthesizing solutions. For example, black-box optimization uses custom strategies to navigate in the parameter space in order to converge to optimal solutions and GANs learn a function that can generate solutions like the ones existing in the dataset.

The consequence of the focus on the final solution is that these techniques often adopt destructive procedures and promote a human-computer interaction based on disruptive turn-taking—i.e., the designer changes the input, then the algorithm executes the synthesis and displays the results as the output. Even in the cases where this is alleviated by some form of interface that partially executes the algorithm, the temporal domain and the repertoire of actions available for the designer are still distinct from the ones of generative procedures.

In this work, we switch the focus of generative systems from automatic generation of design solution to the exploration of design behavior during the generation of partial solutions. The motivation is to incentivize important elements of early and open-ended design exploration by experts, such as the understanding of the consequences of actions over time and the visualization of partial design representations in formation. We intend to enable the designer and the generator to act together on the same design representation over time, which is intrinsic to modeling paradigms that build the solution step-by-step, such as rule-based and agent-based models.

In this research, we focus on agent-based models, because they enable the construction of different parts of the representation in parallel, which results in partial states with more information and, consequently, more opportunities for direct and fine-grained design interactions.

1.2 Multi-agent Space Planning

It is important to be accurate with respect to the term 'agent.' In AI, the term agent is used generally to refer to a construct that, immersed in an environment, uses its program to map percept sequences to actions, in order to solve a certain task rationally [23]. Depending on the task environment, multiple interactive agents can be designed to cooperate, coordinate, and negotiate to achieve a certain goal, forming a multi-agent system (MAS).

Another important referent is agent-based modeling (ABM), where an agent is a unit of representation in the computational modeling of complex systems. Multiple computational agents map percepts to actions, interacting with each other in a shared environment and developing patterns or behaviors that are not necessarily predictable from the perspective of the individual [24]. ABM relies on simulation and fine-grained interactions between agent, environment, and the user to represent complex processes that unfold over time.

We coined the term multi-agent space planning [25] to designate the generative systems based on the simulation of agents that decide how space should be shaped, occupied, or partitioned. In this context, an agent is a spatial entity with local control, interweaving individual perception and action to shape a specific spatial unit. The 'environment' comprehends elements of the space that are independent of the agents. It mediates the relation between the agents (agent-agent interaction), provides global information for the agents (agent-environment interaction), and potentially acts as a game board for designers (meta agent-agent interaction). During a simulation, the agents receive signals from the environment, neighboring agents, or even the designer and must make decisions in order to change the space accordingly.

1.3 Gap in Multi-agent Space Planning

Most of existing agent-based models for spatial synthesis rely on expert knowledge in the form of heuristics or on the adaption of existing models to define how the agent should interact in the environment. This repertoire of methods for interactive spatial synthesis includes swarm algorithms (flocking and pheromone navigation), cellular automata, reaction-diffusion, and physics simulation [25, 26].

Cellular automata (CA) enable the emergence of patterns and have been successfully applied to conceptual form generation, urban morphology [27], and even to building design [28], but the cell-based computation imposes strict restrictions for the satisfaction of architectural requirements. Physics simulation is a more general technique that conciliates simple control with intuitive interaction and can be adapted for different spatial problems and objectives. However, physics-based agents are reactive agents that approximately follow laws of physics. They do not have any sophisticated policy to manage spatial conflicts. Bio-inspired models such as swarm algorithms simulate exogenous phenomena (social navigation) in which the units can move and interact in space. While they enable the incorporation of some architectural requirements, such as area or adjacency, it is usually hard to adapt them to produce conventional architectural forms or satisfy additional requirements.

An alternative to conciliate multiple architectural goals is to hybridize agent-based models with conventional methods, such as heuristics or black-box optimization, for the initialization or the refinement of the global solution. The downside of this hybridization is that it imposes discontinuity in the generation procedure, limiting the opportunities of direct interaction with the spatial solution.

Overall, the challenge for spatial synthesis with agent-based models is twofold: to develop control strategies that incorporate specific architectural requirements and to preserve the fine granularity of the simulation.

1.4 Spatial Synthesis with Self-learning Agents

Reinforcement learning (RL) is a branch of machine learning that addresses the control problem. It is a viable alternative to support domain-specific agents and to preserve fine-grained interaction in spatial synthesis. The goal in RL is not simply computing good trajectories from a given initial state, which could be addressed by methods such as search or optimization. With a proper RL algorithm, the spatial agents can learn how to build spaces in real-time and from varied states.

In CAAD, RL has been recently applied to experiments related to intelligent adaptive building control [29], autonomous robots [30], fire egress evaluation [31], and machine feedback [32]. In the field of spatial synthesis, RL has been adopted for automatic decision-making with shape grammars [33]. Specifically for multi-agent space planning, our literature review [25] identified one experiment with learning agents. In this example [34], teams of agents try to create large clusters of building blocks on the same grid. While this experiment is motivated by recent achievements in the field of RL, in practice, it uses a natural selection mechanism to improve the policies of each competing group of agents between episodes.

Through the proof of concept in this paper, we describe an instance of our method in which custom spatial agents can learn sophisticated policies to conciliate the satisfaction of architectural requirements and the fine-grained interaction. The agents interact in an environment represented as a square grid by controlling a spatial partition. Both the representation of the space and of the agents' action space and goals are defined to take advantage of the properties of the grid. The designer provides the spatial objectives (adjacency, shape, area, etc.), and the agents learn how to behave using reinforcement learning. Particularly, we use custom RL techniques based on deep learning for the multi-agent setting —which is referred to as multi-agent deep reinforcement learning (MADRL) [35].

This proof of concept is part of the first author's larger doctoral research objective. In this paper, we focus on the formulation of the environment, agents, and goals and we will present simulations of trained agents to illustrate it. The learning algorithms and interaction with designers will be presented in future articles.

2 Development

2.1 Grids as Representation of Space and Information

In this proof of concept, we opted to use the square grid as a diagrammatic representation of architectural space (environment and agents). It is a representation that has been widely adopted in the early days of space planning [1] and in the development of recent agent-based models [34, 36–38]. The grid is a very flexible representation that can:

- Simplify spatial queries (neighborhood, adjacency, distance, etc.)
- Structure data discretely and spatially, supporting efficient computer vision, machine learning, and spatial analysis techniques.
- Provide support for an architecture based on an ensemble of discrete entities that encapsulate contextual design information.
- Be extended to 3D version (voxel grid).
- Enable control over multiple resolutions, providing different levels of complexity for design and the grain resolution of the agent.
- Discretize different geometric entities in a homogeneous representation, which supports both conventional room shapes (rectangles, L-shapes, U-shapes, T-shapes, etc.), but also unconventional and irregular shapes (see Fig. 1).

There are, however, disadvantages to using square grids in that that they are anisotropic—i.e., they bias the representation and operations by the horizontal and vertical axes. Additionally, they present a tradeoff between shape information and the design space. High resolution enables complex forms but results in a very large space for exploration. Low resolution simplifies the design space but imposes a considerable loss of shape information.

2.2 Spatial Representation of Environment and Agents

Information is stored in multidimensional arrays. For instance, the environment is an array of shape ($depth_{env}$, w, h). The width (w) and height (h) define the size of the grid, and $depth_{env}$ defines the number of layers of information. The basic layer of information of the environment contains two sets of cells: $empty_{env}$ and $obstacle_{env}$. Empty cells can be occupied by agents while obstacles cannot (see Fig. 2—left).

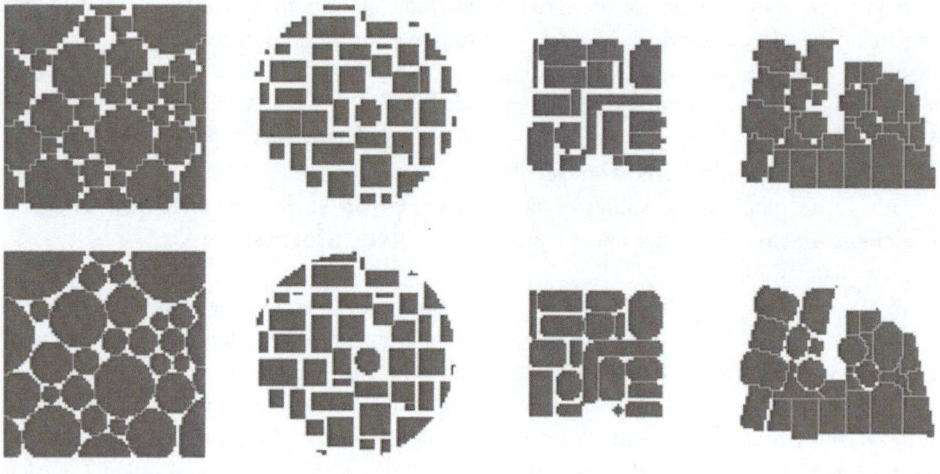

Fig. 1 Examples of real floorplans discretized in grids

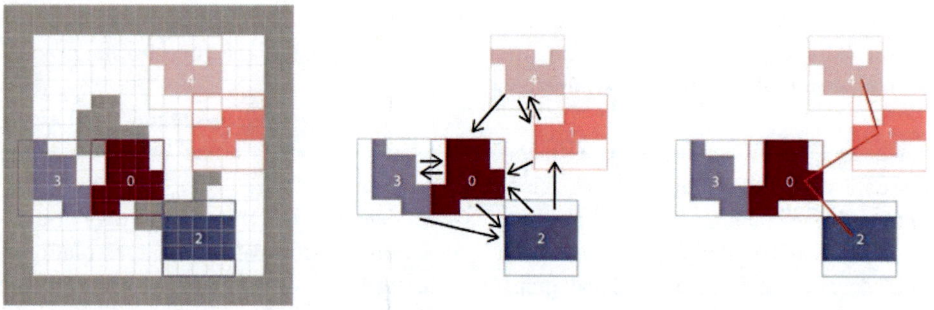

Fig. 2 Left: simplified visualization of agents on the grid with obstacles. Middle: two closest neighbors (m = 2): 0:[3, 2], 1:[4, 0], 2:[0, 1], 3:[0, 2], 4:[1, 0]. Right: adjacency goals (max k = 2): 0:[1, 2], 1:[0, 4], 2:[0], 3:[], 4:[1]

More information can be added to as additional layers, increasing the depth of the environment.

The agent is an array with shape ($depth_{agent}$, w, h), which contains multiple layers of information. Each agent represents a specific spatial partition, such as an activity or room. In this proof of concept, the spatial partition is a polyomino with no holes (PnH). A polyomino is a set of grid cells that share edges. Each PnH_{agent} induces different types of cells (see Fig. 3):

- $adjacent_{agent}$: outside cells in the von Neumann neighborhood [39] of the PnH_{agent}.
- $bridge_{agent}$: adjacent cells that if expanded, create a hole in the PnH_{agent}.
- $non\text{-}adjacent_{agent}$: cells outside of the von Neumann neighborhood of the PnH_{agent}.
- $surface_{agent}$: internal cells that if eliminated will not create a hole in or divide the PnH_{agent}.
- $structure_{agent}$: internal cells that if eliminated will create a hole or divide the PnH_{agent}.
- $mask_{agent}$: square grid placed according to the centroid of the PnH_{agent}.

For the characterization of these cells, we opted for using custom convolution filters to take advantage of parallel computation.

By avoiding the existence of holes, a PnH has certain interesting properties. For instance, it forms a polygon with orthogonal edge boundaries and avoids containment of one agent by another. Therefore, it can form custom diagrams of floorplans, comprehending organizations such as tight packing and loose packing. The shapes can approximate any

Fig. 3 Row 1: simplified representation of agent; row 2: classification of cells: structure (dark blue), surface (medium blue), and adjacent (light blue); row 3: mask with the action space of the agent; row 4: diagram with the legal actions for the current state of the agent; row 5: indication of folding cells: L-folds (light blue), U-folds (dark blue); row 6: representation of ratio area/target area: the agents have the respective (area, target area) pairs: 0: (17, 2), 1: (11, 4), 2: (15, 8), 3: (13, 16), 4: (12, 25); row 7: representation of soft adjacency values; row 8: the current utility of the agent represented in the internal cells of the agent. Color scale for rows 2–8 ▶

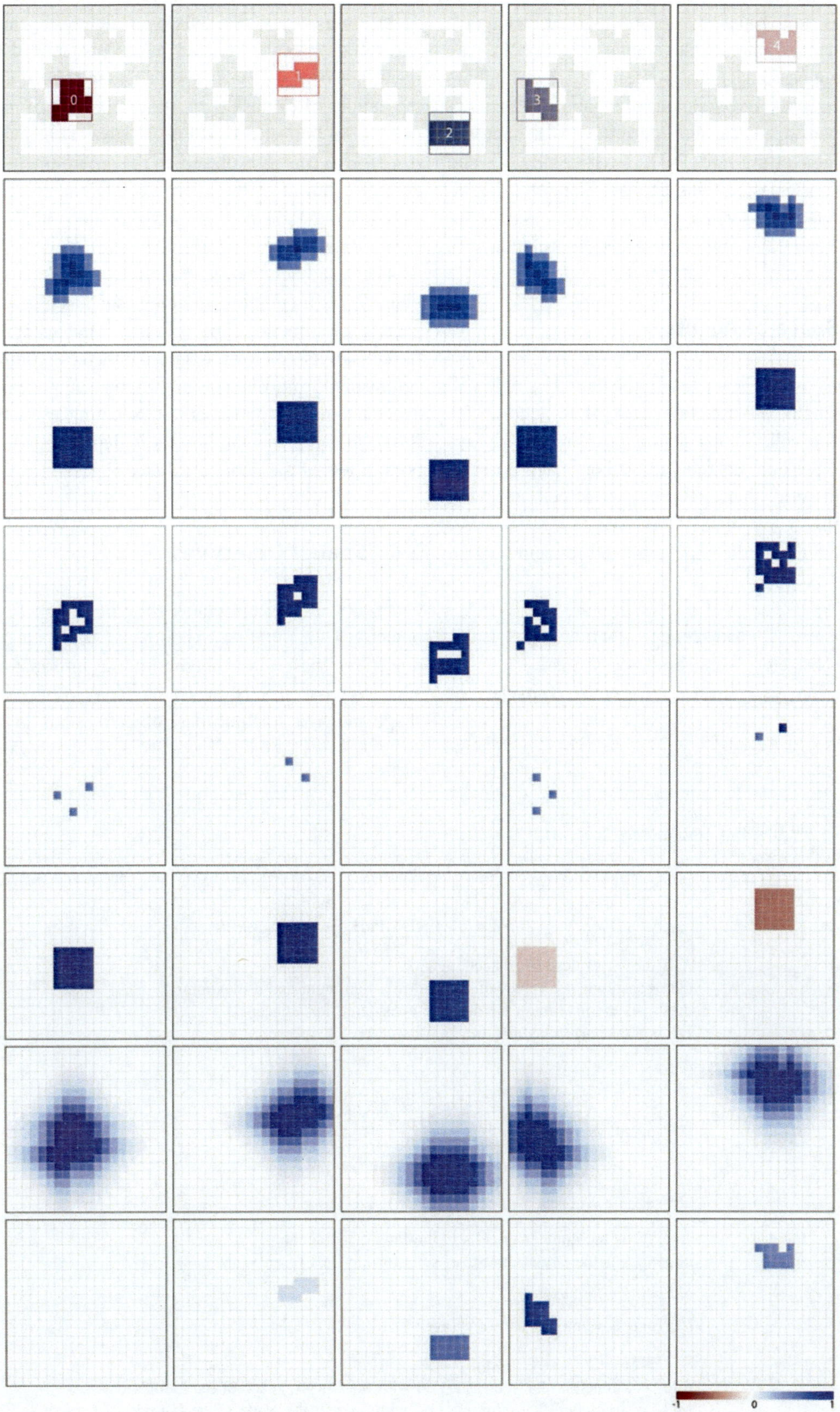
0
1
2
3
4
-1
0
1

polygon, such as orthogonal and non-orthogonal rectangles, or free forms such as amoeba-like shapes (see Fig. 1). The resulting polyominoes can be post-processed to depict bubble diagrams or support the development of a parametric model.

Furthermore, an agent can be interpreted as a microenvironment by converting the PnH_{agent} to $empty_{env}$ and $adjacent_{agent}$ to $obstacles_{env}$. This enables hierarchical relationships and the representation of different aspects of design, such as zoning, building layout, and floorplan layout.

2.3 Action Space of Agents

The basic actions available for the agents are (1) single-cell expansion, (2) single-cell retraction, and (3) no-action.

Notice that the cells of the action grid are not always available for expansion or retraction. For each possible polyomino in this grid, there will be a set of cells that will preserve the consistency of the PnH (see Fig. 3—row 4). The set of legal cells for the expansion and retraction of an agent is defined as

- legal $expansion_{agent}$ = $adjacent_{agent}$—$bridge_{agent}$—$structure_{neighbor}$—$obstacle_{env}$—cells outside of $mask_{agent}$.
- legal $retraction_{agent}$ = $surface_{agent}$—cells outside of $mask_{agent}$.

In other words, the cells by which an agent can expand and retract are not blocked by an obstacle of the environment, preserve the PnH-ness of all the agents, and are inside the action grid ($mask_{agent}$). It is important to notice that the mask or action grid (see Fig. 3—row 3) has shape (w_{action}, h_{action}) < (w, h). It is placed on the environment based on the current centroid of the agent's PnH. Therefore, as an agent expands the PnH in a certain direction, it also moves the centroid and, consequently, the action grid.

The basic actions are building blocks that can be combined to form complex interplays such as blocking, pushing, pulling, or attraction. For example, if the agent eliminates all the cells of its PnH by retraction, it then can jump to any legal cell inside the current action grid, by a single expansion. The agent also has the option of not executing any action this turn. In the grid representation, this is depicted as the cell to the bottom left of the action grid (see Fig. 3—row 4).

2.4 Spatial Objectives

Our proof of concept enables the definition of a variety of objectives. In our first experiments, we focused on simple objectives based on neighborhood information, such as adjacency, or strictly based on local information, such as shape and area. Each of these objectives has an utility (f) and

Table 1 Utility (f) and spatial (g) representations of objectives

Adjacency

$$f^{adj}(\text{agent}) = \begin{cases} \frac{1}{|K|}\sum_{k\in K} \max\left(g^{adj}(k)\,\text{PnH}_{agent}\right) & \text{if}\,|K| > 0 \\ 1 & \text{otherwise} \end{cases}$$

$$g^{adj}(\text{agent}, \text{dist}_{max}, p) = \left(\frac{\text{dist}_{max} - \text{clip}(\text{DT}(\text{PnH}_{agent}) - 1, 0, \text{dist}_{max})}{\text{dist}_{max}}\right)^p$$

K is the set of desired adjacent agents
Clip constrains the values of the grid to a minimum and a maximum
$dist_{max}$ is the maximum L1 distance considered for adjacency
DT is a distance transform function based on L1 distance

Area

$$f^{area}(\text{agent}) = \begin{cases} \frac{\text{area}_{agent}}{\text{target}_{agent}} & \text{if } \frac{\text{area}_{agent}}{\text{target}_{agent}} \le 1 \\ 1 - \frac{\text{area}_{agent} - \text{target}_{agent}}{\text{target}_{agent}} & \text{if } \frac{\text{area}_{agent}}{\text{target}_{agent}} \le 2 \\ 0 & \text{otherwise} \end{cases}$$

$$g^{area}(\text{agent}) = \text{mask}_{agent}\min\left(\frac{\text{area}_{agent}}{\text{target}_{agent}} - 1,\ 1\right)$$

$area_{agent}$ is the number of cells in the PnH of an agent
$target_{agent}$ is the target number of cells for the PnH of an agent

Shape (non-folding)

$$f^{fold}(\text{agent}) = \max\left(1 - \text{sum}(g^{fold}(\text{agent})), 0\right)$$

$$g^{fold}(\text{agent}, \text{fold}_{max}) = \frac{\text{fold}_L(\text{agent})}{\text{fold}_{max}} + \frac{2\text{fold}_U(\text{agent})}{\text{fold}_{max}}$$

$fold_L$ is a function that returns a grid with ones in the place of the L-folds
$fold_U$ is a function that returns a grid with ones in the place of the U-folds
$fold_{max}$ is a parameter that restricts the number of acceptable folds

a spatial representation (g) (see Fig. 3—rows 5–7 and Table 1).

Each goal function (f) has no restriction to any analytical form; it should return a value in the unit interval representing the performance of the agent in a given state with respect to a goal. Each spatial function (g) returns a layer of information of size (w, h) with the spatial hints for the performance of the agent with respect to the goal, which intends to ease training and enable parameterization by the user. For the current proof of concept, we have functions to indicate area, smooth adjacency, and non-folding metrics (stimulates shapes with few folds, such as rectangle, *L*, or *U*).

The utility of the agent state is based on a combination of the selected goals, such as in a weighted average or in the multiplication of the terms. In our experiment, we opted for multiplying the adjacency goal by the average of the other terms. This prevents the agent from getting stuck in local optima based on individual goals, which are generally easier to achieve.

2.5 Learning a Policy

In RL, the goal of the agent is to learn directly or indirectly a policy—i.e., a mechanism that suggests actions in response to its current state in order to maximize the performance of its future trajectory. Learning a policy relies on the interaction of the agent with the environment, which is formalized as a Markov decision process (MDP) [40]. In our proof of concept, the MDP consists of:

- a set of states (S): each state is defined by the agent's percepts, which has multiple layers with the internal and the environmental information.
- a set of actions (A): legal cells for expansion and retraction, induced by the current PnH, plus the extra cell for no-action.
- a set of rewards (R): each reward is a signal with the Δ of the utility between consecutive states s and s' that informs the performance of the agent.
- a function that maps each state-action pair to the resulting reward and state.

At each timestep, an agent can interact with the environment by observing its current state (s) (Fig. 4) and taking an action (a) among the available actions. As a result, the agent moves to the next state (s') and receives an evaluative feedback in the form of a reward (r) from the environment [40] (Fig. 5).

Our setting consists of multiple spatial agents taking actions in the environment. In this case, the architectural spaces are built in parallel, displaying partial information and

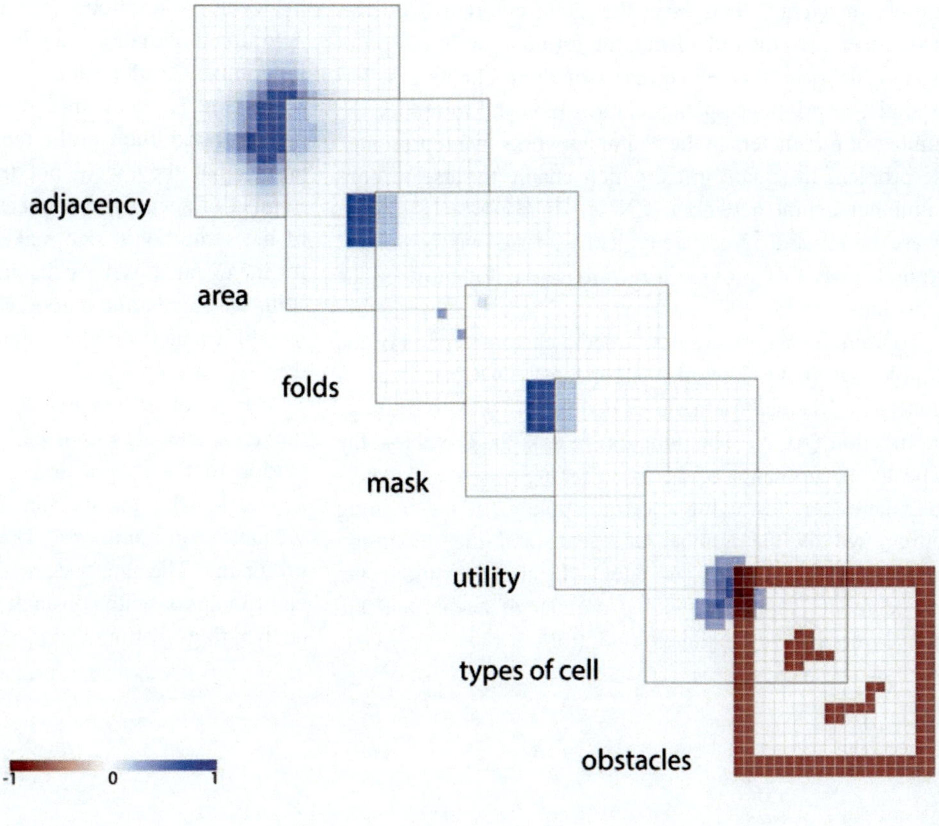

Fig. 4 Example of input state for a single agent: array with shape (7, 20, 20)

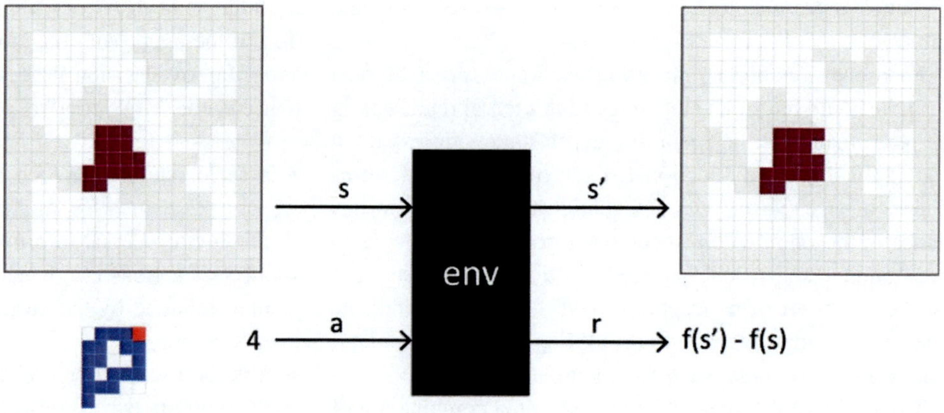

Fig. 5 Interaction between agent and environment. The representation of the states was simplified for visualization

enabling more opportunities for direct, fine-grained interactions. The agents select the actions synchronously, which requires an algorithm to order the execution and solve conflicts. The agents can share information before executing an action, but the possible combination of actions (joint-action space) grows exponentially with the number of agents.

To address this setting, we focus on multi-agent deep reinforcement learning (MADRL), the specific branch of RL that uses models, such as deep neural networks, to estimate the values of state-action pairs (Q-values), or policies for multiple agents. MADRL provides centralized and decentralized strategies to address the problem of multiple interactions and of the exponentially large joint-action space.

On one hand, interactive spatial synthesis presents some non-conventional demands for MADRL, such as varying number of agents, changes in the agent configuration over the time, etc. On the other hand, the formulation of our proof of concept tries to alleviate many of these challenges. For example, to privilege spatial relationships and to reduce the number of parameters in the neural networks, we represented the problem in square grids, which enable the use of convolutional neural networks (CNNs). In contrast to games, where the reward signals are typically sparse, we formulated explicit goals that provide a reward signal for most of the transitions.

To train the agents, we developed a custom version of the Double Deep Q-Networks algorithm (DDQN) [41], a model-free RL method that uses neural networks to estimate the function $Q(s, a)$. This function returns the Q-value—the expected performance of the agent taking action a in state s and, thereafter, following a certain policy. In the training setting, we fix the number of agents and the maximum number of adjacencies per agent. In the execution, the trained model supports a varied number of agents and the input layers provide parametric control.

3 Results

In this initial setting, we trained six agents with a maximum of three adjacencies per agent in an environment with random obstacles. The details of our learning algorithm will be described in a future publication. In this section, we will visually present some of the initial results with agents that mix the learned policy with random actions.

In Fig. 6, we display 20 timesteps of the interaction of the agents in a random environment with the same settings as in the training. The first two rows display the initial 10 timesteps, which show how the trajectory of the agents is easy to grasp visually. In the following two rows, we selected every 100th timestep to show how the behavior of the agent results in local exploration after reaching good configurations. Additional variation can be incentivized by adding more randomness to the policy.

In Fig. 7, we display another simulation with twelve agents. In addition to the fact that the agents will face situations that they were not trained for, the increase in the number of agents and connections as well as the preservation of the same board size makes it significantly more difficult for the agents to satisfy the adjacency and area requirements. Still, the agents find a good configuration and try to improve its performance with local changes, which are constrained by the lack of empty spaces.

Our proof of concept intends to verify whether our formulation and algorithm can result in an effective approximation of the state-action values (Q-values), so that, from any state, the agents can follow a policy that explores valuable configurations. The preliminary results are very promising. The agents learned an effective policy for random environments with obstacles, generalized it for a more challenging setting, and generated variations of the

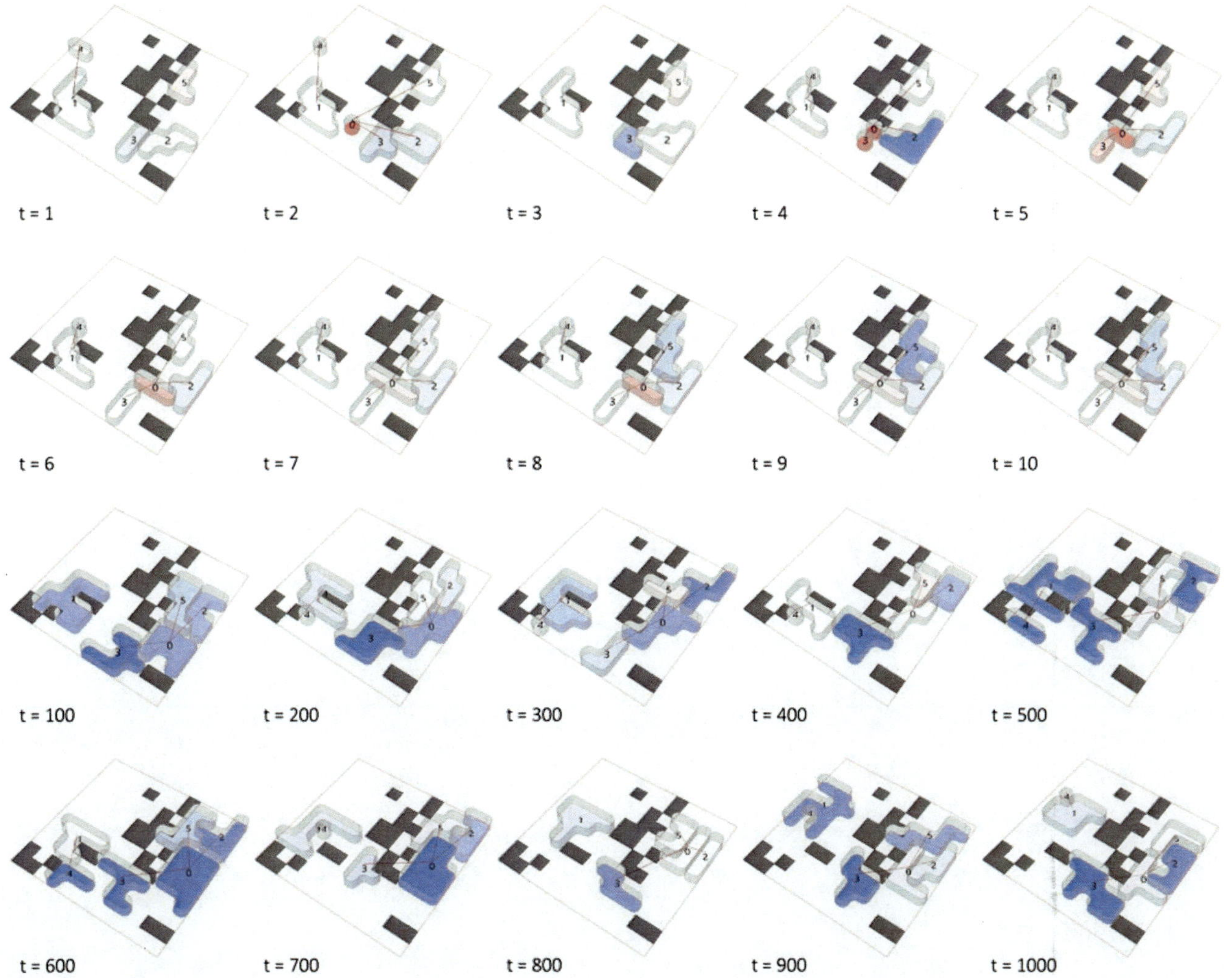

Fig. 6 Simulation of six agents. The goals for the areas are: 0: 5, 1: 7, 2: 4, 3: 3, 4: 1, 5: 5. The goals for adjacencies are: 0: [2, 3, 5], 1: [4], 2: [0], 3: [0], 4: [1], 5: [0]

arrangement in local regions of the design space. Most importantly, they can do all these things by taking increment steps in the environment.

4 Future Steps

In this paper, we have presented a novel framework for interactive space planning with self-learning agents. We describe both the formulation and our proof of concept, which demonstrates that it is possible to teach agents how to behave in large design spaces, addressing specific design goals, and preserving fine-grained interaction.

Future steps of this research will address three topics: learning, representation, and interaction. In terms of learning, we plan to investigate improvements in the learning algorithm, alternative solutions, baselines, and forms of evaluation. The learning algorithm and any improvements will be described in detail in a future publication. We plan to expand this research to the 3D setting and investigate other

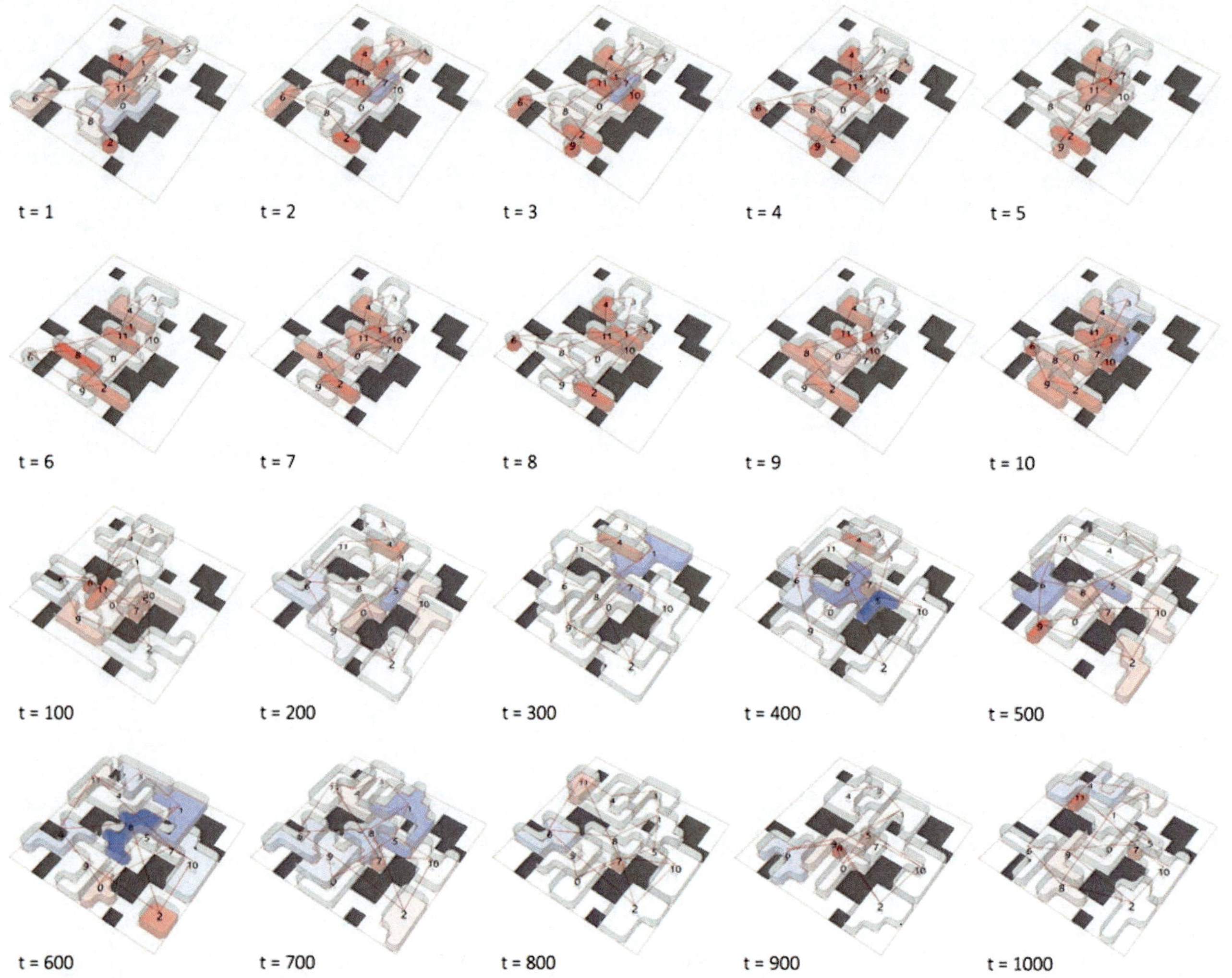

Fig. 7 Simulation of twelve agents. The goals for the areas are [0: 6, 1: 8, 2: 9, 3: 6, 4: 9, 5: 2, 6: 6, 7: 2, 8: 5, 9: 8, 10: 8, 11: 8]. The goals for adjacencies are 0: [5, 7, 9], 1: [3, 5, 8], 2: [8, 9, 10], 3: [1, 4], 4: [3, 11], 5: [0, 1, 10], 6: [8, 9, 11], 7: [0, 10, 11], 8: [1, 2, 6], 9: [0, 2, 6], 10: [2, 5, 7], 11: [4, 6, 7]

forms of design representation that can be integrated into our framework. Finally, our main motivation for developing self-learning agents is to contribute to the development of interactive generative systems that can support observer-dependent and circular processes for custom design generation. We plan to integrate the current proof of concept with a game engine and to explore specific types of interactions and strategies to navigate the design space.

Acknowledgements This research was supported in part by funding from the Carnegie Mellon University Frank-Ratchye Fund for Art @ the Frontier as well as a PhD scholarship granted by the Brazilian National Council for Scientific and Technological Development (CNPq).

References

1. Mitchell, W. (1977). *Computer-aided architectural design*. New York: Mason Charter.
2. Eastman, C. M. (1975). *Spatial Synthesis in Computer-Aided Building Design*. New York: Elsevier Science Inc.
3. Simon, H. A. (1996). The Science of Design: Creating the Artificial. In *The Sciences of the Artificial* (3rd ed.). Cambridge: The MIT Press.
4. Henrion, M. (1978). Automatic space-planning: A postmortem? In: In J. C. Latombe (Ed.), *Artificial Intelligence and Pattern Recognition in Computer Aided Design* (pp. 175–191). New York: North-Holland.
5. Liggett, R. S. (2000). Automated facilities layout: Past, present and future. *Automation in Construction, 9*(2), 197–215.
6. Fischer, T., & Herr, C. M. (2001). Teaching generative design. In: *Proceedings of the 4th Conference on Generative Art*. Milan: Politecnico di Milano.
7. Kolarevic, B. (2005). Digital Morphogenesis. In: B. Kolarevic (Ed.), *Architecture in the digital age: Design and manufacturing* (pp. 12–28). London: Taylor & Francis.
8. Kalay, Y. E. (2004). *Architecture's New Media: Principles, Theories, and Methods of Computer-Aided Design*. Cambridge: The MIT Press.
9. Oxman, R. (2006). Theory and design in the first digital age. *Design Studies, 27*(3), 229–265. https://doi.org/10.1016/j.destud.2005.11.002
10. Gorbman, Y. J., Yezioro, A., & Capeluto, I. G. (2009). Computer-Based Form Generation in Architectural Design—A Critical Review. *International Journal of Architectural Computing, 7*(4), 535–553. https://doi.org/10.1260/1478-0771.7.4.535
11. Vishal, S., & Gu, N. (2012). Towards an integrated generative design framework. *Design Studies, 33*(2), 185–207. https://doi.org/10.1016/j.destud.2011.06.001
12. Henrique, G. C., Bueno, E., Lenz, D., & Sardenberg, V. (2019). Generative Systems: Interwining Physical, Digital and Biological Processes, a case study. In: J. P. Sousa, G. C. Henriques, & J. P. Xavier (Eds.), *Architecture in the age of the 4th Industrial Revolution: Proceedings of the 37th eCAADe and 23rd SIGraDi Conference* (Vol. 1, pp. 25–34). Porto: eCAADe-SIGraDi-FAUP.
13. Nagy, D., Lau, D., Locke, J., Stoddart, J., Villaggi, L., Wang, R., Zhao, D., & Benjamin, D. (2017). Project Discover: An Application of Generative Design for Architectural Space Planning. In: M. Turrin, B. Peters, W. O'Brien, R. Stouffs, T. Dogan (Eds.), *Proceedings of Symposium on Simulation for Architecture and Urban Design* (pp. 59–66). San Diego: Society for Computer Simulation International.
14. Anderson, S. (1966). *Problem-Solving and Problem-Worrying*. Retrieved from http://web.mit.edu/soa/www/downloads/1963-69/TH_AALond-Lect_66.pdf
15. Schön, D. (1983). *The Reflective Practitioner: How Professionals Think In Action*. New York: Basic Books.
16. Corona-Martínez, A. (2003). *The Architectural Project*. (A. Corona-Martínez & M. Quantrill, Trans., M. Quantrill, Ed.). College Station: Texas A&M University Press.
17. Cross, N. (2006). *Designerly ways of knowing*. London: Springer.
18. Lawson, B. (2005). *How designers think: The design process demystified* (4th ed.). Amsterdam: Elsevier.
19. Gänshirt, C. (2007). *Tools for ideas: Introduction to architectural design*. Basel: Birkhäuser.
20. Simon, H. A. (1977). The structure of ill-structured problems. In: *Models of discovery* (pp. 304–325). Dordrecht: Springer.
21. Akin, O., Dave, B., & Pithavadian, S. (1987). Problem structuring in architectural design. *Carnegie Mellon University*. https://doi.org/10.1184/R1/6076109.v1
22. Jones, J. C. (1992). *Design Methods* (2nd ed.). New York: Wiley.
23. Russel, S. J., & Norvig, P. (2010). *Artificial intelligence: A modern approach* (3rd ed.). Upper Saddle River: Prentice Hall.
24. Wilensky, U., & Rand, W. (2015). *An Introduction to Agent-based Modeling: modeling natural, societal, and engineered complex systems with netlogo*. Cambridge: The MIT Press.
25. Veloso, P., Rhee, J., & Krishnamurti, R. (2019). Multi-agent Space Planning: A Literature Review (2008–2017). In J.-H. Lee (Ed.), *Hello, Culture!: Proceedings of 18th CAAD Futures Conference* (pp. 52–74). Daejeon, Korea.
26. Herr, C. M., & Ford, R. C. (2015). Adapting Cellular Automata as Architectural Design Tools. In *Emerging Experience in Past, Present and Future of Digital Architecture: Proceedings of the 20th CAADRIA Conference* (pp. 169–178). Daegu: Kyungpook National University.
27. Koenig, R. (2011). Generating Urban Structures: a Method for Urban Planning Supported by Multi-Agent Systems and Cellular Automata. *Przestrzen I Forma,*(16), 353–376.
28. Araghi, S. K., & Stouffs, R. (2015). Exploring cellular automata for high density residential building form generation. *Automation in Construction, 49*, 152–162.
29. Smith, S. I., & Lasch, C. (2016). Machine Learning Integration for Adaptive Building Envelopes: An Experimental Framework for Intelligent Adaptive Control. In K. Velikov, S. Manninger, M. del Campo, S. Ahlquist, & G. Thün (Eds.), *Proceedings of the 36th ACADIA Conference: Posthumans Frontiers* (pp. 98–105). Ann Arbor: ACADIA.
30. Hosmer, T., & Tigas, P. (2019). Deep Reinforcement Learning for Autonomous Robotic Tensegrity. In: *Ubiquity and Autonomy: Proceedings of the 39th ACADIA Conference* (pp. 16–29). Austin: ACADIA.
31. Jabi, W., Chatzivasileiadi, A., Wardhana, N. M., Lannon, S., & Aish, R. (2019). The synergy of non-manifold topology and reinforcement learning for fire egress. In: *Architecture in the age of the 4th Industrial Revolution: Proceedings of the 37th eCAADE Conference and of the 23rd SIGraDi Conference* (Vol. 2, pp. 85–94). Porto: eCAADe-SIGraDi.
32. Xu, T., Wang, D., Yang, M., You, X., & Huang, W. (2018). An Evolving Built Environment Prototype. In: T. Fukuda, W. Huang, P. Janssen, K. Crolla, & S. Alhadidi (Eds.), *Learning, Adapting and Prototyping: Proceedings of the 23rd CAADRIA Conference* (Vol. 2, pp. 207–215). Beijing: CAADRIA.
33. Ruiz-Montiel, M., Boned, J., Gavilanes, J., Jiménez, E., Mandow, L., & Pérez-de-la-Cruz, J.-L. (2013). Design with shape grammars and reinforcement learning. *Advanced Engineering Informatics, 27* (2), 230–245. https://doi.org/10.1016/j.aei.2012.12.004
34. Narahara, T. (2017). Collective Construction Modeling and Machine Learning: Potential for Architectural Design. In A. Fioravanti, S. Cursi, S. Elahmar, S. Gargaro, G. Loffreda, G. Novembri, & A. Trento (Eds.), *Sharing Computational Knowledge!: Proceedings of the 35th eCAADe Conference*. (pp. 593–600). Rome: eCAADe.
35. Nguyen, T. T., Nguyen, N. D., & Nahavandi, S. (2018). Deep Reinforcement Learning for Multi-Agent Systems: A Review of Challenges, Solutions and Applications. arXiv:1812.11794 [cs, stat]. Retrieved from http://arxiv.org/abs/1812.11794
36. Meyboom, A., Reeves, D. (2013). Stigmergic Space. In P. Beesley, M. Stacey, & O. Khan, (Eds.), *Adaptive*

Architecture: Proceedings of the 33rd ACADIA Conference (pp. 200–206). Toronto: Riverside Architectural Press.
37. Fernando, R. (2014). Space Planning and Preliminary Design Using Artificial Life. In: N. Gu, S. Watanabe, H. Erhan, M. H. Haeusler, W. Huang, & R. Sosa, (Eds.), *Rethinking Comprehensive Design: Speculative Counterculture: Proceedings of the 19th CAADRIA Conference* (pp. 657–666). Hong Kong: CAADRIA.
38. Guo, Z., & Li, B. (2017). Evolutionary approach for spatial architecture layout design enhanced by an agent-based topology finding system. *Frontiers of Architectural Research, 6*(1), 53–62. https://doi.org/10.1016/j.foar.2016.11.003
39. Weisstein, E. W. (n.d.). von Neumann Neighborhood. *MathWorld -A Wolfram Web Resource.* Retrieved from https://mathworld.wolfram.com/vonNeumannNeighborhood.html
40. Sutton, R. S., & Barto, A. G. (2018). *Reinforcement learning: An introduction* (2nd ed.). Cambridge: The MIT Press.
41. van Hassel, H., Guez, A., & Silver, D. (2016). Deep Reinforcement Learning with Double Q-learning. In: *Proceedings of the Thirtieth AAAI Conference on Artificial Intelligence* (pp. 2094–2100). Phoenix: AAAI.

Pedro Veloso Veloso is a computational designer, educator, and researcher interested in design exploration methods supported by different modes of artificial intelligence, such as classical AI, bio-inspired AI, agent-based modeling, and machine learning. He has completed Bachelor of Architecture and Urbanism from the University of Brasilia. He has completed Master of Architectural Design from the University of Sao Paulo. Currently, he is a PhD candidate in Computational Design at Carnegie Mellon University, investigating generative design and the development of interactive systems with learning techniques.

Ramesh Krishnamurti Krishnamurti is a full professor in the School of Architecture at Carnegie Mellon University. He has a BE (Honors) in Electrical Engineering from the University of Madras, a BA in Computer Science from the University of Canberra, and MASc and PhD in Systems Design from the University of Waterloo. His principal area of research is computational design with emphasis on the formal, semantic, and algorithmic aspects of generative construction and development of design as computation via highly coupled parallel explorations of form and description. He is perhaps best known for his work on computational problems in shape grammar theory and algorithms for spatial patterns.

Topologic: Exploring Spatial Reasoning Through Geometry, Topology, and Semantics

Wassim Jabi and Aikaterini Chatzivasileiadi

Abstract

Topologic is a software modelling library that supports a comprehensive conceptual framework for the hierarchical spatial representation of buildings based on the data structures and concepts of non-manifold topology (NMT). Topologic supports conceptual design and spatial reasoning through the integration of geometry, topology, and semantics. This enables architects and designers to reflect on their design decisions before the complexities of building information modelling (BIM) set in. We summarize below related work on NMT starting in the late 1980s, describe Topologic's software architecture, methods, and classes, and discuss how Topologic's features support conceptual design and spatial reasoning. We also report on a software usability workshop that was conducted to validate a software evaluation methodology and reports on the collected qualitative data. A reflection on Topologic's features and software architecture illustrates how it enables a fundamental shift from pursuing fidelity of design form to pursuing fidelity of design intent.

Keywords

Conceptual design • Geometry • Semantics • Spatial reasoning • Topology

1 Introduction

The advent of building information modelling (BIM) systems has revolutionised the architecture, engineering and construction (AEC) sector. By allowing architects, engineers, consultants and contractors to collaborate on a project and share structured and rich information that includes three-dimensional models and federated databases, BIM systems have helped increase coordination and productivity and reduce mistakes [23]. However, while BIM systems have largely served the design development and construction phases well, they have not been as effective in the early phases of conceptual design. This is mainly due to three issues. First, BIM systems focus on creating building components that are highly specified. Designers in the early phases may not know what specific materials, or even forms, they wish to use. They usually employ abstraction as a method to address complexity and postpone specification to a later date. Second, BIM systems, while adept at placing and aggregating components, do not have a formal system of connecting these components. The early phases of design are usually characterised by solving topological issues of adjacency, access, and circulation. Yet, BIM components usually do not have intrinsic adjacency information. Finally, this lack of formal topological information reduces what can be termed the component's spatial awareness—its ability to intelligently react to spatial changes in its surrounding. While BIM components certainly react to user actions and changes in design, they do not always do so correctly and take into account the adjacent spaces because the knowledge of how to react is embedded in the algorithm and not formalised in the data structures. The complexity, lack of formal topology, and resulting difficulties with spatial awareness means that BIM systems struggle to support conceptual design and spatial reasoning. For example, it is usually difficult for BIM systems to conduct shortest path analyses in three dimensions or report on the centrality or isolation metrics of certain spaces. To solve these issues, a plethora of plug-ins have been introduced to help with some of the limitations found in BIM software and improve their functionality. Yet these plug-ins are not generalisable because they must restructure BIM data to satisfy software functionality.

W. Jabi (✉) · A. Chatzivasileiadi
Cardiff University, Cardiff, CF10 3NB, UK
e-mail: jabiw@cardiff.ac.uk

A. Chatzivasileiadi
e-mail: chatzivasileiadia@cardiff.ac.uk

S. Eloy et al. (eds.), *Formal Methods in Architecture*, Advances in Science, Technology & Innovation,
https://doi.org/10.1007/978-3-030-57509-0_25

To address the above issues, the authors have been conducting a funded research project that investigates standardised methods for creating connected three-dimensional entities based on non-manifold topology (NMT) [4, 9, 22]. This has resulted in a comprehensive conceptual framework for the hierarchical spatial representation of buildings implemented in a software modelling library named Topologic [19]. While other features of Topologic have been published elsewhere [3, 11, 18, 20], the focus of this paper is on Topologic's support for conceptual design and spatial reasoning by integrating geometry, topology and semantics.

The remainder of the text is organised into five sections. Section 2 summarises related work on NMT starting in the late 1980s. Section 3 describes the aims of Topologic, its software architecture, methods and classes. Section 4 discusses how Topologic's features support spatial reasoning and conceptual thinking. Section 5 reports on a workshop that was conducted internally to test a software evaluation methodology. In this section, we describe three prototypical workflows in Topologic and some of the qualitative data that we collected from the workshop. Finally, Sect. 6 offers concluding remarks on the role of Topologic in the conceptual design phase.

2 Related Work

The realisation that NMT has benefits to offer in building information modelling goes back to the late 1980s and early 1990s. As part of his Ph.D. dissertation at the Rensselaer Polytechnic Institute, Kevin Weiler invented the winged edge data structures and other topological concepts needed for non-manifold topology modelling [33]. In 1989, Masuda et al. created a non-manifold geometric system that can manipulate entities of different dimensionalities in one architecture [24]. In their paper, they provide the mathematical definition of non-manifold geometric models and introduce the Euler operations needed to manipulate them. In 1991, Crocker and Reinke presented non-manifold topology as a new approach "[…] to overcome editing and coverage limitations associated with the manifold nature of existing boundary evaluation schemes" [12]. Their system extended 3D Boolean operations to include surfaces and wireframes in addition to solids. In 1992, Björk analysed building product models that included the topology of spaces and their boundaries [6]. In anticipation of the issues with which BIM systems currently struggle, Björk posits that building models should contain the appropriate data structures to capture the semantics needed to allow design and construction stakeholders to extract the information they need. In 1997, Cavalcanti et al. introduced a new methodology for subdividing the three-dimensional entities from a set of surface patches [9]. Their system maintained an explicit representation of cells forming a non-homogeneous object and enabled its creation using operations that resemble constructive solid geometry. Similarly in 1997, Chang and Woodbury discussed a non-manifold-based representational scheme for efficient exploration of design spaces [10]. They invented a labelling scheme on top of a non-manifold cell-complex to represent entities of different dimensionalities. Their state representation (*srep*) was bi-partite. The first part contained an exhaustive subdivision of space using 0D, 1D, 2D and 3D entities, and the second part contained attribute sets that listed different collections of cells. This scheme was used to build a generative design system (SEED-Config) based on the ACIS non-manifold 3D modeller [31]. In 2009, Lee et al. used a top-down space subdivision method based on non-manifold topology for the modelling of ship compartments [22]. In 2010, Boguslawski and Gold introduced the dual half edge (DHE) structure that allowed them to represent simultaneously the primal and dual graph of non-manifold entities [7]. As we will see below, dual graphs are essential for querying and visualising topological information and in helping with navigation between topological cells. Boguslawski later used this feature as a representation in escape route planning for tall buildings [8]. In 2017, Furiani et al. at Tre University in Rome, Italy, started the development of a system that uses sparse integer arrays for geometric computing of cells of a very general type [14]. Their system, aimed for biomedical imaging, neural networks and image analysis research, was written using the novel Julia Programming Language. Their project aims to compute the cells of an unknown arrangement and to provide the numerical and topological tools to other algorithms.

3 Topologic

Topologic is intended as a bridge between the conceptual design phase and the BIM-based design development phase. It brings forward building performance analyses including spatial reasoning so that architects and designers can reflect on their decisions and rigorously conceptualise their design before the complexities of BIM set in. Topologic's structured data can then be used in conjunction with intelligent rule-based systems to populate BIM components and to further develop the design within a BIM platform. In this paper, we will focus on Topologic's ability to conduct formal spatial reasoning and semantics-based geometrical and topological operations and illustrate those through detailed workflow descriptions.

Topologic is designed as a core library, written in C++ with additional software layers that expose its methods and data structures to visual data flow programming (VDFP) applications such as Autodesk Dynamo and Grasshopper3D.

At the lowest layer, Topologic uses Open CASCADE, an open-source NMT geometry software development kit that provides data structures and modelling algorithms for 3D solid structures [26]. TopologicCore implements the core topologic classes and methods using an object-oriented programming (OOP) approach with additional support utilities. To enable communication with third-party software, we implemented an interface layer, written in the .NET C++/CLI language [13].

Topologic's classes include: Vertex, Edge, Wire, Face, Shell, Cell, CellComplex, Cluster, Topology, Graph, Aperture, Content, Context and Dictionary (see Fig. 1). A vertex is a point in 3D space with X, Y, Z coordinates. A start Vertex and an end Vertex create an Edge. A set of connected edges creates a Wire. A closed set of one or more wires can form the basis of a Face. A set of connected faces creates a Shell. A closed Shell creates a Cell. A set of connected cells creates a CellComplex. A Cluster is a grouping of topologies of any dimensionality. At this point, it is important to draw a distinction between geometry and topology. An Edge can be used to illustrate this distinction. Topologically, an Edge can be described as a border between two regions or a connection between two locations regardless of its geometric complexity. In Topologic, an Edge can represent a geometry of arbitrary complexity and that geometry can be retrieved on demand.

A Graph in Topologic is an extension of the more classical dual graph [8] and is derived from other topologies. The Graph class and associated methods are based on graph theory [32]. A Graph is composed of vertices and edges that connect them. A Graph in Topologic accepts as input any topology with additional optional parameters and outputs a Graph object. In its simplest form, the dual graph of a CellComplex is a Graph that connects the centroids of adjacent Cells with a straight Edge. The topology of a Graph of a CellComplex is a wire (see Fig. 2). As an option, a graph in Topologic can create edges that connect the centres of cells to the centres of apertures and between cells only if they share not only a face, but also an aperture within that face. An aperture is a special type of face that is hosted by another face (e.g. doors and windows). By building a formal dual graph Topologic can conduct full three-dimensional spatial integration/centrality analysis based on graph theory and space syntax theory [16] (see Fig. 3).

Unrelated topologies can be linked together hierarchically through Topologic's Content/Context system where any topology can store one or more other topologies. Reversely, these content (child) topologies maintain a pointer back to their context (parent) topology. This is useful when one wants to associate topologies that are not linked intrinsically through the boundary representation hierarchy.

Finally, any topology in Topologic can contain a dictionary made of one or more key/value pairs. A key is any identifying string of characters (e.g. "ID"), and its value can be of any data type (e.g. a number, a string). When subjected to geometric operations, operand topologies transmit their dictionaries to resulting topologies. For example, if a cell has a dictionary with key/value pair ["ID", "Cell A"] and is sliced into smaller cells, then each of the resulting cells gains a copy of the original dictionary. In addition, if two cells intersect, the intersection aggregates the key/value pairs of its operands. Furthermore, when a dual graph is created, topologies of the primal graph transmit their dictionary to their counterpart topology in the dual graph. This allows for semantic information to be associated with topologies.

4 Spatial Reasoning

The central premise and current trajectory of BIM systems are to focus on creating highly detailed digital models (digital twins) that attempt to accurately represent the three-dimensional form and inventory of the project [27]. This creates a gap in the semantic content of BIM models. This is because BIM models lack a sophisticated ability to capture design intent and logic—the concept of the design. One of the main goals of Topologic is to address this gap by helping architects and designers reflect on, reason about, represent and encode the spatial and conceptual qualities and logic of their design. The aim is to shift design thinking in the early stages from pursuing a fidelity of design form to pursuing a fidelity of design concept. This can be achieved by supporting the designers' need for abstraction, awareness and conceptual thinking through the tools of geometry, topology and semantics.

Architectural design, by its nature, is a continuous process of creating geometric and non-geometric artefacts that stand in, represent and abstract reality [25, 30, 34]. Architectural construction has traditionally been the process of interpreting and realising these abstracted or idealised models [2]. Topologic's classes of objects such as vertices, edges, wires and faces, etc. act as geometric abstractions and placeholders of more detailed representations. For example, a vertex can be the location of an electric outlet, an edge can be a column's centreline, and a face can be the host surface for a curtain wall. Through a custom dictionary and a content/context system, topologies can hold the information, including more detailed geometries, that are required for future design detailing work.

Through topology (i.e. membership, adjacency and connectivity), these entities transcend mere geometry by possessing and utilising intrinsic awareness of their context. In Topologic, the user can query a cell for its faces, edges, or vertices and a unique, non-overlapping list is generated for each category. These returned entities can in turn, be asked to return their constituent members. For example, a cell can

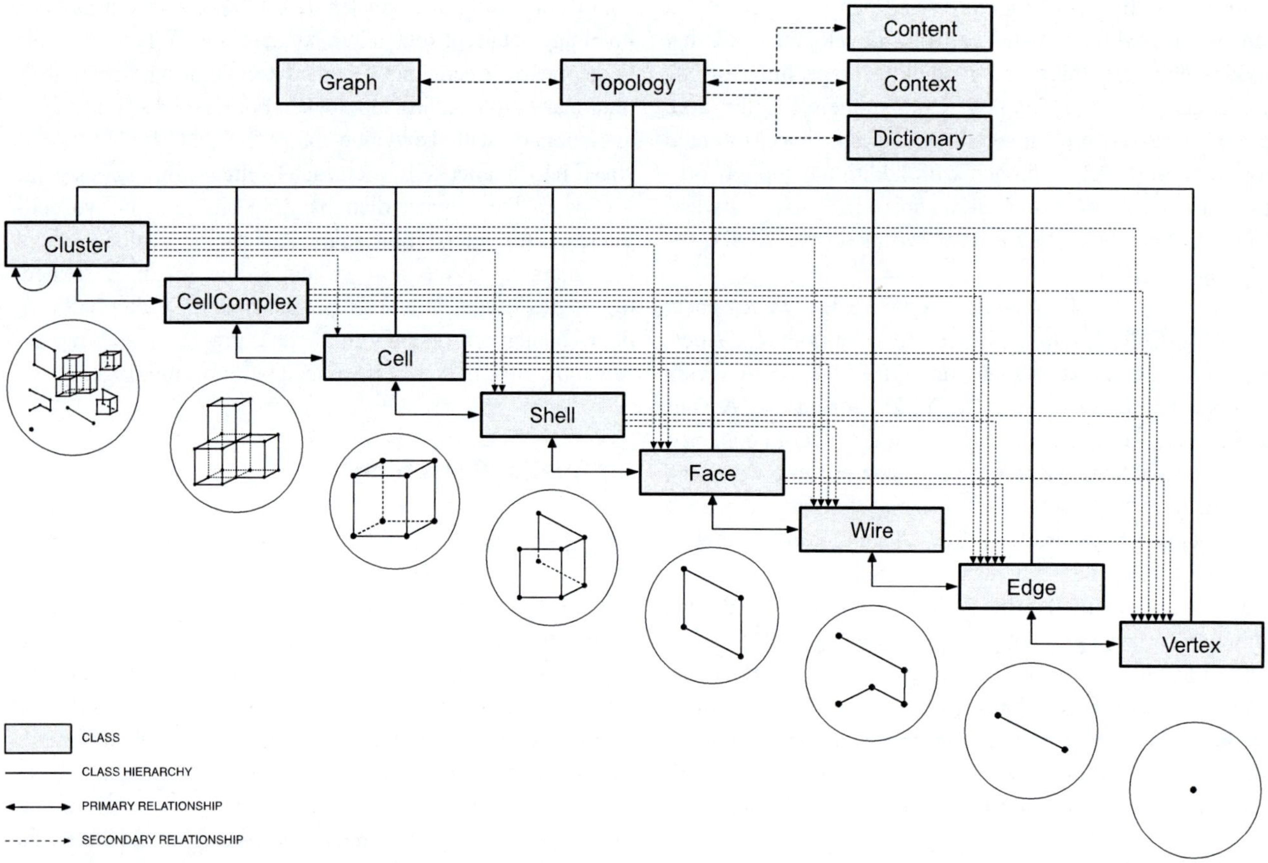

Fig. 1 Topologic's core class hierarchy

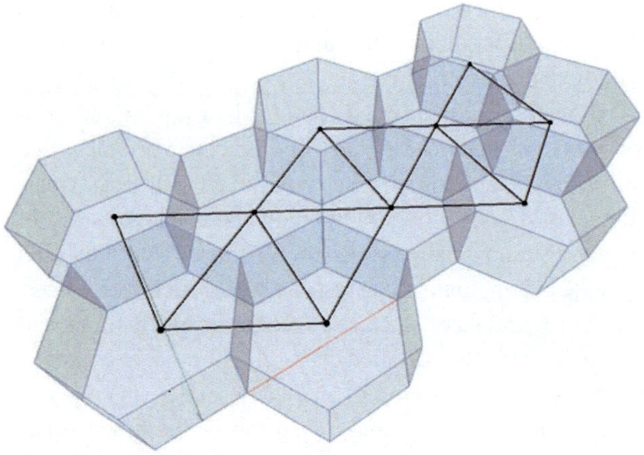

Fig. 2 A primal graph (CellComplex) and its associated dual graph (wire)

return its faces and then ask each returned face for its edges and each returned edge for its vertices. Beyond this top-down query, you can also climb the hierarchy. For example, you can ask a cell to return its vertices, but then you can ask each vertex to return the edges to which it belongs. Similarly, you can ask a cell for its edges and then ask each edge to return the faces that share it.

Conceptual design begins to be supported when you combine these topological queries with other types of semantic queries. As we have seen above, vertices know what edges are connected to them and as such they can query each edge for a certain value in its dictionary or its geometry. Based on the response, the vertex in question can deploy one of several geometric solutions—either by choosing from a list of alternatives in its contents list, or by parametrically modifying its own geometry. One of the interesting features of Topologic is that Boolean operations can manipulate and be manipulated by a topology's semantic information. For example, slicing a cellular building volume with a set of horizontal planes to create floors usually slices all cells/spaces within that building. However, a cell can be assigned a space type (e.g. an atrium) that causes it to resist the slicing and continue to be an undivided volume (see Fig. 4). By simply modifying the semantic assignment, while leaving the remainder of the workflow unchanged, one can achieve a geometrically and topologically different result. Thus, by altering the semantics in the model, the user can experiment with different results and optimise their design. The sum of these abilities enables the designer to think conceptually and strategically about their design.

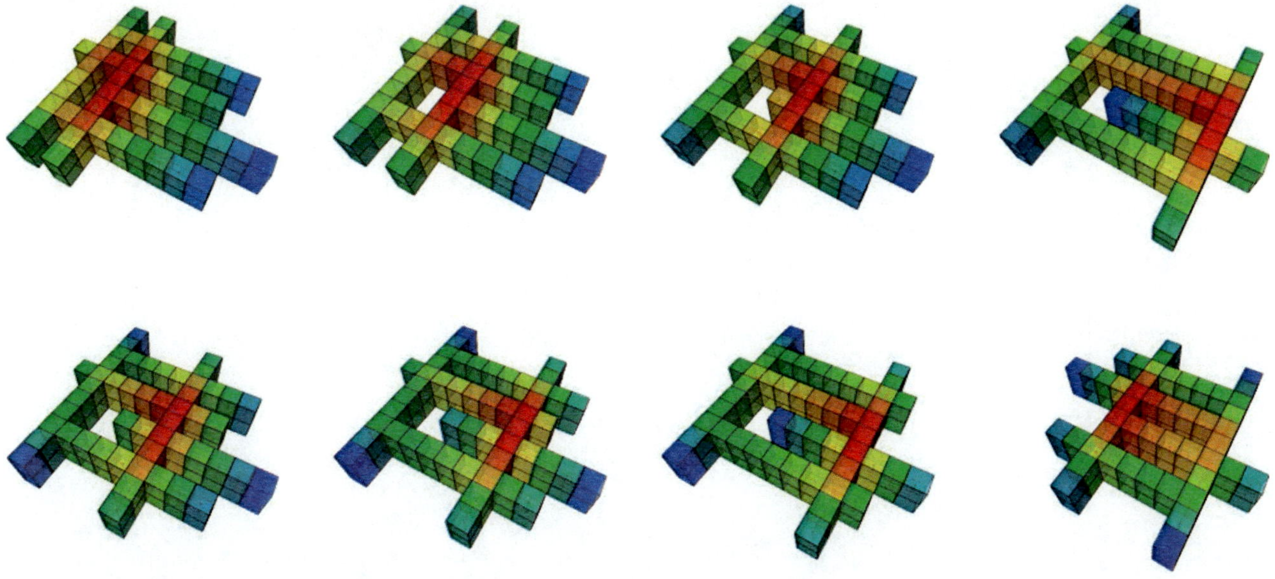

Fig. 3 Graph-based space integration/centrality study conducted in Topologic

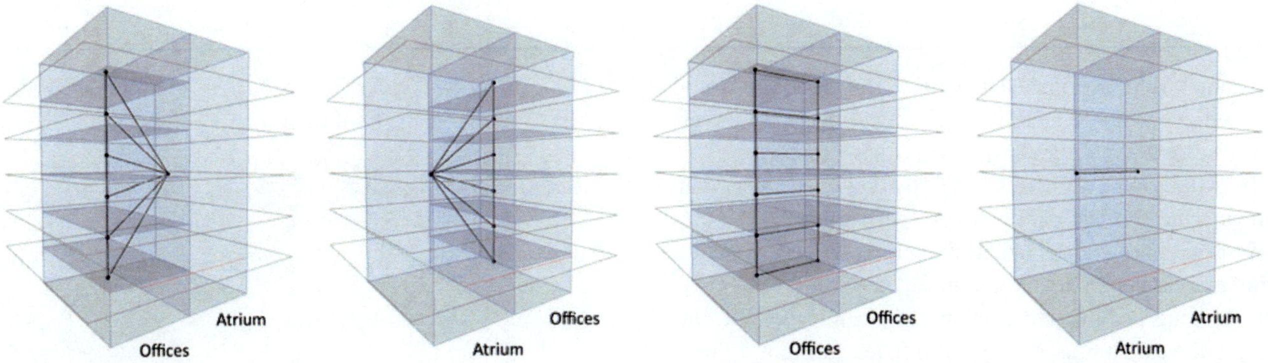

Fig. 4 Semantics-based Boolean operations

They no longer simply aggregate entities in a display list but connect entities that need to be connected and build conceptual models that react logically and provide meaningful information when queried.

5 Usability Workshop

We conducted a software usability workshop where a handful of participants were asked to complete three workflows and then answer a questionnaire and engage in a discussion with the authors. The aim of the workshop was twofold. First, we wanted to evaluate and collect initial feedback on the above concepts and workflows, and second, we wanted to test and verify that the basic evaluation methodology is valid in preparation for conducting a formal and a larger-scale evaluation on which we will report in future publications.

5.1 The Tasks

Three tasks involving creating workflows using Topologic that are relevant to the theoretical concepts described earlier are presented below. In workflow 1 "Geometry and Semantics", the user can explore how information embedded in the dictionaries of topologies can be used to affect the result. Two simple orthogonal intersecting prisms are created. Each prism has a custom dictionary with an identical key ("height") and a custom value assigned to it. These values are independent of the actual geometric height of the prisms. The goal is to use this information later to customise the height of the resulting intersection. The Boolean intersection of these two prisms (indicated below in red) would inherit the custom dictionary values of the two prisms. For example, if one prism has a "height" value of 2 and the other prism has a "height" value of 5, then the resulting intersection will have a dictionary in which the "height" key has a value of {2, 5} (i.e. a list of the

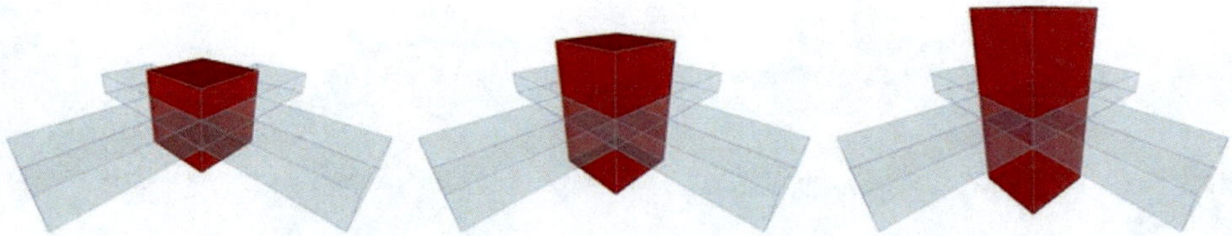

Fig. 5 Two intersecting prisms each with a custom dictionary

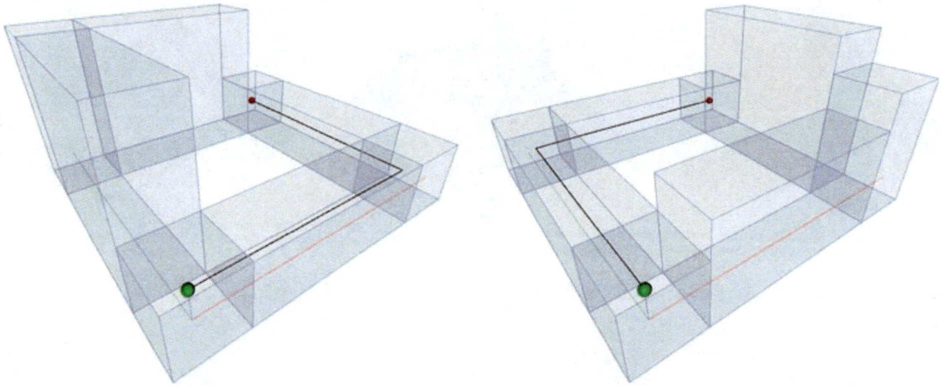

Fig. 6 Shortest path routing through the cells with the smallest overall volume

two values of the original dictionaries). The algorithm would then retrieve and add these two values and use them to vertically scale the intersection geometry (see Fig. 5).

In workflow 2 "Shortest path by key", the user can compute the shortest path between two vertices. The solution takes into consideration the embedded information using graphs and dictionaries. Initially, the user creates Cells with varying heights and combines them into a CellComplex. The next step uses the CellComplex to build a graph and compute the shortest path (see Fig. 6). Each Cell in the CellComplex (and therefore each Vertex) in the graph would have a custom dictionary with an identical key ("volume") and a custom value assigned to it. The goal is to compute the shortest path that goes through the smallest volumes. Input number sliders would be used to change the individual heights of the Cells thus changing their volume. The path would consequently be automatically routed through the Cells with the smallest total volume.

Workflow 3 "Topological distances" enables the user to visualise the integration/centrality of spaces in a building [17], by creating a dual graph of the building. The centrality measure of a space is computed by counting the number of Edges in the shortest path possible between the space in question and every other space in the building. The degree of integration or isolation of a space is indicated using size and colour of spheres placed in each space. The larger the size and the warmer the colour of a sphere the more central/integrated it is and vice versa (see Fig. 7).

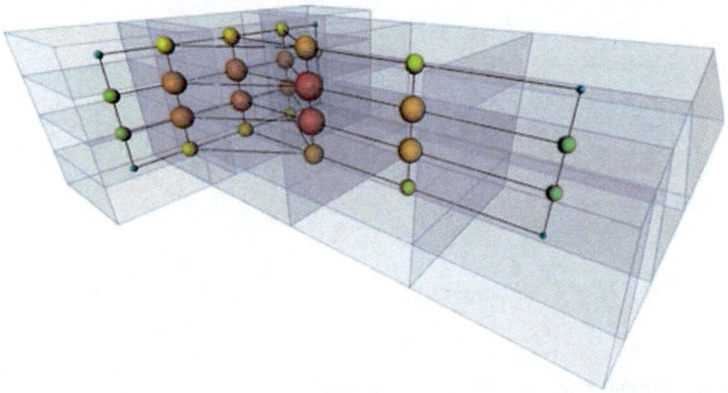

Fig. 7 Visualisation of space centrality

5.2 Evaluation Design and Early Qualitative Results

A usability exercise was conducted to gauge initial perceptions of the user experience regarding the three example workflows presented earlier. These are representative of the type of tasks that a Topologic user might do [1]. A handout including the steps required to implement these was shared with the participants. After completing the workflows, they were asked to share their experience through a questionnaire and a discussion with the authors [15, 21, 28, 29]. This qualitative data, while of moderate reliability, served as a preliminary test bed which the authors will use for the setup of a formal usability experiment in the future.

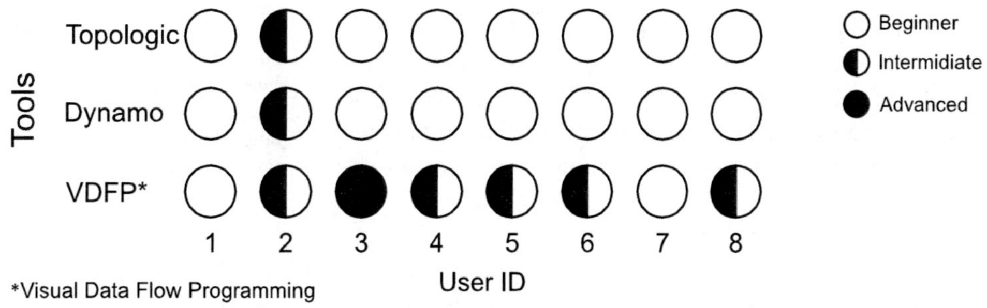

Fig. 8 Level of participants' experience

Participants. Eight postgraduate students who had a similar profile to the targeted end users of the tool participated in the workshop [1, 5]. During the exercise, the authors mainly observed the participants completing the tasks with limited communication [5]. Five of the eight participants were at an intermediate level in using VDFP tools, one was an advanced user, while the remaining two were beginners (see Fig. 8). Seven participants were at a beginner level in Dynamo and Topologic, while one was an intermediate user in both.

Results. All participants were able to complete the given tasks and generally had a positive experience using Topologic. The easy workflows and the fact that scripting can be avoided in some cases were beneficial aspects of Topologic. Some of the collected positive comments were: "pretty straightforward to understand", "this approach of Topologic is an added value and can be pretty useful," and "this function of Topologic would be useful in considering the correlation between different spaces in a building which can help in the design process. This could be used to measure qualitative parameters, such as social interaction […]." On the critical side, users wrote that they "needed to learn the Topologic definitions, e.g. Cells, CellComplex etc., before using the tool in addition to some computer science information, such as the dictionaries concept" and that "without examples to learn the processes and components, [they] would likely find it difficult/timely to [get to know it]. Therefore, links to tutorials/example files would allow users to very quickly pick up Topologic." One user wrote that they are "not sure if [they] would ever undertake such tasks frequently." Other aspects that were confusing were the lack of individual icons for the Topologic nodes and the similar appearance of Dynamo and Topologic nodes, that made it difficult to tell what node belonged to which application. Finally, a limitation of the tool addressed naming conventions and confusion around these, with a typical example being the *Topology.Geometry* and *Topology.ByGeometry* nodes.

Clearly, whether a user would undertake such tasks depends on their interests. It was found that the semantics-based Boolean operations (Task 1) provided the most straightforward approach, while the shortest path analysis (Task 2) required an in-depth study of the theoretical framework prior to engaging with the task. Users seemed to relate more with the spatial integration/centrality analysis through graph theory and space syntax (Task 3), as they were able to link it to practical applications, such as measuring social interactions.

One of the main hurdles of this evaluation was that Topologic is dependent on the host application's interface and naming conventions, so it was difficult to isolate the user experience and focus it solely on the usability of Topologic, instead of the combined usability of Dynamo and Topologic. Moreover, as Topologic is a relatively new tool to be used in conjunction with VDFP applications, most of the participants were at beginner's level with it. The situation was the same regarding the participants' level of experience with Dynamo, yet most of them had an intermediate level of experience with other VDFP applications. Considering the sample of participants, it is anticipated that an intermediate or advanced Dynamo user could potentially give more reliable results in terms of the usability and learnability of the tool. A larger and more diverse sample of participants in terms of experience can yield more reliable results. Additionally, we can give participants the terminology prior to the evaluation so that they familiarise themselves with the concepts. Through the process of conducting this workshop, we had the opportunity to identify parts in the methodology that can be improved regarding the setup of the formal and systematic usability experiment which will allow us to address them in the future and perform a formal and systematic usability experiment with a larger and more diverse group of participants.

6 Conclusion

In this text, we presented a new perspective on supporting conceptual design and spatial reasoning through the integration of geometry, topology and semantics. We described the architecture and features of Topologic, a 3D modelling software library that supports the hierarchical representation of space in architecture and reported on the results from a preliminary evaluation study. Fundamentally, Topologic expands the pursuit of fidelity to design form to include the pursuit of fidelity to design intent. This research has indicated that Topologic can use abstraction as a method to address complexity and provide a formal system for creating, editing and connecting topologies. Components in Topologic have inherent spatial awareness that enables them to intelligently react to requests and changes in their surroundings. Topologic shifts the needed information from ad hoc algorithms to formalised data structures and methods implemented using an object-oriented approach. When combined, these features enable Topologic to support rigorous spatial reasoning and analysis in the conceptual design phase.

References

1. Abdinnour-Helm, S. F., Chaparro, B. S., & Farmer, S. M. (2005). Using the end-user computing satisfaction (EUCS) instrument to measure satisfaction with a web site. *Decision Sciences, 36*(2), 341–364.
2. Aish, R. (1997). MicroStation/J. In A. K. Noor & J. B. Malone (Eds.), Next Generation CAD/CAM/CAE Systems, NASA Conference Publication, 3357. pp. 167–180. National Aeronautics and Space Administration, Hampton, Virginia.
3. Aish, R., Jabi, W., Lannon, S., Wardhana, N. M., & Chatzivasileiadi, A. (2018). Topologic: Tools to explore architectural topology. In L. Hesselgren, A. Kilian, S. Malek, K.-G. Olsson, S.-H. Olga, & C. Williams (Eds.), *Advances in architectural geometry 2018* (pp. 316–341). Klein Publishing GmbH.
4. Aish, R., & Pratap, A. (2013). Spatial information modeling of buildings using non-manifold topology with ASM and DesignScript. In L. Hesselgren, S. Sharma, J. Wallner, N. Baldassini, P. Bompas, & J. Raynaud (Eds.), *Advances in architectural geometry 2012* (pp. 25–36). Vienna: Springer Vienna.
5. Bandi, A., & Heeler, P. (2013). Usability testing: A software engineering perspective. In *2013 International Conference on Human Computer Interactions (ICHCI)* (pp. 1–8). IEEE.
6. Björk, B.-C. (1992). A conceptual model of spaces, space boundaries and enclosing structures. *Automation in Construction, 1*(3), 193–214.
7. Boguslawski, P., & Gold, C. (2010). Euler operators and navigation of multi-shell building models. In T. Neutens & P. Maeyer (Eds.), *Developments in 3D geo-information sciences*. Lecture Notes in Geoinformation and Cartography (pp. 1–16). Berlin: Springer.
8. Boguslawski, P., & Gold, C. (2016). The dual half-edge—A topological primal/dual data structure and construction operators for modelling and manipulating cell complexes. *ISPRS International Journal of Geo-Information, 5*(2), 1–20.
9. Cavalcanti, P. R., Carvalho, P. C. P., & Martha, L. F. (1997). Non-manifold modelling: An approach based on spatial subdivision. *Computer-Aided Design, 29*(3), 209–220.
10. Chang, T. W., & Woodbury, R. (1997). Efficient design spaces of non-manifold solids. In Y. Liu, J.-Y. Tsou, & J. Hou (Eds.), *Second conference on computer-aided architectural design research in Asia* (pp. 335–344). Taiwan: CAADRIA.
11. Chatzivasileiadi, A., Lannon, S., Jabi, W., Wardhana, N. M., & Aish, R. (2018). Addressing pathways to energy modelling through non-manifold topology. In D. Macumber, F. Meggers, & S. Rockcastle (Eds.), *SIMAUD 2018: Symposium for Architecture and Urban Design*. Delft: Society for Modeling & Simulation International (SCS).
12. Crocker, G. A., & Reinke, W. F. (1991). An editable nonmanifold boundary representation. *IEEE Computer Graphics and Applications, 11*(2), 39–51.
13. Ecma International. (2005). C++/CLI language specification. Geneva.
14. Furiani, F., Martella, G., & Paoluzzi, A. (2017). Geometric computing with chain complexes design and features of a Julia package.
15. Hassenzahl, M. (2001). The effect of perceived hedonic quality on product appealingness. *International Journal of Human-Computer Interaction, 13*(4), 481–499.
16. Hillier, B., & Hanson, J. (1984). *The social logic of space*. Cambridge University Press.
17. Hillier, B., Leaman, A., Stansall, P., & Bedford, M. (1976). Space syntax. *Environment and Planning B: Planning and Design, 3*, 147–185.
18. Jabi, W. (2016). Linking design and simulation using non-manifold topology. *Architectural Science Review, 59*(4), 323–334.
19. Jabi, W., Aish, R., Lannon, S., Chatzivasileiadi, A., & Wardhana, N. M. (2019). Topologic, https://topologic.app. Last accessed: April 03, 2020.
20. Jabi, W., Soe, S., Theobald, P., Aish, R., & Lannon, S. (2017). Enhancing parametric design through non-manifold topology. *Design Studies, 52*, 96–114.
21. Laugwitz, B., Held, T., & Schrepp, M. (2008). Construction and evaluation of a user experience questionnaire. In A. Holzinger (Ed.), *Proceedings of the 4th Symposium of the Workgroup Human-Computer Interaction and Usability Engineering of the Austrian Computer Society on HCI and Usability for Education and Work* (pp. 63–76). Springer, Berlin, Heidelberg.
22. Lee, S.-U., Roh, M.-I., Cha, J.-H., & Lee, K.-Y. (2009). Ship compartment modeling based on a non-manifold polyhedron modeling kernel. *Advances in Engineering Software, 40*(5), 378–388.
23. Love, P. E. D., Edwards, D. J., Han, S., & Goh, Y. M. (2011). Design error reduction: Toward the effective utilization of building information modeling. *Research in Engineering Design, 22*(3), 173–187.
24. Masuda, H., Shimada, K., Numao, K., & Kawabe, S. (1989). A mathematical theory and applications of non-manifold geometric modeling. In F.-L. Krause & H. Jansen (Eds.), *International Symposium on Advanced Geometric Modelling for Engineering Applications*. North-Holland, Berlin.
25. McCullough, M. (1997). *Abstracting craft: The practiced digital hand*. Cambridge, MA: MIT Press.
26. Open CASCADE. (2018). Open CASCADE, https://www.opencascade.com. Last accessed: May 15, 2018.
27. Qiuchen Lu, V., Parlikad, A. K., Woodall, P., Ranasinghe, G. D., & Heaton, J. (2019). Developing a dynamic digital twin at a building level: Using Cambridge campus as case study. In M. DeJong, J. Schooling, & G. Viggiani (Eds.), *International Conference on Smart Infrastructure and Construction 2019 (ICSIC)* (pp. 67–75). ICE Publishing.
28. Roberts, A., & Marsh, A. (2001). ECOTECT: Environmental prediction in architectural education. In *Architectural Information Management: 19th eCAADe Conference* (pp. 342–347), Helsinki, Finland.
29. Sabry, H., Sherif, A., Rakha, T., & Fekry, A. (2010). Integration of daylighting simulation software in architectural education. In W. Tizani (Ed.), *The 13th International Conference on Computing in Civil and Building Engineering*. Nottingham University Press.
30. Schön, D. A. (1984). *The reflective practitioner: How professionals think in action*. Basic Books.

31. Spatial Corporation. (1989). 3D ACIS modeler, https://www.spatial.com/products/3d-acis-modeling. Last accessed: March 03, 2020.
32. Voloshin, V. I. (2013). *Introduction to graph and hypergraph theory*. Nova Kroshka Books.
33. Weiler, K. J. (1986). *Topological structures for geometric modeling*. Rensselaer Polytechnic Institute.
34. Woodbury, R., Gün, O. Y., Peters, B., & Sheikholeslami, M. (Roham). (2010). *Elements of parametric design*. Taylor & Francis, Inc.

Wassim Jabi Jabi is a Reader and Director of the M.Sc. Computational Methods in Architecture program at the Welsh School of Architecture, Cardiff University, and holds a Ph.D. in Architecture from the University of Michigan. Dr. Jabi has published widely and is author of the book 'Parametric Design for Architecture' (Laurence King Publishing). Dr. Jabi has secured several grants including a National Science Foundation grant to study design creativity, a Leverhulme Trust grant to develop a conceptual framework for the hierarchical spatial representation of buildings, and an EPSRC grant to research robotic 3D printing of sustainable material.

Aikaterini Chatzivasileiadi Chatzivasileiadi is a Lecturer at the Welsh School of Architecture (WSA), Cardiff University, from which she also holds a Ph.D. in Architectural Science. As part of her Ph.D. research, she served as a visiting researcher at Pennsylvania State University. After completing her Ph.D., she moved to a postdoctoral position at the WSA, where her research focused on two projects funded by the Leverhulme Trust and EPSRC, respectively; the first one was on the development of a conceptual framework for the hierarchical spatial representation of buildings and the second one on robotic 3D printing of sustainable materials.

LBD Server: Visualising Building Graphs in Web-Based Environments Using Semantic Graphs and GlTF-Models

Andrew Malcolm, Jeroen Werbrouck, and Pieter Pauwels

Abstract

Over the last years, Building Information Modelling (BIM) has been increasingly adopted in building projects for its benefits to collaboration and associated risk mitigation. Recent work is trying to make this data available over the web (web-based BIM applications). Most notable in making building data web-ready is the linked building data (LBD) initiative. This group has worked on a number of vocabularies and ontologies, thereby starting from a linked data approach (RDF graphs), aiming at representing building data on the web. Yet, a platform is needed that allows to manage the available linked building data, including 3D geometric visualisation, which is the topic of this article: a LBD server. The LBD server is a web-based application which allows users to upload building data and visualise their geometric representation in a graphical user interface, thereby also enabling to link to this data.

Keywords

Linked building data • IfcOWL • Web-based BIM • Xeogl

1 Introduction

Over the last decade, new advancements in software technologies have enabled the Architecture, Engineering & Construction (AEC) industry to become increasingly more digitised [1]. Such developments, commonly identified as 'Building Information Modelling/Management' (BIM) technologies, encompass many different aspects of the building industry, such as digital visualisation techniques, information exchange, innovative formal design, or collaboration processes. Overall, BIM has been characterised by the shift from interpretative information of two-dimensional plans and paper documents to fully integrated models that incorporate digital building data.

Even though there are still several obstacles to overcome before BIM gets fully adopted into the professional scene and building data becomes fully digitised, the building industry reports economic incentives due to the benefits to collaboration and associated risk mitigation that BIM already brings to the sector at this stage [2]. Building Information Models improve interoperability and efficient exchange between stakeholders through utilising central vendor-neutral formats for building information. Most of these open standards used throughout BIM processes are developed and maintained by BuildingSMART international. The most notable format for AEC information is the IFC or 'Industry Foundation Classes', which aims to describe all architectural construction data in the building lifecycle. The IFC remains legible across all specialised software, facilitating collaboration between project partners [3]. Yet, it remains primarily a data exchange format.

Recent work is trying to make this data available over the web (web-based BIM applications). Having building data available on the web opens up the possibility to exchange building data more openly. Being able to display and share geometric as well as non-geometric building data over accessible web servers and browsers will greatly improve communication between project participants. Furthermore, data can in such case be directly linked to alternative geometric representations (point clouds, different kinds of meshes, simplified CSG models) and to related semantic data (e.g. facility management data, heritage data, geospatial data) in a web-friendly manner, thereby scaling up the usability of building data significantly. This highlights the

A. Malcolm · J. Werbrouck
Ghent University, Jozef-Plateaustraat 22, Ghent, Belgium
e-mail: andrew.malcolm@ugent.be

P. Pauwels (✉)
Eindhoven University of Technology, Groene Loper 3, Eindhoven, Netherlands
e-mail: p.pauwels@tue.nl

S. Eloy et al. (eds.), *Formal Methods in Architecture*, Advances in Science, Technology & Innovation,
https://doi.org/10.1007/978-3-030-57509-0_26

importance of fully digitised and integrated building data. This would not only allow stakeholders to exchange their building data without loss of information, but also allow developers to interact more effectively with the building data, extending the functionalities of BIM using innovative techniques such as machine learning, semantic inference, or querying [4]. In other words, the potential that BIM could offer remains partially untapped as long as we are unable to work with fully interchangeable building data.

The linked building data (LBD) initiative, that started in the W3C LBD Community Group, has focused heavily on making this kind of building data web-ready. This group has worked on a number of vocabularies and ontologies, thereby starting from a linked data approach (RDF graphs), aiming at representing building data on the web. This approach diverges away from the use of IFC as a file-based exchange format. Yet, a platform is needed that allows to manage the available linked building data, including 3D geometric visualisation. Therefore, this paper describes the linked building data server (LBD server), a web-based building data platform that incorporates (semantic) web technologies, which attempts to remedy the aforementioned problems. The LBD server is a web-based application which allows users to upload building data and visualise their geometric representation in a graphical user interface.

At the moment, building data is mainly uploaded in the IFC file format, after which the loaded IFC model is converted to Collada using the IfcConvert application from IfcOpenShell (IfcOpenShell 2019). This in turn is converted into a GL Transmission Format (glTF) using COLLADA2GLTF [5]. These glTF streams are rendered into full 3D scenes of our architectural projects using Xeogl, an open-source JavaScript library from Xeolabs for 3D model visualisation on WebGL, supported by most modern web browsers [6].

In this article, we will first give an overview of the state of the art in representing building data on the web. Then, the LBD server project is presented, including a number of first implementations and tests. The article ends with an overview of conclusions and a list of next steps, which mainly includes a need for an investigation of user interactions with the data and interface.

2 Serving Building Data Online

2.1 Linked Data

The semantic web [7] can be regarded as an extension of the world wide web, enabling users of the web to connect various pieces of digital data in semantic graphs, as a novel alternative to document-based approaches. This method of publishing and connecting data is called 'linked data', referring to the interconnectivity of the data definitions. It utilises the open standardised data model called the 'Resource Description Framework' (RDF[1]). In such an RDF framework, information is organised in the form of 'triples', each time consisting of three data concepts (subject-predicate-object) in which the subject shares a relationship, defined by the predicate, to a certain object. This means that the information stored in graphs can be enriched without limit and remains dereferenceable over the web. This is a significant advantage in comparison to conventional relational databases in the case of highly interconnected data and a very broad scope. Each of these data concepts can be identified separately through a globally unique Uniform Resource Identifier (URI[2]), which keeps similar data concepts distinguishable from one another. The design of the URI also allows the information resource to be semantically enriched with additional data, regardless of discipline. This means that information becomes modular and boundlessly expandable, opening up new possibilities for interdisciplinary communication and collaboration.

This interconnectivity is made possible through the use of semantic schemas that define specific real-world concepts. Inside these schemas, called 'ontologies',[3] abstract data concepts are coherently defined and may refer to other ontologies or definitions inside the same ontology to define it, similar to a dictionary, see Fig. 1. By combining different definitions and relationships, an abstract concept can become fully described and comprehended. By publishing these definitions of data concept openly on the web, digital applications are able to navigate from one concept to another using their URI. Therefore, the RDF representation of data remains structured and legible to computing [8], allowing for semantic reasoning and interpretation of the data as well as advanced rule checking or inference of the data schemas. These benefits in turn make the RDF framework an effective format that can be interpreted and used inside digital applications.

2.2 Linked Building Data

An increasing amount of building data is being published as RDF graphs and AEC industry-specific ontologies are being developed [10]. The benefits of structuring building data using semantic web technologies are the improved interoperability, linking across domains, and additional logical inference and proofs [11] which coincide with many of the ambitions of the building industry such as lossless

[1]https://www.w3.org/RDF/.
[2]https://www.w3.org/wiki/URI.
[3]https://www.w3.org/standards/semanticweb/ontology/.

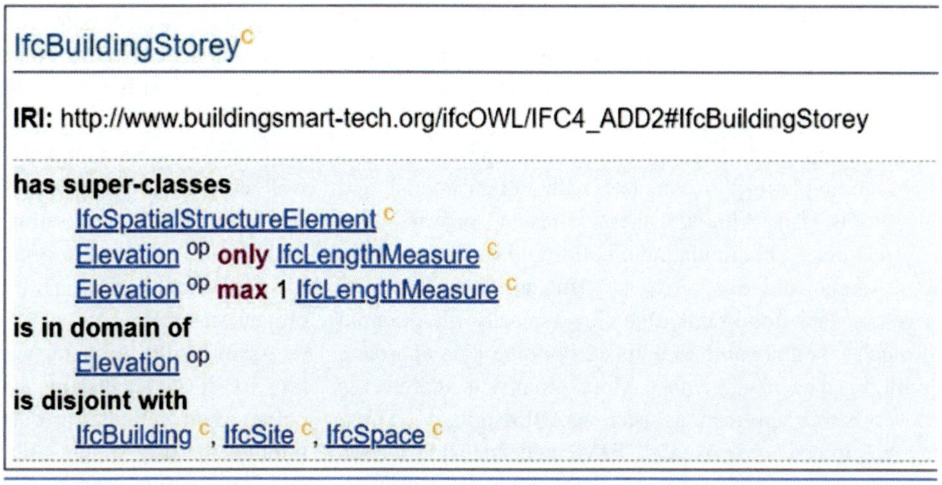

Fig. 1 Definition for ifcBuildingStorey in the ifcOWL ontology [9]

information exchange and complete interoperability. By defining ontologies and data definitions for different disciplines within the AEC sector, a connected and comprehensively legible interdisciplinary practice becomes possible.

Furthermore, the utilised ontologies do not need to be limited to the AEC sector alone, instead building data can be enriched by connecting it to other linked open data from disciplines outside the building industry, further improving the quality and interdisciplinary nature of our building data.

Several ontologies in the Web Ontology Language (OWL)[4] have been introduced to the building industry and expand upon the existing open standards such as the ones developed by BuildingSMART. IfcOWL[5] is a Web Ontology Language representation of the EXPRESS-based IFC schema into a RDF-based schema for linked data [9, 12]. By translating IFC to RDF through the use of ontologies, building information becomes available as an RDF graph while remaining in the structure of the widely used standard of IFC.

However, it is clear that ifcOWL alone will be unable to cover the full dictionary that the AEC sector needs. Rather, the IFC schema is best used as an exchangeable and interoperable format for communicating general information, and therefore serves its purpose as a vendor-neutral exchange format, but not as a fully integrated and all-inclusive description of a project. Fortunately, due to the nature of the semantic web, building data does not need limit itself to a single ontology and can be expanded by multiple, smaller, and modular ontologies that are easier to wield or are able to cover specific areas not included in the IFC schema [13]. Notable ontologies of this sort include the Building Topology Ontology (BOT[6]) and the Building Product Ontology (BPO[7]). BOT is an ontology that simplifies element relationships in construction to more efficiently visualise projects in RDF [14]. The latter, BPO allows users to freely define building element properties while adhering to a semantic web structure. This makes it easier for manufacturers to define their products and still reap the benefits from the semantic web in regards to accessibility and modularity [15].

3 Proposal for an LBD Server Approach

While the benefits of the semantic web technology for the building industry have been made clear, the actual applications on the market today remain limited. This is partially because semantic technologies and applications are characterised by a steep learning curve, but also because the building industry is still catching up with the digital revolution and BIM. Current developments are bringing BIM closer to open platforms to improve interoperability and efficiency of exchange.

There is an opportunity for developers to combine the innovative linked data technologies with BIM visualisation in order to bring the semantic data closer to the user. Giving industry stakeholders an interface to visualise and manipulate linked building data without the required semantic web experience will significantly flatten the learning curve of linked data management and can motivate new developments to take place in the fields of construction industry semantics.

At the same time, if linked building data can be attached to existing BIM models and visualised in web-based clients, the platform for collaboration and exchange of information can shift from standalone vendor software towards open web-based applications (often termed Digital Twins now). This would facilitate lossless information exchange and interdisciplinary communication between stakeholders as the data is more readily available.

[4]https://www.w3.org/OWL/.
[5]https://github.com/buildingSMART/ifcOWL.
[6]www.w3c-lbd-cg.github.io/bot/.
[7]www.github.com/w3c-lbd-cg/product.

Common Data Environments (CDE[8]) for visualising and managing BIM models for collaboration exist in various forms. The most widely used are vendor-proprietary software such as Autodesk BIM360[9] or Graphisoft's Bimcloud.[10] These act foremost as BIM repositories and collaboration platforms, complete with organisational and design tools [16]. Although these kinds of software offer many features to manipulate the building data, they remain closed source and use proprietary models locked behind paywalls. More importantly, they are typically file-oriented and much less amenable to a linked building data approach to building data management. More promising alternatives are web-based applications such as BIMserver.[11] These allow to upload and visualise BIM models inside a web browser and are open source. BIMserver in particular enables users to open up their own BIM server for file sharing. Imported IFC data for each project is stored as objects in a specific database, which allows for analysis of the IFC data. Open-source initiatives such as these give developers the opportunity to create additional plug-ins and tools that expand the application's initial features.

Even though BIMserver permits management and visualisation of BIM data to a large extent, it is unable to cover all aspects of AEC-industry-related information. Also the recent graph-oriented approaches towards building data management are missing, let alone that a fully web-based modular linked data approach is followed. Therefore, it is difficult to enable specific tasks such as managing non-geometric data, linking to external data or computing and analysis can be delivered through modular web applications. Such modular web applications, called BIM bots [17], each provide a single specialised task that admit an input in order to produce a certain output. They need full transactional and web-based data exchange, which is best supported using a linked building data approach. The other option is for current web-based applications to integrate semantic web technologies in order to accommodate for complex building data and at the same time allow interconnectivity with data from other fields, not typically included in the AEC sector.

4 The LBD Server

4.1 Outline and Purpose

This section of the paper describes the Linked Building Data server (LBD server), a proposed web-based building data server that incorporates semantic web technologies, currently in development at the Ghent University, Belgium. The LBD server is a web-based server which allows users to upload building data (e.g. IFC data) and visualise their geometric representation in a graphical user interface, as well as view linked building data attached to the project. By combining BIM visualisation techniques with semantics, the LBD server attempts to bring linked building data closer to the professional scene and improve communication between project stakeholders.

Assimilating semantic web technologies in this application for IFC visualisation has its own merits, as hinted at before. Linked building data can be enriched boundlessly through the use of semantic web ontologies. The semantic data initially imported in the server upon user upload can be enriched with multiple other data sets in alternative vocabularies in order to improve the quality of the data. Linked building data graphs can be interconnected, starting from a single or multiple LBD server nodes, forming a large knowledge network that is able to span across different industries and disciplines.

Additionally, managing building data in linked data graphs provides the possibility of advanced rule checking against rulesets, semantic inference and querying possibilities. Thus, the LBD server acts as an interface (Fig. 2) to upload BIM models and attach relevant linked data to parts of the project and allow users to peruse this linked building data, while still adhering to visually recognisable geometry of the model for navigation. Using semantic web technologies as complementary descriptors of the building data inside a web-based visualisation application will bring the model closer to a complete rendition of the project.

4.2 Implementation Approach

The LBD server is being developed as a JavaScript web application which connects a web GUI with multiple databases that are each optimised for serving specific data (RDF triples, geometry, images). Even though this application promotes the use of graph databases, more conventional types of database management systems still remain useful and can be included because of its modular and web-based system architecture. While RDF graph databases allow for a much larger flexibility for storing highly interconnected building data, conventional document-based databases can be used for storing non-textual documents and files such as images or the IFC model geometry.

Currently, the LBD server makes use of a cloud-based document store to store building data and relevant documents. As of this writing, IFC models and accompanied IFC data are utilised as a use case for development of the LBD server.

[8]https://www.iso.org/obp/ui/iso:std:iso:19650:-1:ed-1:v1:en.
[9]https://www.autodesk.com/bim-360/.
[10]https://www.graphisoft.com/downloads/bimcloud/.
[11]https://github.com/opensourceBIM/BIMserver.

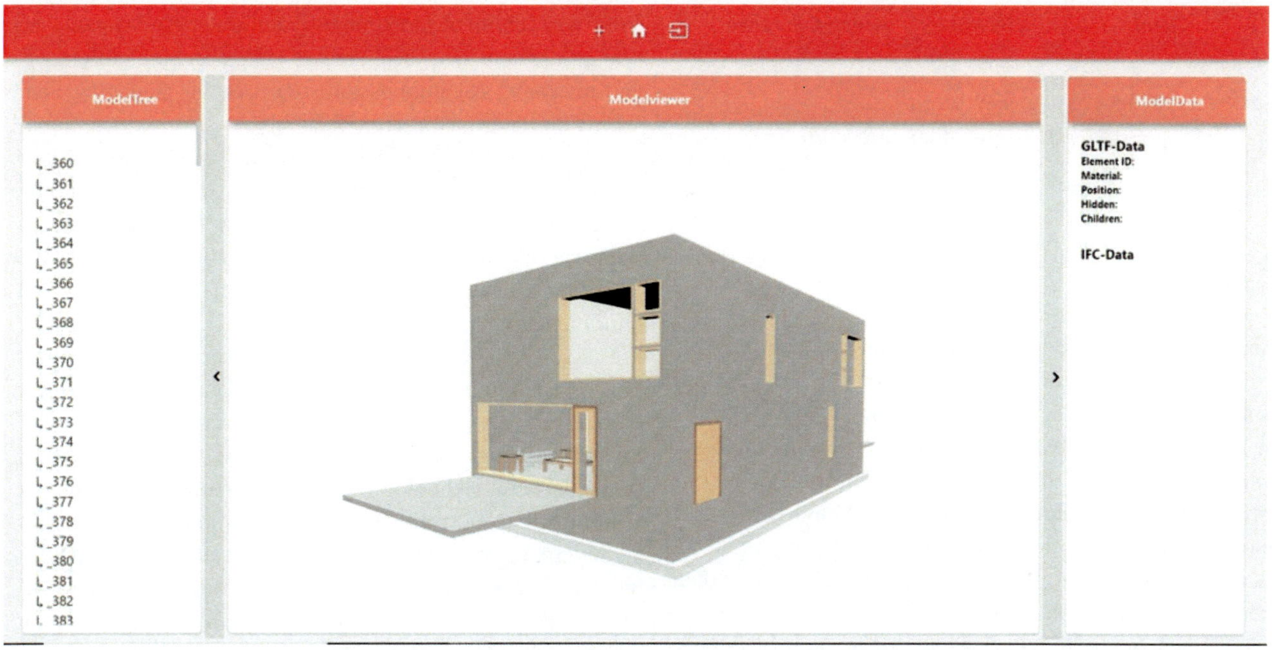

Fig. 2 The LBD server viewer interface

4.3 Geometric Data

Visualisation of the building geometry inside the browser occurs in a similar fashion to most IFC viewer applications. Namely, the geometry is made available into an available standard geometry representation for the web, namely glTF. One can obtain glTF in multiple ways, yet we document below how this glTF representation can be obtained when starting from an IFC file.

First the loaded IFC model is converted to the ‘COLLAborative Design Activity’-format (COLLADA[12]), developed and maintained by the industry consortium KhronosGroup. COLLADA is a standard XML-based schema to make it easy to transport 3D assets between applications and used as an intermediary format (KhronosGroup 2019). The conversion from IFC to COLLADA utilises the IfcConvert[13] application from the open-source software library ifcOpenShell, yet standardised COLLADA files can easily be obtained directly from any 3D editor program. The COLLADA file is afterwards converted into a GL Transmission Format (glTF[14]) using the COLLADA2GLTF convertor [5]. The glTF format utilises JavaScript Object Notation (JSON[15]) schema to organise the geometric information of the model. JSON allows the glTF format ‘*to be cross-platform and web-friendly; and to hold geometry and texture assets in binary blobs for future expandability of asset types that may include streaming and compression*’ [18]. By using these DAE and glTF standards, the LBD server aspires to rely as much as possible on available standards for the representation of data, thereby enabling wide adoption and re-use of embedded data. Also, the JSON and RDF formats play a very important role in this regard. The glTF utilises JSON to divide the 3D model into mesh objects, which include texture information and matrix location. The LBD server allows to upload an IFC file inside a project, after which it automatically goes through the above-mentioned string of conversions, finally to be stored as glTF data. After upload, the IFC file can be discarded, as data exchange is complete. These glTF streams can be rendered in any openGL-enabled browser. To render the converted glTF streams into 3D scenes, the LBD server utilises Xeogl,[16] an open-source JavaScript library from Xeolabs for 3D model visualisation on WebGL, which is a specification supported by most modern web browsers [6]. The Xeogl tool is able to differentiate between the geometries of the various elements present in the converted model.

4.4 Non-geometric Data

Of course, the geometric data needs to be linked to the available semantic linked building data. When starting from IFC data and converting this data using IfcConvert, the mesh

[12]https://www.khronos.org/collada/.
[13]https://ifcopenshell.org/ifcconvert.
[14]https://www.khronos.org/gltf/.
[15]https://www.json.org/json-en.html.
[16]https://xeogl.org/.

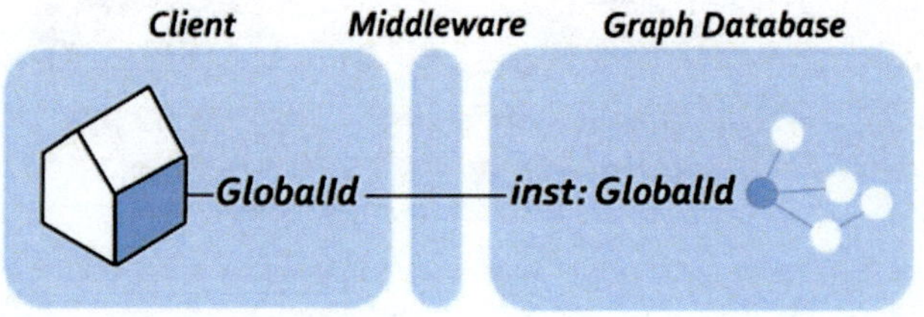

Fig. 3 Schema of connectivity between glTF mesh geometry and graph data

objects in the glTF file also include meta-data from the IFC format; namely an encoded version of the IFC Global Identifier (GlobalId) that remains present in the glTF file. The encoded IFC GlobalId is attached to every mesh object inside the glTF file. This GlobalId can be used as a connection between the geometries inside the Xeogl viewer and the original IFC data schema, as a unique identifier to locate specific IFC elements of the model. Any linked building data cloud typically relies on the full UUIDs that are also used in the glTF data. So, links between data can be made by relying on identical IDs. When relying fully on a linked data approach, it is best to create meaningful links between geometry and semantics, ideally relying on the 'bot: hasComplexGeometry' and/or 'omg:hasGeometry' predicate, instead of using 'sameAs' links. Currently, however, the LBD server aims to utilise this GlobalId as an identifier that connects the geometry to the respective RDF graph (Fig. 3). Links between models are now made directly in the API middleware. This makes it possible to summon Linked Building Data relevant to the selected geometry inside the LBD server user interface.

4.5 User Interaction and Interface

The user is able to navigate the 3D model by interacting with the Xeogl viewer component or by sifting through the 'ModelTree'-tab, depicted in Fig. 3. Clicking on a single element inside the viewer component or inside the ModelTree will highlight the selected element as well as zoom in on the element inside the viewer. Relevant linked building data is queried for and appears inside the 'ModelData'-tab for the selected element. More advanced use cases are currently under development, thereby aiming particularly at enabling BIM Bots that connect directly to the API middleware.

5 Conclusion

This paper proposes the increased combination of semantic web technologies to web-based BIM applications in order to improve integrity and quality of building data. The transition from conventional document-based database systems to semantic graph databases allows rule checking, inference, and reasoning capabilities to assist users in managing their linked building data more effectively.

The concept for the LBD server has been described. The main goal is to allow its users to interact with linked building data very intuitively. By utilising BIM visualisation techniques in combination with semantic web technologies, the LBD server can make it possible to navigate the linked building data in an intuitive manner. The LBD server will be useful for collaboration platforms that necessitate visualisation and communication of building data between project participants. By allowing users to interact more easily with linked building data, web-based BIM applications and semantic AEC technology gain more credence, bringing us closer to an open, web-based and transactional BIM practice.

Several topics for future work remain of interest. First, in the LBD server, the goal is to link building data (RDF nodes) to other RDF graphs to make it possible to enrich the model with additional information. As of this writing, the LBD server viewer displays limited amounts of information about selected elements. Making the attached linked building data readily available in the interface is an important milestone for the LBD server that is not yet reached.

Second, there is a vast amount of building information that is not bound to building geometry or is not meant to be navigated in 3D space or lists. The LBD server project will need to decide how such information such as project data, labour, time, … should be taken into account, if at all.

Lastly, an important topic not yet covered by the LBD server in its current state is the manipulation of semantic data attached to the model. The interface does not yet allow the user to interact with the linked building data found in the graph database. Building an appropriate user interface that displays the semantic data needs to be developed and likely depends on particular use cases.

References

1. Pauwels, P., & Petrova, E. (2018). Information in Construction.
2. Eastman, C., Teicholz, P., Sacks, R. & Liston, K. (2011). *BIM handbook: A guide to building information modeling for owners, managers, designers, engineers and contractors* (2nd ed.). ISBN 978-0-470-54137-1.
3. BuildingSMART. (2020). Industry foundation classes (IFC)—An introduction. https://technical.buildingsmart.org/standards/ifc/. Last accessed January 2020.
4. Zadeh, P. A. (2018). Introduction to machine learning in construction. In *Conference on Construction Management Seminar at UBC*.
5. KhronosGroup. (2020). Collada2gltf. Collada:3D asset exchange schema. https://github.com/KhronosGroup/COLLADA2GLTF and https://www.khronos.org/collada/. Last accessed January 2020.
6. Xeolabs.: Xeogl. https://xeogl.org/. Last accessed January 2020.

7. Berners-Lee, T., Hendler, J., & Lassila, O. (2001). The semantic web. *Scientific American, 284*(5), 34–43.
8. Berners-Lee, T. (2019). Linked Data. https://www.w3.org/DesignIssues/LinkedData.html. Last accessed November 2019.
9. Pauwels, P., & Terkaj, W. (2020). ifcOWL ontology (IFC4x1_RC3). https://users.ugent.be/~pipauwel/ontologies/ifc/20170101/IFC4x1_RC3/. Last accessed February 2020.
10. Pauwels, P., Zhang, S., & Lee, Y.-C. (2017). Semantic web technologies in AEC industry: A literature overview. *Automation in Construction, 73,* 145.
11. Pauwels, P., Krijnen, T., Terkaj, W., & Beetz, J. (2017). Enhancing the ifcOWL ontology with an alternative representation for geometric data. *Automation in Construction, 80,* 77–94.
12. Beetz, J., Van Leeuwen, J., & de Vries, B. (2009). IfcOWL: A case of transforming EXPRESS schemas into ontologies. *Artificial Intelligence for Engineering Design, Analysis and Manufacturing, 23*, 89–101. https://doi.org/10.1017/S0890060409000122
13. Rasmussen, M. H., Pauwels, P., Hviid, C. A., & Karlshøj, J. (2017). Proposal of a central AEC ontology for domain specific extensions. In *Proceedings of the Joint Conference on Computing in Construction* (No. 1, pp. 237–244). Greece: Heraklion.
14. Rasmussen, M. H., Pauwels, P., Lefrançois, M., & Schneider, G. F. (2019). Building Topology Ontology Draft Community Group report 07 January 2019. https://w3c-lbd-cg.github.io/bot/. Last accessed January 2020.
15. Wagner, A., & Rüppel, U. (2019). BPO: The building product ontology for assembled products. In *Proceedings of the 7th Linked Data in Architecture and Construction Workshop—LDAC2019*.
16. Chong, H., Wong, J. S., & Wang, X. (2014). An explanatory case study on cloud computing applications in the built environment. *Automation in Construction, 44,* 152–162.
17. Berlo, L. (2020). Automate your work with BIM bots. https://www.slideshare.net/berlotti/bim-bots, Consulted March 2020. Last accessed February 2020.
18. KhronosGroup. (2020). Khronos COLLADA now recognized as ISO Standard. https://www.khronos.org/news/press/khronos-collada-now-recognized-as-iso-standard. Last accessed March 2020.

Andrew Malcolm Malcom is a Ph.D. student working at the Faculty of Engineering and Architecture at the University of Ghent. He performs research in the field of Open Linked Data in the building environment. He also works on the LBD server project, led by Prof. Pieter Pauwels.

Reconstructing Photogrammetric 3D Model by Using Deep Learning

June-Hao Hou and Chi-Li Cheng

Abstract

In this paper, we propose a deep learning-based method to reconstruct 3D models generated with photogrammetry techniques. The mechanism to reconstruct the model is detecting corners and then segment the polyline contours based on the detected corners. Moreover, we take advantage of the generative design tool (Grasshopper) to build a training gym system, constantly producing unique datasets to train the neural network. In the part of feature engineering, we use series of external angles as the geometrical feature to train the model, which means that the model processes a small number array instead of using images or any kind of huge multi-dimensional data. The result of this research reveals that the neural network model trained by our training gym system can learn how to detect corners on building contours successfully. The details of the training gym, dataset format, dataset generator, the deep learning model, and the potential usages will be elaborated in the following content.

Keywords

Geometry simplification • Deep learning • Polyline segmentation • Mesh reconstruction • Photogrammetry

1 Introduction

Nowadays, data-driven design plays a crucial role in architectural design; using various information to analyze performances of different aspects in design can truly facilitate designers when making decisions. The method to gain data or model for design becomes more and more important. There are several ways to get a model of the site that architects and designers could get access to. For instance, OpenStreetMap provides a quite user-friendly interface for gaining or sharing spatial information; the global terrain data is opened as SRTM format for everyone. However, most of these sources cannot offer the most recent and detailed building models; in order to get more recent models, certain architects begin applying drones with photogrammetry software to build the model around the design site. Nonetheless, these models generated by the photogrammetric method have certain disadvantages, like the redundant faces, defects, and errors, etc. These issues make this kind of model extremely difficult to use in a design process. In our previous research [1], we proposed a contour-based mesh reconstruction with a gaussian mixture model; consequently, the method is able to filter out irrelevant models and then use the most representative contours of buildings to rebuild them (see Fig. 1). In this research, we ponder a method to improve our previous simplification method in the reconstruct process. Moreover, the efficiency of the method is considered because we attempt to make it practical and able to facilitate designers. Thus, there are several aspects of the reconstruction process that should be scrutinized.

1.1 Related Work

This research is to improve our previous research that proposes a contour-based method to reconstruct and simplify the photogrammetric model [1]. Even though the method in our previous research is able to reduce photogrammetric models efficiently, it lacks the capability to simplify curved walls on buildings. The polyline of curved walls remain a series of points after simplification. As a result, the issue in this research is how to rebuild the curved parts as surfaces decently. The difficulty with the issue is that these models

J.-H. Hou
National Chiao Tung University, Hsinchu City, 30010, Taiwan
e-mail: jhou@arch.nctu.edu.tw

C.-L. Cheng (✉)
Graduate Institute of Architecture, National Chiao Tung University, Hsinchu City, 30010, Taiwan
e-mail: micky@arch.nctu.edu.tw

S. Eloy et al. (eds.), *Formal Methods in Architecture*, Advances in Science, Technology & Innovation,
https://doi.org/10.1007/978-3-030-57509-0_27

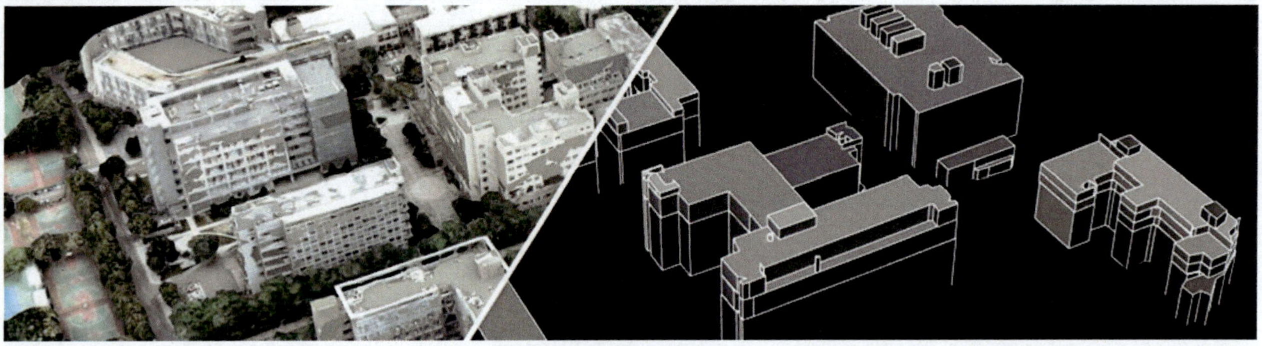

Fig. 1 A method of mesh simplification for photogrammetric models

generated with photogrammetry techniques include different levels of noise depending on the quality and the number of photos, as well as parameters in the photogrammetry software. Accordingly, the method requires the capability to tolerate the variety of factors on the models. In our reconstruction process, the core task is corner detection; the process can detect corners accurately. The number of faces of the reconstructed building model will be reduced because that the nodes on the curved wall will be grouped to form a surface. There are some existing methods to detect corners for different scenarios.

Method for Detecting Corners on Images

For processing contours and detecting corners, Antoni Buades et al. proposed an algorithm to process images, deriving meaningful contours efficiently [2]. In the research, they employed 3D convolution kernels to derive relevant features of the contours; furthermore, their algorithm is able to process images with noise, resulting in distinct contours of objects.

Rosten et al. [3] investigate a way to speed up the process that utilizes machine learning to detect corners on the images. They came up with a way to neglect unnecessary features of the corners but keep the validity. These image-based methods have a kind of disadvantage that the 2D matrix always includes empty value. The scenario in this research can avoid this disadvantage because every external angle on a contour is relevant.

Method for Detecting Corners in 3D Meshes

In addition to deriving distinct corners of contours on 2D images, the topic of how to refine 3D is investigated by many researchers. For instance, research by Yongzhi Wang et al. investigating a method to simplify the mesh model based on the triangle analysis [4]; their goal is to simplify but keep the salient features as much as possible. Another research for processing mesh by Xiao-chao Wang et al. proposed a method to detect corners on noisy meshes via normal tensor voting and neighbor supporting [5]. In this research, we deal with the 3D model as well; nonetheless, we choose polyline contours to reconstruct buildings to reduce the complexity. The reason is that most mesh-based methods seem not to be able to tolerate noise well. Since we reduce the complexity a lot, machine learning could focus more on dealing with noise on a series of angles on a contour.

Method for Processing Polylines

Wolin et al. proposed a simple method to find corners by the short-straws concept, dealing with polylines that are supposed to be polygon efficiently [6]. Zhang et al. presented another method to detect corners with an arc length-based algorithm. It addresses the issue of measuring discrete curvature [7]. For simplifying and combining polylines, Joachim Spoerhase et al. suggest the Simplification of Polyline Bundles. In their research, they deal with polyline simplification and the issues that a bunch of polylines are adjacent to each other [8]. It is quite inspiring because they are different from other similar simplification research that only focuses on a single polyline. Besides, there is an approach utilizing drawing speed data to enhance the validity of corner detection by Tevfik Metin Sezgin. In the research, the detection process is based on scale-space analysis, processing deeper information so that it is able to tolerate the noise and derive the corners preciously [9].

Recently, machine learning was introduced to solve tasks for processing polylines. Long Zeng et al. proposed a machine learning-based method and used a huge number of Human-labelled datasets to train the model to detect corners on drawing lines [10]; however, using numerous human-labeled datasets is too time-consuming to operate. A case about using polyline contour simplification for reconstructing the building model was presented by Biao Xiong et al.; two algorithms in their research depend on footprint maps and LiDAR to segment contour in order to construct the models appropriately [11]. The research illustrates practical methods of constructing high-quality models. Although these mentioned methods are able to derive corners for different scenarios, they still include some parts to be improved when using them to simplify building models from photogrammetry.

First of all, mesh reduction methods are quite inefficient because they require sophisticated algorithms to process complex geometrical data. On the other hand, in our previous research, we focus on polyline contours because most variations of building models on vertical are slight. Therefore, a few contours of a building are adequate to represent the major features of the building. The slicing-based method that we have presented is able to simplify the building model significantly. The performance of our method is more efficient than other methods because we implement neural networks to deal with the relatively simple data (a series of angles on polyline contours); which means that 2D geometry is adequate to facilitate the reconstruction process. Hence, this research will focus on dealing with 2D contours as well.

Second, some polyline-based methods are able to segment polyline accurately but they still have difficulties in segment curved part or tolerance noisy lines. In modern architectural design, a curved form is inevitable to face; as a result, this issue is necessary to solve. There are some approaches requiring data other than geometrical features such as speed data and footprint maps. This information requires special equipment or approach, making it difficult to use by designers or architects. In this research, we propose a method using relatively simple data form, but the results would be accurate.

Third, it is great that machine learning has been utilized in certain related researches; Nonetheless, their methods are processing unnecessary complex information and requiring a lot of human-labelled data. In this research, we propose a training gym system to address the issue.

The purpose of doing this research is to improve a reconstruction method in our previous research. In the previous research, we were able to simplify complex photogrammetrically models significantly yet the contour of its' results are polylines. It might remain certain issues; for one thing, when the user wants to explode the model to get a curved wall, the result will be a lot of segments. And when a user views the results by wireframe or shaded mode, there will be a bunch of irrelevant edges messing up the visualization. Additionally, if users want to generate a pattern to design the façade on the model, these wall surfaces that are represented as NURBS format will make them easier to use in the parametrical environment, facilitating designers further.

2 Methodology

To simplify the model generated with photogrammetry efficiently, three strategies are crafted and employed. The first thing that we are dealing with is simplifying the dataset; most of the machine learning-based method use image or complex feature data to train their model. Instead, we use a series of angles on a polyline as features to detect corners. Then, the labelled data to be used in our method are generated by computer automatically. Besides, a training gym is built to integrate dataset generator and deep learning model; the training gym will generate and train the model continuously until the result is good enough. The other part in this research is the reconstruction process based on the trained model generated with the training gym (see Fig. 2).This method is different from others' because the model of this method will detect corners according to the context of the entire contour; however, the feature for detection is much simpler than others' methods; the characteristic will bring about the validity and efficiency when detecting corners.

2.1 Contour Generator

Instead of using human-labelled data, the labelling process is computerized in this research by using Grasshopper as a data generator; the program keeps generating various contours (see Fig. 3). An universal training aid is designed for enhancing the versatility when detecting corners; based on our experience when dealing with building contours, there are three kinds of variable usually appear on a contour that might influence the detection, such as angle, curvature, and configuration. In addition, certain tiny parts might be neglected by the detection process; therefore, we design the training aid with a tiny part to enhance the sensitivity of paying attention to the details of a contour (see Fig. 4). In the dataset generator, the training aid will be varied by automatically setting different seeds for random components

Fig. 2 Training gym and reconstruction process

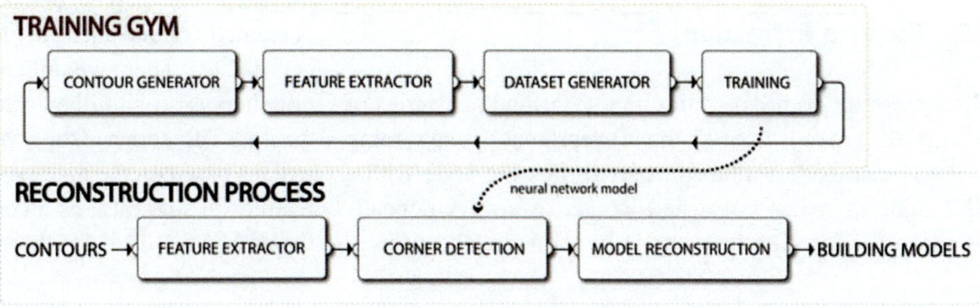

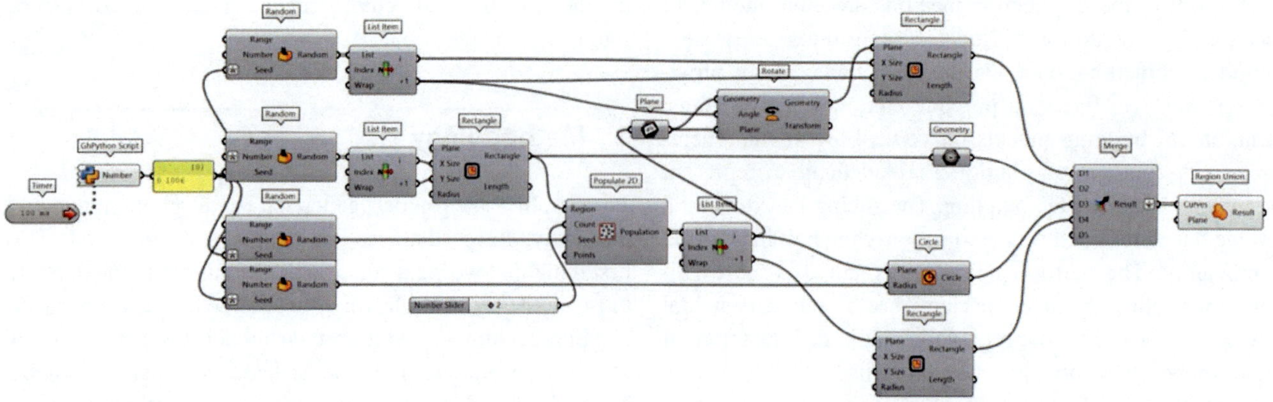

Fig. 3 Dataset generator

Fig. 4 Variable training aid for versatile purpose

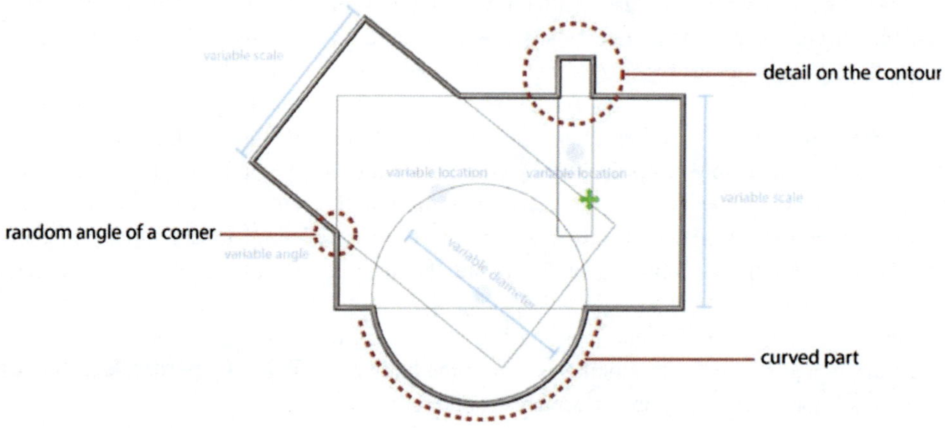

and influencing the result. The results are unified contours of four variable parts. The dataset generator can perform efficiently; hundreds of results can be generated in a second. The image of dataset samples is visualized in Fig. 5.

In addition to these geometric characteristics, noise also plays a crucial role to influence corner detection. To deal with the issue, a post-process to add noise is employed (see Fig. 6). The scale of noise is varied so it can deal with various rugged contours with various scales of noises. After the post-process, the data is quite challenging for computational detection although it might not be an issue to a human. After generating the rugged contours, they will be outputted with the corner labels for the next step.

2.2 Feature Extraction

The feature for corner detection in our method is a series of radians (the range is from $+\pi$ to $-\pi$), representing the turning angle of each node. Although it seems that the node with a high value of radian could be a corner, it always depends because of noise. The first step is to normalize the dataset. Each contour will be divided into 100 nodes then all these 100 nodes are connected as a new polyline. After exploding the polylines, the result will be 100 segments. The means to derive the series radians from it is extending each segment line and measuring the internal radian between the segment and the next one. We define the clockwise turn as positive radian and the anticlockwise turn as negative radian (see Fig. 7).

2.3 Labelling and Processing Dataset

The method to label the dataset is to find the closest corners from the original contour (see Fig. 8). The coordinates of these corners can be derived by exploding the originally generated contour then the vertex will be used as the corner coordinates. Then, we utilize the closest point component in grasshopper to gain the corresponding index number in the dataset. The corner data is derived by shifting the series of radians based on the index number; consequently, the crucial features to suggest it as a corner will appear at the start and the end of the data. On the other hand, these features in the

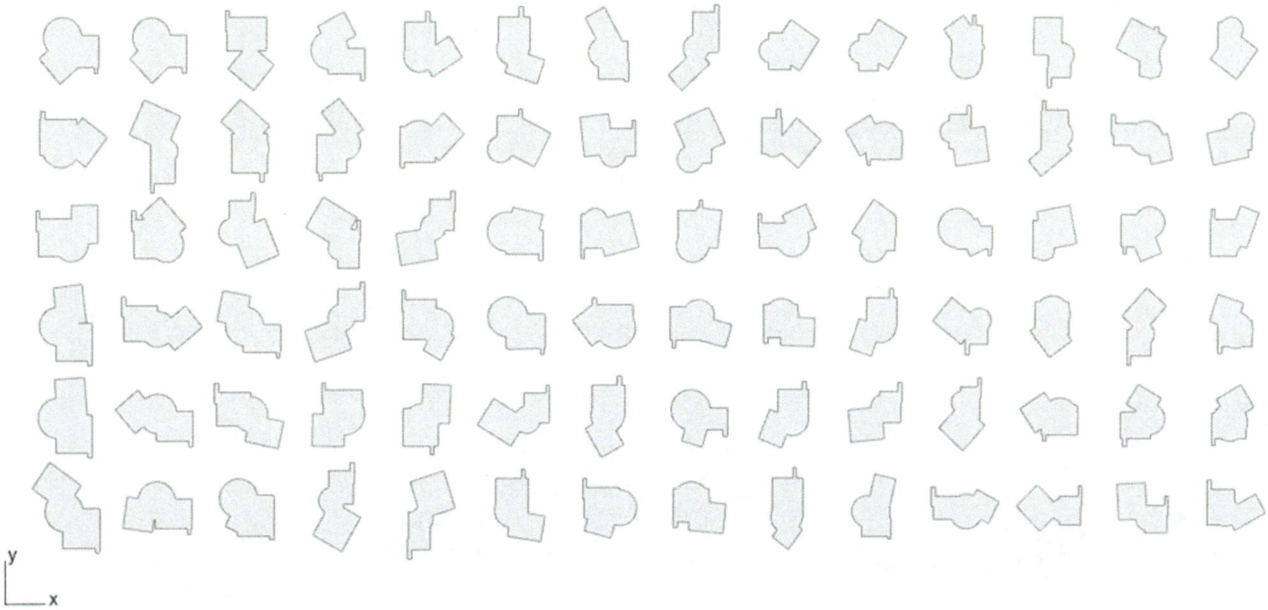

Fig. 5 Various generated contours

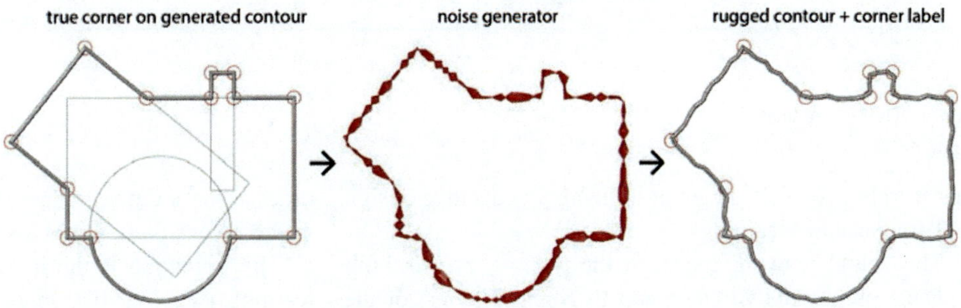

Fig. 6 Adding noise on contours to enhance the tolerance

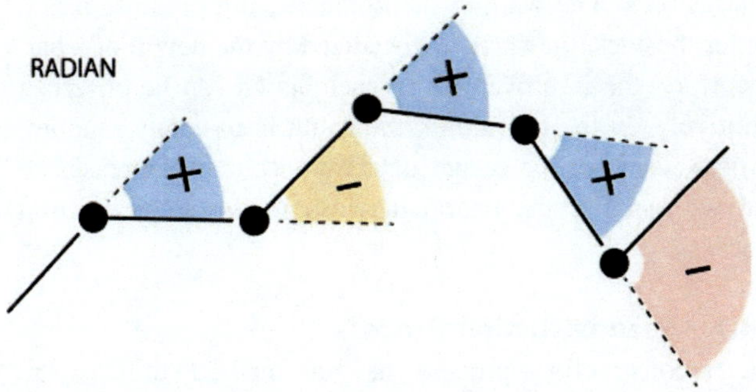

Fig. 7 Using external angle as feature of contour

middle of the data can offer information about noise or inferior features, suggesting it might be a corner. The corner part is labelled as 0 or 1. If it is a corner data, it will be labelled as 0; on the other hand, the data which does not belong to the corner will be labelled as 1. So far, the information for training the neural network is adequate. The neural network will be able to figure out the relationship between the corner number and the series of radians based on the dataset. That is to say that the neural network model will consider the context of the entire contour, making the estimation more thoughtful.

2.4 Training Gym System in Grasshopper

In this research, the neural network model is trained automatically by the training gym in grasshopper. In the training gym, it will keep generating new kinds of contour to challenge the neural network model. There is an issue occurring in the contour that most of the nodes do not belong to the corner. If we use the original distribution of the data it will result in poor sensitivity for corner detection. In order to solve this issue, we separate the corner part and non-corner part into two sets, and then, the non-corner part will be reduced to fit the number of corner parts (see Fig. 9), avoiding the result that the neural network keeps guessing anything as non-corner. Moreover, during training, the dataset will never repeat to avoid overfitting; it is a significant characteristic in our method

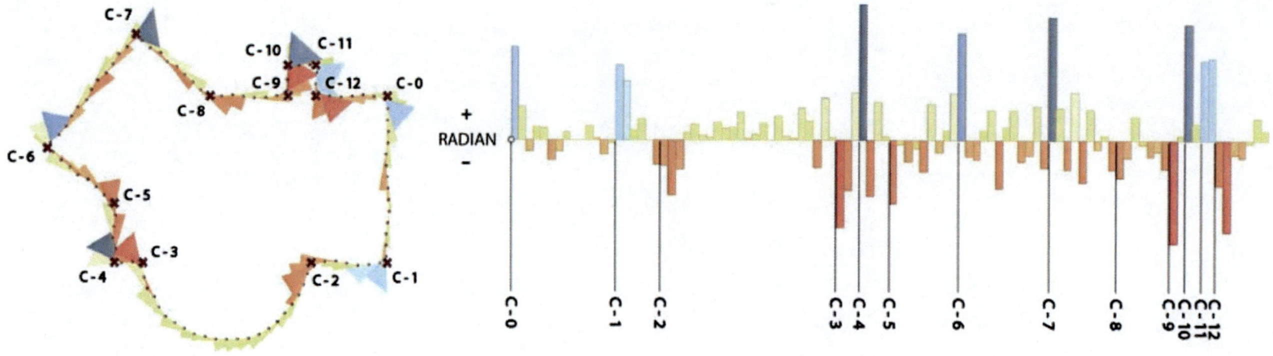

Fig. 8 Contextual dataset

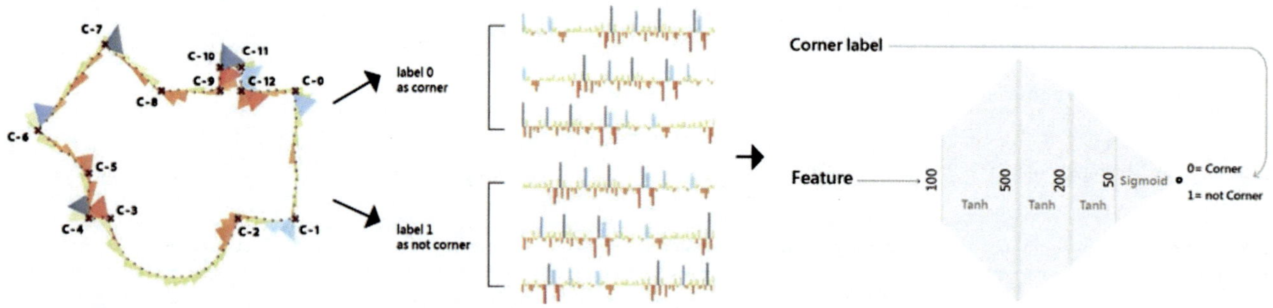

Fig. 9 Training process

that it truly takes advantage of utilizing a generative design tool for machine learning.

The neural network model in the training gym is built with the plugin GH-CPython and PyTorch. The rest of the parts such as dataset generator, noise generator, iterative mechanism, are made of original Grasshopper components. The neural network that we apply is a simple multiple layers with three hidden layers; the output layer has only one node with a sigmoid activity function to restrain the number in the range from 0 to 1. We anticipate that a neural network is adequate for this task because the model needs to focus on the series of radians. The task is much simpler than the multiple layers neural network that is presented by Geofrey E. Hinto; besides, the neural network that Geofrey used is able to classify 10 kinds of digits from 2D images with 784 (28 * 28) pixels [12] while our neural network deals with only 100 input data and classifies 2 kinds of situations (corner or non-corner). As a result, we consider that it is unnecessary to increase the complexity of the neural network like certain fancy methods.

Since the training gym is set, the rest is to let the process progresses toward perfection. In fact, it can get a nice result in a half-hour. During the training process, the neural network model will be saved in standard pkl format (a format to store artificial neural network of PyTorch framework) in order to make the neural network model able to be employed in the reconstruction process independently.

The following is the image showing the progression of learning (see Fig. 10). In each epoch, the neural network model will be tested, visualized with a colour gradient and vertical bars. The warmer the colour is, the possible it is a corner; besides, the value is visualized by the height of a bar. Therefore, the improvement in each epoch can be observed intuitively. In the beginning, the result is basically random; as time went on, the corner detection accuracy increases so that we can use the neural network in the reconstruction process.

2.4.1 Reconstruction Process

The reconstruction process in our method utilizes the neural network model trained by the training gym. The first step is corner detection. The contour to detect should be normalized as a 100-radians format, and then the data will be evaluated by the neural network model. There is only one output of the neural network, whether it is a corner or not. In the defining step, we set a threshold value to decide where to segment (see Fig. 11). The original contour will be separated into individuals; subsequently, each polyline's nodes will be used as control points of individual Bézier curves.

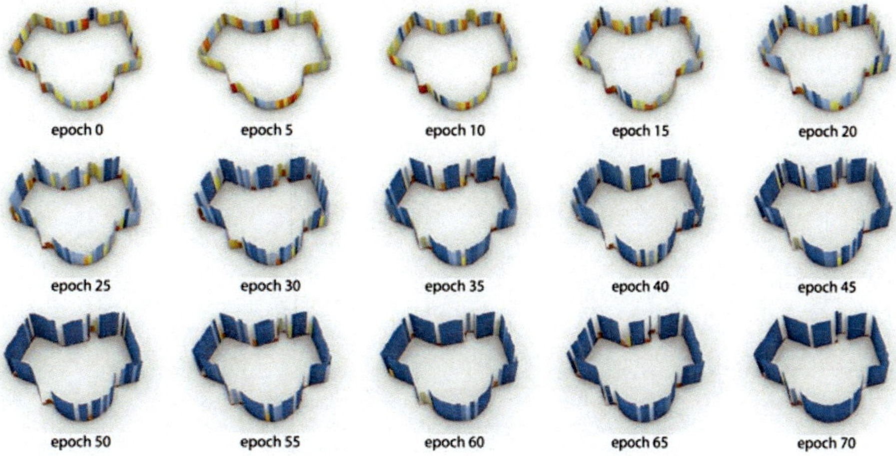

Fig. 10 Learning progression

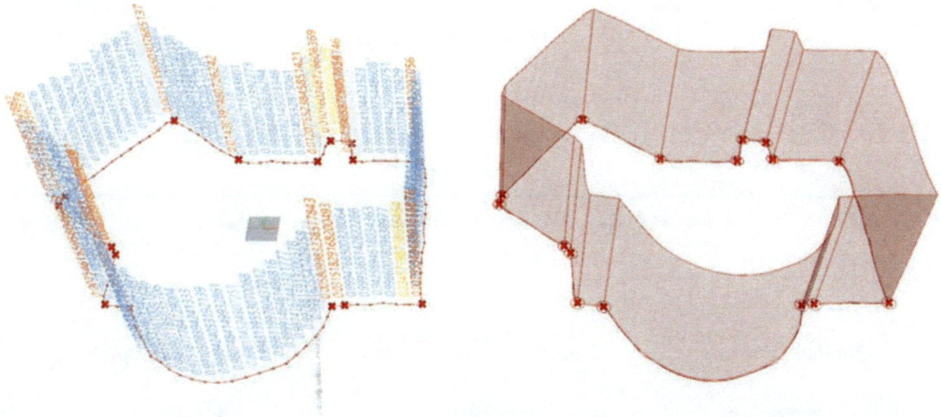

Fig. 11 Segmentation based on corner estimation

3 Results

A training dataset generator and a training gym system are proposed in this research; it can constantly improve the neural network and never repeat the data to avoid overfitting during training. Moreover, a kind of training aid is designed for simulating any circumstance when detecting corners on building contours. In our method, the neural network and data format are designed to be as simple as possible for efficiency.

The capability of our method has been tested and approved that it is able to detect corners on various contours. Even though the neural network is trained by our training gym system instead of real data, it can find corners from real data successfully (see Fig. 12).

Additionally, we have tested them for non-building-like and concave contours (see Fig. 13). The following is the comparison between the other two existing methods to reduce polyline in Grasshopper. Our method can reconstruct the contours neatly, reducing the number of faces significantly. Furthermore, our method can segment the open polylines as well (see Fig. 14); it might be qualified to use as a universal polyline segmentation tool. The main characteristics of the result generated by our method is that it can significantly reduce the number of faces and rebuild the contour as properly segmented surfaces efficiently. A part of the efficiency is caused by the fact that the neural network model in our method is speeded up by GPU.

4 Discussion

Our method simplifies the process of training a neural network for a building model reconstruction process; Although the neural network is trained with the dataset generated by our dataset generator, it is adequate to detect corners on various unseen polylines. It reveals the potential to use a generative design tool to facilitate the training process. Our method is helpful when dealing with models generated through photogrammetry in some aspects. First, our method brings about the neat visualization of building models. Secondly, since the surfaces on the building models are easy to derive, the designer can utilize a parametric tool to analyse or generate façade on these segmented wall surfaces. In addition, these surfaces of the wall make them easier to

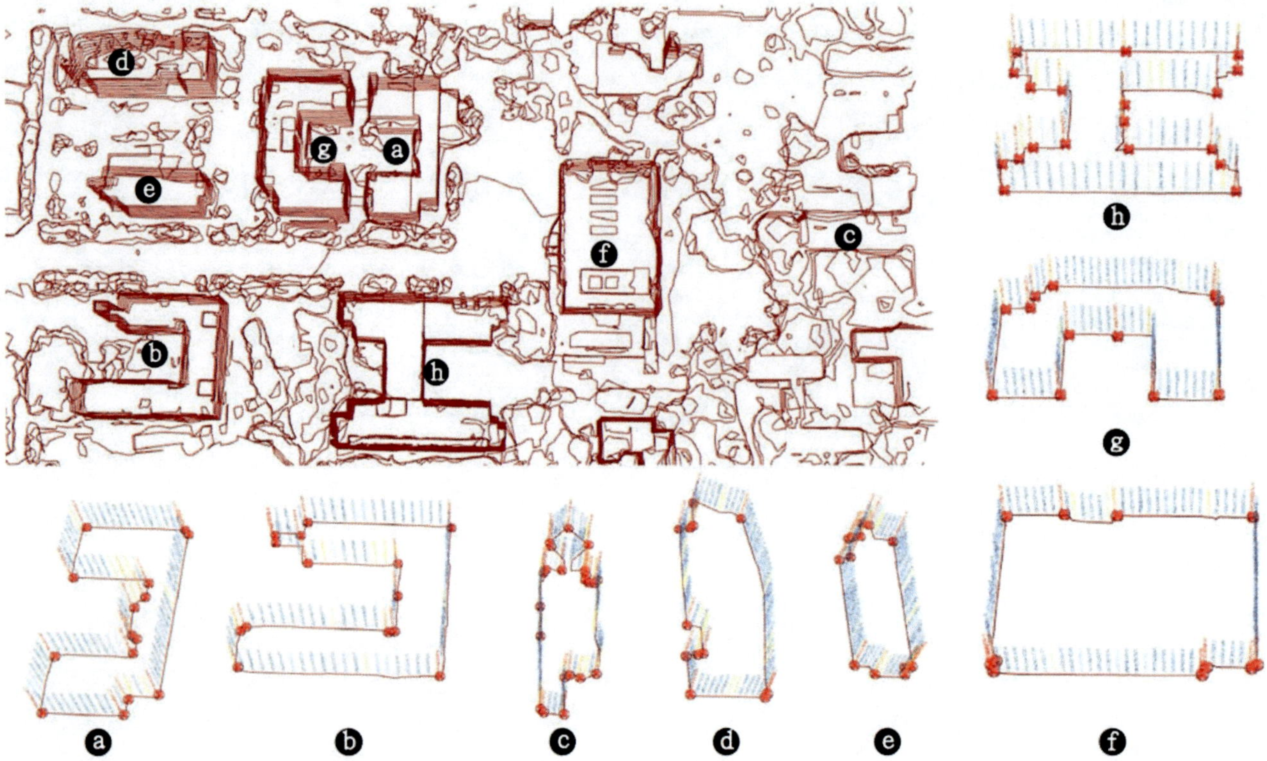

Fig. 12 Finding corners on real data

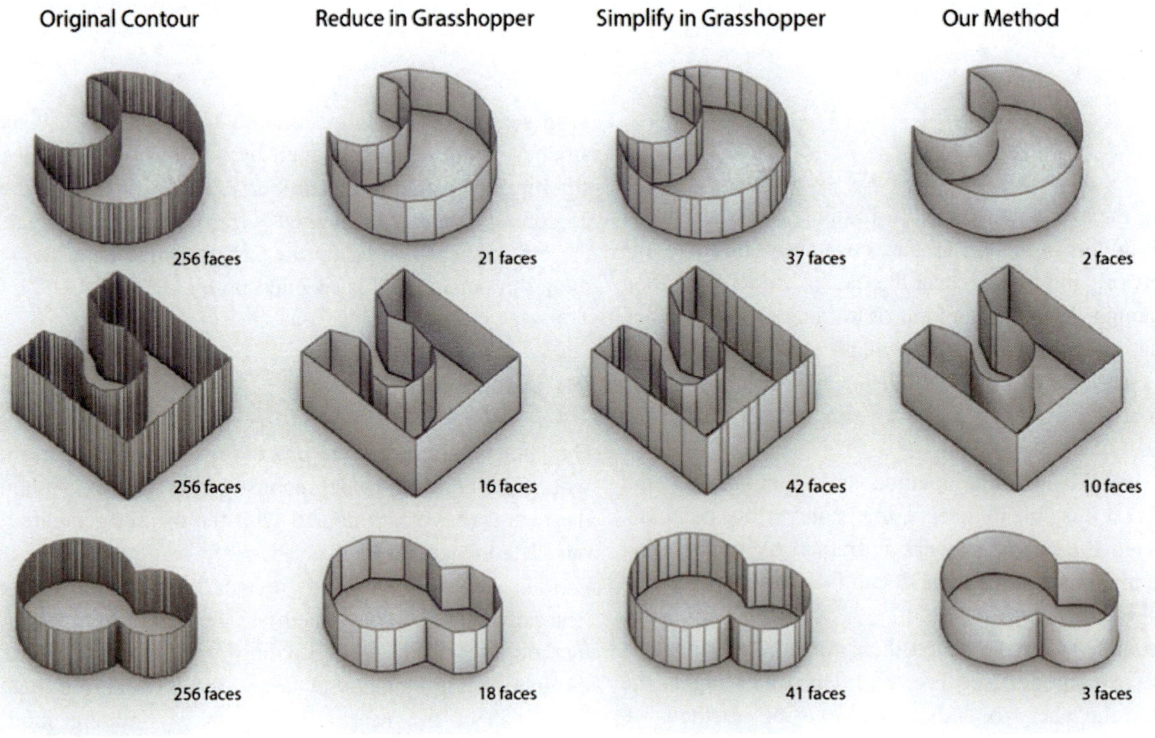

Fig. 13 Using various contours to test the model

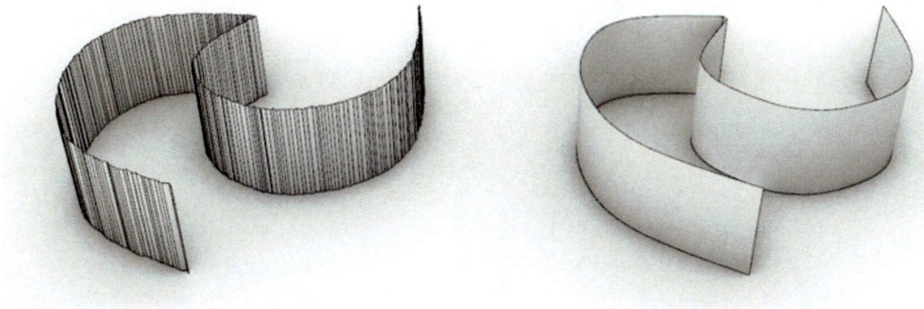

Fig. 14 The result of reconstructing open contour

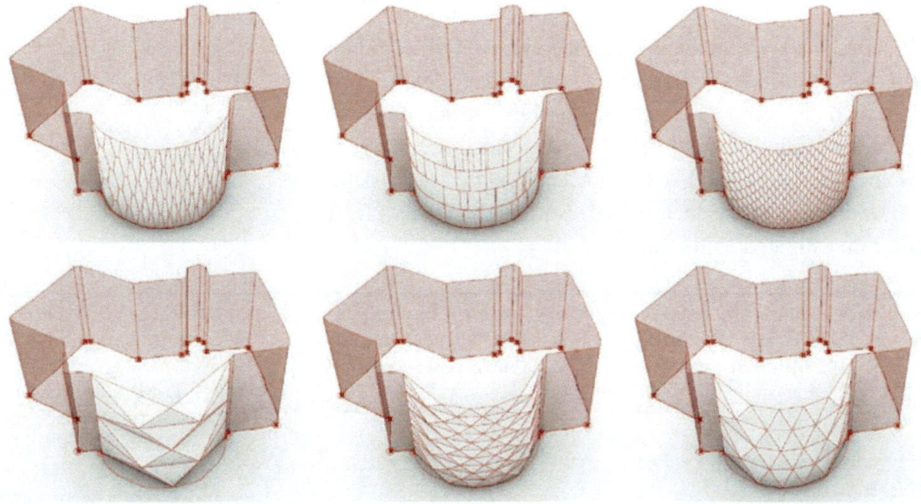

Fig. 15 Taking advantage of our method for parametric design

export to other geometry formats, such as Revit, and AutoCAD; designers are no longer struggling with complex mesh issue. The following photograph is a set of examples of demonstrating how to generate a parametric pattern on these surfaces (see Fig. 15). Despite the fact that our method is suitable for most polylines, it has a reservation about the result; these control points of each surface belong to an unsegmented original polyline. Therefore, when a user is not satisfied with some result, these original nodes can be derived from the control points of the surface. In the future, this method could be utilized as a standard means for reconstructing the photogrammetric mesh model.

5 Future Work

There are certain issues in our method that can be improved in the future. First, there are a few defects might happen around the corner that sometimes remains tiny pieces segments. Second, the resolution of detection in our method is 100 nodes; although it is adequate for most cases, there are some rare cases, might require higher resolution corner detection. The issue might be coped with through the way that makes this method to be an incremental process method. The incremental process is like subdivision; however, the division is based on our deep learning-based segment method; by doing this, the resolution will increase exponentially in a few iterations. On the other hand, in this research, the way to reinforce the sharpness on the corner has not been investigated. It has been considered for future improvement.

There is another application for reverse engineering that could utilize our method (see Fig. 16). It could cope with the issue caused by the fact most simplify and segment methods are dealing with a huge number of triangles on mesh surfaces. Instead, our method is based on the contour; our method is able to detect several corners from one slicing at the same time. In the future work, how to optimize the slicing process will be explored. The similar approach to build a 3D model is presented by Carswell et al. [1].

Although the mesh model generated by the photogrammetry can be significantly simplified through our method, the chaos of texture is not yet solved. The issue might be solved by using these surfaces generated with our method to resample the texture. Once the issue is solved, it will be able to facilitate several usages such as movie and game industries.

Fig. 16 Taking advantage of our method for reverse engineering

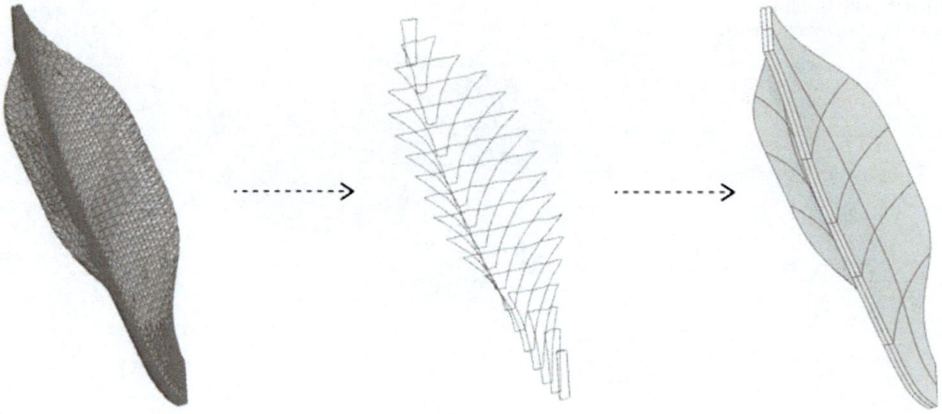

In term of algorithm feature engineering, we utilize radian as the feature in this method and count on deep learning to tackle with the noise. The other possible method that we speculate is using discrete Fourier transform to derive features and filter out noise like the concept of signal processing in pre-treatment phase then using deep learning as well; it will be investigated in the future work.

Acknowledgements The research is partially supported by the Ministry of Science and Technology under Grant Number MOST 109-2634-F-009-015 through Pervasive Artificial Intelligence Research (PAIR) Labs, Taiwan.

References

1. Carswell, D., et al. (2006). 3D solid fin model construction from 2D shapes using non-uniform rational B-spline surfaces. *Advances in Engineering Software, 37,* 491–501.
2. Buades, A., et al. (2018). Contours, corners and T-junctions detection algorithm. *Image Processing on Line,* 24–36.
3. Rosten, E., et al. (2008). Faster and better a machine learning approach. *IEEE Transactions on Pattern Analysis and Machine Intelligence, 32,* 105–119.
4. Wang, Y., et al. (2019). Fast mesh simplification method for three-dimensional. *Scientific Programming, 3,* 1–12.
5. Wang, X., et al. (2012). Feature detection of triangular meshes via neighbour supporting. *Journal of Zhejiang University: Science C, 6,* 440–451.
6. Wolin, A., et al. (2008). ShortStraw: A simple and effective corner finder for polylines, sketch based interfaces and modeling (pp. 33–40).
7. Zhang, S., et al. (2015). Corner detection using arc length-based angle estimator. *Journal of Electronic Imaging, 24.*
8. Spoerhase, J., et al. (2019). Simplification of polyline bundles. 35th European Workshop on Computational Geometry, Utrecht, The Netherlands, arXiv preprint arXiv:1907.05296
9. Sezgin, T. M. (2001). Feature point detection and curve approximation. Department of Electrical Engineering and Computer Science.
10. Zeng, L., et al. (2019). Robust corner and tangent point detection for strokes with deep. Graduate School at Shenzhen, Tsinghua University.
11. Xiong, B., et al. (2016). Footprint map partitioning using airborne laser scanning data. *ISPRS Annals of Photogrammetry, Remote Sensing & Spatial Information Sciences, 3,* 241–247.
12. Hinton, G. E. (2007). Learning multiple layers of representation. *TRENDS in Cognitive Sciences, 11,* 428–434.

Collaboration of Visual Intelligence Between Human and Machinic Actors in Design Processes

Egemen Berker Kızılcan

Abstract

Design is a sensory practice, and for human actors inducing design processes, commonly vision is favored among sensory inputs. The reasons for this preference are prospects provided by abstraction, diagrammatical thinking, and sense-making through mental catalogues that in turn establish conceptual understandings of design organization—rules that represent conception of architectural forms—and representational structures where these rules are anticipated, created, updated, tested, simulated, and maybe discarded. Following three decades of computational design research beginning with methodologies such as shape grammars, followed by CAD technologies emerging increased digital influence on the nature of design, today with advanced techniques such as incorporation of machinic intelligence in design tasks, visual intelligence in architectural design are provided with new potentials. The focus of design research in the scope of visual intelligence should be to locate different potentials in human and machinic actors to form the boundaries of responsibilities, respectively, aiming to form a collective visual intelligence propelled with the best of the both cognitive capacities. This facet is promising for achieving more complex and creative design orders and the paper scrutinizes different potentials in these two different cognitive capacities and argue for the advantages of each in the scope of computational design research.

Keywords

Visual intelligence • Conceptual and perceptual structures • Machine vision • Machine learning • Computational design

E. B. Kızılcan (✉)
Department of Architecture, Middle East Technical University, Ankara, 06800, Turkey
e-mail: egemenkizilcan@gmail.com

1 Introduction

Whether it is in the form of an informed decision over a meticulously collected and catalogued set of data, or mere intuitive marks over a sketch, numerous types of intelligences reside in a spectrum from more formal to more intuitive, and are available for reasoning for any given design task. Still, and perhaps even more, in the context of digital computation, architectural design is foremost a sensory practice that relies on these various types of intelligences and collection of information. Leon Van Schaik discusses different types of intelligences and locates spatial intelligence [29] as the innate type of intelligence necessary for both using and designing space: For Van Schaik Spatial Intelligence is a way to describe how designers address and resolve spatial design problems; it relies highly on sensory input and refers to all formal and intuitive approaches taken during design processes.

Van Schaik illustrates that the more subjective and intuitive parts of designers' mind such as his/her knowledge, past experiences, skills together with spatial intelligence provide for design approaches and solutions. He points out to the concept of 'mental space' that is used to describe where designers create ideas—a cognitive space that depends on and shaped by individual histories of designers. For Van Schaik architecture is necessarily a product of the non-static mental space: 'the accumulated histories of individuals in spaces.' This history is time and place specific hence depends on subjective interpretations of the user/observer of space, therefore, collection and formation of which are ultimately dependent on the sensory input for humans, and rarely escapes dominance of visual information.

In the scope of contemporary formal methodologies inducing architectural design, especially when considering artificial intelligence and machine learning cases broadening their influence, design approaches face challenges. For example, the concept of personal history becomes an issue of

S. Eloy et al. (eds.), *Formal Methods in Architecture*, Advances in Science, Technology & Innovation,
https://doi.org/10.1007/978-3-030-57509-0_28

compiling a consistent dataset that is fed to machine learning systems for training discrimination and/or generation; where innovative ways are sought out to make intuitive approaches more explicit and versatile, both due to limiting constraints of computability [33] and thanks to extensive capabilities of computation. These examples indicate a relocation of vision when considering machine intelligence in parallel with formal design approaches. In other words, it is now possible to form collective intelligences between human and machinic actors that have extended capabilities in both resolving visual complexity and satisfying performative constraints; given that it is possible to compile a consistent dataset/history of actions and experiences for that specific design solution.

2 Decision Making in the Scope of Formal Studies

When considering formal approaches to design problems and rational decision-making, intuitive reasoning seems to be the antagonist that undermines the explicit basis of formal approaches. For some cases that are mostly about solving performative criteria about designs—for example, utilizing genetic algorithms that modify architectural form in order to optimize heat gain concerns—prospects provided by digital computation seem to be so strong, such design processes seem to completely discard intuitive decision making. Indeed, intuitive approaches do pose challenges as reasoning behind intuitive decisions are not always clear, traceable, repeatable or predictable, yet there are ways to reconcile the gap [26] between formal and intuitive approaches. Furthermore, it is difficult to abolish intuitive approaches for mainly two reasons regarding human actors: (1) intuitive reasoning relates more to the already possessed sensory systems, especially vision and (2) intuitive decisions are fast and mostly accurate; mostly because intuitive decisions are open to systematic errors where Daniel Kahneman refers to them as false biases [19]. For design processes such systematic errors may provide for surprises that makes design solutions more varied and arguably creative. In other words, independence from justification may provide for expanding potential spaces. Formal approaches and rational decision making favor a solid ground that consists of predictability and accuracy, arguably for means of justification [34], yet for Kahneman, the least probable and the most random decision is a lucky decision which broadens the potentials for any solution space. Hence, a wider and potent solution space including mistakes is likely to emerge creative solutions.

3 Formal Computational Methodologies in the Scope of Visual Reasoning

Thankfully today, a shift from rule-based to more probabilistic approaches in computational design methodologies is observed [5]. A prominent instance of the more probabilistic and error-tolerant approach would be the work of Roland Snooks: He introduces the concepts of 'strange feedback' and 'messy computation' to intervene with limitations, or namely predefined solution spaces that are solely bound to predictable generation logic aiming to widen the potential of generative processes and to produces 'emergent' results [31]. In other words, Snooks intentionally exploits data and logic, as far as possible from what is more predictable aiming to emerge more creative solutions. This seems counter-intuitive or peculiar, but even with the aim of escaping fixity, Snooks embraces formal techniques and methodologies as a strategy to increase variety, and, therefore, to expand the potential space for design solutions. In the scope of formal methodologies, a larger potential space is not feared but favored since machines and humans together have the capacity to exhaust all potentials and explore all consequences of many available decisions in a faster manner than solely done by humans.

It would be safe to argue that architectural design as a sensory practice establishes a reliance on external information, whether methodologies pursued in order to reach solution are formal or intuitive, computational or conventional; a reliance also exhibited in Snooks' work where gaps between formal and intuitive, as well as computational and conventional are reconciled [26]. So, there is no solution without information and in this case, visual information is perhaps most varied among other formats of information. For human actors inducing design tasks, vision seems to be the most favored among sensory input and similarly, image as the most common means of communication.

For the sake of this paper, visual reasoning is therefore found to be especially curious where a complex process is conducted during making decisions: Visual reasoning spans a variety of faculties making visual reasoning very prospective and rapid; mental catalogues and categories, aesthetic or cultural preferences, geometry and geometrical rules, shape grammars, symbolic thinking, interpretation, representation, abstraction, diagrammatical thinking, sense-making to cite a few. There is ample discussion for each of these faculties so discussing each of them extends well beyond the scope of this paper [5, 8, 10–12, 15, 16, 20–23]. Yet, such faculties provide for both conceptual understandings of design organization—in the context of formal studies, rules and processes which represent the overall organization and conception of the architectural form—and consequent realization of such conceptual representations in the form of an architectural artefact.

Today, the influence of machinic actors establish new possibilities among chronic challenges: Chronic problems are widely regarded as the issues or limitations that are caused by the nature of computational medium: The need for structured logic, enumerability, computability, and management of complexity to cite a major few. Mario Carpo compiled a volume on the whole topic of the 'Digital Turn' by various authors, and it is also beneficial to see Kostas Terzidis's view on 'Algorithmic Complexity' [4, 14, 34]. Concerning visual reasoning, the choice between the more rational data-centrism or the more intuitive sketching and drafting does not present a clear distinction in the current state of the art, instead it presents that a spectrum between two different types of cognitive capacities [13]—machine and human—is available through design processes. Daniel Kahneman characterizes rational decision making as exhausting all potential possibilities to choose the best one where Kahneman's characterization is supported by Herbert Simon's work [19, 30]. Namely, there are two distinct types of decision-making capabilities where the distinction arises from inherent nature of the two different type of actors. Especially, the prospects provided by exciting techniques and methods of artificial intelligence and machine learning provides for an interesting ground of discussion as these approaches can better cope with ambiguity due to their probabilistic and statistical nature, and provides for equal potentials in resolving intricately complex solutions and extending creativity. Namely, formal methodologies now can incorporate ambiguity and vagueness, deeming machinic actors more compatible for humans. An example to this cooperation with machinic actors is the work of Alisa Andrasek that illustrates the possibilities of increased complexity and infinite resolution in design works. Andrasek's AI-based 'design agent' that aims to aid human cognitive capabilities in situations that demand complex decision making exhibit new interfaces to work with machine actors [3].

4 Two Modes of Visual Intelligence

Due to innate strengths and weaknesses of two cognitive capacities and for extensible modes of visual reasoning and creation, the paper argues for an augmented process—a collective intelligence via augmentation—between these two types of reasoning and illustrates curious overlaps together with strengths and weaknesses of both. In other words, both of these visual intelligences have overlaps and distinct advantages, and it is argued in this paper that one should not substitute each other; instead ways of which they can complement each other are pursued. The primary question is that of how to make machine intelligence more relevant to design solutions in the broader sense and here, augmentation becomes an important feature for resolving design problems in real time as the act of design and decision making for design processes are temporal and hermeneutical processes. Main conceptual frames of discussion are found to be abstraction and extensibility of visual reasoning as means of identifying strengths and weaknesses; my research here is based on a discussion entangled with Ernst Gombrich's Sense of Order [15], conceptual and perceptual structures [32] of design in the scope of formal computational processes [24], machine vision [9] and machine learning [1, 2].

For reasoning about design solutions, the relationship of information and conceptualization presents a duality. Pablo Lorenzo-Eiroa discusses intricacies between visual and relational logic in the context of computational design and establishes a distinction between conceptual and perceptual structures for design when in the scope of formal studies. Namely, relational or diagrammatic modes of formal representations of information are possible to be realized in a variety of perceptual structures depending on the preferences of the author—or instances of architectural forms—so within the scope of formal studies resulting form is relatively independent from the underlying conceptual model that initiates conception of form. This is especially relevant for generative design processes and similar to the general understanding of computation where computation consists of data and its formal representation [18, 33]; deeming how information is gathered, interpreted to be formally represented, processed by arithmetic means and then outputted finally is crucial in creating variation over perceptual structures. Therefore, for both machinic and human actors, abstraction and representation are core concepts which help understanding how visual information is acquired, interpreted and processed.

For Gombrich in The Sense of Order, this process of collection and interpretation establishes the basis for creation for humans. To describe what the Sense of Order is, Gombrich illustrates the acquisition of visual information and its assessment by humans through the history of art and architecture with regards to perceptual and psychological effects of such visual information on the mind. The key remark in Gombrich's theory is that human actors reduce variety of the visible world into abstract schematic renderings which are described as 'conceptual images.' The need for such abstraction lies in the biological conditions of which have led to evolution of cognitive capabilities of humans; mainly for a creature to survive in nature, that creature should efficiently and very quickly answer the questions of 'What?' and 'Where?' to visually locate and assess either something is nourishment or threat. Hence, for a creature to survive in the harsh conditions of the wild nature, visual information has to be processed and assessed rapidly, providing for the need for abstraction and accompanying reduction of variety of the information. Gombrich interprets these images as minimal representations/models of the visible world:

> … I emphasized throughout the book that 'making comes before matching'; the minimal schema is first constructed before it is modified or corrected by matching it against reality. If I were asked to sum up the theory underlying this book in a similar formula, it would be that groping comes before grasping or seeking before seeing. In contrast to any stimulus-response theory I would wish to point to the need to regard the organism as an active agent reaching out towards the environment, not blindly and at random, but guided by its inbuilt sense of order. [15, p. 5]

Following Gombrich, vision for humans is highly evolved and efficient to be used for reasoning, and due to its rapid speed, visual reasoning is intuitive. The ability to refer to an inherent mental catalogue seems to be the major upside for achieving this speed. On the other hand, considering machinic actors, and specifically for machine learning models [28], observing how models are trained, the situation is far from different. Learning algorithms utilizing artificial neural networks, specifically of type Generative Adversarial Networks (GANs), or even more specifically Deep Convolutional Generative Adversarial Networks (DCGANs) focus on recognizing repetitions in the training set, and use these patterns recognized during training to generate, for example, faces of people who do not exist [17] where such variation in generated images is partly achieved by randomization and, or as mentioned above in Kahneman's case, luck. Generated images have obvious common traits with regards to the training set (meaning it is possible that the outcomes resembles familiar faces—maybe multiple of them at the same time) but variation is distinguishable as the image is an outcome of an feedback loop between a generator and a discriminator, where constant negotiation between both through numerous epochs among randomization train both of them together simultaneously.

For machines and visual problems, collection of recognized patterns is commonly done by an intense supervised training over a consistent set of data, for instance carefully cropped photos of faces of celebrities in the above-mentioned case. When a model is trained for a specific purpose, such as generating random faces, therefore when regression is achieved and model is optimized, there usually is no need for further intense training because domain-specific tasks, or tasks that maybe even called autistic are unlikely to have changes in their low-level concerns. For face generation, the low-level concern is to find local and global repetitions in the set of photos that establishes biases about what a face is. A very important distinction from human actors here is that, this low-level concern is non-semantical for machines as it does not refer to a mental catalogue and categories of metaphors, symbols, experiences, feelings or beliefs: They are foremost mathematical and statistical features derived from pattern recognition forming up to more convoluted complex relational models that can informedly discriminate concepts within its domain. But unless a higher-order semantical layer is implemented, machinic actors can stay semantically indifferent to interpretations of concepts that they are being trained for: A face will be a face, but the meaning of the face is not questioned, nor it is inferred as that meaning (1) is most probably irrelevant for the survival of the machinic actor, (2) subjective and (3) necessitates a personal history about humans and their facial expressions. So, while a human distinguishes traits of a familiar face in the set of generated faces semantically integrated—for instance by categorizing emotion, age or gender—for the machinic actors those traits are combinations of pixels propagating up to distinction of a concept through recognized recurring patterns, unless a more semantical system is implemented. This independence from meaning is both a strength and a weakness for machinic actors. It is a strength because (1) it makes visual information independent from personal histories and preferences, therefore, forms a neutral basis for visual creation and reasoning, and (2) additional implementations of formal logic are possible for inferring the same categorizations as emotion, age, or gender [7] among numerous other possibilities, making this process configurable, modifiable, and extensible. So, seeing and interpretation are not as integrated as observed in human actors. It is a weakness since there needs to be additional steps in implementation in forming abstract, relational and semantic models or maps over the visual information; in other words, this implementation of inference is not readily available nor integrated, a feature that undermines speed and increases effort in implementation.

An interesting remark is that, for both types of cognitive systems, on collection of visual information, there needs to be an elimination of unnecessary specificity in order to improve the performance, or speed, of decision making. For machinic actors, the important feature is named as low-level seeing, where resolution of the acquired information is configured, and where intensity of regularities or irregularities in the information establish custom sense-making capabilities. But this does not mean that machines cannot compute full specificity: Instead, for [illegible] learning models, the processes of abstraction of information [illegible] form of lowering resolution of images into [illegible] minimal, removing color information, increasing contrast to cite a few: These preprocessing operations are usual prerequisites for low-level seeing for machine vision. So, with this reduction, trained models can refer to a wider set of information to be used for decision making, increasing accuracy of interpretations and categorizations given if an optimal model of inference is implemented.

A wide set of information to refer is crucial since, rational thinking and formal processes require exhaustion of a potential space of decisions to reach the most viable result. For Kahneman to reach the best result, all of the potentials and consequences should be calculated, and this is the only way of rational decision making. Kahneman discusses the game of chess and refers to Von Neumann and Morgenstern's Game Theory [27], providing for an important conceptual example in the frame of formal studies and rational thinking: For finding the best move in the game, all potentials in all the possible branches and subbranches of a decision tree has first to be generated and then all of their consequences are to be calculated with regards to a prospect function to deepest ends, making the most prospective consequence is the best rational decision. Kahneman deems this process of exhaustion impossible for human actors due to an immense mental load. Following Kahneman, formal decisions require rationalization and explicitness, and full rationality embraces complete information, deeming intuition as decision making in the case of incomplete information.

But Kahneman also argues, intuitive decision making is prone to biases, namely systematic errors. These biases are bound to many heuristics, but the main heuristic here is of availability. Repetition and pattern recognition being the main catalyzers for this heuristic in both types of actors. Machine learning relies on pattern recognition, and for patterns to be recognized, there needs to be enough repetition. This is due to the statistical nature of machine learning models. Consistence is a feature of repetition: A given pattern consistently repeating establishes a solidified interpretation in the mental catalogue for human actors or trained models for machinic actors.

For humans, the availability heuristic refers to a mental catalogue of past experiences: Visual information is converted into a conceptual map where meanings and relations are interpreted and where sense-making occurs. As mentioned above referring to The Sense of Order, elimination of unnecessary specificity is for making rapid decisions. Speed of this emerges from a mental catalogue of past experiences recorded in various environments. Similarly, for machines and machine learning specifically, the same bias refers to the training [illegible] regression to a mean forms the bias. Following Kahneman's illustration, regression is [illegible] feature for bypassing biases when supplied [illegible] correct statistics in the course of rational thinking. Therefore, it demarcates the importance of choosing the correct and consistent dataset for training machine learning models. Due to the independence from sense-making mentioned above, and the initial tweaking and configuration in the collection layer, machine vision, and visual reasoning are always more specific and direct. It is not bound to solidified interpretations initially, but biases in machine learning models are formed and modified over time during the course of training. This is an important feature as formation of such biases are mutable.

To summarize up to now, by Gestalt Theory of visual perception [36] and with regards to how design is taught in the more conventional way [20–22], the same catalogue and the ability to form conceptual maps are used for visual reasoning and creation. Designs are formed, tested, and improved hermeneutically within a holistic context, especially when intuitive faculties are more relied on. For Gombrich, seeing is a way of seeking for regularity, and a state of truth where further intentions emerge. So, an important feature for human cognitive capacities is that collection of visual information and creation of representations of conceptual maps are hermeneutical processes. This is a feature that creates varieties in interpretations and genuine conceptions of conceptual maps. The mental catalogue is the initial dataset for humans for formation of intuitions and intentions. For machines, the catalogue is mutable and flexible; therefore, provides for custom ways of seeing, interpretation, reasoning and, therefore, creation. So, the main difference of in human and machine visual cognitive capacities is the boundedness and the level of independence of seeing, interpretation and creation: when boundedness bring out speed and variability as seen in humans, independence provides for extensibility and mutability.

5 Abstraction of Formal Logic and Extensibility

To illustrate extensibility and mutability of formal logic and how semantics come in play in machinic actors, a crucial facet about abstraction lies in software architectures, especially in the case of object-oriented programming: There are certain principles and patterns for object-oriented programming and design; some of the initial ones are named by Robert C. Martin, an influential software engineer, when applied together constitutes maintainability and extensibility of a software [25]. Namely, layers of abstraction for software development briefly means that there are different layers of formal logic added on consecutive layers, deeming each layer added on another is more abstract and less specific; therefore, making that layer more related to interpretations and processing of information in a more semantic way, but less related to raw data collection and output observed in the bottom layer: When navigating upwards on the layers, the logic becomes more semantic and conceptual and the bottom layers are more specific and more related to data layers [35].

Here, it is important that the addition of layers is bound to formal logic. Yet, it should be emphasized that any formal logic is applicable in this case which establishes the basis for custom ways of seeing, or custom senses of order. Namely, higher machine vision requires custom layers of abstractions over low-level seeing. These layers of abstraction contain formal logic that are optimized for different purposes and aims in an extensible manner focusing on carefully limited concerns for each part in the software. How these layers of abstractions are formed are induced by humans aiming to augment their visual reasoning capabilities with various concerns. Humans are bypassing the shortcomings in modifying their inherent, permanent, unmodifiable biological capabilities by augmenting machinic actors to the process even if machine vision is usually domain and concern specific.

For machine vision, as new layers are added, the transition from low-level to higher-order seeing becomes more apparent: For instance, custom formal logic and filters are implemented in order to add on the low-level functions of corner and edge detection, like intermediate vision functions of line detection and boundary pattern analysis and finally to achieve higher seeing regarding 3D Vision and Motion, or even higher and semantical concerns of automated surveillance or in-vehicle vision systems for autonomous cars [9]. So, an important contrast here is the feature of extensibility observed in machinic actors is a strength compared to human actors for whom, seeing and interpretation are integrated processes. The weakness is, as mentioned above, the difficulty to gather a personal history and the inability to refer to the catalogue for faster, more intuitive, if sense-making, therefore, decision making with increased complexity is desired.

In other words, and to summarize, for machines again, vision is built bottom-up from full specificity to custom, abstract, yet formal orders. For humans, seeing is a top-down feature which relies on abstraction, semantics, and conceptualization. But, due to the integrated and embodied features of seeing, it is difficult to extend or modify without augmentation or dedicated modification of the mental catalogue through training. Due to the lack of a catalogue, the intensity of various parts of the visual data have to be pre-processed in order to be emphasized; therefore, there is no room for ambiguity is left. The closest feature to personal history for machine learning is that of the training set for the model, importance of which is discussed thoroughly in this paper.

So, for custom senses of order, the variation in data becomes as important in variation in logic of generation. A good example for this would be DCGAN plan generator which exhibits this variance [6] and the important feature of the plan generator is that how it is structured. It consists of nested neural networks, where responsibilities are encapsulated in different instances of neural networks, where from data to logic layers data is propagated to find its final representation. Each of the layers are aware of the bottom layers but the distributed system is in effect in an autistic manner with regards to each of its component. In this way, the plan generator can provide for differentiating stylistic concerns while keeping the spatial relations similar or vice-versa. The independence between conceptual and perceptual structures observed in this formal methodology allows for an even wider and extensible potential space for design, given that there is a consistent dataset.

It is possible to use any type of consistent visual information aiming to form generative neural networks but inconsistency in the training set almost always undermines machine learning.

Figure 1 shows snapshots from the training process of a DCGAN implemented in different epochs to illustrate features of machinic visual creation. For the sake of experimentation, a Web crawler was first built to gather images from search engines with given keywords. After 42 iterations of modification and optimizations, the final model was trained for 500 epochs with 6000 images relevant to the keywords regarding Modernist architecture and works of famous Modernist architects. But the dataset is never qualitatively assessed and purged to remove inconsistences or *strange* data. So, the main features deviate the results from legible façade images to what might be called digital impressions of Modernist features are: (1) dataset being never qualitatively filtered includes not only images of Modernist buildings but rather odd images as well, and (2) the model is trained with a rather uncommon convolutional neural network stack to cope with the existence of *strange* information in the dataset. This approach incorporated a stable and slower learning rate and prioritized detailing in parts compared to upscaling of wholes which is the most common approach for DCGANs. Upscaling was not expected to work for this model because of the existence of *strange* data and consequent inconsistency of the training set but the purpose of the experiment was to find ways to cope with such inconsistency and to bypass difficulties in forming consistent datasets. The results of the experiment are mildly disappointing in terms of accuracy but greatly satisfactory in dealing with a *dirty* dataset.

Considering Chaillou's plan generator and the experiment done with the dirty dataset for the sake of this paper together, it is safe to assume that visual information in the form of bitmaps of architectural plans and photos of buildings used for training machinic models have personal histories embedded into them. Namely, photos and plans are the outcomes or representations of various subjective past decisions, personal experiences, and spatial intelligences. The variety of subjectivity included in the dataset, or the machinic models' personal histories, makes them independent from semantics, because sense making has already been made and is meticulously included in the dataset. Only task left to the machinic model is to find reoccurrences in various past decisions of various subjects.

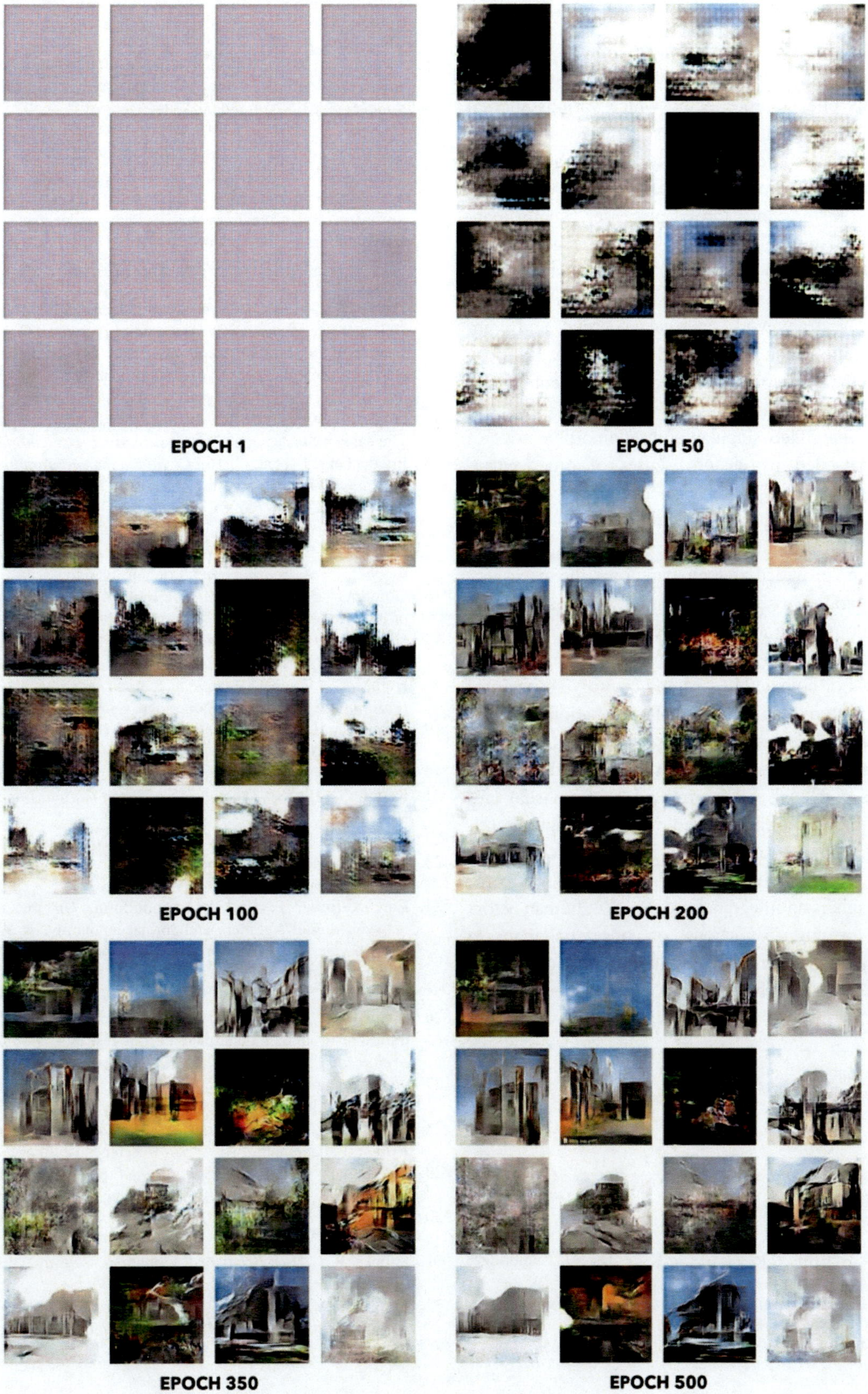

Fig. 1 Process of training a DCGAN that generates modernist facades

6 Conclusion

Augmented processes that is induced by human and machinic actors offer approaches by which various alternatives about a design solution can be simultaneously created and evaluated. But more importantly, they enable visual decision-making processes that are less fixated to complete explicitness—a factor which formal processes usually mandate or favor. Namely, visual decision making becomes more probabilistic that is fed by two types of visual intelligences that have different advantages in achieving different tasks. Within this framework, creative, qualified, and complex design solutions can be achieved easier as by inclusion of machinic actors it becomes possible to implement custom ways of seeing and visual creation.

The innate integrated singular anthropomorphic way of seeing is superseded by custom senses of order where explorative features of design as hermeneutical processes are augmented by the strengths of machinic actors. Designers are now capable of augmenting themselves with custom ways of seeing, provided if they can form the necessary formal logic parallel with their intentions. This allows for intuitive decisions to be tested with full rational and formal basis, and a decision can be both informed and creative at the same time. For both actor types and cognitive capacities, collection and interpretation of visual information is a prerequisite for visual reasoning and creation. This makes it extremely crucial in what way, or sense of order, visual information is collected and interpreted, as for visual creation and reasoning the internalized visual catalogue is utilized. For machine actors, the way of collection, interpretation, and creation is highly modifiable where for human actors this is an integrated process. Yet human actors are better at forming abstractions, or conceptual maps to model and augment themselves with different capacities, inducing the way machinic actors complement design processes.

Acknowledgements This study consists of a part of my own Ph.D. research being conducted in the Middle East Technical University, Department of Architecture.

References

1. Alpaydin, E. (2016). *Machine learning the new AI*. Cambridge, MA: The MIT Press.
2. Alpaydin, E. (2020). *Introduction to machine learning*. Cambridge, MA: The MIT Press.
3. Andrasek, A. (2018). *High resolution fabric of architecture*. Ph.D. Thesis, RMIT.
4. Carpo, M. (2013). *The digital turn in architecture 1992–2012*. Chichester: Wiley.
5. Carpo, M. (2015). The new science of form-searching. *Special issue: Material synthesis: Fusing the physical and the computational* (pp. 22–27). London: Wiley.
6. Chaillou, S. (2019). AI & architecture. Retrieved from https://towardsdatascience.com/ai-architecture-f9d78c6958e0.
7. Chauhan, N. (2019). Predict age and gender using convolutional neural network and openCV. Retrieved from https://towardsdatascience.com/predict-age-and-gender-using-convolutional-neural-network-and-opencv-fd90390e3ce6.
8. Chomsky, N. (2005). *Rules and representations*. New York: Columbia Univ. Press.
9. Davies, R. (2018). *Computer vision* (5th ed.). S.l.: Academic Press.
10. Deleuze, G., et al. (2019). *A thousand plateaus: Capitalism and schizophrenia*. London: Bloomsbury.
11. Derix, C., et al. (2014). *Empathic space: The computation of human-centric architecture*. London: Wiley.
12. Dogan, F., et al. (2010). Generic abstraction in design creativity: The case of Staatsgalerie by James Sterling. *Design Studies, 31*(3).
13. Friedenberg, J., et al. (2016). *Cognitive science: An introduction to the study of mind*. Thousand Oaks, CA: SAGE Publications, Inc.
14. Gleiniger, A., et al. (2008). *Complexity: Design strategy and world view*. Basel: Birkhäuser.
15. Gombrich, E. (1998). *The sense of order: A study in the psychology of decorative art*. London: Phaidon.
16. Gombrich, E. (2014). *Art and Illusion: A study in the psychology of pictorial representation*. Phaidon.
17. Gsurma. (2019). gsurma/face_generator. Retrieved April 7, 2020, from https://github.com/gsurma/face_generator.
18. Huth, M., et al. (2018). *Logic in computer science: Modelling and reasoning about systems*. Cambridge: Cambridge University Press.
19. Kahneman, D. (2015). *Thinking, fast and slow*. New York: Farrar, Straus and Giroux.
20. Kepes, G. (1965a). *Education of vision*. London: Studio Vista.
21. Kepes, G. (1965b). *The nature and art of motion*. London: Studio Vista.
22. Kepes, G. (1966). *Sign, image, symbol*. New York: G. Braziller.
23. de Landa, M. (1998). Deleuze, diagrams and the genesis of form. *Any Magazine Issue, 23*.
24. Lorenzo-Eiroa, P. (2013). Form:in:form, On the relationship between digital signifiers and formal autonomy. In *Architecture in formation: On the nature of information in digital architecture*. London: Taylor & Francis.
25. Martin, R. (2000). *Design principles and design patterns*.
26. Mennan, Z. (2014). Mind the gap: Reconciling formalism and intuitionism in computational design research. *Footprint, Delft Architecture Theory Journal, 15*.
27. von Neumann, J., et al. (2007). *Theory of games and economic behavior*. Princeton: Princeton University Press.
28. Rosebrock, A. (2019). *Deep learning for computer vision with Python*. Philadelphia, PA: Adrian Rosebrock, PyImageSearch.com.
29. van Schaik, L. (2008). *Spatial intelligence: New futures for architecture*. West Sussex: Wiley.
30. Simon, H. A., et al. (2019). *The sciences of the artificial*. Cambridge, MA: The MIT Press.
31. Snooks, R. (2014). *Behavioral formation: Multi-agent algorithmic design strategies*. Ph.D. Thesis: RMIT.
32. Sprecher, A., et al. (2013). *Architecture in formation: On the nature of information in digital architecture*. London: Taylor & Francis.

33. Sudkamp, T. (2007). *Languages and machines: An introduction to the theory of computer science*. S.l.: Pearson Education.
34. Terzidis, K. (2006). *Algorithmic architecture*. Amsterdam: Elsevier.
35. Ullman, J. D. (1976). The mechanization of abstraction. In *Fundamental concepts of programming systems* (pp. 1–23). Reading, MA: Addison-Wesley.
36. Wertheimer, M. (2020). *Productive thinking*. S.l.: Birkhauser.

Egemen Berker Kızılcan Kızılcan graduated from Middle East Technical University (METU), Department of Architecture. He completed his master's degree in the same department; His master thesis titled Complexity Management and Mutability in Architectural Form Conception: Form-blindness and Softform was awarded the METU Thesis of the Year award. He is working on his Ph.D. studies as a Research Assistant in Design Theory and Design by Artificial Intelligence, and he is Ph.D. candidate in Visual Intelligence. He is teaching various courses in METU, among which are design studios, technical drawing and graduate courses and labs.